Energy, Environment, and Climate

//

Second Edition

Energy, Environment, and Climate

Second Edition

RICHARD WOLFSON
Middlebury College

W. W. NORTON & COMPANY
NEW YORK • LONDON

W. W. Norton & Company has been independent since its founding in 1923, when William Warder Norton and Mary D. Herter Norton first published lectures delivered at the People's Institute, the adult education division of New York City's Cooper Union. The Nortons soon expanded their program beyond the Institute, publishing books by celebrated academics from America and abroad. By mid-century, the two major pillars of Norton's publishing program—trade books and college texts—were firmly established. In the 1950s, the Norton family transferred control of the company to its employees, and today—with a staff of more than four hundred and a comparable number of trade, college, and professional titles published each year—W. W. Norton & Company stands as the largest and oldest publishing house owned wholly by its employees.

Printed in the United States of America

Editor: Erik Fahlgren
Project editor: Carla L. Talmadge
Editorial assistants: Mina Shaghaghi, Mary Lynch
Production manager: Chris Granville
Managing editor, College: Marian Johnson
Design director: Rubina Yeh
Photo editor: Stephanie Romeo
Associate editor, emedia: Matthew A. Freeman
Composition: TexTech
Manufacturing: Courier

Library of Congress Cataloging-in-Publication Data

Wolfson, Richard.
Energy, environment, and climate / Richard Wolfson.—2nd ed.
p. cm.
Includes bibliographical references and index.
ISBN 978-0-393-91274-6 (pbk.)
1. Climatic changes. 2. Energy consumption—Climatic factors. 3. Energy consumption—Environmental aspects. 4. Power resources—Climatic factors. I. Title.
QC981.8.C5W645 2012
551.6—dc23 2011030173

W. W. Norton & Company, Inc., 500 Fifth Avenue, New York, N.Y. 10110
www.wwnorton.com

W. W. Norton & Company Ltd., Castle House, 75/76 Wells Street, London W1T 3QT

1 2 3 4 5 6 7 8 9 0

BRIEF CONTENTS

CONTENTS

FOCUS BOXES

PREFACE

Many behaviors distinguish the human species from our fellow inhabitants of Planet Earth. Of these, our use of energy in amounts far exceeding what our own bodies can produce affects the environment in an unprecedented way. Centuries ago, pollution from coal burning was already a serious urban problem. Despite regulatory and technological progress in pollution control, diminished air and water quality continue to be major consequences of our ever-growing energy consumption. Further environmental degradation results as we scour the planet for fuels that contain the stored energy of which we demand an unending supply. Our energy-intensive society also enables other environmentally damaging developments such as sprawl, large-scale mechanized agriculture, and massive deforestation. At the same time, energy brings us higher standards of living and allows our planet to sustain a larger population.

In recent decades, a new and truly global impact of humankind's energy consumption has overshadowed the long-standing and still significant consequences associated with traditional pollution, resource extraction, and energy-enabled development. That impact is global climate change, brought about largely by the emissions from fossil fuel combustion. Climate change is a problem that knows no national or even continental boundaries. It will affect us all—although not all equally. It won't have the civilization-ending impact of an all-out nuclear war or a major asteroid hit, but climate change will greatly stress an already overcrowded, divided, and combative world.

Achieving a healthier planet with a stable, supportive climate means either using less energy or using energy in ways that minimize adverse environmental impacts. Here we have choices: To use less energy, we can either deprive ourselves of energy's benefits or we can use energy more intelligently, getting the same benefits from less energy. To minimize environmental and especially climate impacts, we can shift from fossil fuels to energy sources that don't produce as much pollution or climate-changing emissions. Or we can learn to capture the emissions from fossil fuels and sequester them away from Earth's surface environment.

Earth's energy resources are limited to relatively few naturally occurring stores of energy—the fuels—and to energy flows such as running water, sunlight, wind, geothermal heat, and tides. A realistic grasp of our energy prospects demands that we understand these energy resources. We need to know, first and foremost, if a given resource or combination of resources is sufficient to meet humankind's energy demand. For fuels, we need a good estimate of the remaining resource and a time frame over which we can expect supplies to last. We need to understand the technologies that deliver useful energy from fuels and flows, to assess their environmental impacts, and to recognize that none is without adverse effects. And we need to be realistic about the near-term and long-term prospects for different energy sources in the economic context.

The oil shortages of the 1970s spawned a serious exploration of energy alternatives. Governments and industries sponsored research programs, while tax credits encouraged the installation of alternative energy systems. Vehicle mileage and other measures of energy efficiency increased significantly. At the same time, colleges and universities developed specialized courses in energy issues and the relationship between energy and environment. These courses emerged in traditional departments such as physics, chemistry, and engineering; in interdisciplinary programs dealing with technology and society; and in the burgeoning new programs in environmental studies and environmental science that have sprung up with the emergence of a widespread environmental conscience in the last decades of the twentieth century. Textbooks written for such courses addressed the science and policy issues surrounding energy and the environment.

Energy, Environment, and Climate also focuses on energy and its impact on the environment. Unlike its predecessors, it's built from the ground up on the premise that climate change is the dominant energy-related environmental issue of the twenty-first century. More traditional concerns such as pollution and energy resources remain important, and they, too, are covered here. But a full five chapters—about one-third of the book—are devoted to climate and the energy-climate link.

Energy, Environment, and Climate begins with a survey of Earth's history and the origin of the planet's energy resources. A quantitative look at past and present patterns of human energy consumption follows, including a discussion of the link between energy, economic development, and human well-being. Chapters 3 and 4 provide an introduction to the science of energy, including the all-important role of the second law of thermodynamics. Chapters 5 through 10 describe specific energy sources and their resource bases, the role each plays in today's global energy system, their associated technologies and prospects for future technological development, and their environmental impacts. This section of the book is organized around fundamental resources, including fossil fuels, nuclear energy, geothermal and tidal energy, and direct and indirect solar energy. Because fossil fuels dominate today's energy supply, there are two chapters dealing, first, with the fossil resource and fossil fuel technologies and, second, with the environmental impacts of fossil fuels. Whereas other textbooks have separate chapters on such energy-related issues as transportation, *Energy, Environment, and Climate* includes these topics in the appropriate energy chapters. For example, hybrid vehicles and combined-cycle power plants appear in the fossil-fuel chapters; fuel cells are discussed in the chapter that covers hydrogen as an energy carrier; and wind turbines are included in the chapter on indirect solar energy.

Completely redone in this second edition is Chapter 11, which now provides a detailed look at the energy carriers electricity and hydrogen. The electric power grid, smart-grid technology, distributed generation, AC and DC transmission, load management, integration of renewable electricity sources, energy-storage technologies, and other issues relevant to our growing use of electricity are covered, either for the first time or in much more detail than in the first edition. Material on nuclear fusion, formerly in Chapter 11, has been moved to Chapter 7 on nuclear energy. Hydrogen as an energy carrier remains, appropriately, in the revised Chapter 11.

Four chapters on climate follow the section on energy. Chapter 12 describes the scientific principles that determine planetary climates, including the natural greenhouse effect in the context of planets Venus, Earth, and Mars. The chapter ends with a discussion of the nature of scientific theories and of certainty and uncertainty in science. Chapter 13 details the so-called "forcings"—both natural and anthropogenic—that can upset the energy balance that ultimately establishes Earth's climate. Chapter 14 documents the observations that suggest Earth is now undergoing unusually rapid climate change, and shows why scientists believe much of that change is attributable to human activities. Chapter 15 outlines projections of future climate, and includes a look at the workings of computer climate models and the role of climate feedbacks. The final chapter brings together the two main themes of the book—energy and climate—and explores how humankind might continue to enjoy the benefits of energy while minimizing climate-changing impacts.

Energy, Environment, and Climate is written primarily from a scientific perspective. However, questions of policy and economics are never far behind the science of energy and climate. The text therefore ventures occasionally into policy and economic considerations—although to a far lesser extent than a policy-oriented book would do. In particular, many chapters end with a section specifically dedicated to a policy-related issue that grows out of the science covered in the chapter.

Any serious study of energy and the environment has to be quantitative. We need to understand just how much energy we actually use and how much energy is available to us. It makes little sense to wax enthusiastic about your favorite renewable energy source if it can't make a quantitatively significant contribution to humankind's total energy supply. Assessment of environmental impacts, too, requires quantitative analysis: *How much* pollution does this energy source emit? *At what rate* are we humans increasing the atmospheric CO_2 concentration? *What's the maximum* CO_2 concentration we can tolerate without incurring dangerous climate change? *How long* will nuclear waste remain dangerous? *How much* waste heat does this power plant dump into the river? *How much* CO_2 results from burning a gallon of gasoline? *What's exponential growth* and what are its consequences for future levels of energy consumption, environmental pollution, or carbon emissions? In dealing with such questions, this book doesn't shy away from numbers. At the same time it isn't a heavily mathematical text with equations on every page. Rather, the text attempts to build fluency with quantitative information—a fluency that means being able to make quick order-of-magnitude estimates, work out quantitative answers to simple "how much" questions, and to "read" numerical information from graphs. The book doesn't require higher mathematics—there's no calculus here—but it does demand your willingness to confront quantitative data and to work comfortably with simple equations. Anyone with a solid background in high-school algebra can handle the material here. As for a science background, the text assumes only that the reader has some familiarity with high-school level chemistry and/or physics. Despite its scientific orientation, this book is written in a lively, conversational style that students have welcomed in my other textbooks.

Energy, Environment, and Climate helps reinforce qualitative and quantitative understandings with its end-of-chapter activities. **Chapter Reviews** summarize the big ideas presented in each chapter, invite the student to consider the meaning of new terms introduced in the chapter, and recap important quantitative information and equations. **Questions** probe the concepts behind energy sources, environmental impacts, and climate issues. **Exercises** involve calculations based on the material introduced in each chapter. Answers to the odd-numbered exercises are provided at the back of the book. **Research Problems** send the student to sources of contemporary data—usually Web-based—and allow for more detailed exploration of questions that may be related to energy and environmental issues in the student's home state or country. Given the discoveries quickly unfolding in this growing field, research problems may also ask the student to update data presented in the book or to look more deeply into quantitative data on global energy use and its impacts. **Argue Your Case** questions ask you to formulate an authoritative argument for or against propositions relevant to each chapter's topics.

Energy, Environment, and Climate is illustrated with photos, line drawings, and graphs. Line drawings describe the workings of energy technologies, the flows of energy and material throughout the Earth system, climate models and feedback effects, pollution control and waste storage systems, and a host of other content that's best seen to be understood. Photos are largely of actual energy systems, presented to give a sense of the technologies and their scales. Graphs quantitatively describe everything from the breakdown of our energy use by source or by economic sector to projections of future global temperatures. Every graph is traceable to an authoritative source, and a list of credits and data sources documents these sources.

Energy, Environment, and Climate deals with rapidly changing fields, and this second edition is as up-to-date as possible. Nearly every graph and item of numerical data have been updated through 2010 or later. Wind and solar energy have seen great advances since the first edition was published in 2008, including much expanded deployment of large-scale grid-connected systems, and these developments are reflected here. New understandings of climate change have advanced beyond the 2007 Intergovernmental Panel on Climate Change Fourth Assessment Report, and the climate chapters have been appropriately updated—including the first results from the new Representative Concentration Pathway approach to future climate projections that will feature in the 2014 IPCC Fifth Assessment Report. Energy and climate legislation, both national and international, have been thoroughly updated; an example is an expanded discussion of renewable portfolio standards.

Other improvements in the second edition include a thorough rewrite for clarity and brevity, and the addition of 12 new focus boxes with titles that include "350: The Science behind the Number," "End-use and Primary Energy," "The Warmest Year?," "Being Graphically Literate," and "Converting Units." The second edition also has a more international emphasis, while remaining particularly relevant to students in the United States. The end-of-chapter questions and exercises have been revised in light of instructor and student experience, and the number of exercises has increased significantly. Finally, the second edition incorporates recent environmental incidents relevant to the text; these include such events as the *Deepwater Horizon* oil spill; the natural gas explosion in San Bruno, California; and the Fukushima nuclear accident.

Also complementing the main text are tables displaying important energy- and climate-related quantities; some of the most useful also appear on the inside covers. An appendix tabulates relevant properties of materials, ranging from insulation *R* values of building materials to half-lives of radioactive isotopes to global warming potentials of greenhouse gases. A glossary defines all key terms that appear in the book, and includes acronyms as well as symbols for physical units and mathematical quantities. A list of suggested readings and authoritative websites is also provided. In addition, instructors teaching from *Energy, Environment, and Climate* will find supplementary resources at wwnorton.com/instructors, a password-protected website including images of the figures appearing throughout the text.

Energy, Environment, and Climate is not a book of environmental advocacy or activism; it's much more objective than that. I have my own opinions, and I acknowledge that many—although not all—are in line with the views of the broader environmental movement. But I pride myself on independent thinking based on my own study of others' writings and research, and I'd like to encourage you to do the same. I'm also keenly aware that there is a stronger scientific consensus on some issues, particularly climate change, than either the popular media or the general public may realize. I've been careful to base my scientific statements on the consensus of respected scientists and on peer-reviewed literature that's available to you and everyone else for direct examination. At the same time, I understand the uncertainties inherent in science, especially in an area as complex as the interconnected workings of the global environment. I openly state those uncertainties and quantify them whenever possible. That being said, I would be pleased if the knowledge you gain from this book inspires you to work toward change in our collective patterns of energy consumption. I and the majority of my fellow scientists believe such actions are essential in the coming decades if we're to avoid disruptively harmful environmental impacts.

No individual can be an expert on all the topics covered in a book like this one, and during the writing process I've been fortunate to be able to call on specialists in many fields. They've contributed to making this book more authoritative and timely than I, working alone, could have done. With appreciation, I acknowledge the individuals who have given their expert opinion, read drafts of individual chapters, or otherwise contributed advice and encouragement to this project:

Climate experts Gavin Schmidt (NASA Goddard Institute for Space Studies) and Michael Mastrandrea (Stanford University) reviewed the climate chapters and made many helpful suggestions. Dr. William Glassley (Lawrence Livermore National Laboratory and California Energy Commission) reviewed the sections on geothermal energy in Chapter 8; Dr. JoAnn Milliken (Acting Program Manager, U.S. Department of Energy Hydrogen Program) reviewed Chapter 11; Roger Wallace (Vermont Wood Energy) and Greg Pahl (Vermont Biofuels Association) reviewed sections of Chapter 10 on biomass. Others who offered advice include Dr. William Ruddiman (University of Virginia), Dr. Irina Marinov (University of Pennsylvania), the late Dr. Stephen Schneider (Stanford University), Dr. Michael Mann (Pennsylvania State University), Dr. Peter Vitousek (Stanford University), Dr. Robert Romer (Amherst College), Dr. Mark Heald (Swarthmore College), Dr. Gary Brouhard (McGill University), Elizabeth Rosenberg (Argus Media), and George Caplan (Wellesley College). For this second edition, Dr. James Williams of the Monterey Institute provided a thorough review of Chapter 11's new material on electricity. My Middlebury colleagues Sallie Sheldon (biology), Steve Sontum (chemistry) Jon Isham (economics), Jeffrey Munroe (geology), Chris Watters (biology), Bill McKibben (environmental studies), and Grace Spatafora (biology) were kind enough to share their expertise and encouragement. Finally, I thank my former student Peter Mullen for a thorough reading of the manuscript, and I thank both Peter and Wendy Mullen for their support of this and other projects.

In addition to those acknowledged above, I am grateful to the following instructors of energy and/or climate courses who contributed reviews at the request of W. W. Norton. Their comments, many based on instructor and student experiences with the first edition, offered a blend of pedagogical and scientific expertise that has enhanced the readability, teachability, and authority of this textbook.

First Edition Reviewers

Cecilia Bitz (University of Washington)
Robert L. Brenner (University of Iowa)
F. Eugene Dunnam (University of Florida)
Dorothy Freidel (Sonoma State University)
Jonathan P. Mathews (Pennsylvania State University)
James Rabchuk (Western Illinois University)
Ljubisa R. Radovic (Pennsylvania State University)
Sunil V. Somalwar (Rutgers University)

Second Edition Reviewers

Ani Aprahamian (University of Notre Dame)
Anand Balaraman (Georgia Southern University)
Mona Becker (McDaniel College [Western Maryland College])
David Branning (Trinity College)
Andrew J. Friedland (Dartmouth College)

Douglas Kurtze (St. Joseph's University)
Gaytha Langlois (Bryant College)
Qi Lu (St. John's University)
David Marx (Illinois State University)
Gary Pajer (Rider University)
Zoran Pazameta (Eastern Connecticut State University)
Doug Pease (University of Connecticut)
Alvin Saperstein (Wayne State University)
J. Scofield (Oberlin College)
S. Ismat Shah (University of Delaware)
Sam Tanenbaum (Claremont McKenna College)
Scott Wissink (Indiana University Bloomington)
Eric Woods (State University of New York at Albany)

Finally, special thanks to E. J. Zita (Evergreen State College) for her efforts in checking answers to the end-of-chapter exercises and preparing the online Solutions Manual for this text.

I'm honored to be publishing this book with W. W. Norton, and I am indebted to former Norton editor Leo Wiegman for inviting me to write the first book in what we hope will become a substantial Norton list in environmental studies. Inspired in part by this project, Leo left publishing to start his own environmental work; this second edition was then in the able hands of editor Erik Fahlgren and his assistants Mary Lynch and Mina Shaghaghi. Project editor Carla Talmadge and copyeditor Philippa Solomon spearheaded an efficient and successful production process. I am grateful for all their efforts, and it has been a pleasure to work with all of them.

Finally, I thank my family for their support and patience through the long process of bringing this project to fruition.

ABOUT THE AUTHOR

Richard Wolfson is Benjamin F. Wissler Professor of Physics at Middlebury College, where he also teaches environmental studies. He holds a BA from Swarthmore, an MS in environmental studies from the University of Michigan, and a PhD in physics from Dartmouth. His research involves solar astrophysics and terrestrial climate, and he has published nearly 100 papers in the scientific literature. Wolfson's other books include several introductory physics texts as well as *Nuclear Choices: A Citizen's Guide to Nuclear Technology* (1993) and *Simply Einstein: Relativity Demystified* (2003). He has produced four video courses for The Teaching Company's Great Courses series: *Einstein's Relativity and the Quantum Revolution: Modern Physics for Nonscientists* (1999); *Physics in Your Life* (2004); *Earth's Changing Climate* (2007); and *Physics and Our Universe: How It All Works* (2011).

Chapter 1

A CHANGING PLANET

Earth was born some 4.6 billion years ago, and our planet has been changing ever since. Earth's evolution is driven by an interplay between **matter**—the physical "stuff" that makes up the universe—and **energy**, an equally important universal "stuff" associated with motion, heat, and the fundamental forces of nature. It's energy that makes everything happen; without energy, the universe would be a static, unchanging, lifeless place. In Earth's case, agents of change are astrophysical, geological, chemical, and biological. Astrophysical events formed our planet, and occasionally alter its history. Geological events build mountains and wear them down, move continents, shake the solid Earth, and spew gases into the atmosphere. Chemical reactions change the composition of rocks, soils, atmosphere, and oceans. Life appeared on Earth billions of years ago, and soon biological processes were radically altering the planet's atmosphere and chemistry. Hundreds of millions of years ago, life emerged from the oceans to colonize the land. Just a few million years ago, our human species evolved and began the process of **anthropogenic** (i.e., human-caused) environmental change. We've since become sufficiently plentiful and technologically advanced that we're now having a global impact on Planet Earth.

1.1 Earth's Beginnings

The time is some 4.6 billion years ago; the place a vast cloud of interstellar gas and dust about two-thirds of the way out from the center of the Milky Way galaxy. Most of the material in the cloud is hydrogen and helium, the latter formed in the first 30 minutes after the universe began in a colossal explosion we call the Big Bang. But there are smaller amounts of oxygen, carbon, nitrogen, silicon, iron, uranium, and nearly all the other elements. These were formed by nuclear reactions inside massive stars that exploded several billion years earlier and spewed their contents into the interstellar medium.

Gravitational attraction between the gas and dust particles that make up the cloud causes it to shrink, and—like ice skaters who come together, join hands, and spin—the shrinking cloud begins to rotate. As it rotates, it flattens into a disk, with all the matter in essentially the same plane. The collapse is remarkably rapid, taking only about 100,000 years.

A massive accumulation develops at the disk's center, and under the crushing pressure of gravitational attraction, its temperature rises. Eventually the central mass becomes so hot that hydrogen nuclei—protons—join, through a series of nuclear reactions, to produce helium. This process liberates vast amounts of energy, much of which escapes in the form of light. The Sun is born! The newborn Sun is about 30% fainter than it is today, but its energy output is still equivalent to some 300 trillion trillion 100-watt lightbulbs (that's 3×10^{26} watts). Nuclear "burning" in the Sun's core can sustain the star for 10 billion years, during which time it will grow slowly brighter.

Farther out in the disk, dust particles occasionally collide and stick together. Mutual gravitation attracts more material, and small clumps form. These, too, collide and the clumps grow; in a mere million years the largest have reached kilometer sizes. The more massive clumps exert stronger gravitational forces, so they attract additional matter and grow still larger. After another 100 million years or so, the nascent Solar System has planet-size accumulations—protoplanets—including the newborn Earth. But large numbers of smaller chunks persist in the mix, and they bombard the protoplanets mercilessly, cratering their surfaces and heating them.

In the young Earth, heavier elements sink toward the center, forming Earth's core, and lighter elements float to the surface, eventually forming a solid crust. Gases escape from the interior to form a primitive atmosphere of mostly carbon dioxide and nitrogen, although hydrogen may have been abundant as well. Chunks of interplanetary matter continue their relentless bombardment, heating and reheating the planet. When Earth is a mere 50 million years old, a Mars-size object plows into the young planet, leaving it molten and so hot that it glows for a thousand years like a faint star. Material ejected in the collision condenses to form Earth's Moon. But eventually Earth and its fellow planets have swept up much of the interplanetary stuff, and the bombardment tapers off, although occasional Earth-shaking impacts occur throughout the planet's history. About 4 billion years ago the planet cools, and water vapor condenses to form primeval oceans.

The structure of Earth today reflects the basic processes from those early times. Its center is a solid inner core, mostly iron, at a temperature of many thousands of degrees Celsius. Surrounding this is an outer core of liquid iron, whose motions generate the magnetic field that helps protect us surface dwellers from high-energy cosmic radiation. Covering the core is the mantle—a hot, thick layer that's solid on short timescales but fluid over millions of years. On top of the mantle sits the thin solid crust on which we live. Thermally driven motions in the mantle result in continental drift, rearranging the gross features of Earth's surface over hundreds of millions of years, and giving rise to volcanic and seismic activity. Figure 1.1 takes a cross-sectional look at our planet.

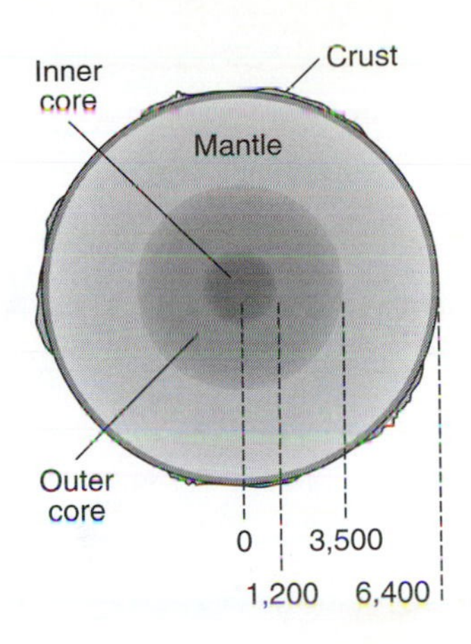

FIGURE 1.1
Structure of Earth's interior. The crust is not shown to scale; its thickness varies from about 5 to 70 km. The distances indicated are measured in kilometers from Earth's center.

1.2 Early Primitive Life

Sometime between 4.2 and 3.5 billion years ago, interactions among natural chemical substances, driven by available energy, led to the emergence of primitive life. The oldest unambiguous evidence for life consists of fossil algae from 3.5 billion years ago, with less direct evidence going back 3.8 billion years. Some would argue that life arose within a few hundred million years of Earth's origin, perhaps as early as 4.2 billion years ago, but the violent bombardment and geological activity of Earth's first few hundred million years have obliterated any firm evidence for very early life. Nevertheless, even the 3.5-billion-year age of the earliest fossils shows that life has been a feature of Planet Earth for most of its history.

We don't know for sure how life developed. Today some biogeologists regard the formation of life as a natural continuation of the processes that differentiated Earth into its core, mantle, and crust. In this view, the first life probably arose deep underground and was fueled by a chemical disequilibrium resulting from Earth's internal heat. Primitive bacterial life of this sort could be common in the universe, with habitats ranging from Earth-like planets to the satellites of distant worlds. The more advanced forms of life we know on Earth, however, probably require specialized conditions, in particular a habitable planetary surface.

The earliest life-forms didn't change much over the billions of years. In fact, some early fossil bacteria and algae are strikingly similar to their modern counterparts. There's a good reason for this: Such simple organisms are generalists, capable of surviving under a wide range of environmental conditions. In that sense they're highly successful, and there's little pressure on them to evolve. In contrast, most highly evolved organisms are specialists, surviving in narrow ecological niches and subject to continuing evolutionary pressures or, worse, extinction as the environment changes.

PHOTOSYNTHESIS

The earliest organisms extracted energy from their chemical surroundings, energy that was ultimately **geothermal**, meaning it came from Earth's interior heat. Some of those organisms, called chemotrophs, are still at work in thermal vents on the ocean floor and at other subsurface locations. But at some point, organisms near the ocean's surface developed the ability to capture sunlight energy and store it in organic molecules built from from carbon dioxide (CO_2) and water (H_2O). This is the process of **photosynthesis**. The stored energy in these molecules is available to the photosynthesizing organisms and to others that prey on them. But there is—or rather was, at first—a downside to photosynthesis: The process released a new chemical compound, the gas oxygen (O_2), into Earth's atmosphere. Oxygen is highly reactive, destructive of many chemical compounds, and therefore was toxic to the early life that had begun, inadvertently, to pollute its environment by creating this new substance.

Pinning down when photosynthesis first began is almost as hard as timing the origin of life itself. Solid fossil evidence shows photosynthetic bacteria dating to 2.7 billion years ago, but the process may have originated as much as a billion years earlier.

1.3 Evolution of Earth's Atmosphere

The histories of life and the atmosphere on Planet Earth are inextricably intertwined, so it's appropriate to pause here and focus on the atmosphere itself. Gases released from Earth's interior gave the young planet an atmosphere that was largely carbon dioxide (CO_2) and nitrogen (N_2), with trace amounts of methane (CH_4), ammonia (NH_3), sulfur dioxide (SO_2), and hydrochloric acid (HCl). Earth's gravity was not sufficient to hold onto hydrogen and helium, so these lighter gases escaped to space. Water vapor (H_2O) was probably a significant atmospheric component very early on, before most of it condensed to form the oceans.

Over roughly Earth's first 2 billion years, the levels of methane, ammonia, and carbon dioxide declined slowly. The details of this early atmospheric history are sketchy, but it's believed that geochemical and biological removal accounted for the decline in atmospheric CO_2. In the geochemical process, CO_2 dissolves in atmospheric water droplets to form carbonic acid (H_2CO_3). Rain carries the acid-laden droplets to the planet's surface, where the carbonic acid reacts with exposed rocks. The effect is to remove CO_2 from the atmosphere and **sequester** it—that is, to store and isolate it—in Earth's crust. In biological removal, early photosynthetic organisms at the ocean surface took CO_2 from the atmosphere and, when they died and sank to the deep ocean, sequestered the carbon in sediments that eventually became sedimentary rocks. The relative importance of geochemical versus biological CO_2 removal is not clear, and scientists are still debating these and other mechanisms. But it's clear that over billions of years CO_2 went from being a major atmospheric component to a gas present only in trace amounts.

Atmospheric nitrogen in the form N_2 is largely nonreactive, so it did not experience significant removal. As a result, Earth's atmosphere from about 3.5 to 2.5 billion years ago was largely nitrogen.

Even as CO_2 declined, atmospheric oxygen was increasing as photosynthetic organisms released O_2 gas. At first the rise in oxygen was slow because the highly reactive O_2 combined with iron and other substances in the oceans and surface rocks, a process called oxidation. But by around 2 billion years ago Earth's exposed surface had become almost fully oxidized, and atmospheric oxygen began to increase significantly. By 1.5 billion years ago, atmospheric oxygen may have reached its current concentration of around 21%. Nearly all the rest was, and still is, nitrogen.

A planetary atmosphere containing free oxygen is unusual. Of all the bodies in our Solar System, only Earth shows significant atmospheric oxygen. Because it's so reactive, oxygen in the form of O_2 soon disappears unless it's somehow replenished. On Earth, that replenishment occurs through photosynthesis. Both the origin and continued existence of our oxygen-rich atmosphere are the work of living organisms. Surely this global modification of Earth's atmosphere ranks as one of life's most profound impacts on our planet. Incidentally, many astrobiologists believe that finding the signature of oxygen in a distant planet's atmosphere might strongly suggest the presence of life.

STRUCTURE OF THE ATMOSPHERE

Like the planet itself, Earth's atmosphere has several distinct layers. At the bottom, extending from the surface to an altitude that varies between about 8 and 18 kilometers (km), is the **troposphere**. Some 80% of the atmosphere's total mass lies within the troposphere, and it's in the troposphere that most phenomena of weather occur. The temperature generally declines with increasing altitude, although particular meteorological conditions may alter this trend in the lower troposphere. A fairly sharp transition, the **tropopause**, marks the upper limit of the troposphere. Above this is the **stratosphere**, which extends upward to some 50 km. The stratosphere is calmer and more stable than the troposphere; only the tops of the largest thunderstorms penetrate into its lowest reaches. The stratosphere contains the well-known **ozone layer** that protects us surface dwellers from harmful ultraviolet radiation. The formation of ozone (O_3) requires life-produced oxygen, so here's another way in which life helped modify Earth's environment, in this case making the land surface a safe place to live. The absorption of solar ultraviolet radiation causes the stratospheric temperature to increase with altitude. Only the troposphere and stratosphere suffer significant impacts from human activity, and these two layers also play the dominant role in Earth's climate. Above the stratosphere lie the **mesosphere** and **thermosphere**, where the atmosphere thins gradually into the near vacuum of space. There is no abrupt endpoint at which the atmosphere stops and space begins. Figure 1.2 shows the structure of Earth's atmosphere, including a typical temperature profile.

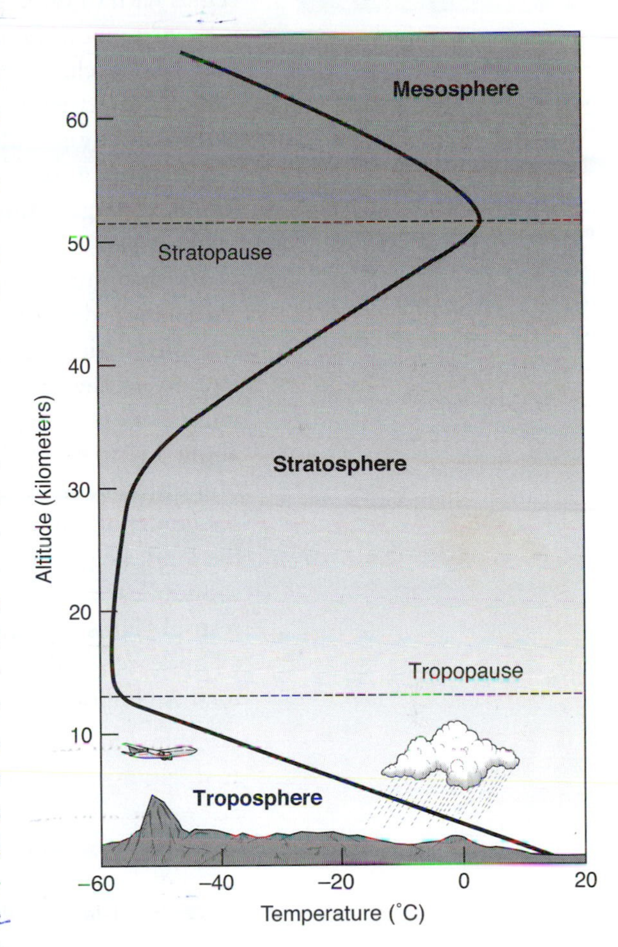

FIGURE 1.2
Structure of Earth's atmosphere, showing a typical temperature profile. Nearly all weather occurs in the troposphere, whereas the stratosphere is important in absorbing solar ultraviolet radiation. Tropopause altitude varies between about 8 and 18 km. Not shown is the thermosphere, a region of high temperature but near vacuum that lies above the mesosphere.

FIGURE 1.3
The Cambrian period, from about 550 to 490 million years ago, produced an enormous diversity of marine life-forms, shown here in an artist's conception.

1.4 Aerobic Life

Back to the discussion of life—because, again, the evolution of life and atmosphere are inextricably linked. Although oxygen was toxic to the life-forms that originally produced it, evolution soon led to new life-forms that could use oxygen in their energy-releasing metabolic processes. In the new oxygen-based metabolism, the process of **aerobic respiration** combines organic molecules with oxygen, producing CO_2 and water and releasing energy. The result is a cycling of oxygen back and forth between life and atmosphere, alternating between the chemical forms CO_2 and O_2.

Because oxygen is so reactive, aerobic respiration releases energy at a greater rate than the **anaerobic respiration** that took place—and still occurs today—in the absence of oxygen. Aerobic respiration therefore helped facilitate the evolution of larger, more complex, and more mobile life-forms exhibiting new behaviors.

Another important behavior that emerged about a billion years ago was sexual reproduction—the organized intermingling of genetic information from two distinct individuals. This led immediately to much greater diversity of life and an acceleration of the evolutionary process. Soon thereafter, about 850 million years ago, the first multicelled organisms appeared. The period from about 550 to 490 million years ago then produced a tremendous diversification of multicelled life-forms (Fig. 1.3). At this point, life was still a strictly marine phenomenon, but around 400 million years ago plants had begun to colonize the land, beginning another of life's major alterations to the planet. Animals soon followed and could take advantage of the food source represented by terrestrial plants. Amphibians, reptiles (including dinosaurs), birds, and mammals all appeared in the last 400 million years or so of Earth's 4.5-billion-year history. An important era in the context of this book is the Carboniferous period,

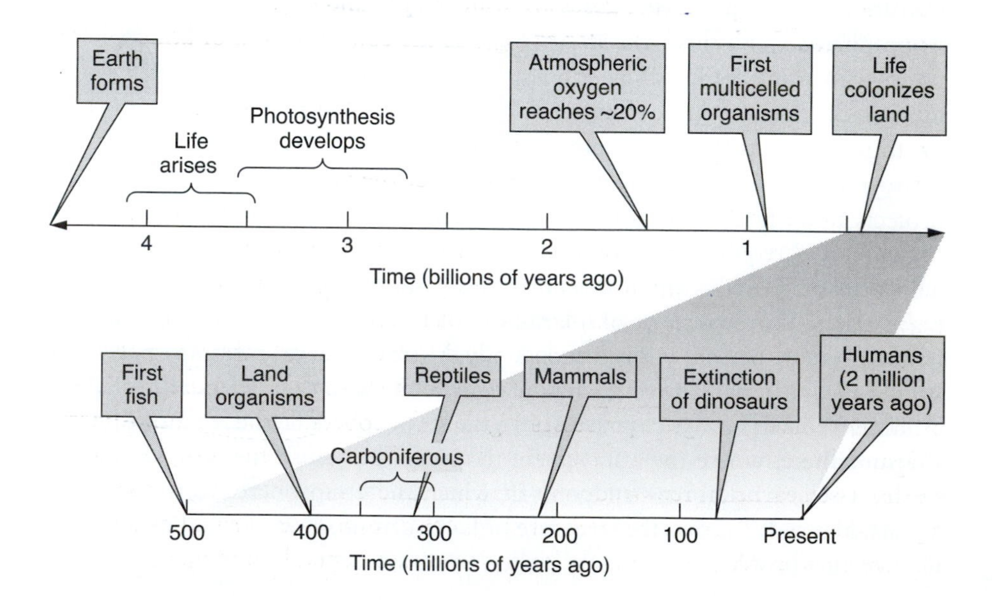

FIGURE 1.4
Some major events in the history of life on Earth. The origins of life and photosynthesis are uncertain, and the dates given for the last 500 million years represent the earliest definitive fossil evidence.

some 360 to 300 million years ago, when prolific growth of the forests led to the burial of carbon that eventually became **fossil fuels**. Finally, human ancestors of the genus *Homo* evolved in about the last 2 million years, and we *Homo sapiens* have been around for only 200,000 years. Figure 1.4 is a timeline of life's evolution on Earth.

1.5 Earth's Changing Climate

Climate describes the average conditions that prevail in Earth's atmosphere—temperature, humidity, cloudiness, and so forth—and the resulting conditions at Earth's surface. Climate is distinct from weather, which describes immediate, local conditions. Weather varies substantially from day to day and even hour to hour, and changes regularly with the seasons. Climate, being an expression of average weather, changes on longer timescales. But change it does, and natural climate change has been a regular occurrence throughout Earth's history.

We've already seen that life and atmosphere are linked through Earth's life-produced atmospheric oxygen and the aerobic organisms that evolved to take advantage of it. Climate, too, is obviously linked with life because the climate of a region determines the kind of life that can survive there. Many factors go into determining Earth's climate, but the two most important are light from the Sun and the composition of Earth's atmosphere. Sunlight provides the energy that warms our planet and drives the circulation of atmosphere and oceans.

Sunlight and atmosphere interact to establish Earth's climate; sparing the details for now, here are the big ideas: (1) Sunlight brings energy to Earth, warming the planet. (2) Earth returns that energy to space, establishing an energy balance that maintains a fairly constant average temperature. (3) **Greenhouse gases** in the atmosphere act like a blanket, blocking the outgoing energy and making Earth's surface temperature higher than it would be otherwise. The most important of these gases are water vapor and CO_2. Change either the rate at which Earth receives solar energy, or the concentration of atmospheric greenhouse gases, and you change Earth's climate.

We know from well-established theories of stellar evolution that the newborn Sun was some 30% fainter than today. We have a very rough record of the average temperature at Earth's surface over the past 3 billion years, and it shows, remarkably, that the temperature hasn't tracked the gradual increase in the Sun's energy output. Take a look at this long-term temperature record shown in Figure 1.5: Despite a fainter young Sun, throughout much of Earth's history our planet has been warmer than it is today.

Many factors influence climate, and they act on timescales ranging from years to billions of years. But over the billions of years' time shown in Figure 1.5, scientists believe that geological processes regulate the concentration of atmospheric CO_2 and thus establish a relatively stable climate. The basic idea is simple: CO_2, as we discussed earlier, is removed from the atmosphere by weathering of rocks; it's replenished by CO_2 escaping from Earth's interior, especially through volcanoes. The chemical reactions that constitute weathering depend on temperature; the

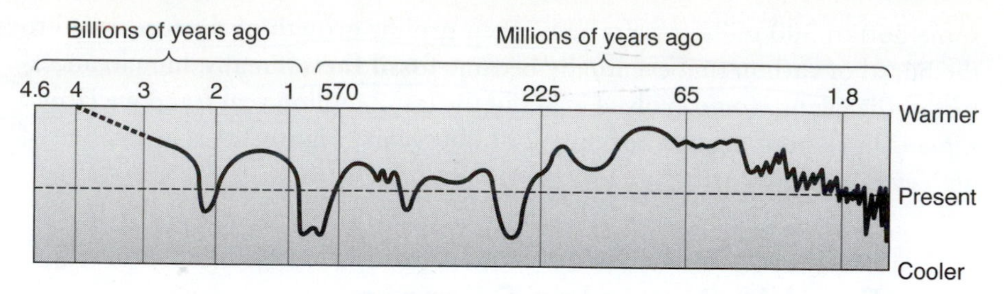

FIGURE 1.5

A rough estimate of Earth's temperature history over the past 3 billion years shows that much of the time it has been warmer than the present, despite the Sun's steadily increasing energy output. The temperature scale is only semiquantitative, with the overall variation shown being around 30°C—comparable to winter–summer differences in today's temperate climates. Note that the horizontal timescale is not uniform.

higher the temperature, the greater the weathering rate. And it's precipitation that brings CO_2 to Earth's surface in the form of the weathering agent carbonic acid. Precipitation, in turn, depends on how much water evaporates into the atmosphere—and that also increases with temperature. With increased temperature, then, both the rate of weathering reactions and the amount of precipitation increase. Those increases promote greater weathering, and thus remove more CO_2 from the atmosphere. With less CO_2 the atmosphere acts less like an insulating blanket, and Earth's surface cools. This drop in temperature decreases the rate of weathering, and the continuing CO_2 emission from volcanoes gradually increases the atmosphere's CO_2 concentration. These two conditions enhance the insulating blanket, and Earth's surface warms.

What I've just described is a process of **negative feedback**. Earth warms, and the Earth–atmosphere system responds in a way that counters the warming. Earth cools, and the system responds to counter the cooling. This is *feedback* because a system, in this case the Earth and atmosphere together, responds to changes in itself. It's *negative* feedback because the response opposes the initial effect.

Scientists believe that the negative feedback process of CO_2 removal by rock weathering has acted over geologic time much like a household thermostat, regulating Earth's temperature within a fairly narrow range, even as the Sun's energy output gradually increased (Fig. 1.6).

SNOWBALL EARTH

Geological temperature regulation hasn't been perfect. Changes in volcanic activity, continental drift, variations in Earth's orbit, and other factors have led to excursions toward warmer or cooler conditions, as suggested in Figure 1.5. Scientists have found evidence of dramatic climate swings that plunged Earth into frozen "snowball" states that were followed by rapid warming. Perhaps as many as four such snowball episodes occurred in the period between about 750 and 580 million years ago. During that time Earth's continents were probably clustered near the equator, and the warm equatorial precipitation made for especially rapid removal of atmospheric CO_2 by rock weathering. Atmospheric

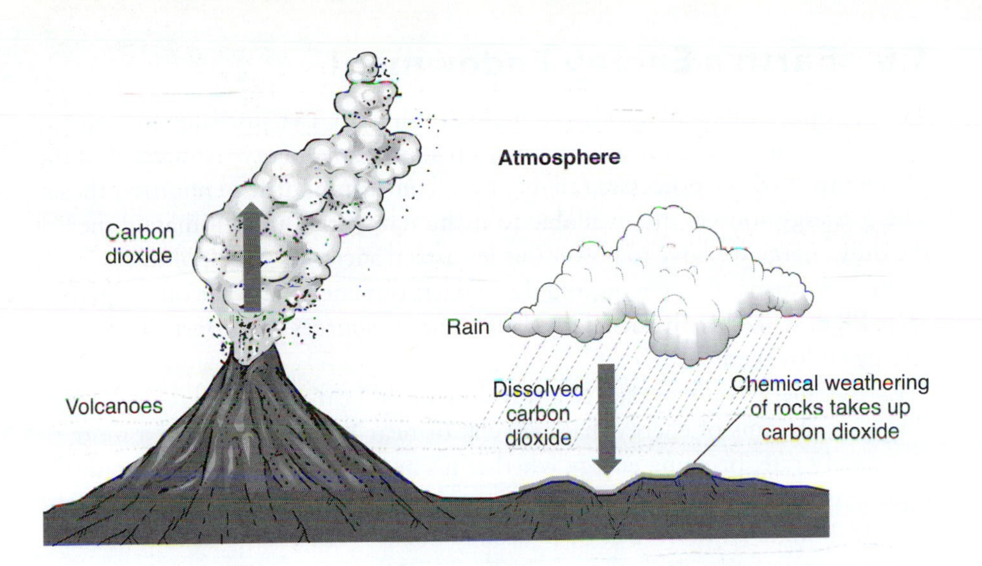

FIGURE 1.6
Over geological time, removal of CO_2 by precipitation and chemical weathering of rocks balances volcanic CO_2 emissions. Carbon dioxide removal increases with temperature, providing a feedback that regulates Earth's temperature.

CO_2 plunged, and ice advanced across the landless northern and southern hemispheres. The larger expanses of ice reflected more of the incoming solar energy, cooling the planet further. The process overwhelmed the natural weathering thermostat, and soon the entire ocean was covered with ice. Starved of precipitation, land glaciers couldn't grow and thus some of the land remained ice-free (Fig. 1.7).

There's a problem here: Ice reflects most of the sunlight incident on it, so once Earth froze solid it would seem impossible for it ever to warm up and thaw again. But remember those volcanoes, which would continue to spew CO_2 from Earth's interior. Normally the atmospheric CO_2 concentration remains fairly constant, with CO_2 removal by weathering occurring at roughly the same rate as volcanic CO_2 emission. With the oceans frozen, however, there was no water to evaporate, precipitate, and cause rock weathering. But volcanism continued, driven by the planet's internal heat, so atmospheric CO_2 increased rapidly—and with it the insulating greenhouse effect and therefore Earth's surface temperature. Eventually equatorial ice melted, exposing dark ocean water to the strong tropical sunlight. The warming rate increased. Both theory and geological evidence suggest that the climate swung from extreme cold to stiflingly hot and wet in just a few centuries. Eventually the geological weathering feedback got things under control, and the climate returned to a more temperate state.

What happened to life during Earth's snowball episodes? Life, still entirely aquatic at this time, hunkered down beneath the kilometer-thick ice that covered the oceans, living off energy escaping from Earth's interior. Many single-celled organisms probably went extinct. Others, however, clustered around geothermal heat sources on the ocean floor, where they evolved in isolation and thus increased the overall diversity of living forms. In fact, some scientists credit snowball Earth episodes and their subsequent hot spells with engendering the huge proliferation of life during the Cambrian period around 500 million years ago (see Fig. 1.3).

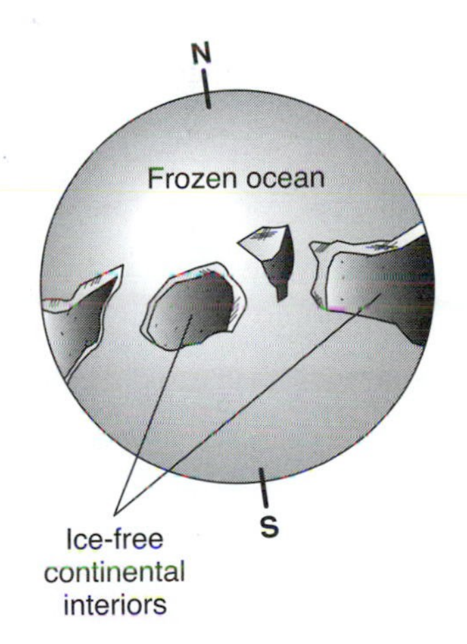

FIGURE 1.7
Earth at the height of a snowball episode. The continents are clustered near the equator, and the entire ocean is frozen to a depth of 1 km. The lack of precipitation arrests glaciers and leaves some land ice-free.

1.6 Earth's Energy Endowment

The astrophysical, geological, and biological history I've just summarized has left Planet Earth with a number of natural sources of energy—sources that the planet taps to drive processes ranging from continental drift to photosynthesis. These energy sources are available to us humans, too, and ultimately they're the only energy we have to power our industrial societies. Much of this book is about how we use that energy and the impacts our energy use has on the planet. Here I'll describe briefly the relatively few energy sources that constitute Earth's energy endowment.

Earth's energy sources take two forms, which I'll call *flows* and *fuels*. **Energy flows** are streams of energy that arrive at or near Earth's surface at a more or less steady rate, bringing energy whether it's needed or not. **Fuels**, in contrast, represent energy that's stored in one form or another—most commonly in the chemical bonds of molecules or in atomic nuclei. Fuel energy usually remains stored energy until a deliberate act liberates it.

ENERGY FLOWS

By far, the dominant energy flow is sunlight, streaming across the 93 million miles of space between Sun and Earth in a mere 8 minutes. Sunlight energy impinges on Earth's atmosphere at the rate of about 10^{17} watts—the rate at which 1,000 trillion 100-watt lightbulbs would use energy. This flow of solar energy represents 99.98% of all the energy arriving at Earth.

What happens to all that solar energy? Except for a nearly infinitesimal portion, it's soon returned to space. By "soon" I mean on timescales ranging from nearly instantaneous to the thousands of years it's stored in the longest-living plants; in any event, nearly all the solar energy enters into the Earth system and goes back out on very short timescales compared with Earth's geological history. This balance between incoming and outgoing energy determines Earth's climate.

Solar energy performs some useful roles between its arrival at Earth and its return to space. Figure 1.8 shows the big picture of Earth's energy flows, including solar and some additional flows. As the figure shows, some 30% of the incident solar energy is reflected immediately back to space, most of it from clouds and reflective particles in the atmosphere and some from ice, snow, light-colored deserts, and other surface features. Another 46% turns directly into thermal energy—what we call, loosely, "heat." That energy gets back to space almost immediately, through the process of radiation that I'll describe in Chapter 4. Another 23% of the solar energy goes into evaporating water, where it's stored in the atmosphere for periods of typically a few days as so-called latent heat—energy that's released when water vapor condenses back to liquid. A mere 1% ends up as the energy of moving air and water—the winds and ocean currents. Chapter 10 shows how we humans capture energy from these indirect solar flows associated with evaporation, wind, and currents. Finally, a tiny 0.08%—less than 1 part in 1,000—of the incident solar energy is captured by photosynthetic plants and becomes the energy that sustains nearly all

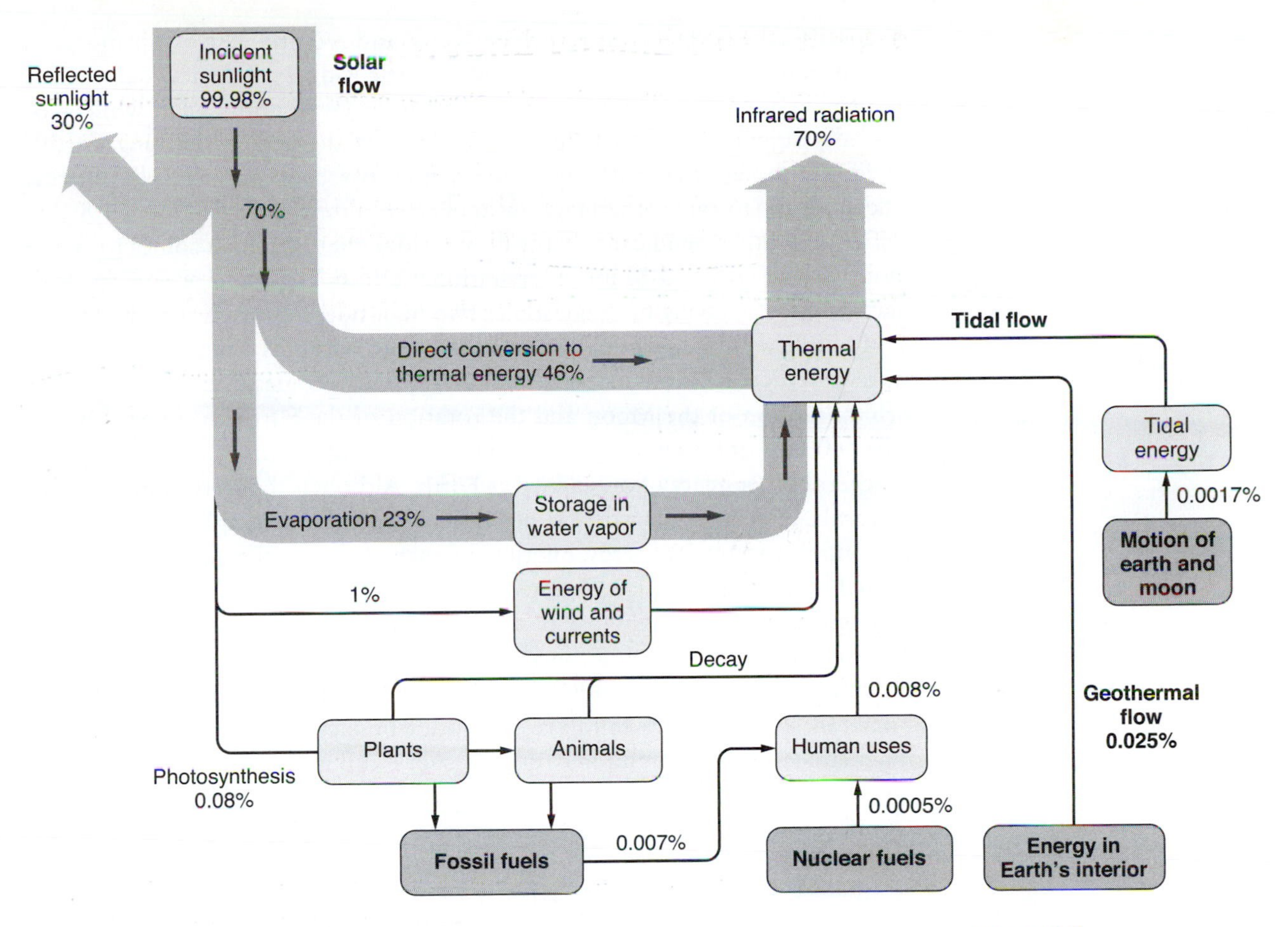

FIGURE 1.8
Earth's energy flows, nearly all of which are from the Sun. Other flows are geothermal and tidal energy. In addition, energy is stored in fossil and nuclear fuels. The 70% of incoming sunlight that's absorbed in the Earth–atmosphere system is almost perfectly balanced by the same amount of infrared radiation returned to space.

life. (I say "nearly all" because some isolated ecosystems thrive on geothermal energy released at undersea vents.) Organisms store photosynthetically captured sunlight as chemical energy in their bodies, and it's this energy that may stick around a while until it all returns to space after death and decay. Actually, it's not quite all. That infinitesimal portion I mentioned a couple of paragraphs ago gets buried without decay and becomes the energy stored in fossil fuels.

Our planet's interior is hot—thousands of degrees Celsius at the core. The source of this geothermal energy is a mix of primordial heat liberated in the gravitational accretion of matter that built our planet over 4 billion years ago, and ongoing energy release in the decay of long-lived natural radioactive elements, particularly uranium and thorium. Geologists aren't sure just how much of Earth's thermal energy is primordial and how much is due to radioactivity, but both sources are probably significant. The internal thermal energy causes the temperature to increase with depth into the planet; near the surface the rate of rise is typically about 25°C per kilometer, or roughly 70°F per mile.

This temperature difference between Earth's interior and surface drives a steady flow of energy to the surface. On average, that flow amounts to some 0.087 W arriving at every square meter of Earth's surface. The rate is much greater in regions of concentrated geothermal activity, where hot molten rock is closer to the surface. These include active and recently active volcanoes, as well as undersea rift zones like the Mid-Atlantic Ridge, where new ocean floor is forming as the continents move slowly apart. Geothermal heat, while geologically significant, contributes just 0.025% of the energy reaching Earth's surface, although the geothermal energy flow is second only to solar energy. The modest natural geothermal flow appears as a thin arrow in Figure 1.8.

The only other significant energy flow is **tidal energy**, which comes from the motions of Earth and Moon. Gravitational effects cause the oceans to slosh back and forth, giving most coastlines two high tides a day. The energy associated with the tides is dissipated, eventually ending up as heat, as the moving water interacts with the shores. That energy, which comes ultimately from the orbital motion of the Moon and the rotation of the Earth, accounts for less than 0.002% of the energy flowing to Earth (see Fig. 1.8).

Several other energy flows arrive at Earth. Although they have physical and cultural significance, they deliver negligible energy. For example, starlight is an insignificant source of energy, but the presence of stars in the night sky has long stirred human imagination, inspired art and literature, and given astronomers throughout the ages a glimpse of the rich universe beyond our Solar System. The Sun itself puts out additional energy in the form of a wind of particles, flowing past Earth at some 400 kilometers per second (km/s). Especially strong bursts of solar particles can have significant impact on Earth's electrical and magnetic environment and may exert a subtle influence on climate by affecting the chemistry of the upper atmosphere. Cosmic rays—high-energy particles from beyond the Solar System—are yet another source. They play a role in the spontaneous mutations that drive biological evolution. Finally, our planet—and indeed the whole universe—is bathed in microwave radiation that originated some 400,000 years after the Big Bang, when the first atoms formed. Energetically, this radiation is totally insignificant, but it has given us profound insights into the origin of the universe itself.

So although Earth is connected to the greater universe through a variety of incoming energy flows, only sunlight, geothermal energy, and tidal energy are at all significant in terms of the energy that drives Earth's natural systems and human society. Of these three, sunlight is by far the dominant source of energy arriving at Earth.

FUELS

Fuels are energy sources that Earth acquired long ago, in the form of substances whose molecular or nuclear configurations store energy. Most familiar are the fossil fuels, which supply the vast majority of the energy that drives human society. These substances—coal, oil, and natural gas—are relative newcomers to Earth's energy endowment. They formed over the past few hundred million years when once-living organic matter was buried before it had a chance to decay. Chemical and physical action then changed the material into fossil fuels, which contain the solar energy that ancient photosynthetic organisms had captured. Today, Earth's endowment of fossil fuels is considerably smaller than it was even a few decades ago. That's because we humans have already consumed somewhere around half of the readily available fossil fuel resources. Although fossil fuels continue to form today, they form millions of times more slowly than we consume them, so they're effectively a nonrenewable energy resource. Figure 1.8 includes arrows showing both the natural formation of fossil fuels and the human consumption of fossil fuel energy.

In contrast to fossil fuels, nuclear fuels have always been present in the Earth. Natural uranium and thorium formed in stellar processes, and today their radioactive decay provides part of Earth's internal heat and thus part of the geothermal energy flow. Through nuclear reactions, uranium and thorium also constitute a more potent energy source. Today, we use uranium to generate some 15% of humankind's electrical energy. Earth's nuclear fuels formed before the planet itself, and there's no terrestrial process that can create them, so these are truly nonrenewable energy resources. Therefore Figure 1.8 shows a one-way arrow representing human consumption of nuclear fuels.

In principle, every atomic nucleus except that of iron represents a source of stored energy—a nuclear fuel. But so far we've only learned to tap that energy from a few very massive nuclei—most significantly, uranium—having the exceptional property that they readily split in two with a great release of energy. But we're exploring other approaches. At the other end of the nuclear mass scale, the hydrogen in seawater represents a vast nuclear energy source. Fused together to make helium, hydrogen releases so much energy that even a rare form of the element could supply humankind's present-day energy needs for 40 billion years—some 10 times as long as the Sun will continue to shine. We know the process works because it's what powers the Sun itself. On a less positive note, we've also learned to release hydrogen's energy explosively in our thermonuclear weapons, or "hydrogen bombs."

Hydrogen represents a potentially enormous resource in Earth's energy endowment, but one that we simply haven't yet learned to use. Hydrogen as nuclear fuel is not to be confused with the chemical fuel hydrogen (H_2), which would power a so-called hydrogen economy. There isn't any significant amount of H_2 in Earth's endowment, which is a big problem for the hydrogen economy. The hydrogen I'm talking about here is already "burned"—combined with oxygen to make water (H_2O). There's no more chemical energy to be extracted. But there's plenty of that harder-to-get-at nuclear energy. Figure 1.8 shows neither this hydrogen-stored nuclear energy nor any associated flow, because at present it isn't being released at all.

Since I'm being exhaustive about the energy sources available to us on Earth, I should mention that there might be sources outside of Earth's own endowment that humankind could someday use. For example, we might mine fuel substances from the Moon, where, among other things, we could find "unburned" hydrogen gas to use as a chemical fuel. Or we might harvest liquid methane (natural gas) on Saturn's moon Titan. But for now any thoughts of getting to such extraterrestrial energy reserves without expending far more energy than we would extract are pure fancy. That may change someday, but probably not any time soon.

A realistic picture of Earth's energy endowment, then, is this: We have available a substantial, continuous energy flow from the Sun, and much lesser flows from Earth's interior heat and from the tidal energy of the Earth–Moon system. Inside the Earth we have fossil fuels, which we're quickly depleting, and the nuclear fuels uranium and thorium that we know how to exploit. We also have a vast nuclear fuel resource in the hydrogen of seawater, but when and even if we'll learn how to use that one is a wide open question. That's it. When we talk about "energy alternatives," "renewable energy," and other popular solutions to energy shortages and energy-related environmental problems, there's no new, hidden, as-yet-undiscovered source. We either have to turn to one of the known sources that comprise Earth's energy endowment, or we have to use less energy.

1.7 The Human Era

Sometime around 5 million years ago, in eastern Africa, creatures evolved that had relatively large brains and walked erect on two feet. A number of human-like variants developed, some of which died out while others evolved further. About 2 million years ago—a drop in the bucket of geological time—humans of the genus *Homo* first appeared. We have firm evidence that our hominid ancestors were making primitive tools as long as 2.6 million years ago and that they had harnessed fire somewhere between 1.6 million and 790,000 years ago. Many factors distinguish human beings from other animals, but for the purposes of this book, the single most important distinction is that we make significant use of energy beyond what our own body's metabolism provides. In that context, the harnessing of fire is a seminal event in human history.

For much of our species' time on Earth, wood-fueled fire and the power derived from domesticated animals were our only energy sources beyond our own bodies. More recently we added oils extracted from animal bodies, tars and oils naturally seeping to Earth's surface, wind energy, water power, coal, and then oil, natural gas, and uranium. In rare instances we've also tapped the direct flows of solar, geothermal, and tidal energy.

Energy consumption is unquestionably a major factor in the advancement of human civilization. It has enabled us to manufacture goods, from the most primitive iron and bronze implements to the most sophisticated computer chips, spacecraft, and synthetic materials. It has enabled us to live comfortably far beyond the warm climates where our species evolved. Energy helps us transport goods for trade and ourselves for exploration, commerce, education, recreation, and cultural enrichment. Today, energy consumption enables the high-speed communication that binds us into one global community. Not surprisingly, measures of human well-being often correlate closely with energy consumption.

Perhaps most significantly, energy-intensive agriculture enables our planet to support a human population far beyond what would have been possible as recently as a few centuries ago. For most of our species' history, resource limitations and the hardships of the hunter–gatherer lifestyle kept the total world population under 10 million. By the beginning of the common era (C.E.; the year 1), the number of humans had reached about 200 million. By 1750, with the Industrial Revolution underway, that number had climbed to some 800 million; around 1800 it reached 1 billion. The entire nineteenth century added another half billion, and a full billion joined the world's population in the first half of the twentieth century, making the population just over 2.5 billion in 1950. By 2000 the total had passed 6 billion and in 2012 it reached 7 billion. Figure 1.9 summarizes human population growth over the past 12,000 years. Fortunately for our planet, the steep upswing shown in Figure 1.9 won't continue. As the figure's inset shows, the annual growth rate as a percentage of population peaked in the 1960s at around 2%. It has been declining since, and by 2010 had dropped to just over 1.1%. But the population itself continues to increase and will do so for some decades to come. Although some projections suggest population growth might continue beyond the year 2100, many show Earth's human population peaking at about 9 billion in the middle of the present century, then beginning a gradual decline to more sustainable levels. Many factors contribute to this welcome trend, but most are related to higher standards of living and particularly to education. Energy consumption helps make possible those higher standards.

A GLOBAL IMPACT

Ironically, the same energy consumption that enables us to enjoy an unprecedented level of comfort and material well-being also threatens natural Earth systems that maintain a supportive environment. Although our ancestors did

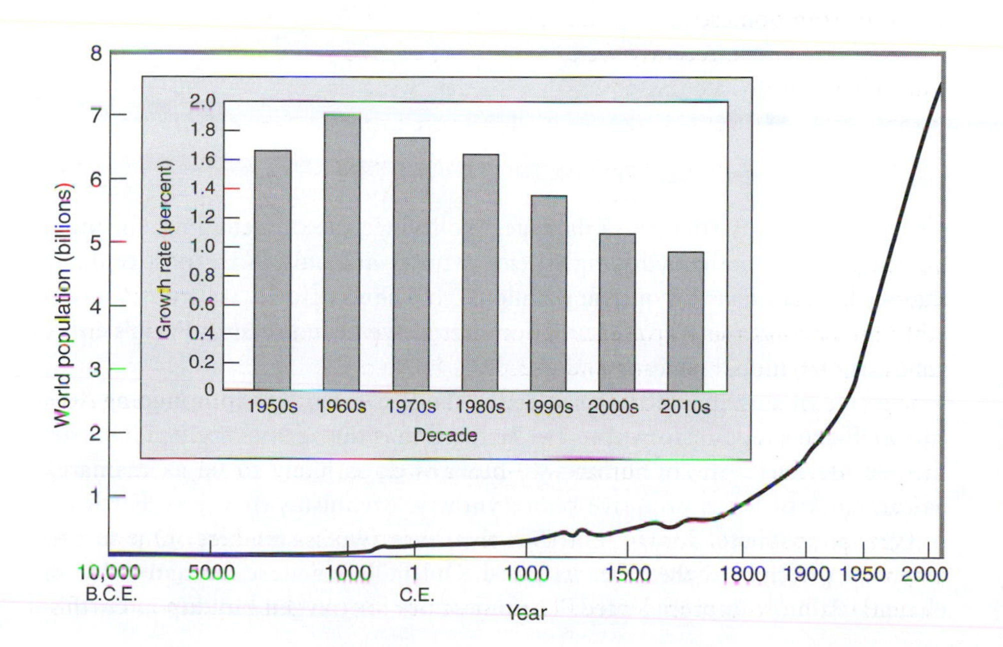

FIGURE 1.9
Human population growth over the past 12,000 years and projected to 2020. Although the world's population continues to grow, the inset shows that the growth rate as a percentage of population peaked in the 1960s. In absolute numbers, the maximum yearly growth occurred in the 1980s.

BOX 1.1 | Being Graphically Literate

A picture, it's said, is worth a thousand words. A graph—a kind of mathematical picture—can be worth thousands of numbers. I'll be using graphs throughout this book, and it's important to know how to "read" them.

Graphs often display *time series*—sequences of data arranged by time. *Line graphs* are commonly used for time series, displaying the data as a continuous line connecting individual data points. The data points themselves may or may not appear. The main part of Figure 1.9 is a line graph. Its horizontal axis is time, measured in years—in this case with a highly nonlinear scale, where a fixed distance on the axis does *not* represent a fixed time interval. Its vertical axis is population, measured in billions of people, and it's linear—meaning that the distance between any two tick marks corresponds to a fixed number, here a billion people. The purpose of the graph isn't just to present a lot of numbers—that could have been done more accurately with a table—but, more importantly, to show a trend. Here that trend is a dramatic increase in population.

To understand a graph, you need to know exactly what it's showing. That means whoever prepared the graph should have labeled the horizontal and vertical axes, in most cases giving numerically labeled tick marks, and indicating units. Is time measured in seconds, or years? Is that energy or power on the vertical axis, and if it's power, is it in watts, kilowatts, or what?

Look carefully at the scaling on the axes. I've already noted that Figure 1.9 has a highly nonlinear horizontal axis. Note that its vertical axis begins at zero population and goes up to 8 billion. Graphs don't always begin at zero, and when they don't the trends they show can appear exaggerated. Beginning a graph at a point other than zero isn't a mistake, but it can be deceptive unless you're sure you understand the vertical scale.

Not all graphs are line graphs. The inset in Figure 1.9 is a *bar graph*, showing population growth rates for distinct decades. In the next chapter you'll meet *pie graphs*, which show how a quantity divides among several categories. You'll also see more complicated line graphs with multiple curves plotted. Sometimes these will share a common vertical axis, but sometimes—as when I plot temperature and carbon dioxide concentration over time—I'll put two distinct vertical axes on the same graph.

In many of the end-of-chapter Research Problems you'll be asked to make your own graphs. An intelligent choice of the graph type, along with careful labeling and scaling of the graph axes, will help communicate the point you're making.

plenty of local environmental damage—polluting waters, denuding hillsides, burning forests, perhaps driving species extinct—it's only in the past century that we humans have become so populous, so technologically vigorous, and so extravagant in our energy consumption that we've begun to alter Earth's environment on a global scale.

The title of this chapter suggests that change is a natural and ongoing feature of Earth's long history. And I've stressed that the agents of global change include life itself. Surely nothing we humans do is likely to be as dramatic as the work of those primitive photosynthetic organisms that gave Earth its oxygen atmosphere. Maybe not. But there are two issues here: One is the extent of the change, the other its speed. On the latter score, human-induced change is almost unprecedented. The timescales for oxygen buildup in Earth's atmosphere and for the tectonic rearrangement of the continents are measured in hundreds of millions of years. The timescale for climate swings into and out of ice ages is thousands of years. But the timescale for the substantial human-caused increase in atmospheric CO_2 is about 100 years. As I've outlined already in discussing the long-term history of Earth's climate, atmospheric CO_2 is a key factor in determining Earth's temperature. So here's a human-caused environmental change, resulting largely from combustion of fossil fuels, that has a direct bearing on global climate. And it's a change that's occurring on an unnaturally rapid timescale.

That last remark needs qualification: Although most natural changes in the global environment occur much more slowly than recent anthropogenic changes, there are exceptions. Climatologists have detected extremely rapid global temperature fluctuations—on timescales as short as a few decades—as Earth came out of the last Ice Age. And as we saw earlier, the ending of a snowball Earth episode may have taken the planet from the deep freeze to a superhot, moist climate in a matter of a century or so. But none of those sudden changes occurred in the presence of a human civilization stretching the planet's resources in an effort to feed, transport, educate, and provide material goods for a burgeoning population. Our advanced human society can ill afford rapid environmental change, and yet such change is just what our very advancement has made increasingly likely.

YOUR ROLE

Given that you're reading this book, it's a good bet that you belong to one of the better-fed, better-educated, and more affluent subsets of human society. If that's true, then no matter how carefully you might try to maintain an environmentally sustainable lifestyle, you're responsible for substantial rates of energy consumption, environmental contamination, CO_2 emission, resource depletion, and other affronts to the global environment. On the other hand, you probably also have the education, wealth, political savvy, and influence to help improve humankind's relationship with Planet Earth.

CHAPTER REVIEW

BIG IDEAS

1.1 Earth is 4.6 billion years old.

1.2 Life developed around 4 billion years ago. Photosynthetic life arose about 3.7 to 2.7 billion years ago.

1.3 Earth's early atmosphere consisted of carbon dioxide, nitrogen, methane, and other gases from the planet's interior. By about 2 billion years ago, photosynthetic plants had added significant amounts of oxygen. Today's atmosphere is largely nitrogen and oxygen.

1.4 Oxygen led to new, more complex, **aerobic** life-forms that could utilize food energy at a greater rate.

1.5 Earth's climate has varied, and is governed by factors including solar luminosity and atmospheric CO_2. Rock weathering and volcanism regulate CO_2 over the long term.

1.6 Earth's energy resources include **fuels** that store energy and **energy flows** that deliver continuous streams of energy. Sunlight is the dominant energy flow, and a small fraction of sunlight is trapped and stored in **fossil fuels**.

1.7 Humankind's earliest ancestors evolved about 5 million years ago, and we are now sufficiently numerous and technological that our impact on Earth is comparable to that of natural processes. Much of that impact stems from our use of energy far in excess of what our own bodies can produce.

TERMS TO KNOW

aerobic respiration (p. 6)
anaerobic respiration (p. 6)
anthropogenic (p. 1)
climate (p. 7)
energy (p. 1)
energy flow (p. 10)
fossil fuel (p. 7)
fuel (p. 10)
geothermal energy (p. 3)
greenhouse gas (p. 7)
matter (p. 1)
mesosphere (p. 5)
negative feedback (p. 8)
ozone layer (p. 5)
photosynthesis (p. 4)
sequester (p. 4)
stratosphere (p. 5)
thermosphere (p. 5)
tidal energy (p. 12)
tropopause (p. 5)
troposphere (p. 5)

GETTING QUANTITATIVE

Earth forms: 4.6 billion years ago

Oxygen-containing atmosphere: ~ 2 billion years ago

Fraction of Earth's energy from the Sun: 99.98%

Human species evolves: ~ 5 million years ago

Humans harness fire: ~ 1 million years ago

Current human population: ~ 7 billion

QUESTIONS

1 What is the most significant change that life has made in Earth's environment?

2 Why can't we tell much about the possible origin of life on Earth before about 4 billion years ago?

3 Why did atmospheric oxygen concentration not rise immediately after life "invented" photosynthesis?

4 What's the difference between an energy flow and a fuel?

5 When Earth was in a snowball state, it would have reflected nearly all the incoming solar energy. So how was our planet able to warm from the snowball state?

6 The huge energy-consumption rate of modern industrial societies poses serious challenges to the environment. Yet energy consumption may help mitigate the growth of world population, which poses its own environmental challenges. Explain.

7 Although the percentage growth rate of world population peaked in the 1960s, it was 1988 when the world added the greatest number of people. Why the difference?

EXERCISES

1 Solar energy is incident on the top of Earth's atmosphere at the rate of about 1,364 watts on every square meter (W/m^2). This energy effectively falls on the cross-sectional area of the planet. From this fact, calculate the total rate at which solar energy arrives at Earth. You'll need to find Earth's radius; see the Physical Constants table on the inside back cover.

2 Use the answer to Exercise 1, along with the percentage of Earth's total energy supplied by the geothermal flow shown in Figure 1.8, to estimate the total geothermal energy flow.

3 In 1965 the world's population was about 3.4 billion and was growing at about 2% annually. In 1985 the population was 4.9 billion, growing at 1.7%, and in 2000 it was 6.1 billion, growing at 1.2%. In which of these three years did the actual number of people increase by the greatest amount? Show by calculating the number in each case.

4 If the 2012 world population growth rate of 1.1% were to continue, what would be the population in 2050? The 2012 population was about 7 billion.

5 Figure 1.8 shows that approximately 0.008% of the energy flow to Earth's surface is associated with humankind's consumption of fossil and nuclear fuels. Use the answer to Exercise 1 to determine the order of magnitude of that flow in watts.

RESEARCH PROBLEMS

(With research problems, always cite your sources!)

1 Prepare a plot of population versus time, like the one in Figure 1.9, but for your own country. Go back as far as you can find data. What factors might affect an individual country's population but not the world's?

2 Find the atmospheric composition of Earth and four other planets or planetary satellites in the Solar System. For each, list at least three of the most abundant gases and their percentage concentration.

3 Washington State's Mount Saint Helens volcano was active during the fall of 2004. Find the dominant gases emitted during this activity and give quantitative estimates of the gaseous emissions if you can find them.

ARGUE YOUR CASE

1 A friend claims that the "human uses" flow in Figure 1.8 is so small compared with natural flows that our energy consumption can't significantly affect the global environment. How do you reply?

2 A friend who's skimmed this chapter thinks it's missing a fundamental energy flow, namely waterpower. Formulate an argument to counter this claim, in the process identifying where waterpower fits into the energy flows of Figure 1.8.

Chapter 2

HIGH-ENERGY SOCIETY

You've just picked up this book, and now I'm going to ask you to put it down. That's because I want to give you a feel for what it means to be a member of a modern, industrial, energy-intensive society. I mean "a feel" literally—a sense, in your own body, of the quantities of energy that you consume directly or that are consumed to support your lifestyle. In the next two chapters we'll develop more rigorous understandings of energy and its measures, but an appreciation for energy and its role in human society is conveyed best by a literal feel for the energy associated with your own body.

So put down your book, and stand as you are able. Put your hands on your hips and start doing deep knee bends. Down and up, down and up—about once a second. Keep it going for a few minutes so you get a real sense of just how vigorously your body is working.

During this short exercise, your body's **power** output—the rate at which it's expending energy—is about 100 **watts** (W). This is the rate at which you're converting energy stored in chemical compounds derived from food into the mechanical energy associated with the motion and lifting of your body. If you're big or exercising more vigorously, the figure is a little higher. If you're small or exercising more slowly, the figure is a bit lower. But it won't be lower than several tens of watts, nor more than several hundred. One hundred watts is a nice round number, and we'll consider it typical of the power output of the average human body.

There are other paths to our 100-W figure. The calories that characterize food are in fact a measure of energy, so calories per day measures power—and you can show (Exercise 3.1) that a typical human diet averages about 100 W. If you've ever turned a hand-cranked or foot-powered electric generator (Fig. 2.1), you probably found that you could comfortably keep a 100-W lightbulb lit for a moderate time. During strenuous activity, your body's power output might be higher—some athletes can put out many hundreds or even thousands of watts for short periods—but 100 W remains a good figure for the average rate at which the human body expends energy.

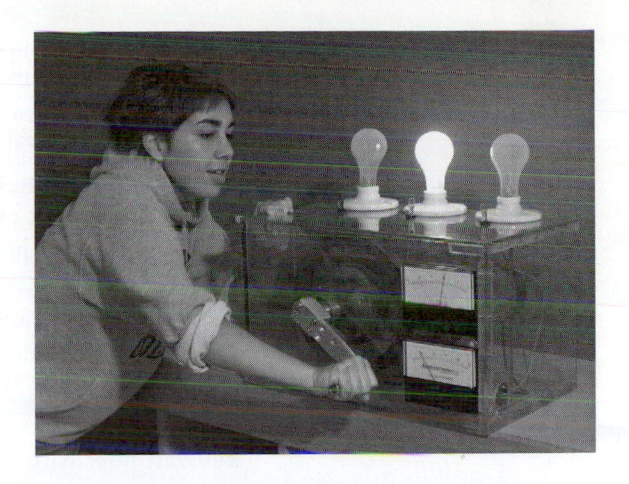

FIGURE 2.1
Turning a hand-cranked electric generator, the average person can sustain a power output of about 100 W. Here the energy ends up lighting a 100-W lightbulb.

2.1 Energy and Power

I've quantified the power output of the human body as being about 100 W. So what's a watt? It's a particular *rate* of energy use. That's a *rate* as opposed to an *amount* of energy. What distinguishes our high-energy society is not so much the total amount of energy we use, but the amount we use each year, or each day, or each second, or whatever—in other words, the *rate* at which we use energy. *Power* is the term for that rate, so the watt is a unit of power. So how big is a watt? The answer is in your muscles: From our knee-bend example, 1 W is about one-hundredth of the average power output of the human body. One kilowatt (kW), or 1,000 W, is then the equivalent of 10 human bodies.

Power and energy are often confused, and it's not uncommon to read newspaper reports of a new power plant that produces "10 million watts every hour" or some such nonsense. Why is that nonsense? Because "every hour" or, more generally, "per time" is built into the meaning of power. Energy is the "stuff" we're talking about, and power—measured in watts—is the rate at which we use that "stuff."

Think of a car as something that produces "stuff," namely, miles of distance traveled. You're much more likely to ask how fast the car can go—that is, the *rate* at which it produces those miles—than how far it can go. If you know the rate, you can always figure out how far you can go in a given time.

Similarly, it makes sense to characterize energy-using devices and even societies by their power consumption—the rate at which they use energy. How much total energy a device or society actually uses depends on the time, so total energy is a less useful characterization.

If I say "This car can go 70 miles per hour," you're not inclined to ask "Is that 70 miles an hour each hour?" because "each hour" is built into the speed designation of 70 miles per hour. Similarly, if I say "This TV uses energy at the rate of 150 watts," it makes no sense to ask "Is that 150 watts every hour?" because the "per time" is implicit in the phrase 150 watts. How far the car goes, and how much energy the TV uses, are determined by how much time you choose to operate these devices. The only subtlety here is that we explicitly identify speed as a rate by using expressions such as "miles per hour" or "meters per second" or whatever, whereas the "per time" is built right into the meaning of the term *watt*.

If watts measure an energy rate, what measures the actual amount of energy? I'll introduce several energy units in the next chapter, but for now here's one that relates easily to the rate in watts: the kilowatt-hour (kWh). One kilowatt-hour is the amount of energy you use if you consume energy at the rate of 1 kW (1,000 W) for one hour. You would also use that much energy if you consumed one-tenth of a kilowatt—that is, 100 W—for 10 hours. So to produce 1 kWh of energy with your own body, you would have to do those knee bends for about 10 hours. You would use 1 kWh if you left a 100-W lightbulb on for 10 hours, or if you used a 1,000-W hair dryer for one hour. Electric bills generally show your monthly consumption of electricity in kilowatt-hours, so we tend to associate the kilowatt-hour with electrical energy. But it's a unit of energy, period, and therefore quantifies any kind of energy. If you understand that watts and kilowatts measure an energy rate, then the name *kilowatt-hour*—a rate multiplied by a time—makes it clear that the kilowatt-hour is a unit of energy, not rate.

Throughout this book I talk about both energy and power. Don't get them confused! Energy is the "stuff," power is the rate.

By the way, I'll be employing phrases such as *using energy, producing energy, expending energy, importing energy, losing energy,* and the like. Because these terms refer to processes rather than energy totals, when quantified they're usually expressed as a rate, measured in watts or other energy-per-time units. You may be aware that energy is a conserved quantity, so terms such as *producing energy* or *energy loss* aren't strictly accurate. What such terms actually describe is the conversion of energy from one form to another. When we "produce" electrical energy, for example, we're converting some other form of energy—usually the stored energy of fossil or nuclear fuels—into electricity. When we "lose" energy, it's usually being converted to the less useful form called, loosely, "heat." We'll deal extensively with energy conversion in subsequent chapters.

EXAMPLE 2.1 | Oil Heat

My home is heated with oil, and when it's on, my oil burner consumes 0.75 gallon of oil per hour. Very roughly, a gallon of heating oil, gasoline, diesel fuel, or other liquid petroleum product contains about 40 kWh of energy. What's the rate at which my oil burner consumes fuel energy?

SOLUTION

That 0.75 gallon per hour amounts to (0.75 gal/h) × (40 kWh/gal) = 30 kWh/h. But what's 1 kilowatt-hour per hour? Simple: Since 1 kWh is the energy consumed in one hour when using energy at the rate of 1 kW, 1 kWh/h—a rate of energy use—is exactly the same thing as 1 kW. So my oil burner uses fuel energy at the rate of 30 kW. By the way, that's far greater than the rate at which typical household electrical appliances use energy.

2.2 Your Energy Servants

You now have a feel for the rate at which your own body can produce energy, about 100 W. So here's the big question for this chapter: At what rate do you, as a citizen of twenty-first-century industrialized society, use energy? More specifically, how many human bodies, producing energy at the rate of 100 W each, would it take to keep you supplied with energy?

Before you answer, think for a minute about your total energy consumption, including the energy that's used for you throughout our complex society. There's energy you use directly—the lightbulb you read by, the heat and air conditioning that keep you comfortable, the energy that runs your computer, the energy of the gasoline that powers your car, the electrical energy converted to sound in your stereo, the elevator that delivers you to your dorm room or apartment, the energy that boils water for a cup of tea or coffee, the refrigerator that stores your cold drinks, and so forth. Then there's energy used indirectly to support your lifestyle. All those trucks on the highway are delivering goods, some of which are for you. Those energy-gobbling banks of open coolers at the supermarket hold frozen food, some of which is for you. And your food itself most likely comes from energy-intensive agriculture, with fuel-guzzling tractors, production and application of pesticides and fertilizers, and processing and packaging of the food. Those airplanes overhead burn a lot of fuel; occasionally they transport you to distant places, and more regularly they supply fresh produce from across the planet or merchandise you ordered with overnight delivery. And speaking of merchandise, think of the factories churning out the products you buy, all of them using energy to process raw materials—to fabricate steel, cloth, silicon, and plastics; and to stamp, melt, mold, weld, rivet, stitch, and extrude the pieces assembled into finished products. Then there's the organizational infrastructure—the office buildings housing commercial, educational, health care, and government services, many with lights blazing around the clock; their air conditioners and heating systems running continuously; and the copiers, computers, and other office equipment on standby for instant use. There's a lot of energy being used to support you!

So how much? The answer, for the United States in the early twenty-first century, is that each of us is responsible for an average energy-consumption rate of roughly, in round numbers, 10,000 W or 10 kW. (The actual figure is between 11 and 12 kW, but we'll stick with the nice, round number—a slight underestimate.) Note, again, that I didn't really answer the question "How much energy do you use?" but rather the more meaningful question "At what rate do you use energy?" And again, don't ask if that's 10,000 W per day or per year or what, because 10,000 W describes the average *rate* at which energy is used by or for you, around the clock, day in and day out.

I've now answered my first question, about the rate at which you use energy—10,000 W. (This number is for citizens of the United States; later in this chapter we'll look at other countries.) The second question asks how many human bodies it would take to supply that energy. We've established that a single human body produces energy at an average rate of about 100 W. Since 100 × 100 = 10,000, this means it would take about 100 human bodies

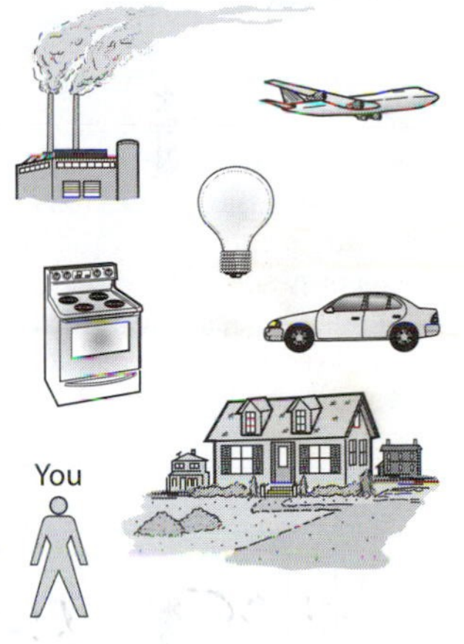

Some of your energy uses

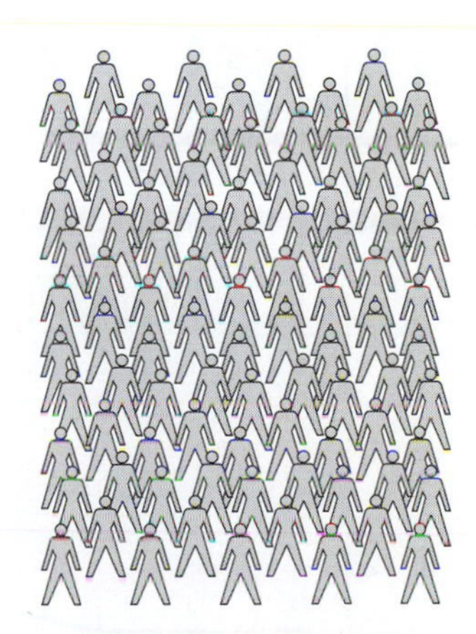

Your energy servants

FIGURE 2.2

It would take 100 energy servants working around the clock to supply the energy needs of the average American.

to supply your energy needs. So picture yourself with 100 "energy servants" working around the clock to supply you with energy (Fig. 2.2). If they work 8-hour shifts instead of all the time, you'd need to employ 300 servants. How much would you be paying those workers if the energy they produced were competitive with fossil fuel energy? The answer is shockingly low—see Exercises 1 and 2 at the end of this chapter.

So this is what it means to be a citizen of a modern, energy-intensive society: In the United States, you have the equivalent of about 100 energy servants working for you all the time. Imagine a room full of 100 people doing those deep knee bends around the clock, or cranking away on electric generators, and you get the picture. Although "energy servant" isn't an official scientific unit, it's a handy way to get a quick feel for the magnitude of your energy consumption. Turn on your 1,000-W hair dryer, and picture 10 energy servants leaping into action. Start up your 250-horsepower SUV and you've unleashed about 2,000 servants. (That's far more than your average of 100, but fortunately your SUV isn't running all the time. You'll have to borrow some servants from friends whose energy consumption is, temporarily, lower than average.) Just leaving your room light on while you're out means there's one poor servant cranking that generator the whole time you're gone.

We haven't always had 100 energy servants, of course. When humankind first evolved, each person had just one servant: him/herself. As our species tamed fire and domesticated animals, that number grew. The discovery of coal and the use of waterpower further increased the number of servants (Fig. 2.3). By the

FIGURE 2.3

Per capita energy consumption rates since 1800. In the early twenty-first century, energy consumption in the United States is about 11 kW per capita, equivalent to some 110 energy servants working around the clock. This rate is about five times the world average. Inset shows recent data for several other countries.

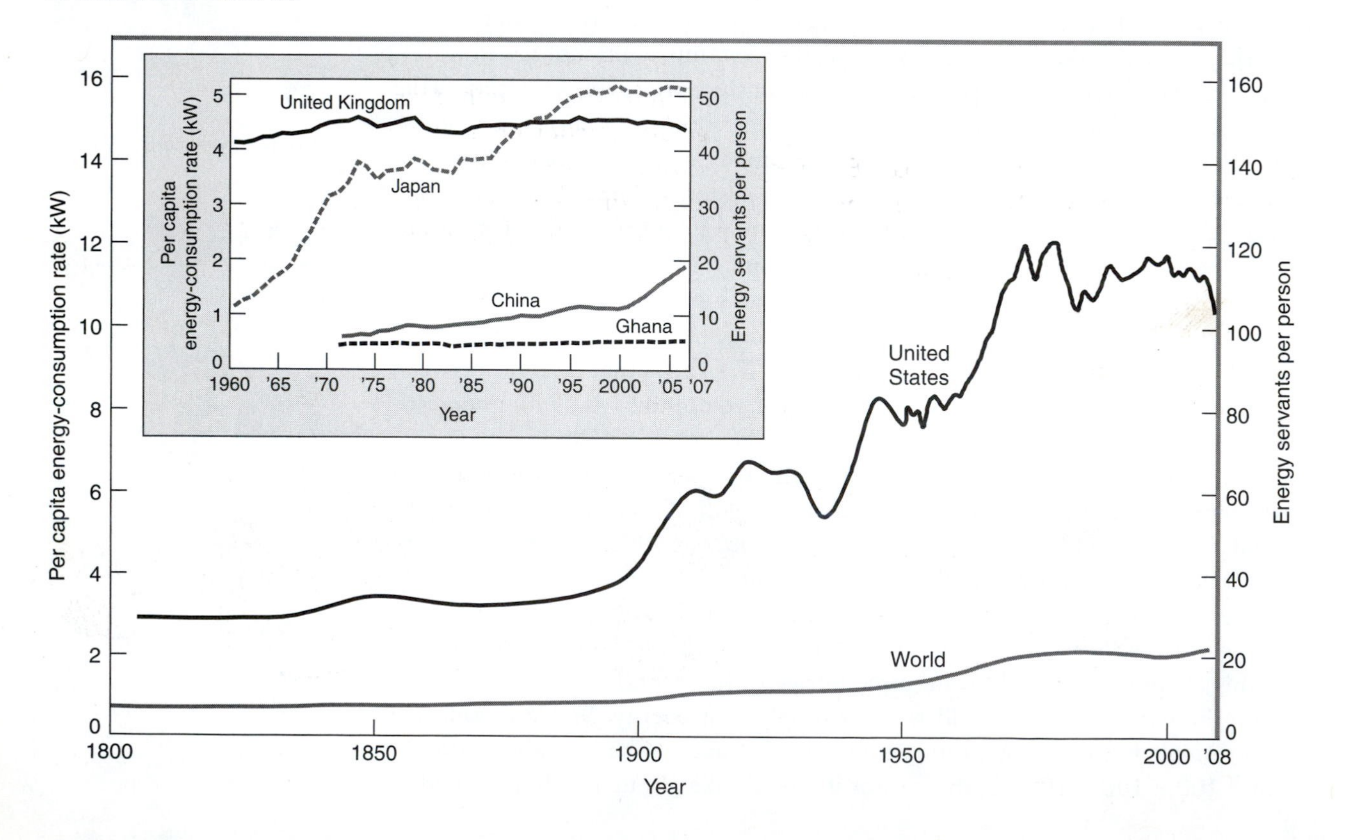

year 1800, the world average was about 5 servants, and by 2000 it had climbed to more than 20. But the distribution is uneven; today, citizens of many developing countries can still claim only a handful of energy servants, while North Americans have more than 100.

2.3 Your Energy Servants' Jobs

What are those 100 energy servants doing for you? It's convenient to assign energy use to four broad sectors: industrial, transportation, residential, and commercial. Figure 2.4a shows the breakdown of energy use among these four sectors for the United States. Since the average U.S. citizen has the equivalent of about 100 energy servants working around the clock, you can think of the numbers in Figure 2.4a as either percentages or as the number of energy servants working in each sector. Thus, in the United States, 21 industrial servants produce the goods you consume, while 45 servants—nearly half—move you and your goods about. An additional 34 commercial and residential servants heat and cool buildings and run lights, computers, stereos, copiers, refrigerators, and all manner of other modern conveniences. Many of your energy servants are working at tasks that may not be obvious to you, such as running machinery in factories or flying fresh produce across the continent. If you want to reduce your own energy consumption, you need to be aware not only of direct actions such as turning off lights or choosing a fuel-efficient car, but also of changes—like buying more local food—that affect the number of energy servants working behind the scenes for you.

How energy servants divide their efforts among sectors varies from country to country. A geographically large country like the United States, with transportation emphasizing private automobiles, uses a greater fraction of its energy for transportation than do compact European countries or Japan. Cold countries like Canada and Norway need more energy for heating. And developing countries, with lower overall energy consumption, devote less of their limited energy to industry and more to residential use. Figure 2.4b and 2.4c uses

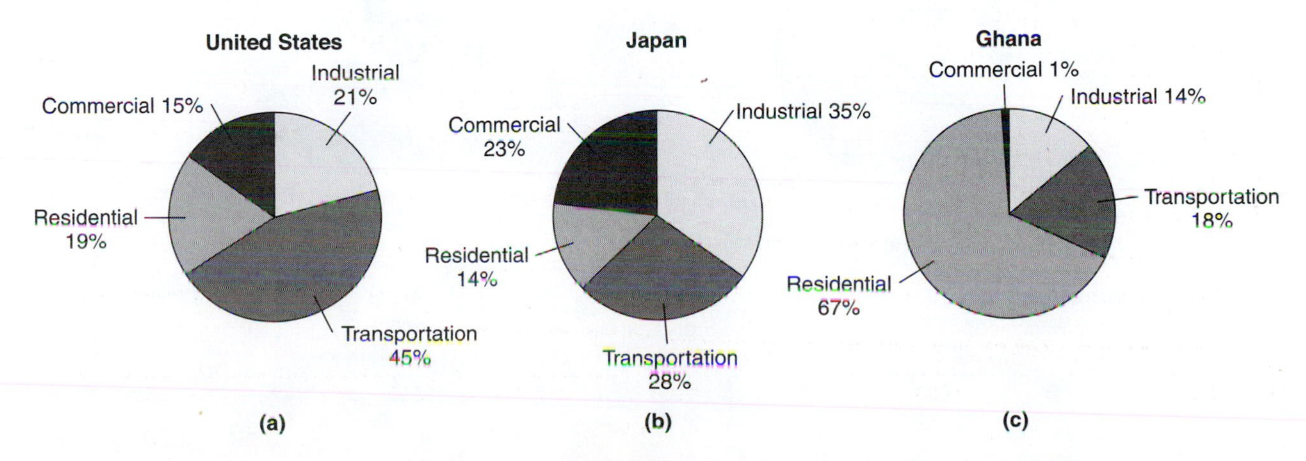

FIGURE 2.4
End-use energy consumption, by sector, for three countries in the early twenty-first century.

Japan and Ghana to illustrate these points. Note that Japan—a compact, highly industrialized country with a state-of-the-art rail network—uses a greater fraction of its energy for industry than does the United States, and correspondingly less for transportation. Japanese houses are smaller than those in the United States, and that's reflected in Japan's lower portion of residential energy consumption. Meanwhile Ghana, a typical developing country, has much smaller industrial and commercial sectors, and devotes most of its limited energy to residential uses.

The data in Figure 2.4 represent **end-use energy**—that is, energy actually consumed in each sector. This contrasts with **primary energy**, which includes the energy of fuels consumed to supply the various sectors. Because of inefficiencies, these two ways of accounting energy may be very different. The distinction between primary and end-use energy will recur throughout this book, and since it's a common source of confusion, I explore it further in Box 2.1.

BOX 2.1 | End-use and Primary Energy

End users of energy are entities—cars, factories, airplanes, stores, homes—that consume energy for useful purposes. End-use energy is measured by the amounts of energy that actually reach the end user. For a car, that's the energy of the gasoline supplied at the pump. For an industry or home, it includes the energy of fuels consumed directly and also of energy supplied as electricity. But electric power plants are typically only about 33% efficient, for reasons you'll learn in Chapter 4. That means their consumption of primary energy—the fuel energy that goes into the power plant—is much greater than their production of electricity. So how do we account for, say, energy consumption in the industrial sector, where much of the energy used is in the form of electricity? Do we count only the energy that's actually consumed, as fuels and as electricity, in factories? Or do we also count the energy that went into making electricity? There's

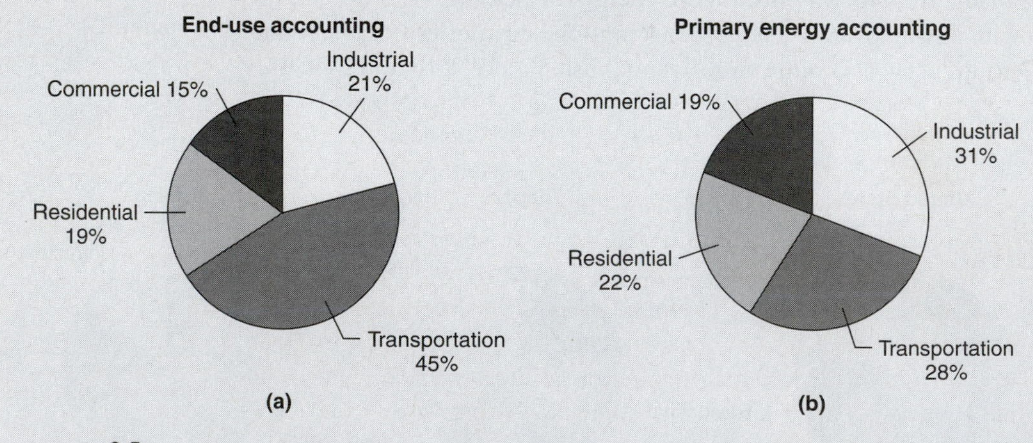

FIGURE 2.5

Energy consumption by sector for the United States, (a) with end-use accounting and (b) accounting for primary energy. Inefficiencies in electric power generation explain most of the difference.

no right answer. To be clear, anyone describing energy consumption needs to be clear as to whether they're talking about end-use or primary energy.

As an example, consider Figure 2.4. The data for all three countries come from the International Energy Agency (IEA; see Credits and Data Sources in the Appendix), and the IEA has chosen to report end-use energy. On the other hand, the United States Department of Energy, in its *Annual Energy Review*, chooses to characterize the four sectors in terms of primary energy consumption. The result, as Figure 2.5 shows, is a very different picture of energy consumption by sector. That industry, rather than transportation, dominates in Figure 2.5b is a consequence of the fact that this figure reflects the total energy consumed to make electricity for industry—thereby greatly increasing industry's share of the total. Since the transportation sector uses relatively little electrical energy, it doesn't get a similar boost. Again, both figures are correct—but they show different things. As you continue your studies of energy—and especially if you seek out new energy statistics—be sure you know whether you're dealing with primary or end-use energy.

2.4 Who Are Your Servants?

You don't really have 100 human servants turning cranks to generate electricity or pedaling like mad to propel your SUV. What you—and society—have are the natural energy resources I outlined in Chapter 1, and the devices that convert them into the forms of energy we find useful. Your car's engine is one such device; others include nuclear or coal-burning electric power plants, hydroelectric generators, wind turbines, boilers that burn fuel to heat buildings and provide steam for industrial processes, solar panels that convert sunlight into electrical energy, and even systems that tap heat deep in the Earth. Behind all these devices are the energy resources themselves—fuels such as coal, oil, natural gas, wood, and uranium, and natural energy flows such as wind, flowing water, and sunlight. To ask "Who are they?" of your energy servants is to ask about these ultimate sources of the energy you use.

The mix of sources varies from country to country, region to region. Places like Norway or the Pacific Northwest of the United States have abundant waterpower, and more of their energy comes from running water. Policy decisions, such as France's choice of a nuclear path to energy independence, give prominence to particular energy sources. But averaged over a country as large as the United States, or for the world as a whole, a striking pattern emerges: The vast majority of the world's energy servants derive from the fossil fuels coal, oil, and natural gas.

Figure 2.6a shows the distribution of energy sources for the United States in the early twenty-first century. It's obvious that some 83% of U.S. energy comes from fossil fuels. Most of the rest is from nuclear and waterpower (hydro). A non-negligible 4% of U.S. energy comes from biomass—wood, corn-based alcohol fuels, and waste burned in "garbage to energy" plants that produce electric power. If you're an advocate of alternative energy sources, Figure 2.6a is sobering. Only 1% of U.S. energy comes from the "other" category that includes the environmentalists' favorites—solar and wind—as well as geothermal energy.

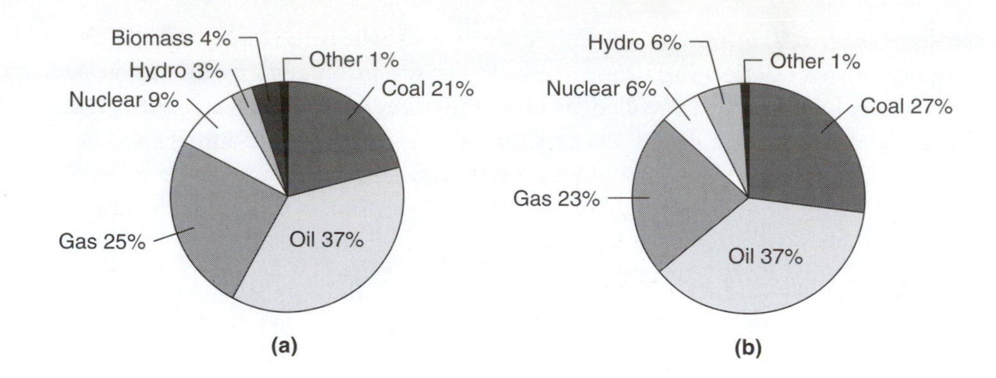

FIGURE 2.6
(a) Sources of U.S. energy in the early twenty-first century. Some 83% comes from fossil fuels. The "other" category includes geothermal, wind, and solar energy. (b) Sources of world energy. More than 85% comes from fossil fuels. Here the "other" category includes biomass, geothermal, wind, and solar energy.

Some two-thirds of the "other" category represents wind, while most of the rest is geothermal; solar energy accounts for only 0.1% of the U.S. energy supply. However, both wind and solar contributions are growing rapidly.

Once again, the fact that energy consumption in the United States is equivalent to about 100 energy servants means that you can translate the percentages in Figure 2.6a directly into numbers of servants. So in the United States, fully 83 of your servants are fossil fueled, 9 are nuclear, and you share a single solar servant with 10 fellow U.S. residents (that's from the one-tenth of the 1% "other" category that's solar).

Things aren't much different for the world as a whole, as Figure 2.6b suggests. The fossil fuel share is a little more than in the United States, at about 87%. The world as a whole gets about twice as great a fraction of its energy from hydropower as does the United States, but less from biomass. And again, the "other" sources currently play a very minor role in world energy consumption, although both wind and solar energy are growing rapidly. Because average per capita energy consumption in the world as a whole is considerably less than in the United States, you can't think of the world percentages as representing equivalent numbers of energy servants.

You might wonder about the actual sizes of the pies in Figure 2.6. We'll quantify those in the next chapter, but for now we'll just compare them. The U.S. share of the world's energy pie is just over 20%, so the U.S. pie is about one-fifth the size of the global pie. China, which in 2010 surpassed the United States to become the world's largest energy consumer, has a similar share—although its per-capita energy consumption rate is far less than that of the United States.

The big picture that emerges from Figure 2.6 is this: The modern world runs on fossil fuels, with more than 85% of our energy coming from coal, oil, and natural gas. This fact is at the heart of the substantial link between energy and climate that's implicit in the title of this book.

2.5 What Energy Servants Buy Us

What do we gain from the efforts of all those energy servants? Whether they're fossil, nuclear, or solar, all are working to provide us with the comforts, conveniences, material goods, and enriched lives that result from use of energy resources beyond our own bodies. Can we measure the benefits that result? Can we establish a relationship between energy consumption and quality of life?

Quality of life is notoriously difficult to measure. Economists have traditionally avoided that difficulty by considering instead measures of economic activity. The presumption is that the higher material standard of living associated with greater economic activity translates for most people into richer, fuller, healthier, more comfortable lives. A commonly used economic indicator is a country's **gross domestic product (GDP)**, a measure of total economic activity. The GDP includes such factors as personal consumption, government expenditures, private and public investment, product inventories, and net exports—the last being an increasingly important factor in today's global economy. Here we'll consider the link between GDP and energy. In the last section of this chapter, we'll look at alternatives to the GDP, which environmentally conscious economists are increasingly adopting as more appropriate measures for quality of life.

ENERGY INTENSITY

To explore the relation between energy and GDP, I've located individual countries on a graph that plots the per capita energy-consumption rate (equivalent to the number of energy servants) on the horizontal axis, and per capita GDP on the vertical axis. Figure 2.7 shows the result for 11 countries. Consider first the United States. Horizontally, it's located at a per capita power consumption of about 11 kW. That's the figure I introduced earlier and approximated as 10 kW, from which our 100 energy servants at 100 W each immediately followed. On the vertical axis, the United States is quite well off, with a GDP around $38,000 per person. Australians use a little less energy each, and they make a little less money. South Korea, Poland, China, Egypt, and the Congo are further down in both per capita energy consumption and GDP (although China's position is changing very rapidly). However, all 7 countries lie pretty

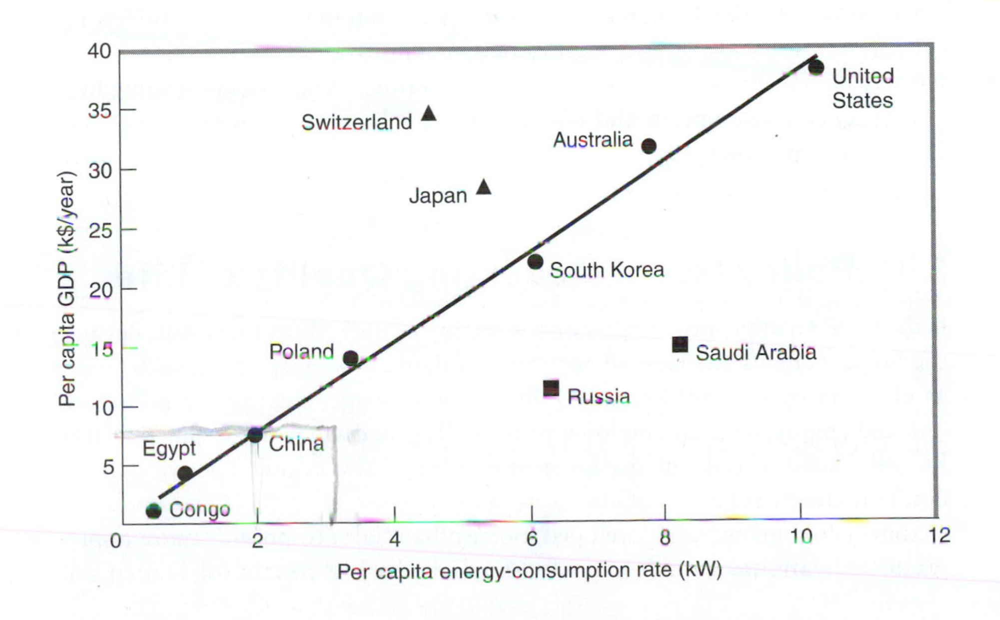

FIGURE 2.7
Per capita GDP (in thousands of U.S. dollars per year) versus per capita energy consumption for 11 countries. The 7 countries that fall near the straight line have approximately the same energy intensity, or energy required per unit of GDP (GDP figures are what economists call GDP *ppp*, for "purchasing power"). Japan and Switzerland are more energy efficient, and Russia and Saudi Arabia less so. Multiplying the numbers on the horizontal axis by 10 gives the number of energy servants per capita.

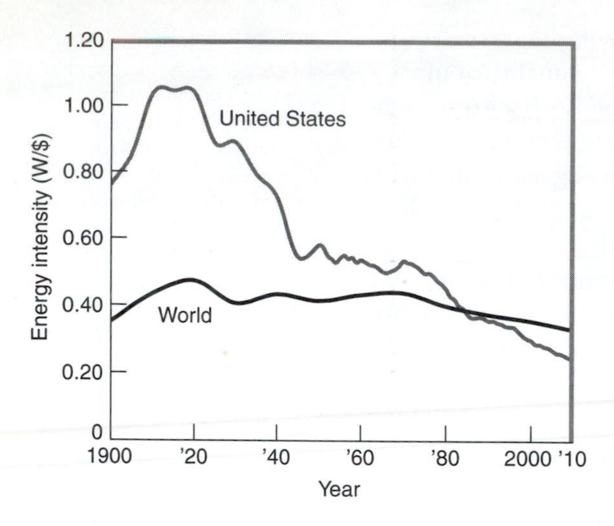

FIGURE 2.8
Energy intensity in the United States (and in many industrialized countries) fell substantially through the twentieth century, with less change in the world as a whole. Intensity is measured in watts of average power needed to produce $1 of GDP each year (based on the value of one U.S. dollar in the year 2000).

much on a straight line, as I've indicated in the figure. This means each uses roughly the same amount of energy to generate each dollar of GDP. This quantity—the ratio of energy use to GDP—is called **energy intensity**. In Figure 2.7, the 7 countries along the straight line have the same energy intensity. Many other countries would fall near this line, too, but some wouldn't. Countries such as Russia and Saudi Arabia lie well below the line. They require more energy to produce a given amount of GDP, as you can see by comparing Saudi Arabia with Poland. Both have about the same per capita GDP, but Saudi Arabia uses a lot more energy to produce it. Therefore Saudi Arabia has the higher energy intensity. Russia's is even higher. On the other hand, Japan and Switzerland lie well above the line. Their citizens get more GDP for every unit of energy consumed. If you think of a country's economy as a machine that converts energy into GDP, then these countries are more efficient in their use of energy. Equivalently, they have lower energy intensity.

There are many reasons for the variations in energy intensity around the world. Small, compact countries need less energy for transportation, so they may have lower energy intensity. Tropical countries need less energy for heating, perhaps resulting in lower energy intensity. Military adventurism may lead to large energy consumption without GDP to show for it. Producing food for export with energy-intensive agriculture (as in the United States) raises the exporter's energy intensity and lowers that of the importer. Taxation, environmental regulation, commitment to public transit, and other policies aimed at improving energy efficiency, reducing air pollution, or mitigating climate change may all lead to lower energy intensity. In fact, energy intensity in many countries has fallen dramatically over the past century (Fig. 2.8). The fact that some countries continue to do better, using less energy per dollar of GDP, is a reminder that others may still have considerable room for improvement. Given that energy use has significant environmental as well as economic impacts, controlling both total energy consumption and energy intensity become important aspects of environmental policy.

2.6 Policy Issue: Measuring Quality of Life

Is the GDP an appropriate measure for quality of life? Many think not. Because the GDP includes virtually all economic activity, money spent on such things as cleaning up oil spills, repairing hurricane damage, treating air pollution–induced emphysema, and building prisons all contribute to increasing the GDP. But oil-fouled waters, intense hurricanes, illness, and criminal activity clearly don't enhance the quality of life.

Some economists have developed indices that claim to provide more appropriate indications of quality of life, especially in the context of human and

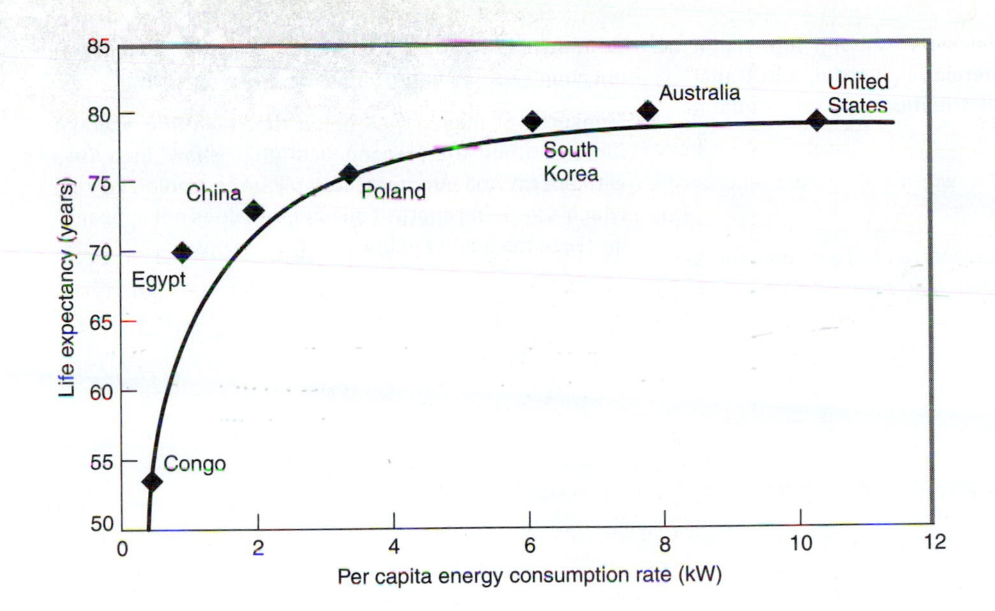

FIGURE 2.9 Life expectancy compared to energy consumption for the seven countries that lie near the straight line in Figure 2.7. Only at very low energy-consumption rates is there a correlation; at higher energy-consumption rates the life expectancy curve saturates. Many other quality-of-life indicators show similar behavior in relation to energy consumption.

environmental health. One such measure is the United Nations' Human Development Index (HDI), which combines the traditional GDP with life expectancy, adult literacy, and school enrollment.

How does energy consumption relate to these broader quality-of-life indices? You can explore this question quantitatively for the HDI in Research Problem 6. For a variety of individual indices, an interesting pattern emerges: At first, increasing per capita energy consumption goes hand in hand with improvements in quality of life; but the improvements eventually saturate, meaning that additional energy consumption doesn't "buy" a higher quality of life. Figure 2.9 shows this effect for life expectancy, a commonly used quality-of-life indicator.

So do all those energy servants buy us a better life? Figure 2.9 suggests that the answer is "yes," up to a point. But beyond that point we don't gain much. And a quality-of-life indicator that accounts for negative factors such as increased pollution or the detrimental effects of climate change might well show a decline in quality of life with excessive energy consumption.

CHAPTER REVIEW

BIG IDEAS

2.1 **Power** measures the *rate* at which we use or produce energy. The standard unit for power is the **watt**, about one-hundredth of the power output of a typical human body.

2.2 Residents of modern industrial societies use far more energy than their own bodies can produce—about 100 times as much in North America and 20 times as much averaged over the globe.

2.3 Energy consumption is distributed among four major sectors: residential, commercial, industrial, and transportation. Details of that distribution vary from country to country.

2.4 Fossil fuels are by far the world's dominant energy source.

2.5 Energy consumption is closely correlated with measures of economic well-being, particularly a country's **gross domestic product**. But there are exceptions, as different countries use energy more or less efficiently.

2.6 Measures of human well-being that take into account factors other than economics also show increases with energy consumption, but only up to a point, beyond which additional energy consumption does not appear to increase the quality of life.

TERMS TO KNOW

end-use energy (p. 26)
end users (p. 26)
energy intensity (p. 30)
gross domestic product (GDP) (p. 29)
power (p. 20)
primary energy (p. 26)
watt (p. 20)

GETTING QUANTITATIVE

Typical power output of the human body: 100 W

Typical per capita energy-consumption rate, global average: just over 2 kW (20 energy servants)

Typical per capita energy-consumption rate, North America: just over 10 kW (100 energy servants)

Portion of world energy supplied by fossil fuels: just over 85%

QUESTIONS

1. A newspaper article claims that a new power plant "produces 50 megawatts of energy each hour, enough to power 30,000 homes." Criticize this statement.

2. Figure 2.3 suggests that global average per capita energy consumption has risen about threefold since the year 1800. You might find that change surprisingly low, given the huge number of new energy-dependent technologies developed since 1800. What factors might account for this apparent discrepancy?

3. Find the energy-consumption rate for five electrical devices you use every day. This value—in watts—is generally listed somewhere on the device or its power adapter, if it has one.

4. For what energy statistic do Egypt and the United States share a common value?

5. Are there enough people in the world to supply all U.S. energy needs, if that energy were to come entirely from human muscle power?

6. What are some factors accounting for the substantial differences between the three pie graphs in Figure 2.4?

7. Identify an occurrence that would raise a country's GDP but lower the quality of life for its citizens.

EXERCISES

1 Suppose you got all your energy from human servants turning hand-cranked electric generators, each producing energy at the rate of 100 W. If you paid those servants $8 per hour, a bit more than the U.S. minimum wage, what would be your cost for 1 kWh of energy? (The typical cost of 1 kWh of electrical energy is actually around 10¢.)

2 What hourly wage would you have to pay your servants from Exercise 1 if the energy they produced were competitive with electrical energy produced by conventional means such as fossil fuels? Use the typical cost of 10¢/kWh.

3 Figure 2.6a shows that 9% of U.S. energy comes from nuclear power plants. The output of those plants is electricity, and this nuclear-generated electricity represents 20% of U.S. electrical energy. Use these facts to determine the fraction of the total U.S. energy supply that is in the form of electricity.

4 For the devices identified in Question 3, estimate the total time you use each device in a day, and use this quantity along with the power consumption to estimate the total energy each device uses in a day. (You can think of this quantity as an average power consumption for each device.)

5 Your author's home uses about 350 kWh of electrical energy each month. Convert this to an average rate of electrical energy consumption in watts.

6 Find the cost per kilowatt-hour of electricity in your area, and the cost of a gallon of gasoline. Considering that gasoline contains about 40 kWh/gal, compare the price per kilowatt-hour for gasoline versus electrical energy.

7 What is the approximate slope (vertical rise divided by horizontal run) of the straight line shown in Figure 2.7? Be sure to include units. What is the meaning of this quantity?

8 A country uses energy at the rate of 3.5 kW per capita, and its citizens enjoy an annual per capita GDP of $7,500. Calculate the energy intensity of this country, and determine whether it uses energy more or less efficiently than the countries on the diagonal line in Figure 2.7.

RESEARCH PROBLEMS

1 Find five countries whose energy intensity is significantly greater than that of the United States. Is there an obvious explanation for their greater energy intensity? Do any of these countries have anything in common? Note: You may find territories of the United States listed as separate countries, but don't include them in your list.

2 China has undergone rapid industrialization in recent decades, and its energy consumption has increased accordingly. What has happened to its energy intensity? Support your answer with a table or graph.

3 Find your state or country's per capita energy consumption and compare it with the U.S. average of roughly 11 kW per capita. What are some factors that might account for any difference?

4 Which of the U.S. or world energy-source categories in Figure 2.6 has changed most significantly over the past few decades? Support your answer with data.

5 France has pursued energy independence through reliance on nuclear power. Prepare a table or graph showing France's energy use by category, like the ones in Figure 2.6, and compare it with those graphs.

6 Find values for the United Nations' Human Development Index (HDI) for the countries listed in Figure 2.7 and prepare a similar graph showing HDI versus energy consumption. Is there any correlation between the two? Is a saturation effect obvious? If you can't discern a trend, add more countries to your plot. You can find the data on the UN Development Programme's web site, where the annual *Human Development Report* is available.

ARGUE YOUR CASE

1. Box 2.1 describes two different ways to measure patterns of energy consumption, using either end-use or primary energy. Decide which you think is the more appropriate means of energy accounting, and formulate an argument to support your view.

2. Figure 2.8 shows that world energy intensity hasn't changed much since 1900, while U.S. energy intensity has fallen substantially. A politician points out that the graph also shows that current U.S. energy intensity is about the same as the world average, and argues that therefore we in the United States can't expect to decrease our energy intensity much further. Formulate your own argument either to support or to counter the politician's statement.

Chapter 3

ENERGY:
A CLOSER LOOK

Chapter 1 began by distinguishing matter and energy, and I defined energy as "the stuff that makes everything happen." That's not a very scientific definition, but it captures the essence of what energy is. You probably have your own intuitive sense of energy, which you reveal when you speak of a "high-energy" person or performance, when you find an event "energizing," or when, at the end of a long day, you're "low on energy." In all these cases, *energy* seems to be associated with motion, activity, or change—hence, energy as "the stuff that makes everything happen." In this chapter, I'll gradually lead you to a more scientific understanding of energy.

The distinction between matter and energy is a convenience in describing our everyday world, but fundamentally the two are manifestations of a single, basic "stuff" that makes up the universe. As Einstein showed with his famous equation $E = mc^2$, energy and matter are interchangeable. The equation shows that matter with mass m can be converted to energy E in the amount mc^2, where c is the speed of light. You can turn matter into energy, and vice versa, but the total amount of "stuff"—call it *mass–energy* for want of a better term—doesn't change. In our everyday world, however, the interchange of matter and energy is a very subtle effect, essentially immeasurable. So for us it's convenient to talk separately of matter and energy, and to consider that each is separately conserved. That's the approach I'll take almost everywhere in this book.

Energy can change from one form to another. For example, a car's engine converts some of the energy stored in gasoline into the energy of the car's motion, and ordinary friction brakes turn that energy into heat as the car slows. But energy can't disappear into nothingness, nor can it be created. In addition to changing form, energy can also move from one place to another. For example, a house cools as thermal energy inside the house flows out through the walls. But again the energy isn't gone; it's just relocated.

Although this treatment of energy and matter as distinct and separately conserved substances is justified in our everyday world, you should be aware that there are times and places where the interchange of matter and energy is so blatantly obvious that it can't be ignored. During the first second of the universe's existence, following the Big Bang, the average energy level was so high that matter particles could form out of pure energy, and matter could annihilate with antimatter to form pure energy. Neither energy nor matter was separately conserved. I'll have more to say about the interchangeability of matter and energy when we explore nuclear energy in Chapter 7.

3.1 Forms of Energy

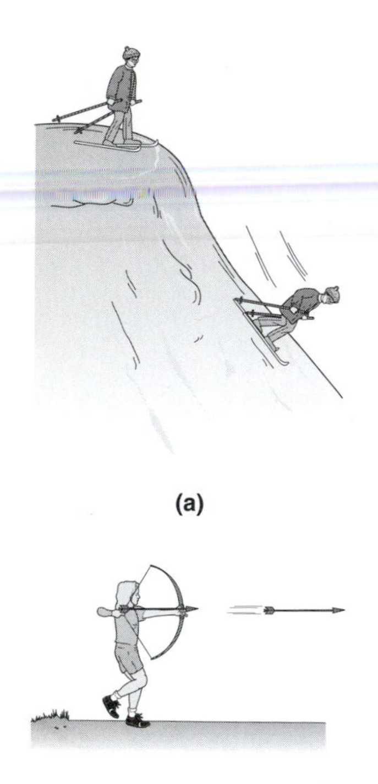

FIGURE 3.1
Potential and kinetic energy. (a) Gravitational potential energy of the skier becomes kinetic energy as she heads down the slope. (b) Elastic potential energy stored in the bow becomes kinetic energy of the arrow.

Stand by the roadside as a truck roars by, and you have a gut sense of the vast energy associated with its motion. This energy of motion, **kinetic energy**, is perhaps the most obvious form of energy. But here's another, more subtle form: Imagine climbing a rock cliff, a process you can feel takes a lot of energy. You also know the danger you face, a danger that exists because the energy you put into the climb isn't gone, but could reappear as kinetic energy were you to fall. Or imagine that I lift a bowling ball and hold it over your head; again you're aware of a danger from an energy that isn't visibly obvious but that you know could reappear as kinetic energy of the ball's downward motion. This energy, associated with an object that's been lifted against Earth's gravity, is **potential energy**; specifically, **gravitational potential energy**. It's potential because it has the potential to turn into the more obvious kinetic energy, or for that matter into some other form. Stretch a rubber band or bungee cord, draw a bow or a slingshot, or compress a spring, and again you sense that there's stored energy. This is called **elastic potential energy** because it involves changing the configuration of an elastic substance. Figure 3.1 shows simple examples of potential and kinetic energy.

FORCE AND ENERGY

Are there other forms of energy? In the context of this book and its emphasis on the energy that powers human society, you might think of answers such as "coal," "wind energy," "solar energy," "waterpower," and the like. But here we'll take a more fundamental look at the different types of energy available to us. There's the kinetic energy associated with moving objects—essentially the same type of energy whether those objects are subatomic particles, trucks on a highway, Earth orbiting the Sun, or our whole Solar System in its stately 250-million-year circle around the center of the Milky Way galaxy. Then there's potential energy, which is intimately related to another fundamental concept—that of **force**. At the everyday level, you can think of a force as a push or a pull. Some forces are obvious, such as the pull of your arm as you drag your luggage through the airport, or the force your foot exerts when you kick a soccer ball. Others are equally evident but less visible, such as the gravitational force that pulls an apple from a tree or holds the Moon in its orbit, the force of attraction between a magnet and a nail, or the frictional force that makes it hard to push a heavy piece of furniture across the floor.

Today, physicists recognize just three fundamental forces that appear to govern all interactions in the universe. For our purposes I'm going to discuss the **gravitational force**, the **electromagnetic force**, and the **nuclear force**, although a physicist would be quick to point out that the electromagnetic and nuclear forces are instances of more fundamental forces. You'll sometimes see the *weak force* mentioned as well. It's important in some nuclear reactions we'll see in Chapter 7, but physicists now understand it as a close cousin of the electromagnetic force. The "holy grail" of science is to understand all these forces as aspects of a single interaction that governs all matter and energy, but we're probably some decades away from achieving that understanding. Figure 3.2 suggests realms and applications in which each of the fundamental forces is important.

Gravity seems familiar, since we're acutely aware of it in our everyday lives here on Earth. Earth has no monopoly on gravity; it's the Sun's gravity that tugs on Earth to keep our planet on its yearlong orbital journey. Actually, gravity is universal; it's a force of attraction that acts between every two pieces of matter in the universe. But it's the weakest of the fundamental forces and is significant only with large-scale accumulations of matter such as planets and stars.

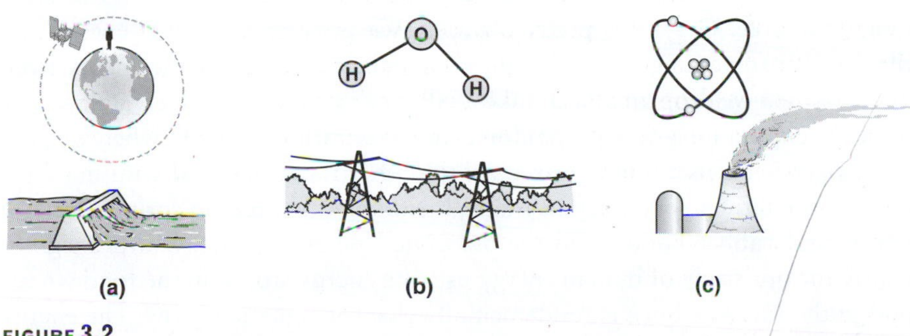

FIGURE 3.2

The fundamental forces and some applications relevant to this book. (a) Gravity governs the large-scale structure of the universe. It holds you to Earth and keeps a satellite in orbit. Gravitational potential energy is the energy source for hydroelectric power plants. (b) The electromagnetic force is responsible for the structure of matter at the molecular level; the associated potential energy is released in chemical reactions such as those that occur in burning fuels. Electromagnetism is also involved in the production and transmission of electrical energy. (c) The nuclear force binds protons and neutrons to make atomic nuclei. The associated potential energy is the energy source for nuclear power plants.

The electromagnetic force comprises two related forces involving electricity and magnetism. The electric force acts between matter particles carrying the fundamental property we call electric charge. It's what binds atoms and molecules, and is involved in chemical reactions. The magnetic force also acts between electric charges, but only when they're in relative motion. In this sense magnetism is intimately related to electricity. The two are complementary aspects of the same underlying phenomenon, which we call electromagnetism. That complementarity has much to do with the ways we generate and transmit electrical energy; thus electromagnetism is vitally important to our energy technologies.

The nuclear force binds together the protons and neutrons that form atomic nuclei. It's the strongest of the three forces—a fact that accounts for the huge difference between nuclear and chemical energy sources, as I'll describe in Chapter 7. We can thank nuclear forces acting deep inside the Sun for the stream of sunlight that supplies nearly all the energy arriving at Earth.

Forces can act on matter to give it kinetic energy, as when an apple falls from a tree and the gravitational force increases its speed and hence its kinetic energy, or when, under the influence of the electric force, an electron "falls" toward a proton to form a hydrogen atom and the energy ultimately emerges as a burst of light. Alternatively, when matter moves against the push of a given force, energy is stored as potential energy. That's what happened when I lifted that bowling ball over your head a few paragraphs ago, or when you pull two magnets apart, or when you (or some microscopic process) yank that electron off the atom, separating positive and negative charge by pulling against the attractive electric force. So each of the fundamental forces has associated with it a kind of potential energy.

What happened to all those other kinds of forces, like the push of your hand or the kick of your foot, or the force in a stretched rubber band, or friction? They're all manifestations of one of the three fundamental forces. And in our everyday lives, the only forces we usually deal with are gravity and the electromagnetic force. Gravity is pretty obvious: We store gravitational energy any time we lift something or climb a flight of stairs. We gain kinetic energy from gravity when we drop an object, take a fall, or coast down a hill on a bicycle or skis. We exploit the gravitational force and gravitational energy when we generate electricity from falling water. All the other forces we deal with in everyday life are ultimately electromagnetic, including the forces in springs, bungee cords, and rubber bands, and the associated potential energy. More significantly for our study of human energy use, the energy stored in the food we eat and in the fuels we burn is fundamentally electromagnetic energy. The energy stored in a molecule of gasoline, for example, is associated with arrangements of electric charge that result in electromagnetic potential energy. (In these cases the energy is essentially all electrical energy; magnetism plays no significant role in the interactions among atoms that are at the basis of chemistry and chemical fuels.) In this book, we're also concerned with one form of energy that isn't either gravitational or electromagnetic—namely, the nuclear energy that we use to generate electricity and that our star uses to make sunlight.

3.2 Electrical Energy: A Closer Look

You probably know that electricity, as we commonly think of it, is a flow of electrons through a wire. That flow is called **electric current**, and what's important here is that the electrons carry **electric charge**, a fundamental electrical property of matter. Electrons carry negative charge, and protons carry equal but opposite positive charge. Because electrons are much lighter, they're usually the particles that move to carry electric current. You might think that the energy associated with electricity is the kinetic energy of those moving electrons, but this isn't the case. The energy associated with electricity, and with its close cousin, magnetism, is in the form of invisible **electric fields** and **magnetic fields** created by the electric charges of the electrons and protons that make up matter. In the case of electric current, electric and magnetic fields surround a current-carrying wire and act together to move electrical and magnetic energy along the wire. Although a little of that moving energy is inside the wire, most is actually in the space immediately around it! Although we call it *electrical* energy, the energy associated with electric current actually involves magnetism as well as electricity, so strictly speaking it's *electromagnetic* energy. However, I'll generally stick with common usage and call it *electrical energy* or just *electricity*.

MAKING ELECTRICITY

Electricity is an important and growing form of energy in modern society. In the United States, for example, about 40% of our overall energy consumption goes toward making electricity, although only about a third of that actually ends up as electricity, for important reasons that I'll discuss in the next chapter. Electrical energy plays an increasingly important role for two reasons. First, it's versatile: Electrical energy can be converted with nearly 100% efficiency to any other kind of energy—mechanical energy, heat, light, or whatever you want. Second, electrical energy is especially easy to transport. Thin wires made from electrically conducting material are all it takes to guide the flows of electrical energy over hundreds and even thousands of miles.

How do we produce electrical energy? In principle, any process that forces electric charges apart will do the trick. In **batteries**, chemical reactions separate positive and negative charge, transforming the energy contained within individual molecules into the energy associated with distinct regions of positive and negative charge—the two terminals of the battery (Fig. 3.3). Hook a complete circuit between the terminals, such as a lightbulb, motor, or other electric device, and current flows through it, converting electrical energy into light, mechanical energy, or whatever.

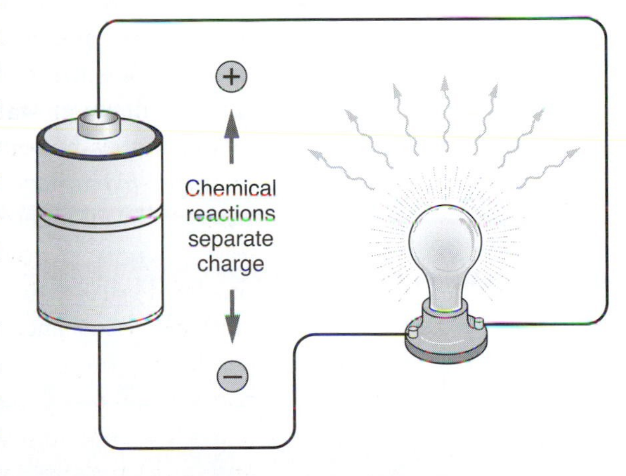

FIGURE 3.3
Chemical reactions in a battery separate positive and negative charge. Connecting an external circuit, such as the lightbulb shown here, allows the battery to deliver electrical energy as charge flows through the circuit.

In some batteries, the chemical reactions go only one way, and when the chemicals have given up all their energy, the battery is "dead" and must be discarded (or, better, recycled). In other batteries, the chemical reactions can be reversed by forcing electric current through the battery in the direction opposite its normal flow. In this mode, electrical energy is converted into chemical energy, which is the opposite of what happens in the battery's normal operation. Such batteries are, obviously, **rechargeable**. Rechargeable batteries power your cell phone, laptop, and cordless tools, and it's a large rechargeable battery that starts your car or, if you have a gas-electric hybrid, provides some of the car's motive power. Batteries are great for portable devices, but they're hardly up to the task of supplying the vast amounts of electrical energy that we use in modern society. Nor are most batteries available today a serious alternative to gasoline and diesel fuel as sources of energy for transportation, although advances in battery technology are closing that gap.

A cousin of the battery is the **fuel cell**, which, conceptually, is like a battery whose energy-containing chemicals are supplied continually from an external source. Today fuel cells find use in a number of specialized applications, including on spacecraft and as backup sources of electrical energy in the event of power failures. Fuel cells show considerable promise for transportation, although issues of cost and fuel storage currently limit that use.

So where does most of our electrical energy come from, if not from batteries? It originates in a phenomenon that reflects the intimate connection between electricity and magnetism. Known since the early nineteenth century and termed **electromagnetic induction**, this fundamental phenomenon entails the creation of electrical effects from *changing* magnetism. Wave a magnet around in the vicinity of a loop of conducting wire, and an electric current flows in the wire. Move a magnetized strip past a wire coil and you create a current in the coil, which is what happens when you swipe your credit card to pay for your groceries. And on a much larger scale, rotate a coil of wire in the vicinity of a magnet and you have an electric generator that can produce hundreds of millions of watts of electric power. It's from such generators that the world gets the vast majority of its electricity (Fig. 3.4).

Remember, however, that energy is conserved, so an electric generator can't actually *make* energy; rather, it converts mechanical energy into electrical energy. Recall Figure 2.1, which shows a person turning a hand-cranked electric generator to power a lightbulb. The person turning the generator is working hard because the electrical energy to light the bulb comes from her muscles. The whole sequence goes something like this:

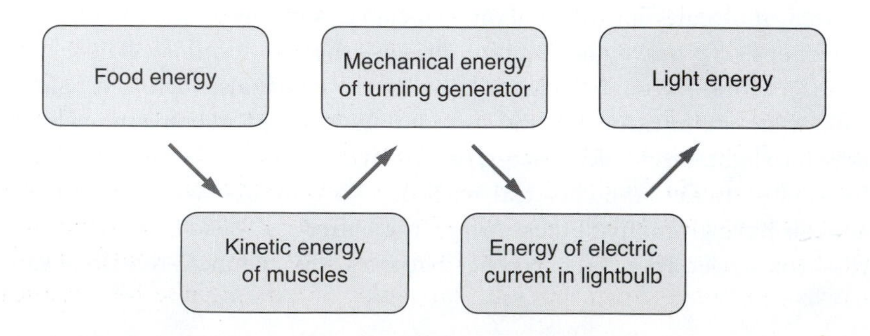

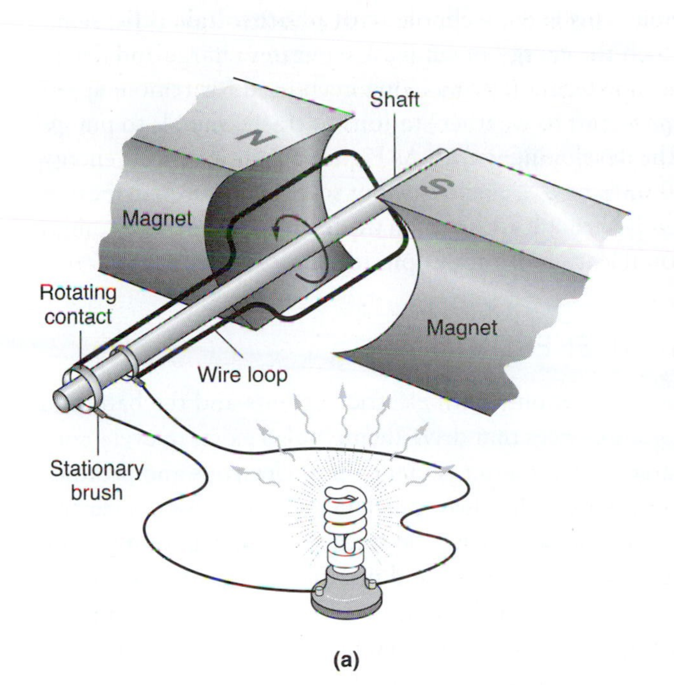

FIGURE 3.4
(a) A simple electric generator consists of a single wire loop rotating between the poles of a magnet. The stationary brushes allow current to flow from the rotating contacts to the external circuit. A practical generator has many loops, each with many turns of wire. (b) This large generator at an electric power plant produces 650 MW of electric power.

Had the lightbulb been off, and not connected in a complete circuit to the generator, the generator would have been very easy to turn. Why? Because turning it wouldn't produce electrical energy, so the turner wouldn't need to supply any energy. So how does the generator "know" to become hard to turn if the lightbulb is connected? Ultimately, the answer lies in another fundamental relationship between electricity and magnetism: Electric current—*moving* electric charge—is what gives rise to magnetism; that's how an electromagnet works. So when current is flowing, there are *two* magnets in the generator—the original, "real" magnet that was built into the generator, and the electromagnet arising from current flowing through the rotating coil. These two magnets repel each other, and that's what makes the generator hard to turn.

I'm going into this level of physics detail because I want you to be acutely aware of what happens every time you turn on a lightbulb, a stereo, a TV, a hair dryer, or an electric appliance: More current flows, and somewhere an electric generator gets harder to turn. Despite my fanciful energy servants of Chapter 2, the generators that produce our electric power aren't cranked by human hands. Most are turned by high-pressure steam created using heat from the combustion of fossil fuels or the fissioning of uranium, while some are turned by flowing water, geothermal steam, or wind (Fig. 3.5). So when you turn on that light or whatever, a little more fossil fuel has to be burned, or uranium fissioned, or water let through a dam, to satisfy your demand for energy. The electric switches you choose to flip on are one of your direct connections to the energy-consumption patterns described in Chapter 2, and the physical realization of that connection is in the electromagnetic interactions that make electric generators hard to turn.

There are a handful of other approaches to separating electric charge and thus generating electrical energy, but only one holds serious promise for large-scale

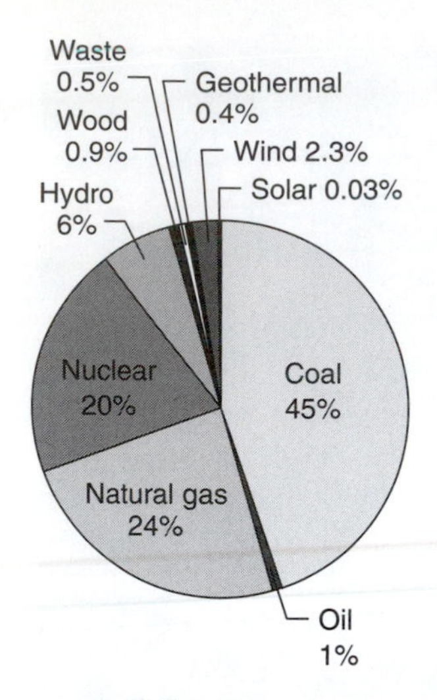

FIGURE 3.5
Sources of electrical energy in the United States.

electric energy production. This is the technology of **photovoltaic cells**, semiconductor devices in which the energy of sunlight separates charge and drives electric current. Photovoltaic technology has long been used for remote applications, ranging from spacecraft to weather stations to traffic signals to pumping water in villages of the developing world. As Figure 3.5 shows, solar energy today generates a mere 0.03% of U.S. electricity. But solar energy's contribution is increasing rapidly, largely through advances in the technology and economics of photovoltaic cells. You'll learn more about photovoltaics in Chapter 9.

STORED ELECTRICAL ENERGY

Electrical energy is associated not only with electric currents and the batteries, generators, and photovoltaic devices that drive them; it also exists in every configuration of electric charges. Since matter is made up of electrons and protons, all matter contains electrical energy. Just how much depends on how the electric charges are arranged. We can create stored electrical energy, for example, by putting positive charge on one metal surface and negative charge on a separate, nearby metal surface. Configurations such as this one store the energy that powers a camera flash or represents information saved in your computer's memory. A larger but similar example is a lightning storm, in which electrical energy is stored in the electric fields associated with layers of charge that build up in violent storm clouds. That energy is released—converted to heat, light, and sound—by lightning discharges. In these and the cases I'll describe next, by the way, the term *electrical energy* is more appropriate than *electromagnetic energy*; magnetism plays a very small role in the energy storage associated with configurations of electric charge, especially when the movement of charge is not significant.

Far more important than the energy of a lightning storm or energy-storage technology is the electrical energy stored at the microscopic level in the configurations of electric charge that we call molecules. Molecules, made up of anywhere from a few to a few thousand atoms, are arrangements of the electric charges that make up their constituent atoms. How much energy is stored in a given molecule depends on its physical structure, which changes in the chemical reactions that rearrange atoms. Most fuels, including in particular the fossil fuels that I introduced in Chapter 1, are substances that store energy in the electric fields associated with their molecular arrangements of electric charges. Gasoline, for example, has molecules consisting of carbon and hydrogen. When gasoline burns, its molecules interact with atmospheric oxygen to form carbon dioxide (CO_2) and water (H_2O). The total electrical energy stored in the CO_2 and H_2O molecules is less than what was in the original gasoline and oxygen, so the process of burning gasoline releases energy.

So fossil fuels store energy, ultimately, as electrical energy of molecular configurations. We generally call that stored energy **chemical energy**, because chemistry is all about the interaction of atoms to make molecules. But the origin of chemical energy is in the electrical nature of matter. By the way, not all fuels are chemical. In Chapter 1, I mentioned the nuclear fuels that power our reactors and bombs; they derive their energy from fields associated with electrical and nuclear forces that act among particles within the atomic nucleus. I'll continue to use the term *fuel* broadly to mean a substance that stores energy in its microscopic structure, at either the atomic/molecular level (chemical fuel) or at the nuclear level (nuclear fuel).

Magnetism is intimately related to electricity and it, too, can store energy. Magnetic energy plays a lesser role than electricity in our everyday lives, although it is associated with the conversion of mechanical energy to electrical energy in electric generators, and vice versa in electric motors. Magnetic energy storage is important in some technological devices, and it plays a major role in the giant eruptive outbursts that occur on our Sun and sometimes hurl high-energy particles toward Earth. The brilliant auroral displays visible at high latitudes are the result of these particles interacting with Earth's own magnetic field and atmosphere. Although stored magnetic energy is of less practical importance than stored electrical energy, be aware that any flow or movement of what you might want to call electrical energy must necessarily involve magnetism, too. This is the case for the current-carrying wire that I discussed at the beginning of this section (Section 3.2), and it's especially true of another means of energy transport that I'll describe next.

ELECTROMAGNETIC RADIATION

The Sun is the ultimate source of the energy that powers life on Earth. As shown in Figure 1.8, our star accounts for 99.98% of the energy reaching Earth's surface. We'll see in Chapter 7 how the Sun's energy arises from nuclear reactions deep in the solar core, but here's the important point for now: Sunlight carries energy across the 93 million miles of empty space between Sun and Earth, so light itself must be a form of energy. Light—along with radio, microwaves, infrared, ultraviolet, X rays, and the penetrating nuclear radiation known as gamma rays—is a kind of energy-in-transit called **electromagnetic radiation.** Light and other electromagnetic radiation are made up of **electromagnetic waves**, structures of electric and magnetic fields in which change in one field regenerates the other to keep the wave moving and carrying energy through empty space (Figures 3.6 and 3.7). I'll occasionally talk of "light energy" or

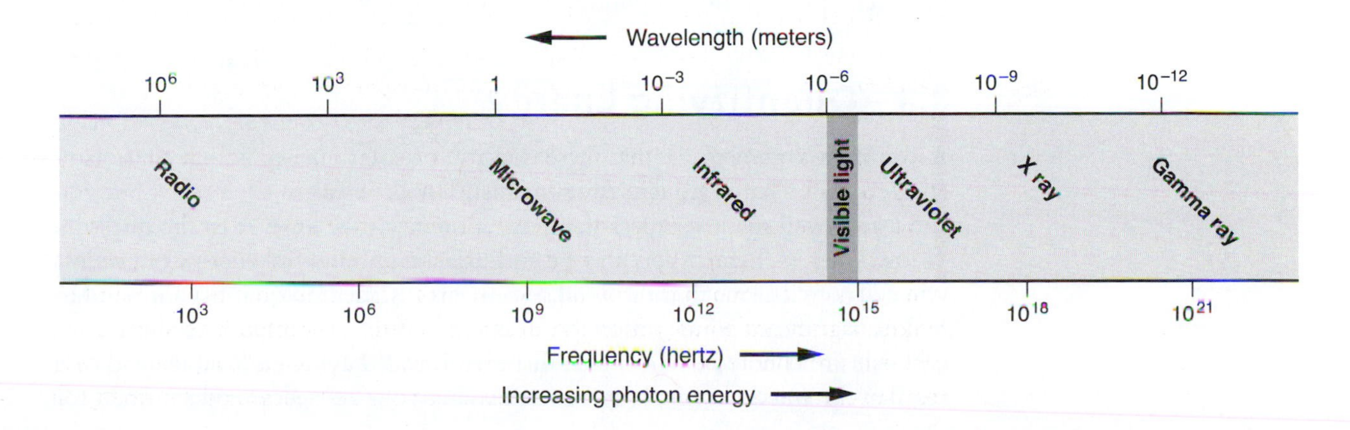

FIGURE 3.6
The electromagnetic spectrum. Electromagnetic waves are characterized by their frequency in hertz (wave cycles per second) or their wavelength in meters (the distance between wave crests). The two are inversely related. Note the nonlinear scale, with each tick mark representing a factor of 1,000 increase or decrease in frequency or wavelength. The Sun's energy output is mostly in the visible and adjacent infrared portion of the spectrum, with a little ultraviolet.

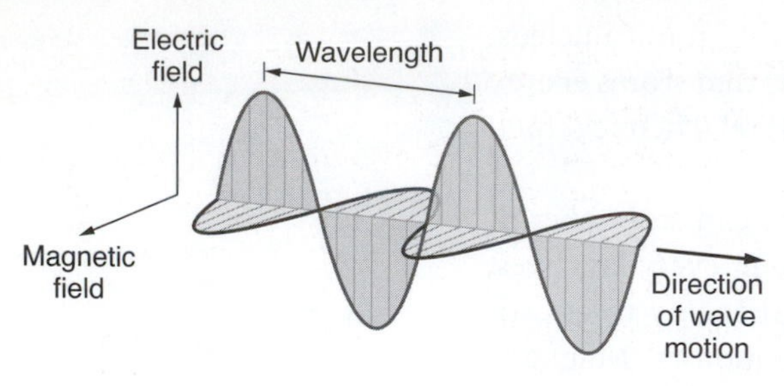

FIGURE 3.7
Structure of an electromagnetic wave. The wave carries energy in its electric and magnetic fields, which are at right angles to each other and to the direction in which the wave is traveling.

"the energy of electromagnetic radiation" as though these were yet another form of energy, but ultimately, light and other electromagnetic radiation are a manifestation of the energy contained in electric and magnetic fields.

Electromagnetic waves originate in the accelerated motion of electric charges—electric currents in antennas producing radio waves; the jostling of charges in a hot object, causing it to glow; atomic electrons jumping between energy levels to produce light; atomic nuclei rearranging themselves and emitting gamma rays. And when electromagnetic waves interact with matter, it's ultimately through their fields producing forces on electric charges—a TV signal moving electrons in an antenna; light exciting nerve impulses in the receptor cells of your eye; sunlight jostling electrons in a solar collector to produce heat, or in the chlorophyll of a green plant to produce energy-storing sugar.

Although it usually suffices to think of electromagnetic radiation in terms of waves, at the subatomic level the interaction of radiation with matter involves quantum physics. Specifically, the energy in electromagnetic waves comes in "bundles" called **photons**, and it's the interaction of individual photons with electrons that transfers electromagnetic wave energy to matter. For a given frequency f of electromagnetic wave, there's a minimum possible amount of energy, corresponding to one photon. That photon energy is directly proportional to the frequency f: $E = hf$, where proportionality constant h is **Planck's constant.** In SI units, $h = 6.63 \times 10^{-34}$ J·s, and it's a measure of *quantization*—the existence of discrete values for many physical quantities at the atomic and subatomic level. For any wave, the product of frequency f and wavelength λ is the wave speed. For electromagnetic waves, that's the speed of light, c (3.00×10^8 m/s). Therefore we can write the photon energy in terms of either frequency or wavelength:

$$E = hf = \frac{hc}{\lambda} \qquad \text{(photon energy)} \qquad (3.1)$$

We'll find the concept of quantized energy and photons especially useful in Chapter 9's description of photovoltaic solar energy conversion.

3.3 Quantifying Energy

A common argument against increased use of solar energy is that there isn't enough of it. That argument is wrong, as I'll make clear in Chapter 9, but you can't very well rebut it unless you have a quantitative answer to the question "How much is there?" You may be enthusiastic about wind energy, but unless you can convince me that the wind resource is sufficient in quantity for wind to make a significant contribution to our energy needs, I'm unlikely to share your enthusiasm. A noted environmentalist claims that buying an SUV instead of a regular car wastes as much energy as leaving your refrigerator door open for

7 years. Is this right? Or is it nonsense? You can't make a judgment unless you can get quantitative about energy.

As these examples suggest, we often need to use numerical quantities to describe amounts of energy and rates of energy use. In Chapter 2, I introduced a basic unit of power, or rate of energy use—the watt (W). I explained what a watt is by quantifying the rate at which your own body can produce energy, about 100 W. I also introduced the prefix *kilo*, meaning 1,000, hence the kilowatt (kW, or 1,000 W). I emphasized the distinction between power—the rate of energy use—and actual amounts of energy. And I introduced one unit that's often used to describe amounts of energy, namely the kilowatt-hour (kWh). Again, here's the relationship between power and energy: Power is a rate and can be expressed in energy units per time units. Equivalently, energy amounts can be expressed in power units multiplied by time units; hence the kilowatt-hour as an energy unit.

Suppose you use energy at the rate of 1 kW for 2 hours. Then you've used $(1 \text{ kW}) \times (2 \text{ h}) = 2 \text{ kWh}$ of energy. Or suppose you consume 60 kWh of energy in 2 hours. Then your energy-consumption rate is $(60 \text{ kWh})/(2 \text{ h}) = 30 \text{ kW}$. Notice how the units work out in this second calculation: The hours (h) cancel, giving, correctly, a unit of power, in this case the kilowatt. Even simpler: If you use 1 kWh of energy in 1 hour, your rate of energy consumption is 1 kWh/h, or just 1 kW. That is, a kilowatt-hour per hour is exactly the same as a kilowatt.

ENERGY UNITS

Still, the kilowatt-hour, although a perfectly good energy unit, seems a bit cumbersome. Is there a simple unit for energy itself that isn't expressed as a product of power with time? There are, in fact, many. In the International System of Units (SI, as it's abbreviated from the French *Système international d'unités*), the official energy unit is the **joule** (J), named for the British physicist and brewer James Joule (1818–1889), who explored and quantified the relationship between mechanical energy and heat. SI is the standard unit system for the scientific community (and for most of the everyday world beyond the United States). If this were a pure science text, I would probably insist on expressing all energies in joules from now on. But energy is a subject of interest to scientists, economists, policymakers, investors, engineers, nutritionists, corporate executives, athletes, and many others, and as a result there are a great many different energy units in common use. I'll describe a number of them here, but I'll generally limit most of the discussion in this book to the joule, the kilowatt-hour, and only occasionally other units.

So what's a joule? In the spirit of your muscle-based understanding of the watt, consider that this book weighs around 2 pounds. Lift it about 4 inches and you've expended about 1 J of energy, which is now stored as gravitational potential energy. In Section 3.4 I'll explain how I did this calculation, and I'll use a similar calculation to verify the 100-W figure from the knee-bend exercise in Chapter 2.

More formally, a joule is a watt-second: $1 \text{ J} = 1 \text{ W·s}$. That is, if you use energy at the rate of 1 W for 1 second, then you've used 1 J of energy. Use energy at the rate of 1 kW for 1 second, and you've used (1,000 W) × (1 s) = 1,000 W·s = 1,000 J, or 1 kilojoule (kJ). Use energy at the rate of 1 kW for a full hour—3,600 seconds—and you've used (1,000 W) × (3,600 s) = 3.6 million W·s, or 3.6 megajoules (MJ; here the prefix *mega* stands for million). But 1 kW for 1 hour amounts to 1 kWh, so 1 kWh is 3.6 MJ.

The joule is the official energy unit of the scientific community, but there are plenty of others. Table 3.1 lists a number of them, along with their equivalents in joules. A unit often used in biology, chemistry, and nutrition is the **calorie** (cal). One calorie is defined as the amount of energy it takes to raise the temperature of 1 gram of water by 1 degree Celsius (°C). The calorie was named before the connection between energy and heat was understood (more on this in Chapter 4). It was Joule himself who established this connection; today, we know that 1 cal is 4.184 J. (There are actually several definitions for the calorie, which vary slightly. I'm using what's called the *thermochemical calorie*.) You're probably most familiar with the calorie as something to be conscious of if you don't want to gain weight. The caloric value of food is indeed a measure of the food's energy content, but since your body stores unused food energy as chemical energy in molecules of fat, it's also an indication of potential weight gain. The "calorie" you see listed on a food's nutritional label is actually 1,000 cal, correctly called a *kilocalorie* (kcal) or sometimes *food calorie* or *large calorie* and written with a capital C: Calorie. Obviously, 1 kcal is then 4,184 J or 4.184 kJ. By the way, "low-calorie soda" might be a meaningful description for an American, but in other countries it's "low-joule soda," making obvious the point that calories and joules measure the same thing—namely, energy (Fig. 3.8).

Since the calorie is a unit of energy, calories per time is a unit of power, convertible to watts. I'll let you show (in Exercise 1 at the end of this chapter) that the average human diet of 2,000 kcal per day is roughly equivalent to 100 W, thus providing another confirmation of Chapter 2's value of 100 W for the typical power output of the human body.

FIGURE 3.8
"Low joule" describes this diet soft drink from Australia. Joules and calories measure the same thing, namely energy.

In the English system of units, no longer used in England but only in the United States and a very few other countries, the analog of the calorie is the **British thermal unit** (Btu), defined as the amount of energy needed to raise the temperature of 1 pound of water by 1 degree Fahrenheit (°F). Table 3.1 shows that 1 Btu is 1,054 J, or just over 1 kJ. You could also calculate this conversion knowing the relationship between grams and pounds, and the Celsius and Fahrenheit scales (see Exercise 2). A power unit often used in the United States to describe the capacity of heating and air conditioning systems is the Btu per hour (Btu/h, but often written, misleadingly, as simply Btuh). As Table 3.1 shows, 1 Btu/h is just under one-third of a watt. My household furnace is rated at 112,000 Btu/h; Example 3.1 shows that this number is consistent with its fuel consumption rate of about 1 gallon of oil per hour. The British thermal unit is at the basis of a unit widely used in describing energy consumption of entire countries, namely the **quad** (Q). One quad is 1 quadrillion Btu, or 10^{15} Btu. The world's rate of energy consumption in the early twenty-first century is nearly 500 Q per year, with the United States and China each accounting for about 100 Q annually.

TABLE 3.1 | ENERGY AND POWER UNITS

Energy units	Joule equivalent*	Description
joule (J)	1 J	Official energy unit of the SI unit system; equivalent to 1 W·s or the energy involved in applying a force of 1 newton over a distance of 1 meter.
kilowatt-hour (kWh)	3.6 MJ	Energy associated with 1 kW used for one hour. (1 MJ = 10^6 J)
gigawatt-year	3.16 PJ	Energy produced by a typical large (1 gigawatt) power plant operating full-time for one year. (1 PJ = 10^{15} J)
calorie (cal)	4.184 J	Energy needed to raise the temperature of 1 gram of water by 1°C.
British thermal unit (Btu)	1,054 J	Energy needed to raise the temperature of 1 pound of water by 1°F, very roughly equal to 1 kJ.
quad (Q)	1.054 EJ	Quad stands for quadrillion Btu, or 10^{15} Btu, and is roughly equal to 1 exajoule (10^{18} J).
erg	10^{-7} J	Energy unit in the centimeter-gram-second system of units.
electron volt (eV)	1.6×10^{-19} J	Energy gained by an electron dropping through an electric potential difference of 1 volt; used in atomic and nuclear physics.
foot-pound	1.356 J	Energy unit in the English system, equal to the energy involved in applying a force of 1 pound over a distance of 1 foot.
tonne oil equivalent (toe)	41.9 GJ	Energy content of 1 metric tonne (1,000 kg, roughly 1 English ton) of oil. (1 GJ = 10^9 J)
barrel of oil equivalent (boe)	6.12 GJ	Energy content of one 42-gallon barrel of oil.
Power units	**Watt equivalent**	**Description**
watt (W)	1 W	Equivalent to 1 J/s.
horsepower (hp)	746 W	Unit derived originally from power supplied by horses; now used primarily to describe engines and motors.
Btu per hour (Btu/h, or Btuh)	0.293 W	Used primarily in the United States, usually to describe heating and cooling systems.

*See Table 3.2 for SI prefixes.

BOX 3.1 | Converting Units

With many different energy units, it's important to know how to convert from one unit to another. The trick is to multiply by a ratio—which you can find or work out from Table 3.1—that gives the final unit you want. For example, to convert the U.S. annual energy consumption of about 100 Q to exajoules (EJ), note Table 3.1's conversion factor for the quad, which can be written as a ratio: 1.054 EJ/Q. Then our conversion for 100 Q becomes

$$(100\ \text{Q})(1.054\ \text{EJ/Q}) = 105\ \text{EJ}$$

Note how the Q "upstairs"—in the numerator of the term 100 Q—cancels with the Q "downstairs" in the term 1.054 EJ/Q. That leaves the final unit as EJ, which is what we want. Since that 100 Q was a yearly energy consumption, we can write the U.S. energy consumption rate as 105 exajoules per year (EJ/y).

That 105 EJ/y describes energy per time, so it's a measure of *power*. Suppose I want to convert it to horsepower: Table 3.1 lists 1 horsepower (hp) as 746 watts (W). A watt is a joule per second, so we'll first convert 105 EJ/y to joules per second (J/s). A year is about 3.15×10^7 seconds (you can get this by multiplying 365 days/year by 24 hours/day by 3,600 seconds/hour), and an exajoule, as Table 3.1 implies, is 10^{18} joules. So our 105 EJ/y becomes

$$\frac{(105\ \text{EJ/y})(10^{18}\ \text{J/EJ})}{3.15 \times 10^7\ \text{s/y}} = 3.33 \times 10^{12}\ \text{J/s}$$

$$= 3.33 \times 10^{12}\ \text{W}$$

Now we convert that to horsepower (hp), using 746 W/hp from Table 3.1. Since we want horsepower "upstairs" in our final answer, we need to *divide* by this conversion factor, in which horsepower appears "downstairs." So we have

$$\frac{3.33 \times 10^{12}\ \text{W}}{746\ \text{W/hp}} = 4.46 \times 10^9\ \text{hp}$$

That makes the United States a nearly 5-billion horsepower country!

EXAMPLE 3.1 | Home Heating

Assuming home heating oil contains about 40 kWh of energy per gallon, determine the approximate rate of oil consumption in my home furnace when it's producing heat at the rate of 112,000 Btu/h.

SOLUTION

We have the oil's energy content in kilowatt-hours, so we need to convert that 112,000 Btu/h into compatible units, in this case kilowatts. Table 3.1 shows that 1 Btu is 1,054 J, or 1.054 kJ, and one hour is 3,600 seconds, so the heat output of the furnace becomes

$$(112{,}000\ \text{Btu/h})(1.054\ \text{kJ/Btu})(1\text{h}/3{,}600\ \text{s}) = 33\ \text{kJ/s} = 33\ \text{kW}$$

Notice how I was careful to write out all the units and to check that they multiplied together to give the correct final unit:

$$(\text{Btu/h})(\text{kJ/Btu})(\text{h/s}) \rightarrow \text{kJ/s} = \text{kW}$$

In any numerical problem like this one, it's essential that the units work out correctly. If they don't, the answer is wrong, even if you did the arithmetic correctly. Note in this case that the unit for hour (h) is in the denominator, so I had to multiply by hours per second (h/s) to convert the time unit to seconds.

So my furnace produces heat at the rate of 33 kW, or 33 kWh/h. Since there are 40 kWh of energy in a gallon of oil, the furnace would need to burn a little over three-quarters of a gallon per hour (33/40) if it were perfectly efficient. But it isn't, so that 112,000 Btu/h heat output requires close to 1 gallon of oil per hour.

Table 3.1 lists several other energy units that we'll have little use for in this book, but that often appear in the scientific literature. The erg is the official unit of the centimeter-gram-second system of units (as opposed to the SI

BOX 3.2 | SI Prefixes

Units for energy and other quantities are often modified with prefixes that indicate multiplication by powers of 10; every three powers gets a new name. You've already met kilo (k, meaning 1,000 or 10^3) and mega (M, 1 million or 10^6). Others you've probably heard are giga (G, 1 billion or 10^9), milli (m, 1/1,000 or 10^{-3}), micro (μ, the Greek "mu," one-millionth or 10^{-6}), and nano (n, one-billionth or 10^{-9}). Table 3.2 lists others, all part of the SI unit system. The large multipliers are useful in quantifying the huge rates of energy consumption in the industrialized world. The world total of 470 Q per year, for example, can be expressed in SI as about 500 exajoules (EJ) per year; converting to watts (Exercise 3) gives a world energy-consumption rate of about 16 terawatts (16 TW, or 16×10^{12} W). I'll use SI prefixes routinely throughout this book, and you can find them here in Table 3.2 and also inside the back cover. Note that the symbols for SI prefixes that multiply by less than 1 are in lowercase, while those that multiply by more than 1 (except for kilo) are capitalized.

TABLE 3.2 | SI PREFIXES

Multiplier	Prefix	Symbol
10^{-24}	yocto	y
10^{-21}	zepto	z
10^{-18}	atto	a
10^{-15}	femto	f
10^{-12}	pico	p
10^{-9}	nano	n
10^{-6}	micro	μ
10^{-3}	milli	m
10^0 (= 1)	—	—
10^3	kilo	k
10^6	mega	M
10^9	giga	G
10^{12}	tera	T
10^{15}	peta	P
10^{18}	exa	E
10^{21}	zeta	Z
10^{24}	yotta	Y

meter-kilogram-second units). The electron volt is a tiny unit (1.6×10^{-19} J) used widely in nuclear, atomic, and molecular physics. Finally, the foot-pound is an English unit, being the energy expended when you push on an object with a force of 1 pound as the object moves 1 foot.

The last two energy units in Table 3.1 deserve special mention because they're based not directly on energy but on oil. The **tonne oil equivalent** is the energy content of a metric ton of typical oil, while the **barrel of oil equivalent** is the energy content of a standard 42-gallon barrel. (A **metric ton**, also called a **tonne**, is equal to 1,000 kg or a little more than a 2,000-lb U.S. **ton**.) The values given for the tonne oil equivalent and barrel of oil equivalent in Table 3.1 are formal definitions; the actual energy content of oil varies somewhat, depending on its source and composition.

Table 3.1 also lists several units for power. Among those in common use is **horsepower** (hp), a holdover from the day when horses supplied much of the energy coming from beyond our own bodies. One horsepower is 746 W, or about three-quarters of a kilowatt. So a 400-hp car engine can, in principle, supply energy at the rate of about 300 kW (most of the time the actual rate may be much less, and very little of that energy ends up propelling the car; more on this in Chapter 5).

Fuels—those substances that store potential energy in the configurations of molecules or atomic nuclei—are characterized by their energy content, expressed as energy contained in a given mass or volume. Table 3.3 lists the energy contents of some common fuels. Some of these quantities find their way into alternative units for energy and for energy-consumption rate, as in the tonne oil equivalent and barrel of oil equivalent listed in Table 3.1. Related power units of millions of barrels of oil equivalent per day often describe the production of fossil fuels, and national energy-consumption rates are sometimes given in millions of barrels of oil equivalent per year. At a smaller scale, it's convenient to approximate the energy content of a gallon of petroleum product—oil, kerosene, gasoline—as being about 40 kWh (the exact amount varies with the fuel, and the amount of useful energy obtained depends on the efficiency of the energy-conversion process). Finally, Table 3.3 hints at the huge quantitative difference between chemical and nuclear fuels; just compare the energy per kilogram of petroleum with that of uranium!

You'll find Tables 3.1 through 3.3 sufficiently useful that they're printed inside the front and back covers for easy reference, along with other useful energy-related information.

3.4 Energy and Work

Push a stuck car out of the mud, a lawnmower through tall grass, or a heavy trunk across the floor, and in all cases you're doing a lot of work. **Work** has a precise scientific meaning: It's a quantity equal to the force you apply to an object multiplied by the distance over which you move that object. Work is essentially a measure of the energy you expend as you apply a force to an object. This energy may end up as increased kinetic energy of the object, as when you kick a soccer

TABLE 3.3 | ENERGY CONTENT OF FUELS

	Typical energy content (varies with fuel source)	
Fuel	**SI units**	**Other units**
Coal	29 MJ/kg	7,300 kWh/ton 25 MBtu/ton
Oil	43 MJ/kg	~40 kWh/gallon 138 kBtu/gallon
Gasoline	44 MJ/kg	36 kWh/gallon
Natural gas	55 MJ/kg	30 kWh/100 cubic feet 1,000 Btu/cubic foot
Biomass, dry	15–20 MJ/kg	13–17 MBtu/ton
Hydrogen gas (H_2) burned to produce H_2O	142 MJ/kg	320 Btu/cubic foot
Uranium, nuclear fission:		
Natural uranium	580 GJ/kg	161 GWh/tonne
Pure U-235	82 TJ/kg	22.8 TWh/tonne
Hydrogen, deuterium-deuterium nuclear fusion:		
Pure deuterium	330 TJ/kg	
Normal water	12 GJ/kg	13 MWh/gallon, 350 gallons gasoline equivalent per gallon water

ball and set it in motion, or it may end up as potential energy, as when I lifted that bowling ball over your head. One caveat: You do work only when the force you apply to an object is in the direction of the object's motion. If the force is at right angles to the motion, no work is done, and there's no change in the object's energy. If the force is opposite the motion, then the work is negative and you take energy away from the object. Figure 3.9 illustrates these possibilities.

So work is a measure of energy supplied to an object by mechanical means—that is, by applying a force, such as pushing or pulling. Again, it's the product of force times the distance the object moves:

$$W = Fd \tag{3.2}$$

Here W is the work, F the force, and d the distance. Equation 3.2 applies in the case where the force and the object's motion are in the same direction. In the

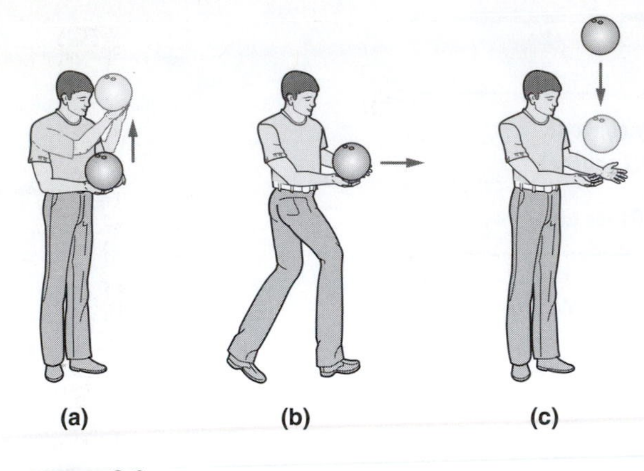

FIGURE 3.9
Force and work. (a) When you lift a bowling ball, you apply a force in the direction of its motion. You do work on the ball, in this case increasing its gravitational potential energy. (b) When you carry the ball horizontally, you apply a force to counter its weight, but you don't do work on the ball because the force is at right angles to its motion. (c) When you apply an upward force to stop a falling ball, the force is opposite the ball's motion, so here you do negative work that reduces the ball's kinetic energy.

English system commonly used in the United States, force is measured in pounds and distance in feet; hence the English unit of work is the foot-pound (see Table 3.1). In the SI system, the unit of force is the **newton** (N), named in honor of the great physicist Isaac Newton, who formulated the laws governing motion. One newton is roughly one-fifth of a pound (1 pound = 4.48 N). The SI unit of distance is the meter, so work is measured in newton-meters (N·m). And what's a newton-meter? Since work is a measure of energy transferred to an object, 1 N·m is the SI unit of energy, namely the joule. So another way of understanding the joule is to consider it the energy supplied by exerting a force of 1 N on an object as the object moves a distance of 1 m.

An object's **weight** is the force that gravity exerts on it. Near Earth's surface, the strength of gravity is such that an object experiences a force of 9.8 N for every kilogram of mass it possesses. We designate this quantity g, the strength of gravity near Earth's surface. The value of g is then 9.8 newtons per kilogram (N/kg), which is close enough to 10 N/kg that I'll frequently round it up. (If you've had a physics course, you probably know g as the acceleration of gravity—the rate at which an object falling near Earth's surface gains speed. My definition here is equivalent, but more useful for thinking about energy.) We can sum all this up mathematically: If an object has mass m, then its weight is given by

$$F_g = mg \tag{3.3}$$

Here I've designated weight as F_g, for "force of gravity," because I've already used the symbol W for work. If m is in kilograms and g in newtons per kilogram, then the weight is in newtons.

To lift an object at a steady rate, you have to apply a force that counters the force of gravity. That is, the force you apply is equal to the object's weight, which Equation 3.3 shows is simply the product mg. Suppose you lift the object a height h. Using h for the distance in Equation 3.2, we can then combine Equations 3.2 and 3.3 to get an expression for the work you do in lifting the object:

$$W = mgh \tag{3.4}$$

This work ends up being stored as gravitational potential energy. Since the gravitational force "gives back" stored potential energy, Equation 3.4 also describes the kinetic energy gained when an object with mass m falls a distance h.

Earlier I suggested that you could get a feel for the size of a joule by lifting your 2-pound book about 4 inches. Two pounds is about 1 kg, so your book's weight is about 10 N (here I multiplied 1 kg by the approximate value for g, namely 10 N/kg). Four inches is about 10 centimeters (cm), or one-tenth of a meter, so Equation 3.2 gives $W = (10\text{ N}) \times (0.10\text{ m}) = 1.0\text{ J}$. Or just derive it from Equation 3.4:

$$W = mgh = (1\text{ kg})(10\text{ N/kg})(0.10\text{ m}) = 1\text{ J}$$

So there it is: Lift your 2-pound book 4 inches and you've done one 1 J of work, giving the book 1 J of gravitational potential energy.

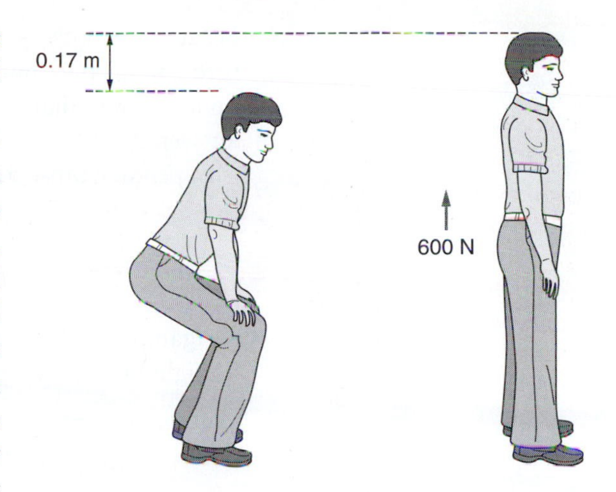

FIGURE 3.10
Rising out of a knee bend requires that I apply a 600-N force over a distance of 0.17 m, resulting in 100 J of work done. Repeating once per second gives a power output of 100 W.

What's your weight? My mass is about 70 kg, so according to Equation 3.3 my weight is (70 kg) × (10 N/kg) = 700 N. With 4.48 N per pound, this is equivalent to (700 N)/(4.48 N/lb) = 156 lb. That's just what my U.S. scale reads. Suppose I start doing those knee bends from Chapter 2. I've marked the position of the top of my head on a chalkboard when standing and again when at the lowest point of my knee bend, and the marks are 17 cm apart (0.17 m) (Fig. 3.10). So as I come up out of each knee bend, I raise most of my weight by 17 cm. Most? Yes: My feet stay on the ground and my legs participate in only some of the motion. So the weight that's actually raised is somewhat less than my actual weight; suppose it's about 600 N. Then each time I rise to my full height, the work I do is $W = (600\ \text{N}) \times (0.17\ \text{m}) \simeq 100\ \text{N·m} \simeq 100\ \text{J}$. If I do those knee bends once every second, that's 100 J/s, or 100 W. Again, this is a rough figure; you could quibble with my choice for just what fraction of my weight I actually raise, and surely the once-per-second is only approximate. But there it is: While doing those knee bends, my body expends energy at the rate of roughly 100 W.

You might wonder what happens as I lower my body when bending my knees. In that downward motion I do negative work, or the force of gravity does work on me. If my muscles were like springs, this energy would be stored as elastic potential energy and be available for the next upward motion, with the result that my overall energy output would be considerably less. But my muscles aren't springs, and most of that energy is lost in heating the muscles and other body tissues. There's a little bit of springlike energy storage in muscles, but it isn't very significant with the relatively slow muscle contractions and extensions of this knee-bend exercise.

EXAMPLE 3.2 | Mountain Run!

(a) How much work do I do in running up a mountain with a vertical climb of 2,500 feet (760 m)? Express the answer in both joules and calories. (b) If the run takes 50 minutes, what's the average power involved?

SOLUTION

My mass is 70 kg, g is 9.8 N/kg, and I'm climbing 760 m; Equation 3.4 then gives

$$W = mgh = (70\ \text{kg})(9.8\ \text{N/kg})(760\ \text{m}) = 521\ \text{kJ}$$

With 1 kcal = 4.184 kJ, this amounts to (521 kJ)/(4.184 kJ/kcal) = 125 kcal. This is the bare minimum amount of work it would take to scale the mountain because my body is far from 100% efficient, and I do work against frictional forces in my own muscles even when walking horizontally. So I probably burn up a lot more than the 125 "calories"—actually kilocalories—implied by this answer.

Expending those 521 kJ over 50 minutes gives an average power of

$$p = \frac{521 \text{ kJ}}{(50 \text{ min})(60 \text{ s/min})} = 0.174 \text{ kJ/s} = 174 \text{ W}$$

Again, the actual rate of energy expenditure would be a lot higher, although it might take rather more than 50 minutes to make a 2,500-foot climb.

3.5 Work and Kinetic Energy

The examples of the preceding section involved our doing work against the gravitational force, resulting in stored gravitational energy. But what if there is no opposing force? Then the work we do on an object goes into kinetic energy. Examples include pushing a fellow skater or hockey puck on ice, kicking a ball on a smooth horizontal surface, or accelerating a car from a stoplight.

An important result, which follows from Newton's laws of motion, is the **work–energy theorem**. It states that the *net work* done on an object—that is, the total work done by all the forces acting—is equal to the change in the object's kinetic energy. The theorem specifically identifies the kinetic energy K with the quantity $\frac{1}{2}mv^2$, where m is the mass and v (for velocity) is its speed:

$$K = \tfrac{1}{2}mv^2 \qquad \text{(kinetic energy)} \qquad (3.5)$$

It's important to recognize that the work–energy theorem applies not to individual forces, but only to the sum of all forces acting on an object. For example, you do work lifting an object at constant speed against gravity, but the downward-acting gravitational force does negative work, and the *net work* is zero. Thus the object's kinetic energy doesn't change—although the work you've done has increased its potential energy. But if you kick a ball horizontally with a force that exceeds the frictional force, then the positive work you do exceeds the negative work done by friction; thus the net work is positive and the ball's kinetic energy $\frac{1}{2}mv^2$ increases—and so, therefore, does its speed v.

Note that the kinetic energy depends on an object's speed *squared*, which means kinetic energy increases rapidly with speed. For example, doubling the speed (a factor of 2) results in quadrupling the energy (a factor of 2^2, or 4). That's one reason why driving at high speeds is particularly dangerous, more so than if energy increased in direct proportion to speed (Fig. 3.11). And it's also why accelerating rapidly to high speeds is a rather inefficient use of fuel.

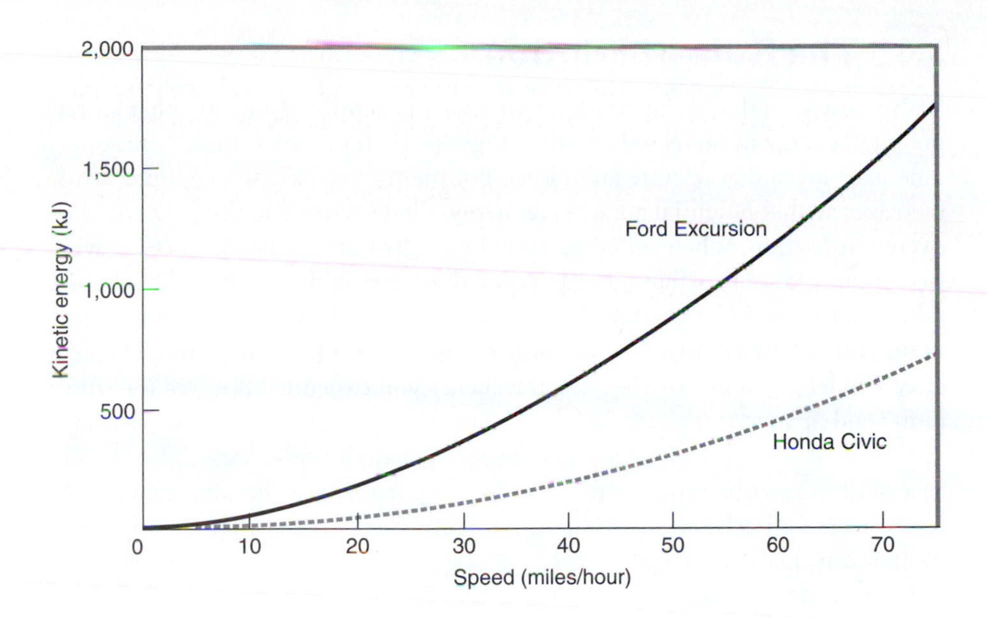

FIGURE 3.11
Kinetic energy increases with the square of the speed. Shown here is kinetic energy versus speed for a Honda Civic compact car (empty weight 1,268 kg) and a Ford Excursion SUV (empty weight 3,129 kg), each with a 68-kg driver. The weight difference accounts for much of the difference in fuel efficiency between these vehicles.

EXAMPLE 3.3 | Takeoff Power!

A fully loaded Boeing 767-300 jetliner has a mass of 180,000 kg. Starting from rest, it accelerates down the runway and reaches a takeoff speed of 270 kilometers per hour (km/h) in 35 seconds. What engine power is required for this takeoff roll?

SOLUTION

The plane gains kinetic energy during the takeoff roll, and we know how long the roll takes, so we can calculate the energy per time, or power. Equation 3.5 gives energy in joules if the mass is in kilograms and speed is in meters per second. So first we convert that 270 km/h into meters per second:

$$(270 \text{ km/h})(1{,}000 \text{ m/km})(1 \text{ h}/3{,}600 \text{ s}) = 75 \text{ m/s}$$

Then the plane's kinetic energy at takeoff is

$$K = \tfrac{1}{2}mv^2 = \left(\tfrac{1}{2}\right)(180{,}000 \text{ kg})(75 \text{ m/s})^2 = 5.1 \times 10^8 \text{ J} = 510 \text{ MJ}$$

The plane is on the runway for 35 seconds, so the engines must be supplying power at the rate of 510 MJ/35 s = 14.6 MJ/s or 14.6 MW. With 1 hp being 746 W, that's about 20,000 hp.

3.6 The Role of Friction

We've now seen that doing work on an object generally results in either stored potential energy or increased kinetic energy (or both if, for example, you simultaneously lift and accelerate an object). But there's one case where doing work increases neither potential nor kinetic energy. That's when you do work solely to overcome friction. When I lifted that bowling ball over your head, you were worried because you knew that the gravitational force would "give back" the stored potential energy if I let go of the ball. But if you push a heavy trunk across a level floor, you don't have to worry about the trunk sliding back and doing damage once you let go. Why not? Because the energy you expended pushing the trunk didn't end up as stored potential energy. But energy is conserved, so where did it go? Most of it went into the form of energy we call, loosely, "heat." Unlike the case of lifting an object or compressing a spring, that energy became unavailable for conversion back to kinetic energy. To convince yourself that friction turns mechanical energy into heat, just try rubbing your hands rapidly together!

The **frictional force** is fundamentally different from forces like gravity or the force in a spring, in that work done against friction doesn't end up as stored potential energy. Rather, it ends up in a form—"heat"—that isn't particularly useful. This inability to recover the energy "lost" to friction is an important limitation on our efforts to use energy wisely and efficiently. For example, despite our best engineering efforts, much of the mechanical energy extracted from gasoline in a typical automobile is lost to friction in the engine, transmission, and tires.

3.7 The Art of Estimation

In this chapter I've thrown around a lot of numbers and a few equations, because it's important to be able to calculate, and calculate accurately, when dealing quantitatively with energy. But it's equally important to be able to make a quick, rough, "back of the envelope" estimate. This kind of estimate won't be exactly right, but it should be close enough to be useful. Many times that's all you need to appreciate an important energy concept or quantity. Below are a couple of examples of estimation. Note how readily I've approximated quantities to nice, round numbers or guessed quantities I wasn't sure of.

EXAMPLE 3.4 | Lots of Gasoline

What's the United States' annual gasoline consumption?

SOLUTION

How many cars are there in the United States? I don't know, but I do know that there are around 300 million people. Americans love cars, so I'm guessing there are around 200 million cars, vans, SUVs, and pickup trucks. How far does a typical car go in one year? More than a few thousand miles, but (for most of us), not 50,000 or 100,000 miles. So suppose it's about 10,000 miles.

And what's the average fuel efficiency in the United States? After falling from the 1980s through 2004, it's risen slightly and now averages about 25 miles per gallon for cars, vans, SUVs, and pickups. Using that figure, the U.S. gasoline consumption rate is

$$(200 \times 10^6 \text{ cars})(10^4 \text{ miles/car/year})\left(\frac{1}{25 \text{ miles/gallon}}\right) = 8 \times 10^{10} \text{ gallons/year}$$

Note my use of scientific notation to make dealing with big numbers easier. I rounded the final result to just one digit because this is an estimate; I could have rounded further to 10^{11} gallons per year because the numbers going into my estimate were so imprecise. How did I know to put the 25 miles per gallon in the denominator? Two ways: First, it makes the units come out right, as they must. I was asking for gallons per year, so I'd better do a calculation that gives gallons per year. Second, and a bit more physically meaningful, I know that higher fuel efficiency should result in lower fuel consumption, so the fuel efficiency had better be in the denominator. More cars and more miles driven increase consumption, so they go in the numerator.

How good is this estimate? According to the U.S. Energy Information Administration (EIA), U.S. gasoline consumption in 2010 was about 380 million gallons per day. Multiplying by 365 days per year gives 1.4×10^{11} gallons. My estimate, just under 10^{11} gallons per year, is a bit low but certainly in the ballpark. And the EIA figure includes gasoline-powered commercial trucks as well as private vehicles. By the way, the number of cars, SUVs, and light trucks registered in the United States in 2010 was about 250 million—above my estimate of 200 million and close to one vehicle for every adult and child!

EXAMPLE 3.5 | A Solar-Powered Country?

The average rate at which Earth's surface absorbs solar energy is about 240 W/m²—that is, 240 watts on each square meter of surface area (this figure accounts for night and day, reflection off clouds, and other factors). If we had devices capable of capturing this energy with 100% efficiency, how much area would be needed to supply the total U.S. energy demand?

SOLUTION

At what rate does the United States use energy? The U.S. population is about 300 million, and, as we saw in the preceding chapter, U.S. residents consume energy at the rate of about 10 kW per capita. So the area needed would be:

$$(300 \times 10^6 \text{ people})(10^4 \text{ W/person})\left(\frac{1}{240 \text{ W/m}^2}\right) = 10^{10} \text{ m}^2$$

where I rounded 300/240 to 1, and put the watts per square meter in the denominator for the same reason I did the miles per gallon in the previous example. There are 1,000 meters or 10^3 m in 1 km, so there are 10^6 m^2 in 1 km^2 (picture a square 1,000 m on a side; it contains a million little squares each 1 m × 1 m). So that 10^{10} m^2 is 10^4 km^2. For comparison, that's roughly one-sixth the area of California's Mojave Desert. Of course we don't have solar energy conversion devices that are 100% efficient, so considerably more area would actually be needed—perhaps 10 times as much with efficiency and infrastructure taken into account. Still, the required land area is remarkably small. See Research Problem 4 for more on this point.

3.8 Wrapping Up

This chapter examines basic energy concepts at the most fundamental level; in that sense it's both essential for and yet most removed from the main themes of this book, namely human energy use and its environmental impacts. You now understand that energy manifests itself either as the kinetic energy associated with motion, or as potential energy associated with forces such as gravity, electromagnetic forces, and nuclear forces. You understand conceptually how energy is stored as gravitational potential energy or as chemical or nuclear energy in fuels. And you know how to talk fluently about energy and power in a variety of unit systems. Finally, you can calculate potential and kinetic energies in simple systems, determine power consumption, and make quick estimates of energy-related quantities.

Still, something may seem missing. Have we really covered all forms of energy? In one sense, yes. But in another, no: We haven't said much about that form of energy called, loosely, "heat." That's a topic with big implications for our human energy consumption, and it's the subject of the next chapter.

CHAPTER REVIEW

BIG IDEAS

3.1 Many forms of energy are instances of **kinetic energy** or **potential energy**. Kinetic energy is the energy of moving objects, while potential energy is stored energy.

3.2 Electricity and magnetism play a crucial role in energy storage and energy technologies. The energy of chemical fuels is ultimately stored in the electric fields associated with molecular configurations. The electric power we use is produced through the process of **electromagnetic induction**. Electromagnetic energy is also carried by **electromagnetic radiation**, of which light is an important example.

3.3 The watt (W) is the standard unit of power, and the **joule** (J) is the corresponding energy unit. One watt is 1 joule per second (J/s). Other common energy units include the kilowatt-hour (kWh), **calorie** (cal), and **British thermal unit** (Btu). Multiples of units are expressed with standard prefixes, and conversion factors relate energy measurements in different units.

3.4 You do **work** when you apply a force to an object as it moves, provided the force acts in the direction of the object's motion. Work is the product of the force and the distance the object moves. Doing work on an object increases its energy. For example, applying a force to lift an object results in an increase in its gravitational potential energy.

3.5 An object of mass m moving with speed v has kinetic energy $K = \frac{1}{2}mv^2$. The **work–energy theorem** states that kinetic energy changes only when nonzero *net work* is done on the object. The net work includes work done by *all* forces acting on the object; if the net work is positive, then kinetic energy increases.

3.6 Friction is a force that dissipates energy, turning the energy of motion into less useful forms. Friction can limit our ability to use energy efficiently.

3.7 Energy is a quantitative subject. However, you can learn a lot about energy with quick, simple estimates of numerical quantities.

TERMS TO KNOW

barrel of oil equivalent (p. 50)
battery (p. 39)
British thermal unit (p. 46)
calorie (p. 46)
chemical energy (p. 42)
elastic potential energy (p. 36)
electric charge (p. 39)
electric current (p. 39)
electric field (p. 39)
electromagnetic force (p. 37)
electromagnetic induction (p. 40)
electromagnetic radiation (p. 43)
electromagnetic wave (p. 43)
force (p. 37)
frictional force (p. 56)
fuel cell (p. 40)
gravitational force (p. 37)
gravitational potential energy (p. 36)
horsepower (p. 50)
joule (p. 45)
kinetic energy (p. 36)
magnetic field (p. 39)
newton (p. 52)
nuclear force (p. 37)
photon (p. 44)
photovoltaic cell (p. 42)
Planck's constant (p. 44)
potential energy (p. 36)
quad (p. 46)
rechargeable battery (p. 40)
ton, tonne, metric ton (p. 50)
tonne oil equivalent (p. 50)
weight (p. 52)
work (p. 50)
work–energy theorem (p. 54)

GETTING QUANTITATIVE

Energy of a photon: $E = hf = \frac{hc}{\lambda}$ (Equation 3.1; p. 44)

Planck's constant: $h = 6.63 \times 10^{-34}$ J·s

Speed of light: $c = 3.00 \times 10^{8}$ m/s

Energy and power units: see Table 3.1

Energy content of fuels: see Table 3.3

Work: $W = Fd$ (Equation 3.2; p. 51)

Force of gravity: $F_g = mg$ (Equation 3.3; p. 52)

Work done lifting object of mass m a distance h: $W = mgh$ (Equation 3.4; p. 52)

Kinetic energy: $K = \frac{1}{2}mv^2$ (Equation 3.5; p. 54)

QUESTIONS

1 Why is it harder to walk up a hill than on level ground?

2 Describe qualitatively the relationship between force and potential energy.

3 Table 3.3 shows that hydrogen has a higher energy content per kilogram than natural gas, but a lower energy content per cubic foot. How can this be consistent?

4 You jog up a mountain and I walk. Assuming we weigh the same, compare (a) our gravitational potential energies when we're at the summit, and (b) the average power each of us expends in climbing the mountain.

5 How many (a) megajoules are in 1 exajoule; (b) petagrams in 1 gigatonne (1 tonne = 1,000 kg); (c) kilowatt-hours in 1 gigawatt-hour?

6 How is the frictional force fundamentally different from the gravitational force?

EXERCISES

1 The average daily human diet has an energy content of about 2,000 kcal. Convert this 2,000 kcal per day into watts.

2 Using appropriate conversions between pounds and grams, and degrees Celsius and Fahrenheit, show that 1 Btu is equivalent to 1,054 J.

3 Express in watts the world energy-consumption rate of approximately 470 Q per year.

4 There are two ways to calculate the power output of a car when you know that (1) the car has a 250-horsepower engine and (2) the car gets 20 miles per gallon when traveling at 60 miles per hour: (a) Convert the horsepower rating into watts. (b) Calculate the gasoline consumption rate and, using Table 3.3's energy equivalent for gasoline, convert the result to a power in watts. Comparison of your results shows that a car doesn't always achieve its engine's rated horsepower.

5 An oil furnace consumes 0.80 gallons of oil per hour while it's operating. (a) Using the approximate value of 40 kWh per gallon of petroleum product, find the equivalent power consumption in watts. (b) If the furnace runs only 15% of the time on a cool autumn day, what is the furnace's average power consumption?

6 The United States imports about 12 million barrels of oil per day. (a) Consult the tables in this chapter to convert this quantity to an equivalent power, measured in watts. (b) Suppose we wanted to replace all that imported oil with energy produced by fission from domestic uranium. How many 1,000-MW nuclear power plants would we have to build?

7 Assuming that 1 gallon of crude oil yields roughly 1 gallon of gasoline, estimate the decrease in daily oil imports (see preceding question) that we could achieve if the average fuel efficiency of U.S. cars and light trucks, now around 23 miles per gallon, were raised to the 50 miles per gallon typical of a hybrid car. Assume the average vehicle is driven about 10,000 miles per year and that there are about 250 million cars and light trucks operating in the United States.

8 In the text I cited an environmentalist's claim that buying an SUV instead of a regular car wastes as much energy as leaving your refrigerator door open for 7 years. Let's see if that's right. A typical refrigerator consumes energy at the rate of about 400 W when it's running, but it usually runs only about a quarter of the time. If you leave the door open, however, the refrigerator will run all the time. Assume that an average SUV's fuel efficiency is 20 miles per gallon, an average car gets 30 miles per gallon, gasoline contains 40 kWh of energy per gallon, and you drive the vehicle 15,000 miles per year. Calculate how long you would have to leave your refrigerator door open to use as much extra energy as the difference between the car and SUV in the first year you own the vehicle.

9 I want to expend 100 "calories" (that is, 100 kcal) of energy working out on an exercise machine. The readout on the machine says I'm expending energy at the rate of 270 W. How long do I need to exercise to expend those 100 kcal?

10 You buy a portable electric heater that claims to put out 10,000 Btuh (meaning 10,000 Btu/h). If it's 100% efficient at converting electrical energy to heat, what is its electrical energy-consumption rate in watts?

11 A car with a mass of 1,700 kg can go from rest to 100 km/h in 8.0 seconds. If its energy increases at a constant rate, how much power must be applied to achieve this magnitude of acceleration? Give your answer in both kilowatts and horsepower.

12 You're cycling on a level road at a steady 12 miles/hour (5.4 m/s), and your body's mechanical power output is 150W. Because your speed is constant, you're applying a force that's just enough to overcome the forces of friction and air resistance. What is the value of your applied force?

13 You're cycling up a hill, rising 5 feet for every 100 feet you move along the road. You're going at a steady 4.3 m/s, and you need to overcome frictional forces totalling 30 N. If you and your bicycle together have a mass of 82 kg, what's the minimum power you need to supply to overcome both friction and gravity?

14 An energy-efficient refrigerator consumes energy at the rate of 280 W when it's actually running, but it's so well insulated that it runs only about one-sixth of the time. You pay for that efficiency up front: It costs $950 to buy the refrigerator. You can buy a conventional refrigerator for $700. However, it consumes 400 W when running, and it runs one-fourth of the time. Calculate the total energy used by each refrigerator over a 10-year lifetime and then compare the total costs—purchase price plus energy cost—assuming electricity costs 10¢ per kilowatt-hour.

15 In 2010 the *Deepwater Horizon* oil well in the Gulf of Mexico blew out, spilling some 5 million barrels of oil into the Gulf over 3 months. (a) Find the energy equivalent of the spilled oil, in joules or suitable multiples, and (b) estimate how long this oil could have supplied the entire U.S. energy demand.

RESEARCH PROBLEMS

1 Find a value for the current population of the United States, and use your result along with the approximate U.S. annual energy-consumption rate of 100 Q per year to get an approximate value for the per capita U.S. energy-consumption rate in watts.

2 Choose a developing country and find the values for its current population and its total annual energy consumption in quads. Using these numbers, calculate the per capita energy-consumption rate in watts.

3 Find the official EPA (Environmental Protection Agency) fuel efficiency for the car you or your family drive. Estimate your yearly mileage, and determine the amount of fuel you would save each year if you switched to a 50-mile-per-gallon hybrid.

4 Make the assumption that the solar-collector area calculated in Example 3.5 should be increased by a factor of 10 to account for the inefficiency of the solar cells and to allow room for infrastructure. What fraction of the total area of (a) the state of New Mexico and (b) the continental United States would then be needed? (c) Compare your answer to part (b) with the fraction of land that's now under pavement.

ARGUE YOUR CASE

1 A friend claims that a metric tonne is more than twice the mass of an English ton, since a kilogram is more than twice the mass of a pound. Make a quantitative argument in support of or against your friend's claim.

2 Upgraded fuel economy standards released in 2010 call for a 34.1 mile-per-gallon average for new cars and light trucks in 2016. A politician opposing this increase claims it will save "only a drop in the bucket" compared with the 2010 fleet average of about 23 MPG. Formulate a counterargument showing that the increase—once the entire fleet reaches 34.1 MPG—could significantly reduce the United States' gasoline consumption rate from its 2010 value of about 380 million gallons per day. State any additional assumptions you make.

Chapter 4

ENERGY AND HEAT

Huge cooling towers loom over an electric power plant; their steamy plumes carry away some two-thirds of the energy released from the fuel consumed at the plant. Only the remaining one-third gets transformed into electrical energy. You fill your car's gas tank, pay, and drive off—yet only about 20% of the energy in the gas ends up actually moving your car. Why are power plants and cars so inefficient? The answer lies in the nature of what we commonly call "heat," and in its relationship to other forms of energy. That relationship is so important to humanity's energy appetite that we need to explore it in detail.

4.1 Heat and Internal Energy

You probably think of heat as the energy contained in a hot object. Brimming with many gallons of water at 120°F, your hot-water heater surely contains a lot of heat. But strictly speaking, that's not what "heat" means. The energy in the water heater or any other hot object is **internal energy**, also called **thermal energy**. Internal energy consists of kinetic energy associated with the random motions of the atoms and molecules that make up the object. So it's a form of energy we've already discussed. The only difference between thermal energy and the kinetic energy of, say, a moving car is that thermal energy is random, with the individual molecules moving in different directions at various speeds. In contrast, all parts of the moving car participate in the same motion. Both are examples of kinetic energy, but the difference between random and directed motion is a big one, with serious consequences for human energy use.

So what's heat? Strictly speaking, **heat** is energy that is flowing as a result of a temperature difference. The term *heat* always refers to a *flow* of energy from one place to another, or one object to another. And not just any flow, but a flow that occurs because of a temperature difference. If I take a can of gasoline out of the garage and carry it to my lawnmower, that's a flow of energy. But it's not heat, because the flow of energy from garage to lawnmower is caused not by a temperature difference but by my carrying the can. Similarly, and perhaps surprisingly, the flow of energy from your microwave oven to the cup of water you're boiling is not heat. There's no temperature difference involved, but rather a carefully engineered flow of electromagnetic energy. On the other hand, the loss of energy from your house on a cold day, associated with the flow of energy through the walls and windows, is heat. That's because this flow is driven by the temperature difference between inside and outside. If the inside and outside of the house were at the same temperature, there would be no energy loss. If it were warmer outside, the flow would be into the house. In or out, that flow is properly described as *heat* or *heat flow*. Other examples include the heat radiating from a hot woodstove or lightbulb filament, the transfer of energy from a hot stove burner into a pan of water, and the movement of hot air from Earth's surface to higher levels of the atmosphere. All are heat flows because they occur as a result of a temperature difference.

You might think I'm splitting hairs with this distinction. After all, a microwave oven heats water just as well as a stove burner, so is there really a significant difference between the two processes? There is, and that's precisely the point. You can raise temperature with heat flow, but there are other ways to do it that don't involve a temperature-driven energy transfer. The microwave is just one example; I could also heat the water by agitating it vigorously with a spoon. The spoon is no hotter than the water, so this energy flow isn't heat. Instead, it's a transfer of kinetic energy as the moving spoon does work on the molecules of water. Another example: Push on the plunger of a bicycle pump, and you raise the temperature of both the pump and the air inside it. This doesn't happen because of heat transfer from a hotter object, but because you supplied mechanical energy that ended up as internal energy of the pump and the air. An experiment involving the agitation of water is what allowed James Joule to determine the quantitative relationship between heat and mechanical energy. That relationship is further developed in the so-called first and second laws of thermodynamics.

The distinction between heat and mechanical energy flows underlies the **first law of thermodynamics**, which says that the change in an object's internal energy is the sum of the mechanical work done on the object and the heat that flows into it. The first law is essentially a generalization of the law of conservation of energy to include heat. In other words, it doesn't matter where an object's internal energy comes from, whether by heat transfer or mechanical energy or a combination of the two. A cup of hot water is just that; whether its temperature increased because of heat flow or because someone shook it violently makes no difference. That heat or mechanical energy flow can go either way, into or out of the object, respectively raising or lowering its internal energy. The first law handles both cases using positive and negative signs, respectively.

Low temperature

High temperature

FIGURE 4.1
Temperature measures the average molecular kinetic energy. At lower temperatures, the molecules of a gas have lower kinetic energy and move more slowly; at higher temperatures, the molecules have higher kinetic energy and move more quickly.

4.2 Temperature

If heat is energy flowing because of a temperature difference, then what's temperature? The answer is simple: **Temperature** is a measure of the average thermal energy of the molecules in a substance. That is, temperature measures the average of the quantity $\frac{1}{2}mv^2$ (kinetic energy) associated with individual molecules. The word *average* is important, because thermal motions are random in both direction and speed. Figure 4.1 suggests this kinetic-energy meaning of temperature.

You're probably familiar with several temperature scales. The Fahrenheit scale, used mainly in the United States, has water freezing at 32°F and boiling at 212°F. The Celsius scale, used in science and in everyday life throughout most of the world, has water freezing at 0°C and boiling at 100°C. But neither scale is truly fundamental, because neither corresponds to zero energy at the zero-degree mark. The **Kelvin scale**, the official temperature scale of the SI unit system, has the same size degree gradations as the Celsius scale, but its zero is where molecular kinetic energy—at least in classical physics—would be precisely zero. This is **absolute zero**, and it occurs at about –273°C. The temperature unit on the Kelvin scale is the kelvin (not degrees kelvin or °K, but just K). A temperature change of 1 K is the same as a change of 1°C, but the actual values on the Kelvin and Celsius scales differ by 273. Another absolute scale, used primarily by engineers in the United States, is the Rankine scale. Its degrees are the same size as Fahrenheit degrees, but its zero, like that of the Kelvin scale, is at absolute zero. We'll find the absolute property of the Kelvin scale useful in some areas of energy and climate studies. On the other hand, the Celsius scale is more familiar and often more appropriate for expressing everyday temperatures, so I'll be going back and forth between the two. Again, where temperature *differences* are concerned, there's no distinction between the Kelvin and Celsius scales. Figure 4.2 compares the four temperature scales.

4.3 Heat Transfer

Several different physical mechanisms produce the temperature-difference-driven energy flow that we call heat. We refer to these as *heat-transfer mechanisms*, although strictly speaking, "heat transfer" is redundant because heat *means* energy that's flowing from one place to another.

CONDUCTION

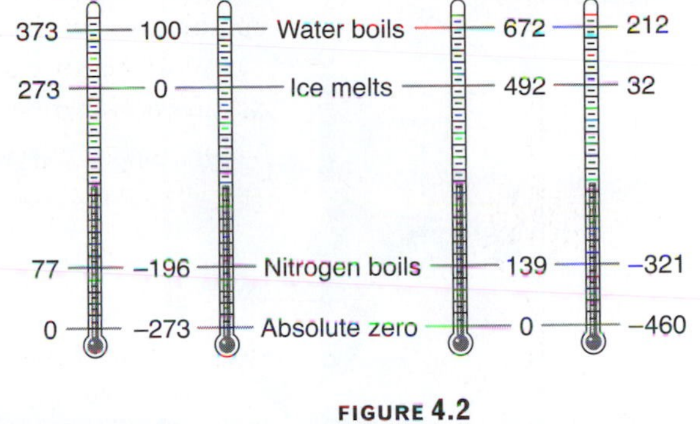

FIGURE 4.2
Four temperature scales compared: Kelvin, Celsius, Rankine, and Fahrenheit.

Think of a cool pan sitting on a hot stove burner. The burner's higher temperature means the molecules in the burner have greater kinetic energies. When these molecules collide with those in the pan, they transfer kinetic energy. As a result, the pan warms. That's a microscopic description of heat transfer by **conduction**, the direct sharing of kinetic energy by molecular collisions between materials in contact. Another example is the flow of heat through the walls of a house. Air molecules in the house collide with the wall, transferring kinetic energy. Molecules at the indoor edge of the wall transfer their energy to molecules deeper within the wall, and eventually the energy gets to the outdoor edge of the wall where it's transferred to the exterior air. The temperature difference between the warmer inside and cooler outside ensures that collisions, on average, transfer energy in the direction from warmer to cooler.

Conduction is an important process in the study of energy and climate. Conduction is largely responsible for energy losses from buildings, and the vast energy resources we devote to heating attest to these losses. Conduction also transfers energy from the sunlight-warmed surface of the Earth into the atmosphere; other heat-transfer mechanisms then carry the energy to higher levels of the atmosphere.

In the case of buildings, we would like to avoid energy loss by conduction through walls. We do so with insulating materials, which, by virtue of their structure, are poor conductors of heat. A given material is characterized by its **thermal conductivity**, designated *k*. Table 4.1 shows thermal conductivities of

TABLE 4.1 | **THERMAL CONDUCTIVITIES OF SELECTED MATERIALS**

Material	Thermal conductivity, (W/m·K)	Thermal conductivity, (Btu·inch/hour·ft²·°F)
Air	0.026	0.18
Aluminum	237	1,644
Concrete (typical)	1	7
Fiberglass	0.042	0.29
Glass (typical)	0.8	5.5
Rock (granite)	3.37	23.4
Steel	46	319
Styrofoam (extruded polystyrene foam)	0.029	0.2
Water	0.61	4.2
Wood (pine)	0.11	0.78
Urethane foam	0.22	0.15

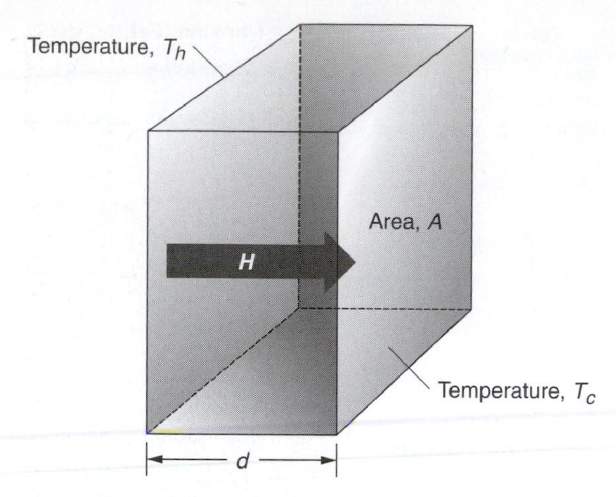

FIGURE 4.3
Conductive heat flow H through a slab of material with conductivity k is given by Equation 4.1. T_h and T_c are the temperatures on the warmer side and the cooler side of the slab, respectively. A is the slab area, and d is its thickness.

some common materials. The heat flow through a given piece of material is determined by its thermal conductivity, its area, its thickness, and the temperature difference from one side of the piece to the other. Figure 4.3 and Equation 4.1 show how to determine this heat flow:

$$H = kA\,\frac{T_h - T_c}{d} \tag{4.1}$$

Here H is the heat flow, measured in watts; k is the thermal conductivity; A and d are the area and thickness, as shown in Figure 4.3; and T_h and T_c are the temperatures on the hot and cold sides of the material, respectively.

EXAMPLE 4.1 | A Better Refrigerator

Assume it's 3°C inside your refrigerator, while the surrounding kitchen is at room temperature (20°C or 68°F). The refrigerator is essentially a rectangular box 150 cm high by 76 cm wide by 64 cm deep. Determine the rate at which heat flows into the refrigerator if it's insulated on all sides using 4-cm-thick urethane foam.

SOLUTION

Heat flows into the refrigerator as a result of thermal conduction. Equation 4.1 describes conduction quantitatively; to determine the heat flow, we need to know the area, the temperature difference, and the thickness and thermal conductivity of the insulating material. We're given the two temperatures and the insulation material (urethane foam), so we can get the thermal conductivity from Table 4.1; it's 0.022 W/m·K. We know the insulation thickness (4 cm), but we need to calculate its area, which is the total of the refrigerator's top and bottom, front and back, and two side areas. You can see from Figure 4.4 that the total area is then

$$2[(76\text{ cm})(150\text{ cm}) + (76\text{ cm})(64\text{ cm}) + (150\text{ cm})(64\text{ cm})] = 51{,}728\text{ cm}^2 \simeq 5.2\text{ m}^2$$

where the last step follows because there are 100 cm in 1 m, and thus $100 \times 100 = 10^4$ cm^2 in 1 m^2. Using this area in Equation 4.1, along with the temperatures, insulation thickness, and thermal conductivity, we get

$$H = kA\,\frac{T_h - T_c}{d} = (0.022\text{ W/m·K})(5.2\text{m}^2)\,\frac{20°\text{C} - 3°\text{C}}{0.04\text{ m}} = 49\text{ W}$$

To keep cold, the refrigerator needs to remove energy from its interior at this rate. That's why it's plugged into the wall and consumes electricity. However, its average electrical power consumption may be less than 49 W, for reasons I'll describe in Section 4.9.

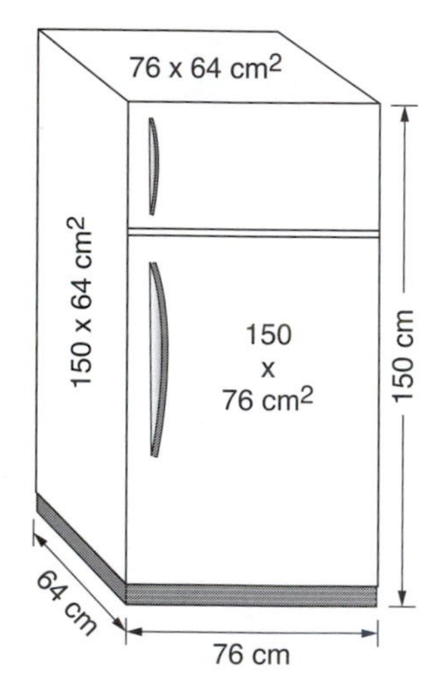

FIGURE 4.4
The refrigerator from Example 4.1 has a total area of 5.2 m^2, and heat flows through its insulated walls at the rate of 49 W.

BUILDING INSULATION AND *R* VALUE

The thermal conductivities of the materials comprising a building's walls and roof determine the rate of heat flow out of or into the building, and therefore its energy requirements for heating and cooling. Building (or retrofitting) with insulating materials that have low conductivity is a critical step in reducing overall energy consumption, especially the nearly 40% that occurs in the commercial and residential sectors.

If you go to your local building-supply outlet and buy a roll of fiberglass insulation or a slab of rigid foam, you'll find it marked with its ***R* value**. This quantity measures the material's resistance to the flow of heat and is closely related to thermal conductivity (k). However, where k is an intrinsic property of a given material, R characterizes a specific thickness of material. The relationship between the two is $R = d/k$ where d is the thickness. The R value is almost always measured in English units; although rarely stated explicitly, those units are ft^2·°F·h/Btu. It's easier to make sense of R by considering its inverse, whose units are British thermal units per hour per square foot per degree Fahrenheit. Take, for example, a 1-inch-thick slab of "blueboard," a foam insulation often used on basement walls (Fig. 4.5). Its R value is 5, which means each square foot of blueboard loses one-fifth of a Btu of energy every hour for every degree Fahrenheit temperature difference across the insulation. So if it's 70°F inside your basement and 45°F in the ground outside, and if your basement wall area totals 1,000 square feet, then the rate of energy loss through the wall is

$$(1/5 \text{ Btu/h/ft}^2/°\text{F})(1{,}000 \text{ ft}^2)(70°\text{F} - 45°\text{F}) = 5{,}000 \text{ Btu/h}$$

Incidentally, that's about 1,500 W (see Exercise 1), so picture 15 energy servants hunkered down in your basement working away just to produce the energy that's lost through your basement wall. Of course, things would be a lot worse without the insulation, and in any event on a cold day there's a lot more energy being lost through the walls and windows upstairs.

FIGURE 4.5
Worker installing "blueboard" rigid foam insulation on exterior foundation walls. This material provides an *R* value of 5 per inch of thickness, some 40 times that of the concrete foundation itself.

One virtue of the R value is that the overall R value for a composite structure can be calculated by simply adding the R values of the component materials. Table 4.2 lists R values for some common materials used in construction, although you could calculate some of these from Table 4.1 using $R = d/k$. Note that an 8-inch-thick concrete wall rates at roughly $R = 1$, so it's a lousy insulator compared with our 1-inch foam. But we would have had a more accurate account of the heat loss through the basement wall had we added the concrete's $R = 1$ to the foam's $R = 5$, for a total R value of 6. Because the R value scales with thickness, you can easily determine the R value for any material appearing in Table 4.2, even if your particular thickness isn't listed in the table.

Calculating the total R value for a composite wall is a bit subtle (Fig. 4.6); in addition to the building materials, there's usually a layer of "dead" air against the wall, which provides additional insulation. Table 4.2 lists the R value of this dead air for both interior and exterior wall surfaces, but be aware that these figures depend on the air being still. A strong wind blowing outside replaces slightly warmed air with cold air, removing that extra insulating

TABLE 4.2 | ***R* VALUES OF SOME COMMON BUILDING MATERIALS**

Material	*R* value ($ft^2 \cdot °F \cdot h/Btu$)
Air layer:	
Adjacent to inside wall	0.68
Adjacent to outside wall, no wind	0.17
Concrete, 8-inch	1.1
Fiberglass:	
3.5-inch	12
5.5-inch	19
Glass, 1/8-inch single pane	0.023
Gypsum board (Sheetrock), 1/2-inch	0.45
Polystyrene foam, 1-inch	5
Urethane foam, 1-inch	6.6
Cellulose, blown, 5.5 inch	20
Window (*R* values include adjacent air layer):	
Single-glazed wood	0.9
Standard double-glazed wood	2.0
Argon-filled double-glazed with low-E coating	2.9
Argon-filled triple-glazed with low-E coating	5.5
Best commercially available windows	11.1
Wood:	
1/2-inch cedar	0.68
3/4-inch oak	0.74
1/2-inch plywood	0.63
3/4-inch white pine	0.96

value. Dead air actually contributes more than glass to the insulating properties of windows and is especially effective when trapped between the panes of a double- or triple-glazed window. Replacing air with argon gas further enhances the insulating qualities of a window, as do "low-E" (low-emissivity) coatings that reduce energy loss by infrared radiation (more on radiation heat loss later in this section). Premium windows cost a lot, but over the lifetime of a building they can pay for themselves many times over in fuel costs, as well as reduce the building's environmental impact. Table 4.2 includes some windows, from the most basic to the most advanced; the difference in *R* value is dramatic. Today's most advanced windows, in fact, are rated at *R*11, equivalent to 4 inches of fiberglass insulation.

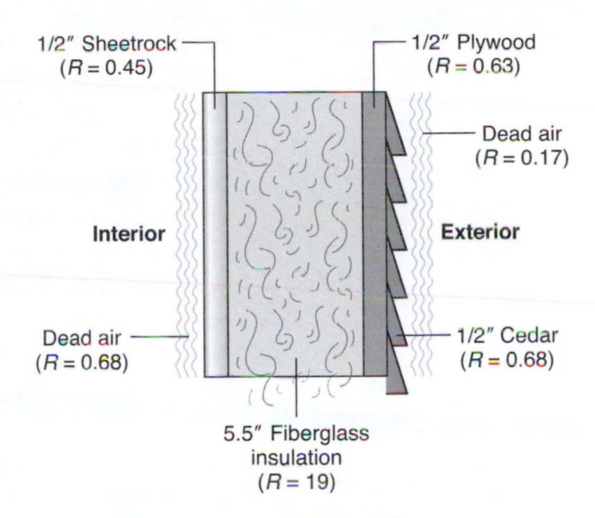

FIGURE 4.6
The *R* value of a composite wall is the sum of the *R* values of its individual components. Here the *R* value is 21.6, nearly all of it from the fiberglass insulation.

EXAMPLE 4.2 | Saving on Heat

Imagine you've bought an old house that measures 36 feet wide by 24 feet deep and is 17 feet from the ground to the second-floor ceiling. The walls have no insulation, and the *R* value of the wall materials and dead air is only 4. The second-floor ceiling has 3.5 inches of fiberglass insulation over 1/2-inch gypsum board. The attic under the sloping roof is uninsulated, a common practice to avoid condensation and rot. This makes our calculations easier, since the house can be considered a rectangular box.

(a) On a winter day when it's 10°F outside and 68°F inside, what's the rate of heat loss through the walls and ceiling? Neglect heat flow through the floor (a reasonable approximation because the ground is much warmer than the outside air) and through the windows (less reasonable, because windows generally account for a lot of heat loss; see Exercise 11). (b) How much oil would you burn in a winter month with this average temperature? Assume an oil furnace that is 80% efficient at converting the chemical energy of oil into heat in your house. (c) At $2.75 per gallon, how much would it cost to heat this house for a month? (d) How much would you save on your monthly heating bill if you put 3.5-inch fiberglass insulation in the walls and raised the ceiling insulation to 11 inches of fiberglass?

SOLUTION

Since they have different *R* values, we need to calculate separately the areas of ceiling and walls to determine the individual heat losses. The ceiling measures (24 ft) × (36 ft) = 864 ft^2. Table 4.2 shows the ceiling's *R* value is 13.3, consisting of the 3.5-inch fiberglass insulation (R = 12) plus the gypsum board (R = 0.45), the inside air layer (R = 0.68), and the outside layer (R = 0.17). This means every square foot of ceiling loses 1/13.3 Btu per hour for every degree Fahrenheit of temperature difference. Here we have 864 square feet of ceiling area and a temperature difference of 68°F – 10°F = 58°F (I'm assuming the unheated, ventilated attic is at the outdoor temperature, which may be a slight underestimate.) Then the heat loss through the entire ceiling is

$$(1/13.3\ \text{Btu/h/ft}^2/°\text{F})(864\ \text{ft}^2)(58°\text{F}) = 3{,}768\ \text{Btu/h}$$

Now for the walls: The total area of the equal-sized front and back and the two identical sides is (2)[(36 ft × 17 ft) + (24 ft × 17 ft)] = 2,040 ft^2. We're given the total R value of 4, so the heat-loss rate through the walls is

$$(1/4\ \text{Btu/h/ft}^2/°\text{F})(2{,}040\ \text{ft}^2)(58°\text{F}) = 29{,}580\ \text{Btu/h}$$

So the total loss rate—walls and ceiling—is 33,348 Btu/h, which I'll round to 33 kBtu/h. (Again, this is a very rough calculation, neglecting windows, losses through the floor, losses due to air infiltration, etc.) There are a total of $24 \times 30 = 720$ hours in a 30-day month, so this amounts to a monthly heat loss of (33 kBtu/h)(720 h/month) = 24 MBtu/month, where again I rounded and changed to the more convenient prefix *mega* (M) instead of *kilo* (k).

On to the oil consumption and cost. With a furnace that's 80% efficient, you need oil with an energy content of 24 MBtu/0.80 = 30 MBtu. Table 3.3 lists oil at 138 kBtu/gallon. This is the same as 0.138 MBtu/gallon, so we'll need (30 MBtu)/(0.138 MBtu/gal) = 217 gallons of oil to heat the house for a month. At $2.75 per gallon, that will cost $597.

Putting 11-inch fiberglass in the ceiling gives twice the insulating value of the 5.5-inch fiberglass listed in Table 4.2, or an *R* value of 38 for the insulation alone; adding the gypsum board and dead air on both sides gives a total *R* of 39.3. Adding 3.5-inch fiberglass to the walls increases the walls' *R* value by 12, to 16. Areas and temperature differences remain the same, so we can simply repeat the above calculations with the new *R* values. The result is a heat-loss rate of 8,670 Btu/h, about one-fourth of the earlier value. Everything else goes down proportionately, so you need only 56 gallons of oil for the month, at a total cost of $154. You save $597 – $154 = $443 on your monthly oil bill, and you produce a lot less carbon dioxide and pollutants.

So the insulation is a good investment. Exercise 11 explores this situation in the more realistic case of a house with windows (when you decide to upgrade those, too).

EXAMPLE 4.3 | Heat Loss and Solar Gain

On a sunny winter day, a window is situated so that during the midday hours it admits solar energy at the rate of approximately 300 watts per square meter of window surface—equivalently, 95 Btu per hour per square foot. For the best and worst windows listed in Table 4.2, find the minimum outdoor temperature for which that window provides the house with a net energy gain during the midday hours. Assume the indoor temperature is fixed at 70°F.

SOLUTION

We know the rate at which each square foot of window gains energy: 95 Btu/h. We want to know when the rate of heat loss is equal to this gain. Since the inverse of the *R* value gives the heat-loss rate in Btu per hour per degree Fahrenheit per square foot, multiplying the temperature difference (call it ΔT) by $1/R$ gives the heat loss per square foot. Now, we want to know when the rate of energy gain exceeds the heat-loss rate. The break point occurs when they're just equal, so we set the heat-loss rate $\Delta T/R$ equal to the rate at which the window gains energy:

$$\frac{\Delta T}{R} = 95 \text{ Btu/h/ft}^2$$

Solving for ΔT gives $\Delta T = (95 \text{ Btu/h/ft}^2)(R)$. The single-glazed window listed in Table 4.2 has an R value of 0.9, giving $\Delta T = (95 \text{ Btu/h/ft}^2)(0.9 \text{ ft}^2\cdot°\text{F}\cdot\text{h/Btu}) = 86°\text{F}$ (note how everything works out when the units of the R value are stated explicitly). With an interior temperature of 70°F, this means the outdoor temperature can be as low as 70°F – 86°F, or a chilly –16°F, and even this relatively inefficient window still provides a net energy gain. (Of course, that's only true during the midday hours on sunny days. Accounting for night and clouds, even the best south-facing windows are likely to be, on average, net energy losers.) I won't bother to repeat the calculation for the better window, since it provides a midday energy gain down to temperatures far below anything you would encounter. By the way, the better-insulated window probably admits a little less solar energy than the single-pane model, but it more than compensates for that with far lower heat loss at night. In fact, many high-quality windows are designed to reduce solar gain as well as heat loss in order to cut energy use for air conditioning during the summer.

CONVECTION

Conduction carries energy from a hot stove burner into a pan of boiling water, but from there the vigorous boiling motion is what carries energy to the water's surface. That bulk motion of the water (as opposed to the random motions of individual molecules) is called **convection**. In general, convection is the bulk motion of a fluid, driven by temperature differences across the fluid. Convection generally occurs in the presence of gravity, because as fluid becomes less dense when heated, it rises, gives up its energy, then sinks to repeat the convective circulation (Fig. 4.7).

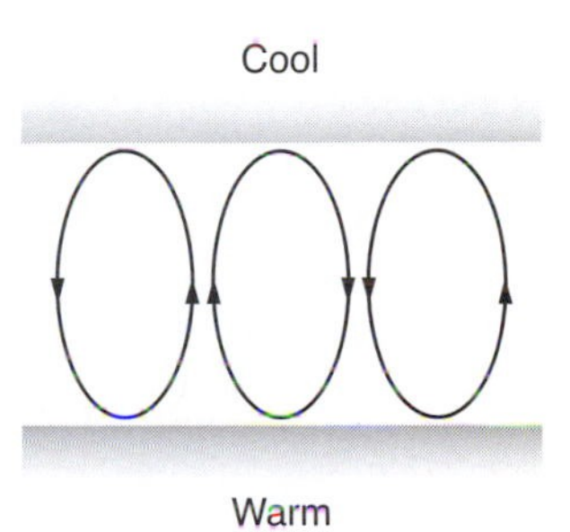

FIGURE 4.7
Convection. Warm fluid rises, gives up its thermal energy, and then sinks to form individual cells of convective motion.

Home heating systems often exploit convection. Baseboard hot water pipes or old-fashioned radiators heat the air around them, which then circulates about the room in a large-scale version of the convection patterns shown in Figure 4.7. Convection carries heated air near Earth's surface higher into the atmosphere. The violence of a thunderstorm results from vigorous convective motions within the storm, and, on a planetary scale, convection produces large-scale atmospheric circulation patterns that transfer energy from the tropics toward the poles (Fig. 4.8). On a much larger scale, convection carries energy from the Sun's interior to just below its surface—energy that becomes the sunlight that ultimately powers life on Earth.

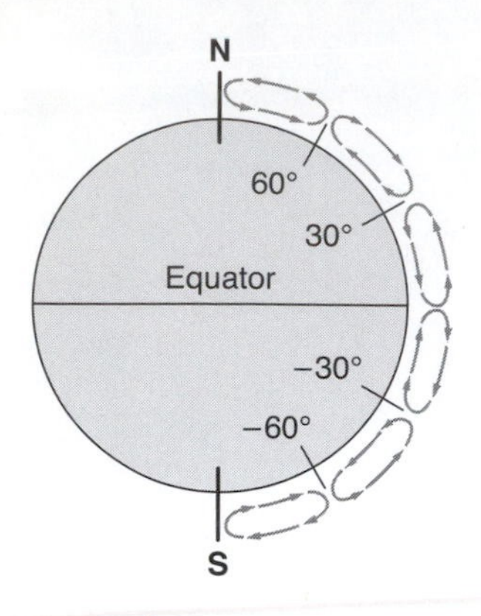

FIGURE 4.8
Large-scale convective motions transport energy from Earth's warm equatorial regions to the cold polar regions. Coupled with Earth's rotation, these convective motions also give rise to the prevailing winds that blow from east to west at low latitudes and from west to east in the temperate zones.

Convection is also responsible for energy loss. Although air is a good thermal insulator, its convective motion between the panes of a double-glazed window can still result in heat loss. For that reason, a narrower air space that inhibits convection is actually better than a wider one. Insulating materials such as fiberglass and foams work because the interstices between glass fibers or tiny bubbles in the foam trap air or gases and prevent convective motions from developing. Energy loss in buildings also results from **infiltration**, as winds (themselves large-scale convective flows) drive cold air in through cracks, especially around windows and doors. Infiltration is exacerbated by mechanical systems that vent air from a building, reducing the interior pressure and encouraging infiltration from outside.

The strength of convective flows, and therefore the rate of energy transfer, depends on the temperature difference across a convecting fluid. But fluid motion is complicated, and there's no simple formula that applies in all cases of convection.

What I've been describing here is **natural convection**, so called because it occurs naturally when there's a vertical temperature difference. There's also **forced convection**, which occurs when we use fans or pumps to move fluids. Hot-air and hot-water heating systems generally use forced convection to move fluid from the furnace or boiler into a living space where natural convection in the air then takes over. Forced convection also moves heat-transfer fluids through solar energy systems and through the boilers of fossil fuel and nuclear power plants. So forced convection is an important process in many energy technologies.

RADIATION

Stand near a hot woodstove, and you feel heat coming from the stove to your skin. Around a campfire, even on a frigidly cold evening, you're warmed by heat from the fire. And when you lie on the beach, your body is heated by energy from the Sun. Conduction or convection aren't involved in these cases, since the air between you and the heat source isn't itself being warmed (and there is no air between Sun and Earth). Instead, the heat transfer occurs by **radiation**—in particular, electromagnetic radiation.

As long as a material is above absolute zero, its atoms and molecules are in random thermal motion, moving about in gases and liquids or vibrating in place in solids. And because atoms and molecules consist of electrically charged particles, those random motions result in electromagnetic radiation. So all objects, as long as they're above absolute zero temperature, emit energy in the form of electromagnetic radiation. They may also gain electromagnetic energy from their surroundings, so there isn't necessarily a net energy loss. If an object is hotter than its surroundings, it emits more energy than it absorbs; if it's cooler, it gains energy. So the net flow of electromagnetic energy to or from an object is driven by the temperature difference between the object and its surroundings. This temperature-driven flow of electromagnetic radiation therefore qualifies as a third mechanism for heat transfer.

The total energy radiated increases rapidly with an object's temperature—specifically, as the fourth power of the absolute temperature measured on the Kelvin scale. It also depends on the object's surface area and on a property called its **emissivity**. Emissivity is a number between 0 and 1 that describes a material's efficiency as an emitter of electromagnetic radiation. A well-known law of physics states that a given material's efficiency as an emitter is equal to its efficiency as an absorber; that is, materials that are good at absorbing radiation are equally good at emitting it. A black object (which looks black because it absorbs all colors of light incident on it) is, in principle, a perfect absorber. Therefore it's also a perfect emitter, and it has an emissivity of 1. A shiny material such as metal reflects nearly all the light incident on it, and therefore it's both a poor absorber and a poor emitter, with an emissivity close to zero. That's why the inside of a thermos vacuum bottle is shiny; the resulting low emissivity cuts down on energy transfer by radiation. The vacuum between the walls of the bottle ensures that there's no conduction or convection either. That's why your coffee stays hot for hours. Those low-E windows I mentioned earlier are actually coated with a thin film that reduces emissivity and thus lowers energy loss by radiation.

The efficiency of absorption and emission are equal at any given wavelength of radiation, but may vary with wavelength. An object that looks black to the eye has high absorption and therefore high emissivity for visible light, but it might behave very differently for infrared, ultraviolet, or other forms of electromagnetic radiation. The difference between emissivity for visible and infrared wavelengths has great practical significance; for example, low-E windows manage to minimize heat loss via infrared radiation without cutting down too much on the visible light they admit. And that same difference between visible light and infrared emissivities plays a major role in climate, especially in the greenhouse effect and the role our fossil-fueled society plays in altering climate (much more on this in Chapter 12).

Putting together fourth-power dependence on temperature with emissivity gives the **Stefan–Boltzmann law** for radiation:

$$P = e\sigma AT^4 \tag{4.2}$$

Here P is the power, in watts, emitted by an object at temperature T; A is the object's surface area; and e is its emissivity. The symbol σ (the Greek letter "sigma") is the **Stefan–Boltzmann constant**, a universal quantity whose value, in SI units, is 5.67×10^{-8} W/m^2/K^4. Here I'm using the symbol P rather than the H of Equation 4.1 because Equation 4.2 gives, strictly speaking, only the power radiated *from* an object. To find the net heat transfer by radiation would involve subtracting any radiation coming from the object's surroundings and being absorbed by the object. In many cases where radiation is important, however, the object in question is significantly hotter than its surroundings. Then that strong T^4 temperature dependence means the energy emitted far exceeds the energy absorbed, and in such cases Equation 4.2 provides an excellent approximation to the net heat transfer by radiation.

EXAMPLE 4.4 | A Red-Hot Stove

A hot stove burner with nothing on it loses energy primarily by electromagnetic radiation, which is why it glows red hot. A particular burner radiates energy at the rate of 1,500 W and has a surface area of 200 cm^2. Assuming an emissivity of 1, what's the temperature of the burner?

SOLUTION

Equation 4.2 describes the rate of energy loss by radiation, here 1,500 W. In this example, the only thing in Equation 4.2 that we don't know is the temperature, which is what we're trying to find. It's straightforward algebra to solve Equation 4.2 for the quantity T^4; raising the result to the 1/4 power (or taking the fourth root) then gives the answer. But one caution: Equation 4.2 works in SI units, which means using square meters for area. We're given the area in square centimeters; since 1 cm is 10^{-2} m, 1 cm^2 is 10^{-4} m^2. Thus our 200-cm^2 area amounts to 200×10^{-4} m^2. So here's our calculation for the burner temperature:

$$T^4 = \frac{P}{e\sigma A}$$

or

$$T = \left(\frac{P}{e\sigma A}\right)^{1/4} = \left(\frac{1{,}500\ \text{W}}{(1)(5.67 \times 10^{-8}\ \text{W/m}^2/\text{K}^4)(200 \times 10^{-4}\ \text{m}^2)}\right)^{1/4} = 1{,}070\ \text{K}$$

or about 800°C.

Equation 4.2 shows that the total radiation emitted by a hot object increases dramatically as its temperature rises. But something else changes, too—namely, the wavelength range in which most radiation is emitted (Fig. 4.9). At around 1,000 K, the stove burner of Example 4.4 glows red, although, in fact, most of its radiation is invisible infrared mixed with some visible light. A lightbulb filament heats up to about 3,000 K, and it glows with a bright yellow-white light, although as Figure 4.9 shows, it's still emitting mostly infrared and is therefore a rather inefficient light source. Put the bulb on a dimmer and turn down the brightness, though, and you'll see its color changing from yellow-white to orange and then to a dull red as the filament temperature drops. The Sun's surface is at about 6,000 K, so its light is whiter than that of the lightbulb (which is why photographs taken in artificial light rather than sunlight sometimes show strange colors). On the other hand, a much cooler object such as Earth itself, with a temperature around 300 K, emits essentially all of its radiation at infrared and longer wavelengths. This distinction between the 6,000-K Sun radiating primarily visible light and the 300-K Earth radiating primarily infrared

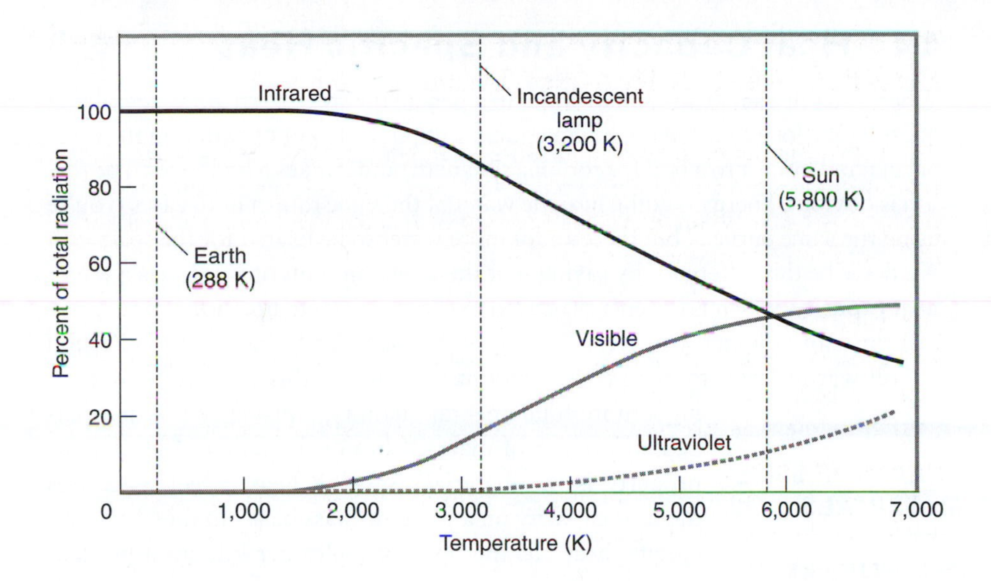

FIGURE 4.9
Percentage of radiation emitted in different wavelength ranges, as a function of temperature. The infrared curve represents longer wavelengths; the ultraviolet curve represents shorter wavelengths. The visible spectrum is defined as wavelengths in the range of 380 to 760 nanometers.

plays a major role in establishing Earth's climate. Finally, the entire universe is at an average temperature of only 2.7 K (about –270°C), with its radiation primarily in the microwave region of the electromagnetic spectrum. Studying that microwave emission gives astrophysicists detailed information about the structure and evolution of the universe.

EXAMPLE 4.5 | Earth's Radiation

Earth's radius (R_E) is 6,370 km, or 6.37 Mm. Assuming a temperature of 288 K and an emissivity of 1, what's the total rate at which Earth loses energy by radiation?

SOLUTION

This is a straightforward application of Equation 4.2, with the surface area of the spherical Earth given by $4\pi R_E^{\ 2}$. So we have:

$$P = e\sigma AT^4 = (1)(5.67 \times 10^{-8}\ \text{W/m}^2/\text{K}^4)(4\pi)(6.37 \times 10^6\ \text{m})^2(288\ \text{K})^4$$
$$= 2.0 \times 10^{17}\ \text{W}$$

Because of Earth's relatively cool temperature, this power is essentially all in the form of infrared radiation. It's no coincidence that the number here is in the same order-of-magnitude ballpark as the 10^{17} W given in Chapter 1 as the total solar energy flow to Earth. In fact, there's an almost perfect balance between incoming sunlight and outgoing infrared, and that's what determines Earth's temperature. It's this balance that establishes Earth's climate, and I'll have much more to say on this topic in Chapters 12 and 13.

4.4 Heat Capacity and Specific Heat

When you put a small amount of water in a pan on the stove to steam vegetables, it doesn't take long to reach the boiling point. But put a big pot of water on the same burner and bring it to a boil for cooking spaghetti, and it takes a lot longer. The reason is obvious: Energy is going into the water at the same rate in both cases (you're using the same burner), but there's a lot more water to be heated for the spaghetti. We describe this difference by saying that the larger amount of water has a larger **heat capacity**, which is the energy needed to cause a given temperature rise.

If, instead of water, you were heating oil for deep-fat frying, you would find the oil warms faster than an equal amount of water. In this case we're describing a more fundamental quantity, a property not of a particular amount of material but of the material itself. This property is the material's **specific heat**, which measures the heat capacity on a per-unit-mass basis. In the SI system, specific heat has the units of joules per kilogram per kelvin (J/kg/K); equivalently, J/kg·K because the double division becomes a multiplication in the denominator. Because 1 K and 1°C are the same as far as temperature *differences* are concerned, this can equally well be written J/kg/°C or J/kg·°C. Water's specific heat of 4,184 J/kg·K is especially large, which is why it takes a long time to heat water, and why large lakes heat or cool slowly and thus exert a moderating effect on local climate. The figure 4,184 should sound familiar, since it's the number of joules in a kilocalorie. The calorie was initially defined as the heat needed to raise the temperature of 1 gram of water by 1°C, so the specific heat of water is simply 1 cal/g·°C. Converting calories to joules and grams to kilograms gives 4,184 J/kg·°C. Table 4.3 lists specific heats of some common materials.

TABLE 4.3 | SPECIFIC HEATS OF SOME COMMON MATERIALS

Material	Specific heat (J/kg·K)
Aluminum	900
Concrete	880
Glass	753
Steel	502
Stone (granite)	840
Water:	
Liquid	4,184
Ice	2,050
Wood	1,400

The definition of specific heat as the energy per kilogram per kelvin (or degree Celsius) shows immediately how to calculate the energy needed to raise the temperature of a given mass m of material by an amount ΔT (here Δ is the capital Greek letter "delta," commonly used to signify "a change in . . ."). We simply multiply the specific heat (commonly given the symbol c) by the mass involved and by the temperature change:

$$Q = mc\Delta T \tag{4.3}$$

Note that I use Q for heat energy, to distinguish it from other forms of energy. The heat Q in Equation 4.3 is in joules.

EXAMPLE 4.6 | Meltdown!

During Japan's 2011 nuclear crisis, cooling systems at three operating reactors of the Fukushima Daiichi Nuclear Power Station were disabled when a massive earthquake cut power to the station and a subsequent tsunami wiped out backup generators. Although the reactors shut down automatically during the earthquake, decay of radioactive fission products within the nuclear fuel continued to release energy at a rate that, during the first day, averaged about 25 MW for each reactor.

In a desperate attempt to avoid full meltdowns, operators pumped seawater into the crippled reactors. If they were able to get 450 tonnes (450,000 kg) of water, initially at 10°C, into each reactor, how long would it take that water to reach the boiling point?

SOLUTION

This example is a good test of your fluency in working quantitatively with energy and power. We're given power—the *rate* at which radioactive material is putting out energy. But Equation 4.3 relates temperature rise ΔT to *energy* Q. That energy is the product of power P with the unknown time t. So Equation 4.3 becomes $Pt = mc\,\Delta T$. We're given $P = 25$ MW, and $\Delta T = 90°C$ from the initial 10°C to water's 100°C boiling point (here I'm neglecting small differences between seawater and fresh water). The mass m is 450,000 kg, and, from Table 4.3, c = 4,184 J/kg·K or 4,184 J/kg·°C since temperature *change* is the same whether measured in K or °C. Solving for the time t then gives

$$t = \frac{mc\,\Delta T}{P} = \frac{(450 \times 10^3\ \text{kg})(4{,}184\ \text{J/kg·°C})(90\ \text{°C})}{25 \times 10^6\ \text{W}} = 6{,}780\ \text{s}$$

or just under 2 hours. Our answer explains the urgency the entire world felt during the Fukushima crisis. Fortunately, as we'll see in Example 4.7, it takes a lot more time to boil all that water away. Nevertheless, Fukushima reactors 1, 2, 3 all suffered meltdowns of their nuclear fuel.

4.5 Phase Changes and Latent Heat

Supplying heat to an object usually raises its temperature, but not always. Sometimes the energy goes into a change of state instead, as when ice melts or water boils. What's happening is that the added energy is breaking the bonds that hold molecules into the rigid structure of a solid or in close contact in a liquid. Different materials change state at different temperatures and require different amounts of energy to make those changes.

The energy required to change the state of a unit mass of material (1 kg in SI units) is the **heat of transformation**—specifically, the **heat of fusion** for the liquid-to-solid transformation and the **heat of vaporization** for the change from liquid to gas. In this book we're interested mostly in water, for which the heats of fusion and vaporization at the melting and boiling points are, respectively, 334 kJ/kg and 2.257 MJ/kg. At more typical atmospheric temperatures where evaporation takes place, water's heat of vaporization is somewhat larger—approximately 2.5 MJ/kg. Those are big numbers, and they tell us that it takes lots of energy to melt ice and even more (nearly seven times as much) to vaporize an equal amount of water.

The adjective *latent* is sometimes added in front of *heat* in the heats of transformation. That's because a substance in, say, the gaseous state contains more energy than the same substance in the liquid state, even if they're at the same temperature. The extra energy is latent in the higher-energy state and can be released in a change of state. Specifically, water vapor contains more energy than liquid water. When water vapor condenses, energy is released. Hurricanes are powered by condensation of water vapor that rises into the atmosphere above warm tropical oceans. On a larger scale, transfer of energy by water vapor rising in the atmosphere plays a significant role in Earth's global energy balance. In the context of weather and climate, we refer to the energy associated with water vapor as **latent heat**, again because the energy is latent and can be released by condensation to liquid water. Latent heat contrasts with **sensible heat**, meaning heat you can feel, because it's associated with molecular kinetic energy—as measured by temperature.

EXAMPLE 4.7 | Back to the Meltdown

Once the emergency cooling water of Example 4.6 reached the boiling point (just under 2 hours into the crisis), how long would it take to boil away completely?

SOLUTION

The heat of vaporization of water is 2,257 kJ/kg, and in Example 4.6 we have 450,000 kg of water. So the energy required is (2,257 kJ/kg) × (450 × 10^3 kg) = 1.0 × 10^9 kJ, or 10^6 MJ. The reactor is putting out energy at the rate of 25 MW—that is, 25 MJ per second. So it's going to take

$$\frac{10^6 \text{ MJ}}{25 \text{ MJ/s}} = 40{,}000 \text{ s}$$

or about 11 hours. All the while, the water stays at a safe 100°C. So water's large heat of vaporization bought precious time during the Fukushima crisis.

4.6 Energy Quality

A 2-ton car moving at 30 miles per hour has, in round numbers, about 200 kJ of kinetic energy. One teaspoon of gasoline contains a similar amount of chemical energy. And the extra internal energy stored in a gallon of water when heated by 10°C is also around 200 kJ. Are all of these energies equivalent? Are they equally valuable? Equally useful?

BOX 4.1 | Higher and Lower Heating Values

With the exception of high-grade coal, the fossil fuels contain substantial amounts of hydrogen as well as carbon. So do wood and other biofuels. When these substances or pure hydrogen are burned, one product is water vapor. Water's heat of vaporization introduces a complication in accounting for the energy content of these substances, because there's latent energy in the water vapor. If that's simply vented to the atmosphere—as it is, for example, with automobile engines, then its energy content is lost. But if the water is condensed as part of the energy-extraction process—as in some advanced boilers—then we've captured the latent energy. Thus the energy content of hydrogen-containing fuels has two values: The **lower heating value** applies when we don't capture the latent energy of water vapor, the **higher heating value** when it does. The difference is on the order of 10% for liquid and gaseous fossil fuels. With biomass and waste, there's a further complication: These materials are often wet, and it takes energy to remove the water. That's why biomass energies are usually stated in terms of dry mass.

The answer to all these questions is no. To understand why, consider what it would take to convert the energy of either the water or the gasoline to the kinetic energy of a car. The energy in the water is the random kinetic energy of individual H_2O molecules. Somehow we would have to get them all going in the same direction and then impart their energy to the car. To use the gasoline, we'd have to burn it, again producing the random kinetic energy of molecules, and then somehow get all those molecules moving in the same direction so they could impart all their energy to the car.

How likely is it that the randomly moving molecules in water or in combustion products of the gasoline will organize themselves so they're all going in the same direction? If there were only two or three molecules, we might expect, on rare occasion, that random collisions among the molecules or their container would result in all of them going in the same direction. But the number of molecules in a macroscopic sample of material is huge—on the order of 10^{23} or more. It's extraordinarily unlikely that these molecules will ever find themselves all moving in the same direction. So unlikely, in fact, that it probably wouldn't happen spontaneously even if you waited many times the age of the universe.

The extreme improbability of those random molecular motions ever being in the same direction means that it is, in practice, impossible to convert all the internal energy of a heated material into kinetic energy. However, it's easy to go the other way. When our car brakes, its 200 kJ of directed kinetic energy ends up as random internal energy associated with frictional heating of the brakes (unless it's a hybrid, which captures the energy for reuse). We have here an **irreversible process**: Once directed kinetic energy of a macroscopic object is converted to heat—that is, to random internal energy associated with an increased temperature—then it's impossible to turn it all back into kinetic ordered energy (Fig. 4.10).

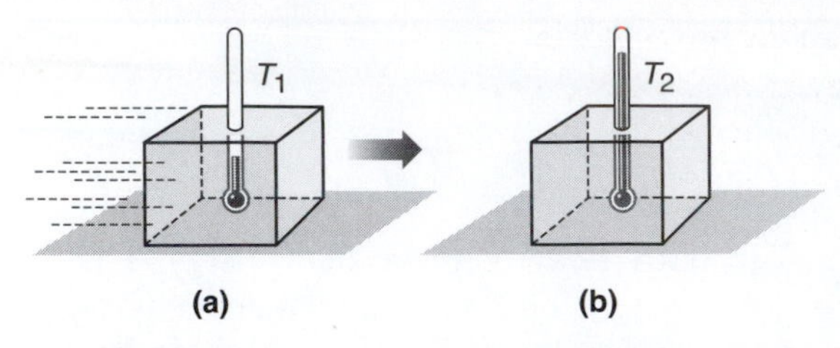

FIGURE 4.10
An irreversible process. (a) A block slides along a surface, gradually slowing. Friction converts the kinetic energy of its directed motion to random thermal energy of its molecules, indicated by the higher temperature when it comes to rest (b). The reverse process, from (b) to (a), is so highly improbable that it's essentially impossible.

Irreversible processes are common. Beat an egg, intermixing white and yolk; even if you reverse the beater, you'll never see the white and yolk separate from the scrambled mix. Take two cups of water, one hot and one cold, and put them together inside an insulated box. A while later you have two cups of lukewarm water. Wait as long as you want and you'll never see them spontaneously revert to one hot and one cold cup. That's true even though the amount of energy in the two lukewarm cups is exactly the same as in the original hot/cold situation.

There's something ordered about the kinetic energy of a moving car, with all its molecules sharing the common motion of the car. There's something ordered about an unbeaten egg, with all its yolk molecules in one place and its white molecules in another. And there's something ordered about cups of hot and cold water, with more of the faster-moving molecules in one cup and more of the slower-moving ones in the other. But these ordered states gradually deteriorate into states with less order. So do many other ordered states you can think of, such as the state of your room after you clean it. It gradually gets messy, and the messy room never spontaneously gets neater.

There's a simple reason for this tendency to disorder—namely, of all the possible ways to arrange the molecules of an egg, or the stuff in your room, or the molecular energies of two cups of water, there are far more ways that correspond to disordered states. You could arrange the books on your shelf in different ways, but if you toss them randomly on the floor, there are many more possible arrangements. Because of the greater number of disordered states, systems naturally tend to end up disordered. Ordered states are few, and therefore special, and it takes an effort to put a system into an ordered state.

That systems tend naturally toward disorder is a general statement of the **second law of thermodynamics**. The second law places serious restrictions on our ability to harness energy, and this is what I'll emphasize from now on. But be aware that the second law is broader, describing the general tendency to disorder—a tendency that occurs simply because there are more disordered states.

Back to our 200 kJ of energy in the moving car, the teaspoon of gasoline, and the gallon of hot water. The second law of thermodynamics says that we can't turn all the internal energy of the water or the gasoline into the directed kinetic energy of a car. We can, on the other hand, turn all the kinetic energy of the car into internal energy. So there's a very real sense in which the kinetic energy of the car is more valuable, more versatile, more useful. Anything we can do with 200 kJ of energy, we can do with the kinetic energy of the car. We can't do just anything with the gasoline or the hot water. In particular, we can't turn it all into kinetic energy, so the energy of the gasoline and hot water is less valuable.

I'm not talking here about different *amounts* of energy (we have 200 kilojoules in all three cases), but about different *qualities* of energy. The kinetic energy is of higher quality, because it's the most versatile. And the gasoline's energy is of higher quality than that of the heated water. Why? Because when I burn the gasoline I can convert its energy to internal energy of far fewer molecules than are in a gallon of water. Those molecules each have, on average, more energy than the molecules in the water, meaning a higher temperature. Because there are fewer of these high-temperature molecules, the number of disordered states, although still fantastically huge, is nevertheless smaller than in the case of the water. This translates into our being able to convert more of the gasoline's energy to directed kinetic energy.

4.7 Entropy, Heat Engines, and the Second Law of Thermodynamics

Entropy is a measure of disorder. Our moving car with its 200 kJ of kinetic energy represents a state of low entropy. Convert the car's kinetic energy to random thermal energy, and entropy has increased, even though the total energy remains the same. The combustion products of our teaspoon of gasoline have higher entropy than the moving car, but less entropy than the gallon of hot water—even though the car, the gasoline's combustion products, and the water all have 200 kJ of energy. For a given amount of energy, there's a hierarchy of energy qualities, from the highest quality and lowest entropy to lower qualities and higher entropy. At the top are mechanical and electrical energy—energies that we can convert to other forms with, in principle, 100% efficiency. A given amount of energy has somewhat lower quality, and correspondingly higher entropy, if it's associated with material heated to a temperature much higher than its surroundings. Still lower quality, and even higher entropy, results when the same amount of energy brings material to a temperature just a little above that of its surroundings. Figure 4.11 shows this hierarchy of energy qualities.

The second law of thermodynamics can be restated in terms of entropy: *The entropy of a closed system can never decrease*. The "closed system" designation is important, for reasons we'll see in Section 4.9. And in almost all cases, entropy in fact increases. So in general, energy quality deteriorates. That's why the fact that energy is conserved doesn't really do us much good; as we use energy, it degrades in quality and becomes less useful.

We're now ready to understand in more detail why it is that we can't extract as mechanical energy all the energy contained in fuels. We *can* burn fuel directly and convert all its energy to heat—that is, to internal energy. But this is low-quality energy. What we can't do is turn it all into high-quality mechanical energy or electricity.

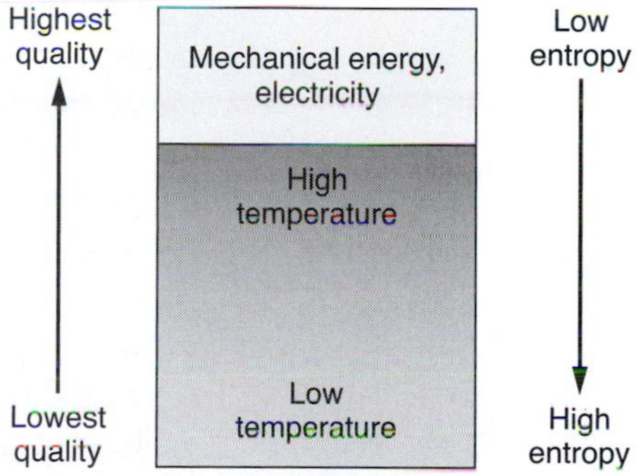

FIGURE 4.11
Quality associated with a given quantity of energy is highest for mechanical energy and electricity, and lowest for low-temperature thermal energy. Low entropy corresponds to high quality, and vice versa.

HEAT ENGINES

A **heat engine** is a device that converts random thermal energy into mechanical energy. Examples include the gasoline engine in your car, the boilers and steam turbines in electric power plants, and jet aircraft engines. Although these engines differ in their design and operation, they're conceptually similar: All extract random thermal energy from a hot substance and use it to deliver mechanical

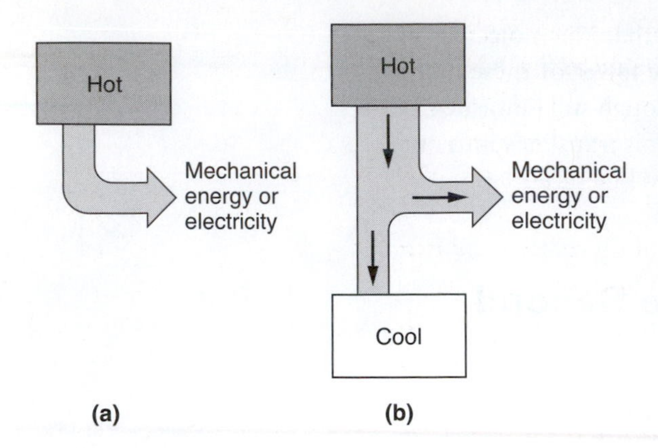

FIGURE 4.12
Conceptual diagrams for heat engines. (a) The impossible case of an ideal engine, which extracts heat from a hot substance and delivers an equal amount of mechanical work or electricity. (b) A real engine converts only some of the extracted heat into work; the rest is rejected to the cool environment.

energy. In an **internal combustion engine,** the fuel burns inside the engine and the combustion products become the hot substance. This is what happens in the cylinders of your car engine. In an **external combustion engine**, burning fossil fuel or fissioning uranium heats a separate substance, usually water. Either way, we can consider that we start with a hot substance whose high temperature is sustained by burning a fuel.

Figure 4.12a shows a conceptual diagram of what an ideal heat engine would be like. It would extract energy from a hot material and deliver it all as mechanical energy. But the second law of thermodynamics tells us that can't happen. In fact, another statement of the second law is this: *It is impossible to build a perfect heat engine.* What a real engine does is to extract energy from a hot substance and deliver only *some* of it as mechanical energy; the rest gets transferred as heat to the engine's surroundings. Figure 4.12b shows a conceptual picture of a real heat engine. The real engine is less than 100% efficient, because the mechanical energy it delivers is less than the energy it extracts from the hot substance (which in turn is kept hot by the burning fuel). The efficiency, e, is just the ratio of mechanical energy delivered to total energy extracted:

$$e = \frac{\text{mechanical energy delivered}}{\text{energy extracted from fuel}} \qquad (4.4)$$

So how efficient can we make a heat engine? Maybe we can't convert all the fuel energy to mechanical energy, but why can't clever engineers get the efficiency of Equation 4.4 close to 1 (i.e., to 100%)? Early in the nineteenth century a young French engineer named Sadi Carnot pondered this question. He conceived a simple engine involving a fluid inside a cylinder closed with a moveable piston. Placed in contact with a hot substance, the fluid would push on the piston and deliver mechanical energy. But then the fluid would have to be cooled to its original state, so the cylinder would be placed in contact with the ambient environment. In the process, heat would flow to the environment—this heat being energy extracted from the fuel but *not* delivered as mechanical energy. Although the engineering details of the **Carnot engine** are not exactly those of a modern gasoline engine or a steam power plant, the general principles are similar. More important, Carnot proved that no engine, no matter how cleverly designed, could be more efficient than his engine.

So what's the efficiency of Carnot's engine? I won't go through the derivation, which is done in most introductory physics texts. The result depends on just two numbers: the temperature of the hot substance (T_h) and the temperature of the cool substance (T_c) to which the waste heat flows. This quantity, which I'll call the **thermodynamic efficiency limit**, is given by

$$e = 1 - \frac{T_c}{T_h} \qquad \text{(thermodynamic efficiency limit)} \qquad (4.5)$$

Here the temperatures must be expressed on an absolute scale (Kelvin or Rankine). I've designated Equation 4.5 as the thermodynamic efficiency *limit* because, as Carnot proved, it represents the *maximum possible* efficiency for *any* engine operating between the temperatures T_c and T_h. Real engines fall below the Carnot ideal, so Equation 4.5 really is the limiting case of the best possible engine.

We're generally stuck with the low temperature; it's that of Earth's ambient environment, or a little higher; say 310 K. And T_h is the hottest temperature we can achieve in our heat engine, which is set by the materials from which the engine is made. As a practical matter, for example, the maximum steam temperature in a fossil-fueled power plant is around 650 K; for a nuclear plant, where water is heated in a single large pressure vessel, it's around 570 K. These lead to thermodynamic efficiency limits of

$$e_{\text{max fossil}} = 1 - \frac{T_c}{T_h} = 1 - \frac{310\text{ K}}{650\text{ K}} = 0.52$$

or 52%, and

$$e_{\text{max nuclear}} = 1 - \frac{T_c}{T_h} = 1 - \frac{310\text{ K}}{570\text{ K}} = 0.46$$

or 46%. T_h is somewhat higher for an automobile engine, but the low temperature of the warm engine block is higher, and the thermodynamic efficiency limit comes out about the same, a little under 50%.

Again, Equation 4.5 gives the absolute maximum efficiency that's theoretically possible; that's why I've put the subscripts "max" on the efficiencies calculated above. Given the temperatures T_h and T_c, no amount of cleverness on an engineer's part can increase the efficiency over the limit expressed in Equation 4.5. Actual efficiencies will be a lot lower because friction and other losses sap energy, and because energy is diverted to pumps, pollution control systems, or other devices essential to the operation of an engine or power plant. By the time those losses are considered, most older electric power plants are somewhere around 30% to 40% efficient, meaning that roughly two-thirds of the energy extracted from fuels is dumped to the environment as waste heat.

EXAMPLE 4.8 | Ocean Thermal Energy Conversion

Ocean thermal energy conversion (OTEC) is a scheme for extracting energy from the temperature difference between the warm surface waters of tropical oceans and the cool water deeper down. Solar heating of the surface water would take the place of a burning fuel. Although there are many engineering details to work out, OTEC has potential as an energy source, particularly for tropical island nations (more on OTEC in Chapter 10). However, as this example shows, OTEC is not particularly efficient.

Typical tropical ocean surface temperatures are around 25°C, while several hundred meters down the temperature remains at about 5°C. What's the maximum possible efficiency for a heat engine operated between these two temperatures?

SOLUTION

The maximum possible efficiency is the thermodynamic limit of Equation 4.5, which requires absolute temperatures. Since 0°C is 273 K, our hot and cold temperatures of 25°C and 5°C become, respectively, 298 K and 278 K. Then Equation 4.5 gives

$$e_{\text{max}} = 1 - \frac{T_c}{T_h} = 1 - \frac{278\ \text{K}}{298\ \text{K}} = 0.067$$

or only 6.7 percent. However, there's no fuel to pay for, so low efficiency is not necessarily unacceptable.

Our understanding of heat engines illustrates a point I made earlier when I described the cups of hot and cold water that gradually became two cups of lukewarm water, despite no overall loss of energy. Given the original cups of hot and cold water, we could have run a heat engine and extracted some useful, high-quality mechanical energy. But if we let the cups become lukewarm we could no longer do that, despite all the original energy still being in the water. It's in that sense that the quality of energy in the lukewarm water is lower; none of it is available for conversion to mechanical energy.

FIGURE 4.13
The cooling tower at this nuclear power plant transfers waste heat to the atmosphere.

THERMAL POLLUTION

We need to dispose of the waste heat produced in our heat engines. In a car, some of the heat goes out the tailpipe as hot gases; much of the rest goes out through the radiator, a device designed explicitly to dump waste heat to the surrounding air. Power plants are almost always built adjacent to large bodies of water, giving them a place to dump waste heat. But the associated temperature rise of the water can affect the aquatic ecology. To avoid such **thermal pollution**, water from the power plant is often cooled through contact with air before being returned to a river, lake, or bay. Those huge concrete cooling towers you probably associate with nuclear power (but which are used in fossil-fueled power plants as well) serve that purpose; the cloudlike plumes emerging from the towers consist of water droplets condensed from water vapor that formed when water was cooled by evaporation inside the towers (Fig. 4.13). So vast is the amount of waste heat produced in power generation that most of the rainwater falling on the continental United States eventually makes its way through the cooling systems of electric power plants.

4.8 Energy Quality, End Use, and Cogeneration

To propel a car, truck, or airplane, we need kinetic energy—that is, mechanical energy of the highest quality. To operate a computer, TV, or washing machine, we also need the highest-quality energy, this time in the form of electricity. But heating water for a shower or producing steam for processing food only requires lower-quality thermal energy. These different **end uses** have different energy-quality requirements. We can produce low-quality energy simply by burning a fuel and transferring the energy as heat to water or whatever. The second law of thermodynamics places no restrictions on this process. So it makes a lot more sense, for example, to heat your hot water with a gas flame than with electricity. That's because each joule of electrical energy you use represents two additional joules that were rejected to the environment as waste heat back at the electric power plant. Energy from the gas, on the other hand, transfers to the water with close to 100% efficiency (practical heaters achieve some 85% efficiency).

So why can't we use all that waste heat for low-quality energy purposes? Increasingly, we do. Whole cities in Europe are sometimes heated with energy that's the byproduct of electric power generation. Increasingly, institutions in the United States are choosing a similar energy path called **cogeneration** or **combined heat and power (CHP)**. In this process, fossil fuels are burned to produce steam, which turns a turbine and generates electricity just as in any electric power plant. But the waste heat is then used for heating buildings or for industrial processes. In the United States today, only about 5% of the waste heat from electricity generation is put to use, most of it in the industrial sector. So there's considerable additional potential for energy savings through cogeneration.

PRIMARY ENERGY

The second law of thermodynamics can lead to confusion when we try to quantify energy consumption. If I ask about the energy used to run your car, do I mean the energy content of the gasoline you burned, or only the actual kinetic energy that was put into the car's motion? In this case the answer is perhaps obvious: You burn gasoline right there in your car, and it's clearly the total energy of the gasoline that you're responsible for consuming, despite the fact that most of that energy was simply dumped to the environment as waste heat.

Electrical energy presents a murkier situation. Your house is miles from the power plant that produces your power, and you tend to think only about what's going on at your end of the wire. When you turn on your air conditioner and consume, say, 5 kWh of electrical energy, should you count that as your total energy consumption? Or should you recognize that the electricity came from a power plant with an overall efficiency of only about 30%, in which case your total energy consumption is more like 15 kWh?

BOX 4.2 | Cogeneration at Middlebury College

Like many institutions, Middlebury College heats its buildings using a central steam-production plant. Steam is delivered through heavily insulated underground pipes to the various campus buildings, where it can be used directly for heating or to operate chillers for summer air conditioning. Wood chips and oil fuel the steam-production boilers.

The temperatures and pressures generated in the boilers are greater than required for heating buildings; thus they constitute a higher-quality form of energy than what's needed. To make use of this high-quality energy, Middlebury has installed turbine-generators to generate electricity and pass the "waste heat" on to the steam distribution system. This is a classic example of cogeneration, in this case installed retroactively on what was a purely thermal energy system. Today the college operates several such generators, with a combined electrical generation capability of 1.7 MW (Fig. 4.14). Although the turbine-generators don't run all the time at capacity, they do produce some 18 percent of the college's total electrical energy consumption.

FIGURE 4.14
This cogeneration unit at Middlebury College produces nearly 1 MW of electric power.

That 15 kWh—the total energy extracted from the fuel used to make your electricity—is called primary energy. Primary energy includes the substantial amount of fuel energy that the second law of thermodynamics keeps us from converting to electricity or other high-quality energy. But primary energy is a more honest accounting of the actual energy consumed to provide us with what's generally a lesser amount of useful energy. In discussions of world energy use, you'll often see the term **total primary energy supply (TPES)** used to describe a country's energy use. Again, *primary* means that we're accounting for energy right from the source, not just the end use. I explored this distinction between primary and end-use energy briefly in Box 2.1, but here you can see how it's ultimately grounded in the second law of thermodynamics.

In the case of electric power plants, we often distinguish the thermal power output—the primary energy, or total energy extracted from fuel, including what's dumped as waste heat—from the electrical energy that is the plant's useful product. You'll sometimes see the designation MWth, for *megawatts thermal,* and MWe, for *megawatts electric,* to distinguish the two. Most power plants are specified by their electrical output; after all, electrical energy is what they're trying to produce and sell. A typical large power plant might have an output of 1,000 MWe (equivalently, 1 GWe). If its efficiency is 33%, then it extracts energy from fuel at three times this rate, so its thermal output is 3,000 MWth. Incidentally, you might want to look back at Example 4.6 and decide whether that 2-GW nuclear plant is a 2-GWe or a 2-GWth plant. In the former case, it would be a very large plant indeed; in the latter, it would be of only modest size.

The electricity picture is further complicated by the mix of different sources of electrical energy, and even of the different efficiencies of thermal power plants. As a result, there's no single conversion between 1 kWh of electrical energy consumption and the corresponding primary energy. In the Pacific Northwest of the United States, for example, much of the electricity comes from nonthermal hydropower. There, 1 kWh of electrical energy corresponds, on average, to only a little more than 1 kWh of primary energy. In New England, which gets much of its electricity from nuclear power plants that are roughly 33% efficient, the ratio is more like a factor of 3 or so. In fact, for the United States as a whole, the total primary energy consumed to generate electricity in the early twenty-first century is some 2.9 times the total amount of electrical energy produced. Accounting for electricity consumed within power plants and lost in electrical energy transmission raises the average primary energy consumed to nearly 3.2 kWh for every 1 kWh of electrical energy delivered to the end user. Again, think of your energy servants: For every one you've got working for you running anything electrical, there are on average two more back at the power plant laboring away just to produce waste heat.

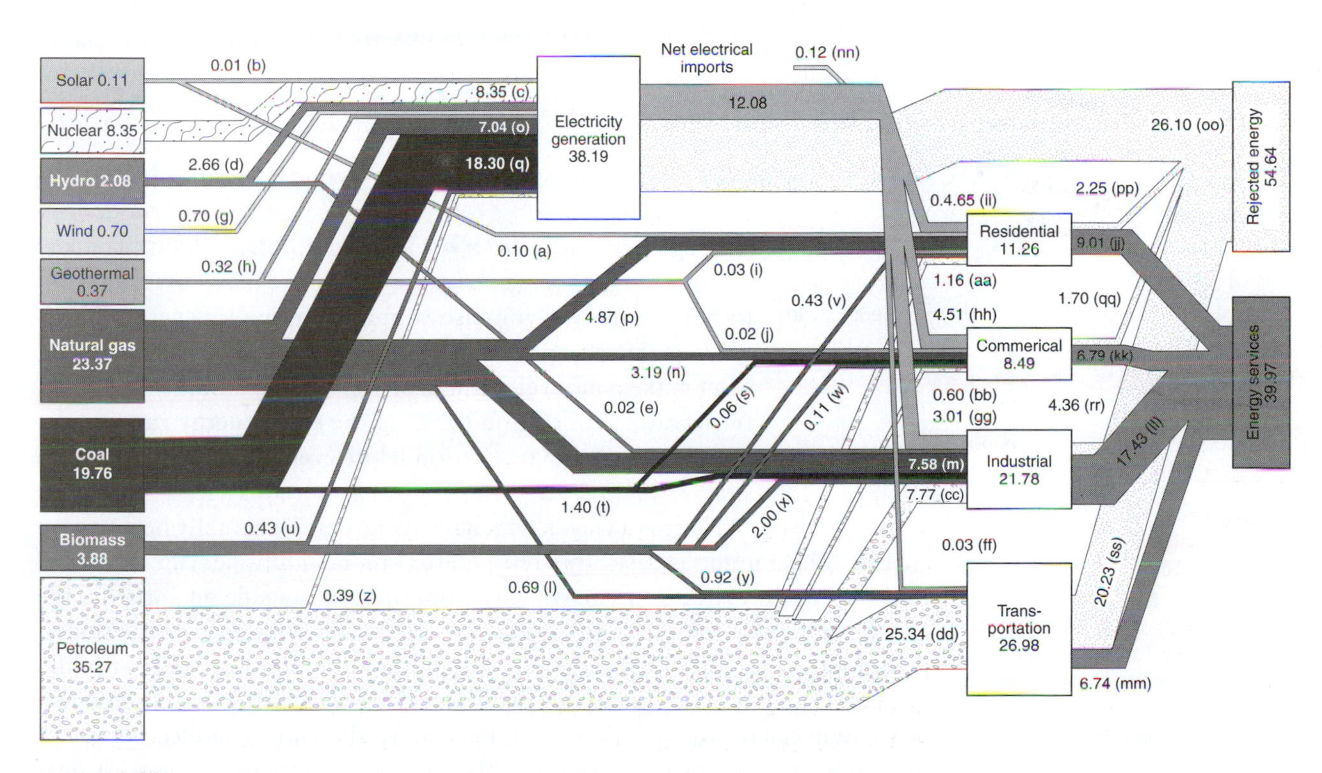

FIGURE 4.15

Energy-flow diagram for the United States in 2009, showing primary energy sources on the left, and end uses and waste ("rejected energy") on the right. Total consumption of 94.6 Q is lower than normal because of the Great Recession.

The distinction between primary and end-use energy is embodied in Figure 4.15, a complex diagram illustrating energy flows from primary to end use in the United States. Note that more than half the primary energy coming in on the left-hand side of the diagram is wasted, ending up as "rejected energy." This includes not only losses associated with the second law of thermodynamics, but also losses to friction, electrical resistance, and other irreversible mechanisms that dissipate useful energy into random thermal energy.

4.9 Refrigerators and Heat Pumps

Equation 4.5 shows that we can increase the thermodynamic efficiency limit of a heat engine either by raising the maximum temperature or by lowering the minimum temperature. The materials that comprise the engine set the maximum temperature. But why can't we just refrigerate the cooling water to lower the minimum temperature? This won't work, and the reason is again the second law of thermodynamics. How do we cool things? With a **refrigerator**, which, conceptually, is like a heat engine run in reverse. With a heat engine, we extract heat from a hot substance, turn some of it into mechanical energy or electricity, and reject the rest as waste heat to a cool substance. With a refrigerator, we extract energy from a cool substance and transfer it to a hot substance (Fig. 4.16). An ideal refrigerator would do exactly that, with no other energy needed. But this is impossible, as my example of the two cups of lukewarm water suggests. They never become, spontaneously, one hot and one cold cup.

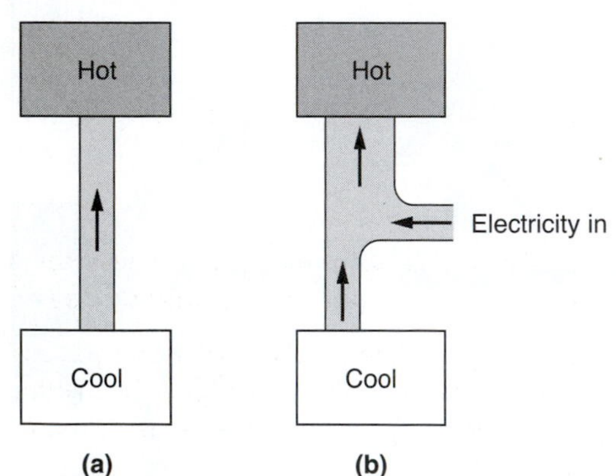

FIGURE 4.16
(a) A perfect refrigerator would transfer energy spontaneously from cool to hot, but this is impossible according to the second law of thermodynamics. (b) A real refrigerator requires a high-quality electrical or mechanical energy input to effect the energy transfer from cool to hot.

In fact, another statement of the second law is this: *It is impossible to build a perfect refrigerator*. Equivalently, *heat doesn't flow spontaneously from a cooler to a hotter object*. The key word here is *spontaneously*. You can make heat flow from cool to hot, but only if you also supply high-quality energy. That's why you have to plug in your refrigerator. And that's why you can't make a more efficient heat engine by teaming it up with a refrigerator; at best you'll use up the extra energy running the refrigerator, and in reality you'll lose even more energy to waste heat.

There is a way in which refrigerators can actually help us save energy. A **heat pump** is basically a refrigerator or air conditioner run in reverse. In the winter it extracts thermal energy from the cool outside air (or from the ground in cold climates) and delivers it to the warmer interior of the house. Of course, heat won't flow spontaneously in that direction, so the heat pump needs a source of high-quality energy, namely electricity. But a well-designed heat pump can transfer a lot more thermal energy than it uses as electricity, and therefore heat pumps have considerable potential for energy savings. A heat pump's effectiveness is quantified by its **coefficient of performance (COP)**, which is the ratio of heat delivered to electrical energy consumed. The best commercially available heat pumps, sourced from groundwater wells, have COPs in excess of 5—meaning that, for every kilowatt-hour of electricity they consume, these pumps deliver more than 5 kWh of heat. (More on heat pumps in Chapter 8.)

4.10 Energy Overview

Energy is a big subject, and Chapters 3 and 4 have just skimmed the surface. But you now have enough background to understand quantitatively how we power our high-energy society, what constraints the fundamental laws of nature place on our use of energy, and what steps we might take to reduce our energy use or mitigate its environmental impact. In particular, you should be fluent in the various units used to measure energy and power, you should recognize the different fundamental kinds of energy, and you should be keenly aware of the restrictions the second law of thermodynamics places on our ability to make use of thermal energy. Along the way, you've learned to calculate the effects of energy flows as they cause changes in temperature or in the physical states of substances, and in that context you've learned to evaluate quantitatively the important role of insulation in making energy-efficient buildings. You're now ready to apply your understanding of energy to the various sources of society's energy and to the complex flows of energy that govern Earth's climate.

CHAPTER REVIEW

BIG IDEAS

4.1 **Heat** is energy that flows because of a temperature difference. Heat usually raises a material's **internal energy**, the kinetic energy of random molecular motion. The **first law of thermodynamics** says that the change in internal energy is the sum of the mechanical work done on an object and the heat that flows into it.

4.2 **Temperature** measures the average thermal energy of molecules. Common temperature scales include Celsius (°C) and Fahrenheit (°F), as well as the **Kelvin scale** (K), which is based on **absolute zero**.

4.3 Heat-transfer mechanisms include **conduction**, important in building insulation; **convection**, important in circulating heat within buildings as well as in energy loss associated with windows; and **radiation**, especially important for objects such as planets and stars that are surrounded by vacuum.

4.4 **Heat capacity** and **specific heat** describe the amount of heat needed to effect a given temperature change.

4.5 **Heats of transformation** describe the energy per unit mass required to change a material's state. On melting or vaporizing, the heat of transformation is stored as **latent heat**.

4.6 Energy has *quality* as well as quantity. The highest quality energies include mechanical and electrical energy. Internal energy is of lower quality. The **second law of thermodynamics** prohibits the transformation of low-quality energy into high-quality energy with 100% efficiency.

4.7 The second law of thermodynamics is ultimately about the tendency toward more disorder. **Entropy** is a measure of that disorder. The second law limits our ability to produce high-quality energy efficiently through methods involving heat. As a result, **heat engines** inevitably reject as waste heat some of the energy they extract from their fuels.

4.8 Energy efficiency dictates that we not use high-quality energy like electricity for low-quality uses such as heating. With **cogeneration**, also called **combined heat and power (CHP),** waste heat from electricity generation is put to use for heating.

4.9 **Refrigerators** and **heat pumps** transfer heat from cooler to hotter systems. Heat pumps can provide efficient heating by using small amounts of high-quality energy to transfer larger amounts of heat.

TERMS TO KNOW

absolute zero (p. 64)
Carnot engine (p. 82)
coefficient of performance (COP) (p. 88)
cogeneration (p. 85)
combined heat and power (CHP) (p. 85)
conduction (p. 65)
convection (p. 71)
emissivity (p. 73)
end use (p. 85)
entropy (p. 81)
external combustion engine (p. 82)
first law of thermodynamics (p. 63)
forced convection (p. 72)
heat (p. 63)
heat capacity (p. 76)
heat engine (p. 81)
heat of fusion (p. 77)
heat of transformation (p. 77)
heat of vaporization (p. 77)
heat pump (p. 88)
higher heating value (p. 79)
infiltration (p. 72)
internal combustion engine (p. 82)
internal energy (p. 62)
irreversible process (p. 79)
Kelvin scale (p. 64)
latent heat (p. 78)
lower heating value (p. 79)
natural convection (p. 72)
radiation (p. 72)
refrigerator (p. 88)
R value (p. 67)
second law of thermodynamics (p. 80)
sensible heat (p. 78)
specific heat (p. 76)
Stefan–Boltzmann constant (p. 73)
Stefan–Boltzmann law (p. 73)
temperature (p. 64)
thermal conductivity (p. 65)
thermal energy (p. 62)
thermal pollution (p. 84)
thermodynamic efficiency limit (p. 82)
total primary energy supply (TPES) (p. 86)

GETTING QUANTITATIVE

Conductive heat flow: $H = kA\,\dfrac{T_h - T_c}{d}$ (Equation 4.1; p. 66)

Stefan–Boltzmann law for radiation: $P = e\sigma AT^4$ (Equation 4.2; p. 73)

Energy required for temperature change: $Q = mc\Delta T$ (Equation 4.3; p. 76)

Efficiency of an engine: $e = \dfrac{\text{mechanical energy delivered}}{\text{energy extracted from fuel}}$ (Equation 4.4; p. 82)

Thermodynamic efficiency limit: $e = 1 - \dfrac{T_c}{T_h}$ (Equation 4.5; p. 82)

QUESTIONS

1 Describe the difference between heat and thermal energy.

2 My neighbor tells me that his new wood stove "puts out a lot of heat." Is this a correct use of the term *heat*? Discuss.

3 Solid glass is not a particularly good insulator, but fiberglass—a loose agglomeration of thin glass fibers—is an excellent insulator widely used in buildings. What gives fiberglass its insulating quality?

4 You can bring water to a boil fairly quickly, but it takes a long time to boil it away. What two properties of water determine this fact?

5 Some modern medical thermometers work by measuring the infrared emission from the eardrum, and they give nearly instantaneous readings. What's the principle behind such thermometers?

6 Large lakes tend to exert a moderating effect on the surrounding climate. Why?

7 What is the essential idea behind the second law of thermodynamics?

8 Explain, in the context of the second law of thermodynamics, why heating water with electricity is not an energy-efficient thing to do.

9 In the United States, a significant portion of the energy we produce is in the form of high-quality, low-entropy electricity. But many end uses of energy tend more toward low- and medium-quality thermal energy, so we end up using some of our high-quality energy for low-quality uses. Why is this an inefficient use of energy resources?

10 My local nuclear power plant is rated at 650 MWe. What does the lowercase *e* mean here?

11 Explain the difference between higher and lower heating values of fuels.

EXERCISES

1 Verify that the 5,000 Btu/h heat flow through the basement wall described in Section 4.3 is equivalent to about 1,500 W and therefore 15 energy servants.

2 Find the rate of energy loss through 10-square-foot windows with the worst and the best *R* values listed in Table 4.2, on a day when it's 70°F inside and 15°F outside.

3 My solar hot-water system has a 120-gallon storage tank with a surface area of 4.3 m². It's insulated with 5-cm-thick urethane foam, and the water in the tank is at 76°C. It's in the basement, where the ambient temperature is 15°C. What is the rate of heat loss from the tank?

4 You're considering installing a 4-foot-high by 8-foot-wide picture window in a north-facing wall (no solar gain) that now has a total *R* value of 23. As a compromise between initial cost and energy savings, you've chosen a standard double-glazed window like the one listed in Table 4.2. As a result of installing this window, how much more oil will you burn in a 6-month winter heating season with an average outdoor temperature of 25°F and an indoor temperature of 68°F? Assume your oil furnace is 80% efficient, as in Example 4.2.

5 The electric heating element in a 40-gallon water heater is rated at 5 kW. After your sister takes a shower, the water temperature drops to 90°F. How long will it take the heater to bring the temperature back up to 120°F? The density of water is 3.79 kg per gallon. Neglect heat loss, which in a well-insulated heater should be a lot less than that 5-kW power input.

6 (a) Calculate the energy needed to bring a cup of water (about 250 g) from 10°C to the boiling point. Then find the time it takes to heat this water (b) in a 900-W microwave oven that transfers essentially all of its microwave energy into the water and (c) in a 1-kg aluminum pan sitting on a 1,500-W electric stove burner that transfers 75% of its energy output to the water and the pan. Assume the pan, too, starts at 10°C and has to be heated to water's boiling point.

7 Assuming your electricity comes from a power plant that is 33% efficient, find the actual primary energy consumed in heating the water as described in both parts (b) and (c) of Exercise 6. Assume that the microwave oven uses electrical energy at the rate of 2.4 kW to produce its 900 W of microwave energy. Remember also that you have to heat the aluminum pan on the stove. Compare your answers with the energy needed solely to heat the water.

8 The total annual electrical energy production in the United States is about 4,000 TWh, while the useful heat produced by cogeneration is about 1,500 TBtu annually. Use these figures to confirm the statement on page 84 that only about 5% of waste heat from electric power generation gets used. Assume 33% average power generation efficiency.

9 Lake Champlain, bordered by New York, Vermont, and Quebec, has a surface area of 1,126 km². (a) If it freezes solid in the winter with an average ice thickness of 0.5 m, how much total energy does it take to melt the ice once spring comes? The density of ice is about 920 kg/m³ and the heat of fusion is 334 kJ/kg. (b) If this energy is provided by sunlight shining on the lake at an average rate of 200 W/m² but with only half that amount absorbed, how long will it take the lake to absorb the energy needed to melt it?

10 A house measures 42 feet wide by 28 feet deep. It sits on a 4-inch-thick concrete slab. The ground below the slab is at 48°F, and the interior of the house is at 68°F. (a) Find the rate of heat loss through the slab. (b) By what factor does the heat-loss rate drop if the slab is poured on top of a 2-inch layer of polystyrene insulation?

11 Rework Example 4.2, now assuming that the house has 14 windows, each measuring 45 inches high by 32 inches wide. When you buy the house, they're single-pane windows, but when you upgrade the insulation, you also upgrade the windows to the best you can buy (see Table 4.2). Remember that the windows replace some of the wall area. Neglect any solar gain through the windows, which would increase your savings even more.

12 The surface of a woodstove is at 250°C (that's degrees Celsius, not kelvins). If the stove has a surface area of 2.3 m^2 and an emissivity of 1, at what rate does it radiate energy to its surroundings?

13 The filament of a typical lightbulb is at a temperature of about 3,000 K. For a 100-W bulb, what is the surface area of the filament? You can assume that the entire 100 W goes into heating the filament, and that the filament has an emissivity of 1.

14 Assuming a low temperature of 300 K, at what high temperature would you have to operate a heat engine for its maximum possible efficiency to be 80%?

15 A 2.5-GWth nuclear power plant operates at a high temperature of 570 K. In the winter the average low temperature at which it dumps waste heat is 268 K; in the summer it's 295 K. Determine the thermodynamic efficiency limit of the plant in each season, and then derive from it the maximum possible electric power output.

16 A jet aircraft engine operates at a high temperature of 750°C, and its exhaust temperature—the effective low temperature for the engine—is a hot 350°C. What is the thermodynamic efficiency limit for this engine?

17 Show that an R value of 1 (that is, 1 ft^2·°F·h/Btu) is equivalent in SI units to 0.176 m^2·K/W (equivalently, m^2·°C/W).

RESEARCH PROBLEMS

1 Find the average temperature for a particular winter month in the region where your home is located. From your family's heating bills, estimate as best you can the amount of oil or gas consumed to heat your house during this month. Estimate the surface area of your house and, on the assumption that all the energy content of your heating fuel escaped through the house walls and roof, estimate an average R value for the house.

2 Locate the nearest thermal power plant to your home or school and describe what type it is, what kind of fuel it uses, and any other details. From this information, determine (a) its electric power output and (b) its thermal power output, and then (c) calculate its actual efficiency. If you're in the United States, you may find useful the U.S. Department of Energy's listings entitled "Existing Electric Generating Units in the United States."

3 Locate a microwave oven and find its rated microwave power and its actual electric power consumption. From these determine its efficiency at converting electrical energy into microwaves. Then time how long it takes to heat a known amount of water from a known initial temperature until it reaches the boiling point. Assuming that all the microwave energy ends up in the water, use the result of your experiment to calculate a value for the microwave power. How does it compare with the oven's rated power?

4 Find specifications on your car or your family's: its mass, its engine's power rating in horsepower, and its acceleration (0 to some speed in how many seconds?). Compare the horsepower rating with the actual average power needed to give the car its kinetic energy at that final speed.

ARGUE YOUR CASE

1. A friend suggests that power plants should be more efficient in the summer, since "there's more heat in the air." In a database of electric power plants, you find that most are actually rated for higher outputs in winter than in summer. Formulate an argument as to why this should be the case.

2. A politician claims that it's just a matter of better engineering before we have power plants that are 100% efficient at converting fuel energy to electrical energy. Counter this claim.

Chapter 5

FOSSIL ENERGY

The time is 100 million years ago, the place is a steamy swamp. Earth is a much warmer planet than today, thanks to a level of atmospheric CO_2 more than three times what it is now, and more than four times the level at the dawn of the industrial era. Our swamp could be almost anywhere, because although the tropics in this ancient world are just a few degrees Celsius warmer than today, the poles are some 20°C to 40°C warmer and therefore ice-free. Dinosaurs thrive even north of the Arctic Circle. However, there are plenty of places in today's world where our swamp couldn't be. That's because sea level is some 200 meters higher than at present, so some of today's land is under water.

The tropical heat and abundant CO_2 promote rapid plant growth. Plants take in CO_2 from the atmosphere and water from the ground. Using energy from sunlight, they rearrange the simple CO_2 and H_2O into sugars and other complex molecules that store this solar energy in the electric fields associated with the new molecular configuration. This, of course, is the process of photosynthesis. The energy-storing molecules produced in photosynthesis become the energy source for the plants that made them, for the herbivorous animals that eat those plants, for the carnivorous animals that eat the herbivores, and for the insects and bacteria that decompose all of these organisms when they die.

5.1 The Origin of Fossil Fuels

Coal, oil, and natural gas are the fossil fuels that power modern society. As I showed in Chapter 2, these fuels provide nearly 90% of the energy we humans consume. A few other fossil products, especially peat and the tarry solid called *bitumen,* are also used for energy or applications such as paving roads. But coal, oil, and gas are by far the dominant fossil fuels. They're also sources of air pollution and of greenhouse gases that are altering Earth's climate. I'll devote two chapters to fossil fuels, beginning in this section with their origins.

COAL

Back to our ancient swamp, where giant ferns surround stagnant, acidic waters. The acidic conditions limit the decay of this organic material, and as more and more generations of dead plants accumulate, they push the older material ever deeper. Eventually the material forms a soft, crumbly, brown substance called **peat**. Peat contains some of the energy that was originally fixed by photosynthesis when the plants were alive. Indeed, peat is still forming today, in places ranging from the Florida Everglades to the bogs of Scotland, Canada, Scandinavia, and Russia. When dried, peat provides fuel for warming houses and for cooking.

But our peat doesn't meet that fate. Rather, geologic processes bury it ever deeper. Temperature and pressure increase, and the material undergoes physical and chemical changes. Much of the oxygen and hydrogen escape, leaving a substance rich in carbon. The final result is the black solid we call **coal**. Coals vary substantially in type, depending on where they form and how far evolved they are. The most evolved—bituminous and anthracite coal—are largely carbon. However, coals contain significant quantities of other elements, including sulfur, mercury, and even uranium.

Coal also contains energy. The carbon atoms in coal aren't combined with oxygen as they are in CO_2. As such this carbon, together with oxygen in the air, represents a state of higher energy than does CO_2. Combining the carbon with oxygen in the process of combustion releases that excess energy. Coal is quite a concentrated energy source; the highest-quality anthracite contains some 36 MJ/kg, or about 10,000 kWh per ton. (The 29-MJ/kg figure in Table 3.3 is an average over different types of coal.)

So coal's energy is ultimately sunlight energy trapped by long-dead plants. This direct link with once-living matter makes coal a fossil fuel. Some lower-grade coals actually contain recognizable plant fossils (Fig. 5.1).

FIGURE 5.1
A fossil leaf in low-grade coal.

Coal is still forming today, although under current conditions that happens much more slowly than it did in the Carboniferous period, some 300 million years ago. But the rate of coal formation has always been agonizingly slow. Nearly all living matter decays promptly after death, releasing the CO_2 that plants removed from the atmosphere. Most of the stored energy is released, too, and is quickly radiated to space. But as I discussed in Chapter 1 when

I described Earth's energy endowment, a tiny fraction of living material escapes immediate decay and is buried to form coal and other fossil fuels. The fossil fuels we burn today are the result of hundreds of millions of years of growth, death, escape from decay, and physical and chemical processes that transform once-living matter into these fuels.

OIL AND NATURAL GAS

Picture another ancient scene, this time a warm, shallow ocean. A steady "rain" of dead organic matter falls to the ocean floor. Most of this consists of tiny, surface-dwelling organisms called, collectively, **plankton**. Plankton includes animal species (zooplankton) and plant species (phytoplankton), the latter using photosynthesis to capture the solar energy that is the basis of marine food chains. The organic material reaching the ocean floor joins the inorganic matter carried into the sea from rivers, resulting in a loose mix of ocean-floor sediments.

In time, the pressure of the overlying water compacts the sediments. Natural cements and the crystallization of inorganic minerals eventually convert the sediments into solid material. Over geologic time, ocean-floor sediments are buried and subject to high pressures and temperatures. The result is the layered solid we call sedimentary rock. During the formation of sedimentary rock, organic material may be "cooked," first to a waxy substance called **kerogen** and eventually to a combination of organic liquids and gases called, collectively, **petroleum** (this term is often used for liquids alone, but formally *petroleum* refers to both liquids and gases). Trapped in petroleum is some of the energy that was captured long before by photosynthetic plankton on the ocean surface.

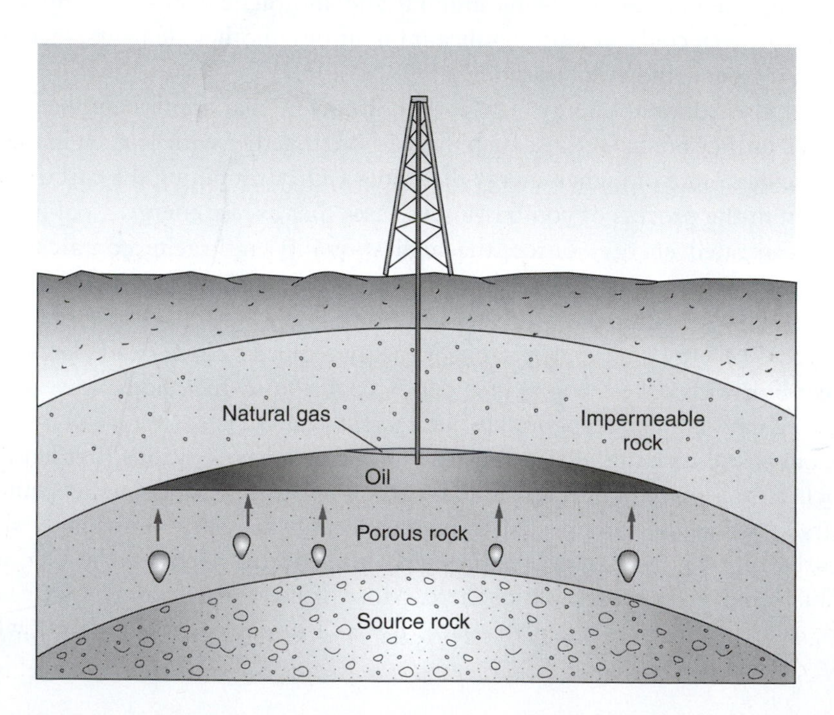

FIGURE 5.2
Formation of an oil deposit in a geological structure called an *anticline*, in which rock layers have bent upward. The oil originally forms in the source rock, then migrates upward through the porous layer to collect at the top, where it is trapped by the overlying impermeable rock. Natural gas may collect above the oil.

Petroleum occurs throughout many sedimentary rock formations, but usually not in sufficient quantities to make its extraction economically feasible (at least not at present). But being fluid, some petroleum migrates to collect in porous rocks such as sandstones or limestones. Overlying layers of impermeable rock may trap the petroleum, resulting in a concentrated source of fossil energy (Fig. 5.2). Fast-forward to the present, and you're likely to find an energy company drilling to extract the trapped fuel.

Like coal, petroleum represents fossil energy. And as with coal but even more so, we're removing petroleum from the ground at a rate millions of times greater than the rate at which it forms.

5.2 The Fossil Fuels

COMPOSITION

We've seen that coal consists largely of carbon, although lower grades contain significant quantities of hydrocarbons (molecules made of hydrogen and carbon) and contaminants such as sulfur and mercury. Petroleum is even more of a mix. As with coal, the exact composition of petroleum varies, depending on where and under what conditions it was formed. Most significant is the ratio of gas to liquid. The gaseous component of petroleum is **natural gas**, which consists largely of methane (CH_4, the simplest organic hydrocarbon) and other light, gaseous hydrocarbon molecules. The liquid component of petroleum is **crude oil**. Crude oils vary widely in composition, but all contain a mix of hydrocarbons. Typically, crude oil is about 85% carbon by weight, and most of the rest is hydrogen. Sulfur, oxygen, and nitrogen are also present in significant quantities.

REFINING

We **refine** crude oil to produce a variety of products ranging from the heaviest oils for industrial boilers, to fuel oil used in home heating, diesel oil that powers most trucks and some cars, jet aircraft fuel, and gasoline for our cars. Figure 5.3 shows the typical distribution of refinery products. Each of these products contains a mix of different molecules, so no single chemical formula describes a given fuel. Molecules typically found in gasoline, for example, include heptane (C_7H_{16}), octane (C_8H_{18}), nonane (C_9H_{20}), and decane ($C_{10}H_{22}$). These molecules consist of long chains of carbon with attached hydrogens; for every n carbons, there are $2n + 2$ hydrogens. Higher grades of gasoline have greater ratios of octane to heptane, which provides more controlled burning in high-performance engines. (Hence the term "high octane" you see at the gas pump.)

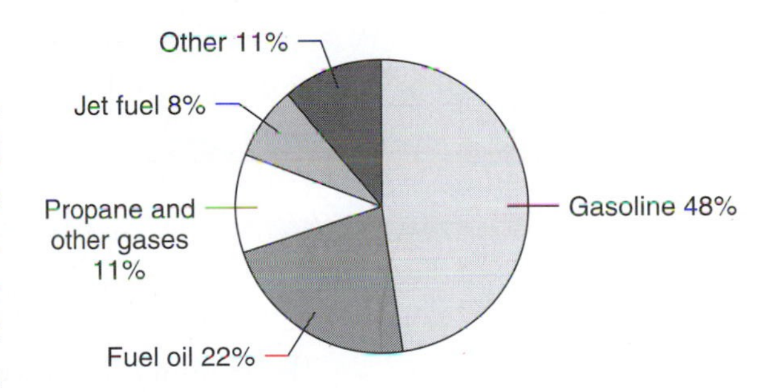

FIGURE 5.3
Typical distribution of refinery products in the United States. The "other" category includes lubricants, asphalt and road oil, waxes, kerosene, solid fuels, and feedstocks for the manufacture of petrochemicals. The product mix varies with seasonal changes in demand.

The main refining process involves **fractional distillation**, in which crude oil is first heated to vaporize it. The vapor then rises through a vertical column,

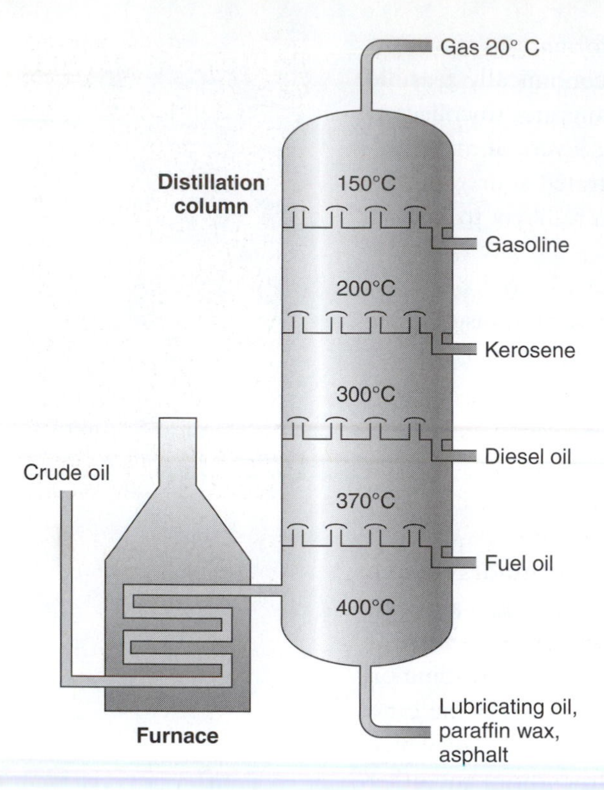

FIGURE 5.4
Simplified diagram of the fractional distillation process used in oil refining, showing temperatures at which different products condense out of the distillation column.

where its temperature decreases with height. Heavier components condense out at higher temperatures, so they're removed near the bottom of the column. The lighter components rise higher before condensing, with gasoline near the top. Gaseous fuels such as propane then exit from the top of the column. Additional steps, including so-called catalytic cracking, break up large hydrocarbon molecules to increase yields of lighter components, especially gasoline. After separation, fuels may undergo further refinement to remove undesirable substances like sulfur, or to have additives mixed in for improved combustion. Figure 5.4 diagrams the fractional distillation process.

The energy content of refined fuels varies somewhat, but a rough figure is about 45 MJ/kg for any product derived from crude oil. Refining is itself an energy-intensive process, with some 7% of the total U.S. energy consumption going to run oil refineries.

OTHER PRODUCTS FROM FOSSIL FUELS

We don't burn all the products we make from petroleum. About half of the "other" category in Figure 5.3, which is around 7% of all crude oil, is turned into road-making products, lubricants, and petrochemical feedstocks. These feedstocks go into making a vast array of products—everything from plastics to medicines, ink to contact lenses, insecticides to toothpaste, perfumes to fertilizers, lipstick to paint, false teeth to food preservatives. As oil grows scarce, it will become increasingly valuable as a raw material for manufacturing, giving us all the more incentive to look elsewhere for our energy. Our descendants a century hence may well look back and ask, incredulously, "Did they actually burn such valuable stuff?"

HISTORY OF FOSSIL FUEL USE

Humankind has used coal as an energy resource, albeit in limited quantities, for as long as 2,000 years. Some evidence suggests that around 1,000 B.C.E. the Chinese were burning coal to smelt copper; archeologists have found coal cinders from 400 C.E. in what was Roman-occupied Britain; and written accounts of coal use begin appearing in the thirteenth century. Coal was a widely used heating fuel in seventeenth-century England, so much so that the famous St. Paul's Cathedral was blackened from coal smoke even before it was completed. Coal consumption in Europe expanded greatly with the beginning of the industrial revolution and James Watt's eighteenth-century improvements to the steam engine. By the 1830s, coal mining thrived in the eastern United

States, supplying coal for industry and for steam locomotives on the newly developed railroads.

Petroleum, too, has a long history. Liquid petroleum seeps naturally to Earth's surface in some locations, and as early as 5,000 B.C.E. humans exploited this resource for a variety of purposes, including medicinal use in ancient Egypt. Oil became a military resource when Persians and Arabs developed oil-soaked arrows and other incendiary weapons. By the twelfth century, use of oil for illumination had spread from the Middle East to Europe. Indians of North America discovered oil seeps in what are now New York and Pennsylvania, and they used the oil for medicines. Whale oil was a favored fuel for lighting purposes until the nineteenth century, but the decline of whale populations drove the need for new liquid fuels, including petroleum. The world's first oil wells were drilled in Russia, near the Caspian Sea, in the mid-1800s; in the United States, Edwin Drake struck oil in Titusville, Pennsylvania, in 1859. These pioneering wells marked the beginning of an ever-expanding quest for oil worldwide. Today, oil is surely the single most important natural resource for modern society.

FIGURE 5.5
Gas flaring at PetroChina's Tazhong oil refinery.

Natural gas is the latest of the fossil fuels to see widespread growth. Gas generally occurs in the presence of oil and, until recently, was often treated as a waste product and *flared*—that is, burned—right at the well (Fig. 5.5). But construction of gas pipelines, development of ships for transporting liquefied gas, and discovery of new gas fields have made distribution and importation easier. These factors, along with gas's lower greenhouse emissions and clean, efficient combustion, have contributed to natural gas's increased popularity, and suggest that this fuel's share of the world energy supply will continue to grow in the coming decades. In the United States, primary energy production from natural gas pulled even with coal around 2009; today, both supply about one-fourth of the country's energy. Worldwide the situation is less clear; use of gas is growing in the West, but coal consumption is also increasing rapidly, especially in China. Figure 5.6 shows quantitatively the trends in fossil fuel consumption.

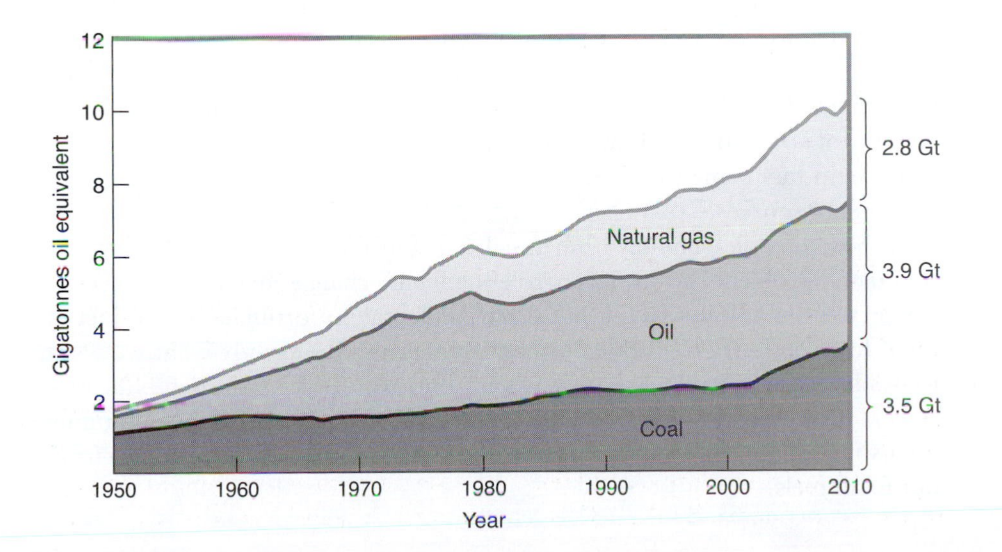

FIGURE 5.6
World fossil fuel consumption since 1950. The height of each shaded area represents the amount for one of the three fuels, so the top curve is the total fossil fuel consumption. Thus the graph shows that, in the final year plotted, fossil fuels equivalent to more than 10 gigatonnes of oil (Gtoe) were consumed globally.

5.3 Energy from Fossil Fuels

When we burn a fossil fuel, we bring fuel into contact with atmospheric oxygen and raise the temperature so that chemical reactions occur between the fuel and oxygen. In sustained combustion, as in a coal-fired power plant or a gas-fired home water heater, heat released in the burning process sustains the combustion. In a gasoline engine, an electric spark ignites the gasoline–air mixture in each cylinder, whereas in a diesel engine, ignition results from compression and consequent heating of the fuel–air mixture as the piston rises in the cylinder. In any event, chemical reactions rearrange the atoms of fuel and oxygen, resulting in new chemical products and simultaneously releasing energy as heat.

PRODUCTS OF FOSSIL FUEL COMBUSTION

Because fossil fuels consist largely of hydrocarbons, complete combustion produces mostly carbon dioxide gas (CO_2) and water vapor (H_2O). Both are nontoxic, commonplace substances already present in the atmosphere, but both are also climate-changing greenhouse gases. Although water vapor is actually the dominant naturally occurring greenhouse gas in Earth's atmosphere, CO_2 produced from fossil fuel is largely responsible for recent changes in atmospheric composition.

Because the various fossil fuels have different chemical compositions, their combustion results in differing amounts of CO_2 emission. Coal is mostly carbon, so CO_2 is the main product of coal burning. Oil consists of hydrocarbon molecules with roughly twice as many hydrogen as carbon atoms. So when oil burns, a good deal of the energy released comes from the formation of H_2O and so, for a given amount of energy, less CO_2 is emitted. Finally, natural gas is largely methane (CH_4), so it contains still more hydrogen and produces even less CO_2. That's why a shift from coal and oil to natural gas would slow but not halt the process of anthropogenic climate change. Figure 5.7 shows quantitatively the carbon dioxide emissions per unit of energy released in burning each of the three main types of fossil fuel. Unfortunately, we're going to run out of natural gas (and oil) long before we run out of coal, so switching to gas isn't a long-term solution. However, it could help buy some time until we start making large-scale use of carbon-free energy sources.

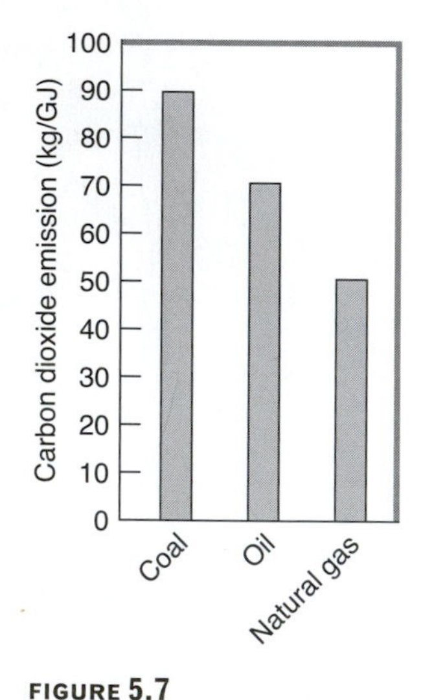

FIGURE 5.7
Carbon dioxide emission per gigajoule of energy released in the combustion of the three fossil fuels. Natural gas produces just over half the CO_2 of coal, making it a more climate-friendly fuel.

Carbon dioxide emission from fossil fuel burning is a major global environmental problem because of its role in climate change (this is the subject of Chapters 12 to 15). But CO_2 is not some incidental, unfortunate by-product of fossil fuel combustion. Rather, in a very real sense, it's exactly what we want to produce (along with water) when we burn fossil fuels. Carbon dioxide represents the lowest possible energy state of carbon in Earth's oxygen-containing atmosphere, so by making CO_2, we've extracted the most energy possible from our fossil fuels.

EXAMPLE 5.1 | Coal versus Gas

An older coal-burning power plant produces electrical energy at the rate of 1 GWe, typical of a large power plant. The plant is 35% efficient, meaning that only 35% of the coal's energy ends up as electricity. Estimate the rate at which the plant emits CO_2, and determine the reduction in CO_2 emissions if the coal plant were replaced with a modern gas-fired facility with 44% efficiency.

SOLUTION

Figure 5.7 shows that coal burning yields CO_2 emissions of 90 kg per gigajoule (GJ) of energy produced. This figure refers to the thermal energy content of the fuel; at 35% efficiency, a 1 GWe power plant would need to produce thermal energy at the rate of 1 GW/0.35 = 2.86 GWth (I've added the *th* to remind you of the difference between electric and thermal power). Since 1 GW is 1 GJ per second, this 2.86-GWth rating, coupled with CO_2 emissions of 90 kg/GJ, gives a CO_2 emissions rate of

$$(2.86 \text{ GJ/s}) \times (90 \text{ kg/GJ}) = 257 \text{ kg/s}$$

for the coal-burning plant. That's more than 500 pounds of CO_2 every second!

A gas-fired plant that is 44% efficient, in contrast, would require a thermal power of 1 GW/0.44 = 2.27 GWth to produce that same 1 GW of electric power. For natural gas, Figure 5.7 shows 50 kg of CO_2 per gigajoule of energy, so the emissions rate for the gas-fired plant becomes

$$(2.27 \text{ GJ/s}) \times (50 \text{ kg/GJ}) = 114 \text{ kg/s}$$

The increased efficiency and, more significantly, the lower CO_2 emissions per unit of energy produced from natural gas combine to give the gas-fired plant a total CO_2 emissions rate less than half that of the coal-fired plant.

Fossil fuels aren't pure hydrocarbon, and even if they were, they wouldn't necessarily burn completely to CO_2 and H_2O. So a host of other substances result from fossil fuel combustion. Many of these are truly undesirable, and they're often unnecessary in that their production isn't required in the energy-release process. These substances comprise pollution. Pollution from fossil fuels is such a major environmental problem that I devote all of Chapter 6 to it.

5.4 Fossil Energy Technologies

Nearly all technologies for extracting energy from fossil fuels involve burning the fuel. Burning releases energy as heat, which we can use directly to heat water, air, or other substances and thus use nearly all the energy contained in the fuel. But heated substances represent low-quality energy, as I described in Chapter 4.

BOX 5.1 | Carbon versus Carbon Dioxide

In describing carbon emissions from fossil fuels, different authors in different contexts may express the emission rate either as an amount of CO_2 per unit time, as I did in Example 5.1, or in carbon (C) per unit time. The latter is in some sense more general, because we do emit carbon in other forms (CH_4 [methane], for example). When you're reading about flows of carbon or CO_2, whether natural or anthropogenic, be sure you know which is being specified.

What's the relationship between CO_2 emissions when specified as CO_2 and as C? Simple: Carbon has an atomic weight of 12, and oxygen's atomic weight is 16, so the weight of the CO_2 molecule is $12 + (2 \times 16) = 44$—that is, 44/12, or 3.67 times that of a carbon atom. This means that a CO_2 emission rate specified as, say, 9 gigatonnes (Gt) of carbon per year is equivalent to $(9\ \text{Gt}) \times (44/12) = 29$ Gt of CO_2. (This 9 Gt per year isn't just an arbitrary figure; it's roughly the annual total of global anthropogenic carbon emissions from fossil fuel combustion.) In short, to convert from carbon to CO_2, just multiply by 44/12.

If we want high-quality energy, such as electricity or energy of motion, then we need to run a heat engine that converts *some* of the heat released from burning fuel into high-quality energy. The second law of thermodynamics prohibits us from converting *all* of that heat to high-quality energy.

The ingenuity of engineers, tinkerers, and inventors from James Watt and his predecessors on to the present has resulted in a bewildering array of heat-engine designs. But, as Sadi Carnot showed, none can escape the fundamental efficiency limitations of the second law of thermodynamics, and none can be more efficient than Carnot's early-nineteenth-century design for a theoretical engine with the maximum possible efficiency. This doesn't mean we can't continue to improve real heat engines, but it does mean that such improvements can at best approach Carnot's ideal limit, not reach it. We won't explore heat engine designs in detail here, but it is helpful to distinguish two general types, external combustion engines and internal combustion engines.

FIGURE 5.8
This steam locomotive, built around 1930, is an example of an external combustion engine. The efficiencies of steam locomotives were low, typically in the single digits.

EXTERNAL COMBUSTION ENGINES

The first steam locomotives had a long, cylindrical boiler heated by a wood or coal fire. The water boiled to steam, and steam pressure drove pistons that turned the locomotive's wheels (Fig. 5.8). Because the fuel burned in an external firebox, which subsequently heated the water, these were called *external combustion engines*. With their efficiencies measured in single digits, steam locomotives were never very efficient, and they came nowhere near Carnot's thermodynamic limit. Steam automobiles were produced in the early twentieth century, but they, too, were not at all efficient. External combustion engines for transportation vehicles are now a thing of the past.

However, the external combustion engine lives on as the prime mover in most electric power plants. In a typical fossil-fueled plant, fuel burns in a boiler consisting of a chamber containing banks of water-filled pipes. This arrangement allows hot combustion gases to surround the pipes, resulting in efficient heat transfer. The water boils to high-pressure steam, which drives a fanlike **turbine** connected to an electric generator (recall Chapter 3's discussion of electromagnetic induction). The spent steam, now at low pressure, still contains energy in the form of its latent heat. It moves through a **condenser**, where the steam-containing pipes contact a flowing source of cooling water, usually from a nearby river, lake, or ocean. The steam condenses back to water and returns to the boiler for reuse. In the process, it gives up its latent heat to the cooling water. This energy is usually dumped to the environment as waste heat, either directly to the cooling water source or to the atmosphere via the large **cooling towers** typically seen at power plants. Figure 5.9 is a diagram of such

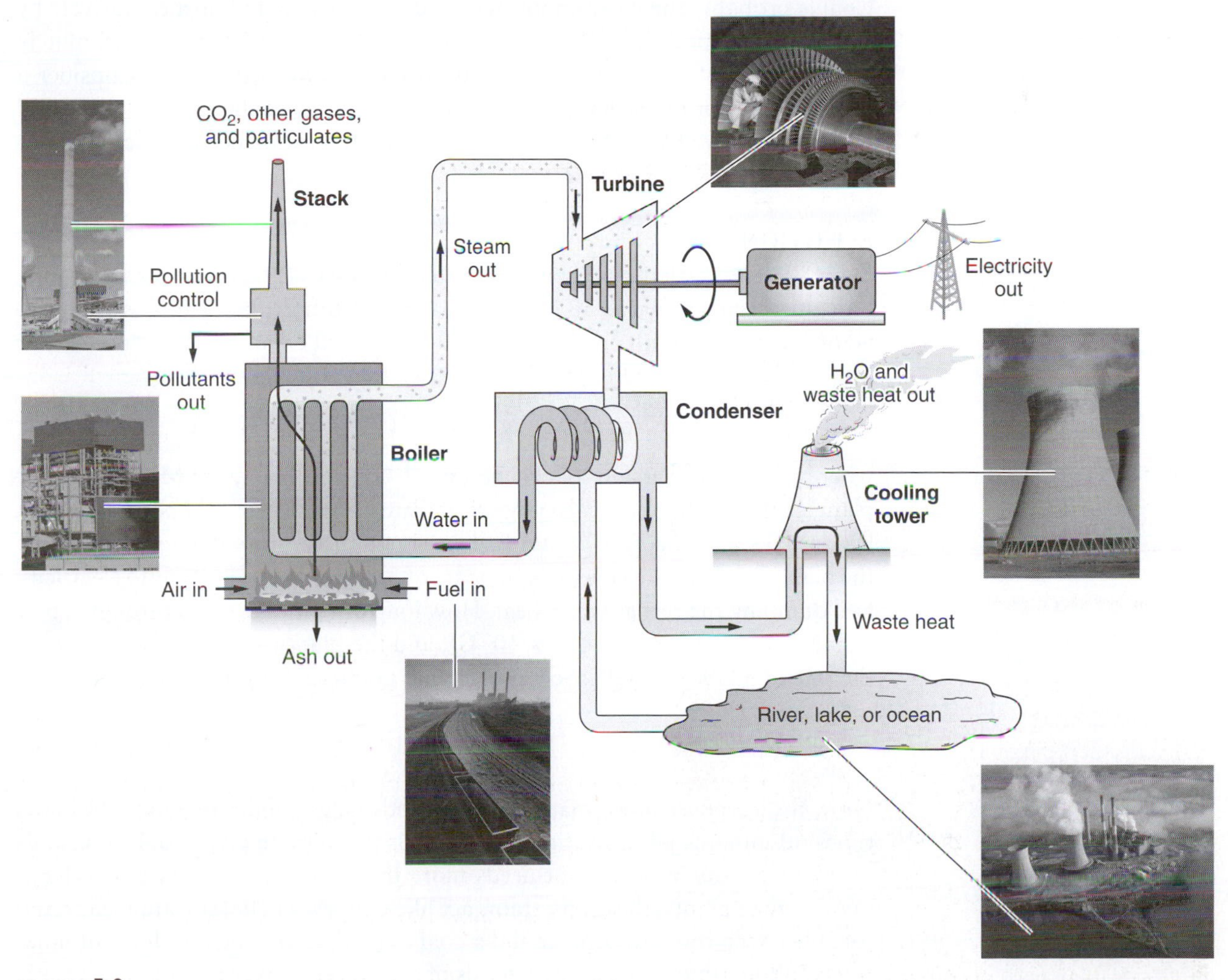

FIGURE 5.9
Diagram of a typical fossil-fueled power plant.

a **thermal power plant**, with photos of several main components. Incidentally, this description applies to nuclear plants as well, with the exception that (in the most common design) the water occupies a single large pressure vessel in which fissioning uranium fuel is immersed.

Thermal power plants are large, stationary structures. Their performance is optimized for a particular turbine speed and power output. For that reason they can be more efficient than vehicle engines, which have to function over a range of speed and power. The fuel requirements for a large fossil fuel power plant are staggering: As Example 5.2 shows, a 1-GW plant may consume many 100-car trainloads of coal each week!

EXAMPLE 5.2 | Coal Train!

Coal is probably the most important commodity carried almost exclusively by rail. A typical modern coal train, called a *unit train,* consists of approximately 100 carloads of coal, each with a capacity of about 100 tonnes. Consider a coal-fired power plant that produces electrical energy at the rate of 1 GWe and is 33% efficient. Estimate the number of unit coal trains needed each week to keep this plant supplied with fuel.

SOLUTION

Table 3.3 shows that coal's energy content averages 29 MJ/kg; for this rough estimate I'll round that to 30 MJ/kg. Recall that 1 tonne (metric ton) is 1,000 kg. So our unit train, with one hundred 100-tonne coal cars, carries 10^4 tonnes, or 10^7 kg of coal. Therefore, the energy carried in a unit train is

$$(10^7\ \text{kg})(30\ \text{MJ/kg}) = 30 \times 10^7\ \text{MJ} = 3 \times 10^5\ \text{GJ}$$

where the last conversion follows because 1 GJ (10^9 J) is 1,000 MJ. Our power plant is putting out electrical energy at the rate of 1 GWe, or 1 GJ per second, but it's only 33% efficient, so it uses fuel energy at three times this rate. Therefore the plant consumes fuel energy at the rate of 3 GW, turning 1 GW into electricity and dumping the rest as waste heat. How long does it take to go through a unit train? The unit train carries 3×10^5 GJ, and the plant consumes fuel energy at the rate of 3 GW or 3 GJ/s. So a unit train is going to last for a time given by

$$\frac{3 \times 10^5\,\text{GJ}}{3\,\text{GJ/s}} = 10^5\,\text{s}$$

Now, for purposes of estimation, 10^5 seconds is just under one day (24 hours times 60 minutes per hour times 60 seconds per minute gives 86,400 seconds per day). So our power plant needs more than seven trainloads a week! For many power plants, these unit trains act like conveyer belts, shuttling back and forth between the power plant and a coal mine that may be hundreds of miles distant. You can explore the fuel needs of a gas-fired power plant in Exercise 2.

In Chapter 4, I showed how the second law of thermodynamics prohibits any heat engine from extracting as useful work all the energy contained in a fuel. In particular, the efficiency of a heat engine depends on the maximum and minimum temperatures within the engine. For a typical fossil-fueled power plant, with maximum steam temperature of 650 K, we calculated this thermodynamic efficiency limit at just over 50%. But real plants fall well short of this theoretical maximum. Friction robs some of the energy, and so do deviations from Carnot's ideal engine cycle. In addition, a significant part of the power plant's output goes into running pumps, pollution control systems, cooling towers, and other essential equipment. All that extra energy eventually ends up as heat, adding still more to the waste required by the second law of thermodynamics. The net result is that a typical coal-fired steam-turbine power plant has an overall efficiency of around 35%. That's where I get the rough rule of thumb that two-thirds of the energy extracted from fuel is discarded as waste heat. We'll see later in this chapter how more advanced systems can achieve higher efficiencies.

INTERNAL COMBUSTION ENGINES

In an internal combustion engine, the products of fuel combustion themselves provide the pressures that ultimately produce mechanical motion. There's no transfer of heat to a secondary fluid, as in the steam boiler of an external combustion power plant. In **continuous combustion engines**, fuel burns continually and generally produces rotary motion directly. Jet aircraft engines are a good example; so are devices used in some advanced power plants. In **intermittent combustion engines**, fuel burns periodically, usually in a cylindrical chamber containing a movable piston.

Intermittent combustion engines are the workhorses of our surface transportation fleet. More than a century of engineering has brought this technology to a high level of sophistication and performance. Nearly all engines used in cars and trucks, as well as in many ships, railroad locomotives, and smaller aircraft, are intermittent combustion engines in which rapid burning of the fuel in a cylinder produces hot, high-pressure gas that pushes on a piston. The piston is connected to a **crankshaft**, with the result that the back-and-forth motion of the piston becomes rotary motion of the crankshaft. Most modern engines, except those used in smaller devices such as lawnmowers, have more than one cylinder. The physical layout of the cylinders, the strategies for initiating and timing the combustion, and the type of fuel all vary among different engine designs. Here I'll mention just a few variations on the basic intermittent combustion engine.

Spark-ignition engines include the gasoline engines used in most cars in the United States. Fuel is injected into the cylinders at an appropriate point in the cycle and then, at the optimum instant, an electric spark ignites the fuel. In modern automobile engines a computer controls the spark timing to maximize performance, or to maximize fuel efficiency, or to minimize pollution (these three goals are generally mutually exclusive). Figure 5.10 shows a cutaway diagram of a typical modern gasoline engine. Used in today's automobiles, such engines

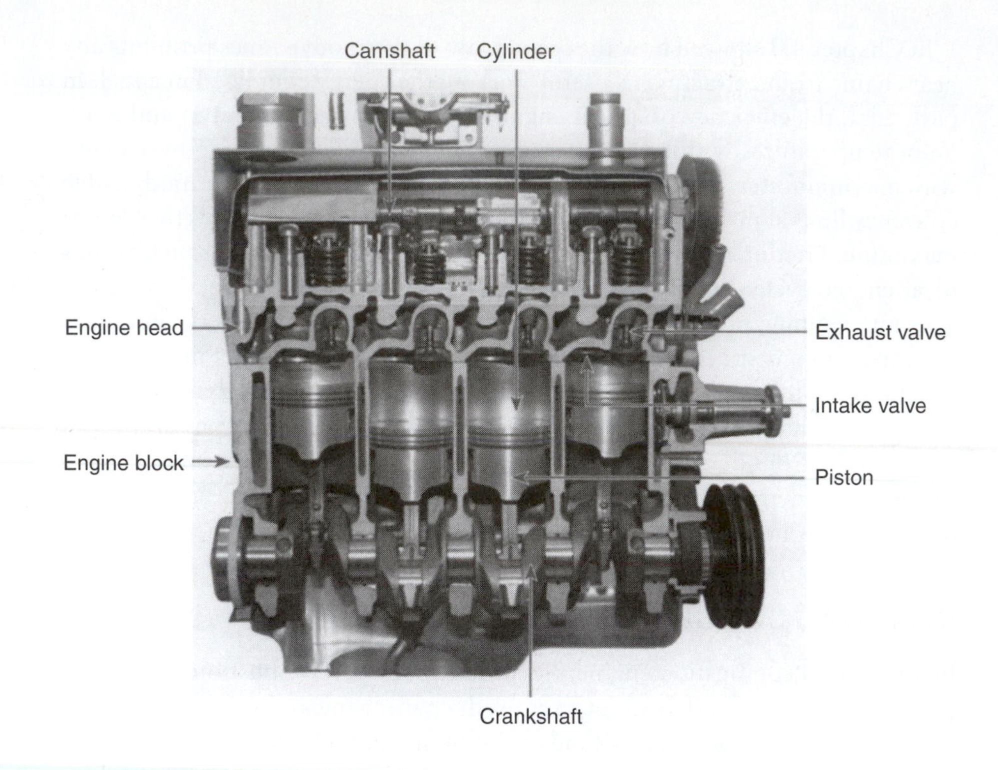

FIGURE 5.10
Cutaway diagram of a four-cylinder gasoline engine. The up-and-down motion of the pistons is converted to rotary motion of the crankshaft. Valves admit the fuel–air mixture and release exhaust gases.

are about 20% efficient at converting the energy content of gasoline into mechanical energy, but some of this energy is lost to friction in other mechanical components, leaving only about 15% of the original fuel energy available at the wheels. Figure 5.11 shows energy flows in a typical automobile.

Compression-ignition engines include the diesel engine, used universally in large trucks, buses, and locomotives, and in just over half the cars sold in Europe today. If you've pumped a bicycle tire, you know that compressing air in the pump makes it hot; what you're doing is turning mechanical work into thermal energy. In a diesel engine, the rising piston compresses air in the cylinder to such an extent that it's hot enough to ignite fuel, which is injected into the cylinder at just the right instant. Diesel engines burn a refined oil that's heavier than gasoline and results in significantly more particulate pollution (more on this in Chapter 6). However, higher compression—implying higher temperature and thus a higher thermodynamic efficiency—along with other design features make the diesel engine considerably more fuel efficient than its gasoline-powered cousin. This is the reason for Europe's ongoing boom in diesel car sales. The few diesel-fueled cars sold in the United States have fuel efficiencies comparable to those of gas–electric hybrids. Greater fuel efficiency means that diesel cars emit less CO_2 per mile of driving and thus have a smaller impact on climate. However, their high compression means that diesel engines must be more ruggedly built, which makes them heavier and more expensive than their gasoline counterparts; that's one reason they're most commonly found in larger vehicles like trucks.

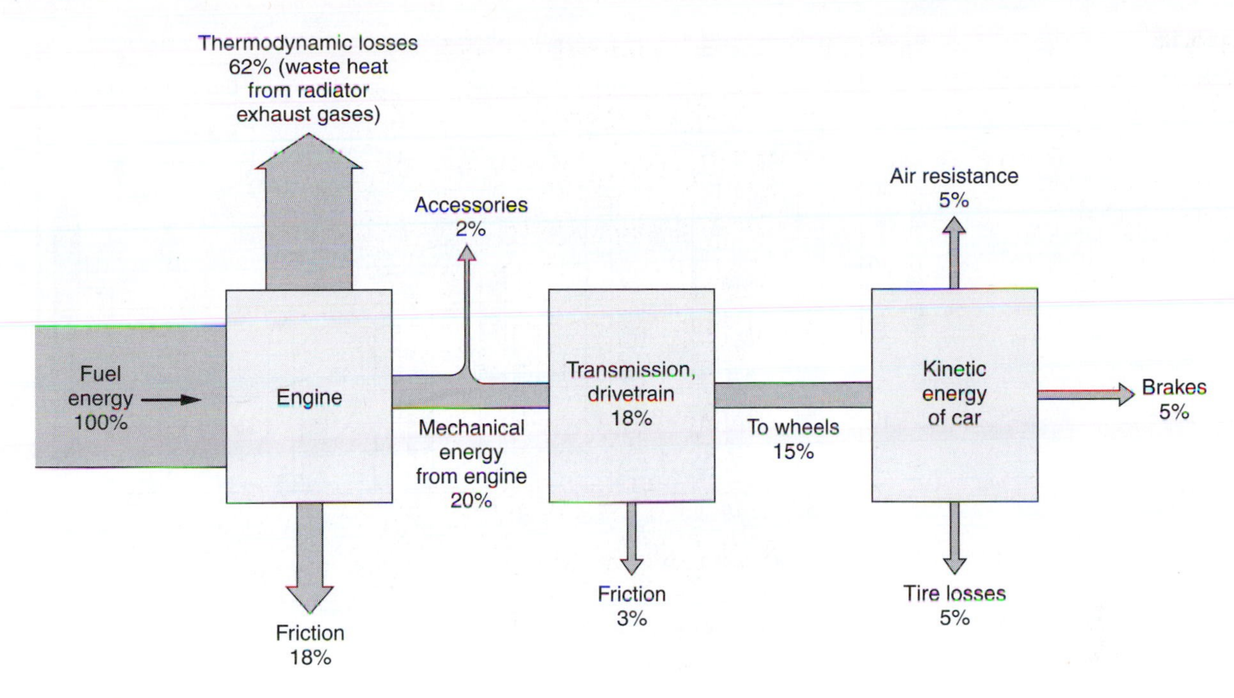

FIGURE 5.11
Energy flows in a typical gasoline-powered car. Thermodynamic losses and friction leave only about 15% of the fuel energy available at the wheels, all of which is dissipated by air resistance, tire friction, and braking. The power needed for accessories runs the air conditioning, lights, audio system, and vehicle electronics.

HYBRID AND ELECTRIC VEHICLES

Gas–electric hybrids combine gasoline engines with electric motors. In these vehicles, motive power to the wheels may come directly from the gasoline engine, from the electric motor, or from a combination of the two. Hybrids are noted for their high fuel efficiency, compared with cars of similar weight and configuration (Fig. 5.12). This results largely from **regenerative braking**, in which the car's kinetic energy is recaptured as it slows, turning an electric generator and charging the battery. The energy is then available to power the car through its electric motor. In a gasoline-only car, this energy would be dissipated as useless heat through friction in the brakes. Also, a hybrid's gasoline engine is usually smaller, can operate closer to its optimum efficiency, and may shut off altogether when the car is stopped or the electric motor is running. All these factors contribute to increased fuel efficiency.

Hybrids come in several flavors. **Parallel hybrids** are the most common; they utilize both their gasoline engine and electric motor for powering the wheels, either separately or in combination. Figure 5.13 diagrams the parallel system in Toyota's Prius hybrid. Within the parallel category are many technological variations. Parallel hybrids that make greater use of electric propulsion tend to get better fuel mileage in city driving than on the highway, a reversal of the usual situation. In **series hybrids**, a gasoline engine drives a generator that

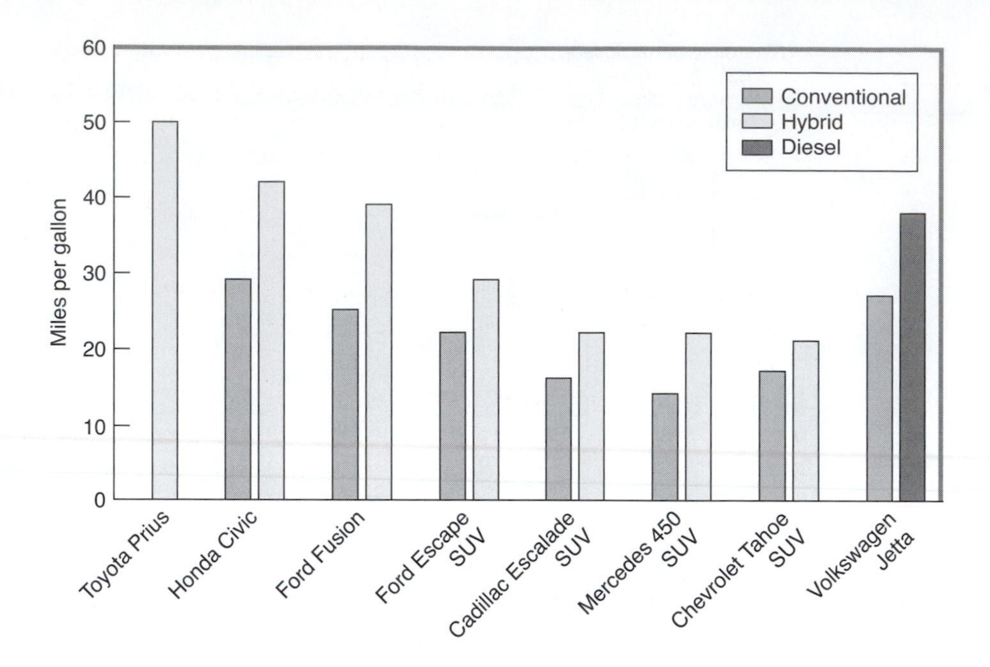

FIGURE 5.12
Fuel efficiencies of conventional and hybrid vehicles for the 2010 model year; also shown are the conventional and diesel Volkswagen Jetta, which was not available as a hybrid in 2010. Note that hybrid SUVs, although more efficient than their conventional counterparts, get fewer miles per gallon than smaller conventional vehicles.

charges a battery; motive power then comes solely from an electric motor powered by the generator. Chevrolet's Volt is an example of a series hybrid.

Some hybrids, like Toyota's original Prius, are entirely gasoline powered, in that the gasoline engine runs a generator that charges the battery, which, in turn, supplies energy to the electric motor. Others, like the Volt and the Prius PHV, are **plug-in hybrids**, whose batteries can be charged either by plugging into the electric power grid or from the on-board gasoline-powered generator.

Plug-in hybrids may or may not be entirely fossil fueled, depending on where the electricity comes from. The same goes for pure **electric vehicles**, like Nissan's Leaf, which have no gasoline engine and are powered solely by grid-charged batteries and electric motors. Electric vehicles are often touted as "cleaner" than conventional gasoline-powered vehicles or gas-electric hybrids,

(a)

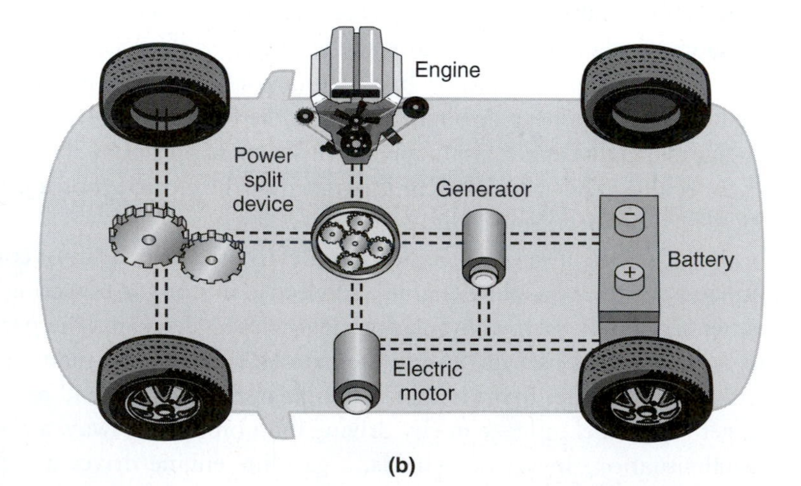

(b)

FIGURE 5.13
(a) Engine and electric motor for the Toyota Prius. Gasoline engine is at left; electric motor and generator at right. Engine is rated at 73 kW (98 horsepower), motor at 60 kW (80 horsepower).
(b) Diagram of Toyota's Hybrid Synergy Drive. Energy flows among the different components depending on the driving situation.

but that depends on where the electrical energy comes from. A sobering report by the U.S. Department of Energy suggests that most of the energy powering plug-in hybrids and electric cars comes from fossil-fueled power plants, rather than "greener" sources—even if the latter are available on the power grid.

GAS TURBINES

The second major class of internal combustion engine is the continuous combustion engine, of which the **gas turbine** is the most widely used example. In a gas turbine, fuel burns to produce hot, pressurized gas that rushes out of the engine exhaust and, on the way, drives a rotating turbine. The turbine may, in turn, drive an electric generator, airplane propeller, or other device. In a jet aircraft engine, the turbine compresses air and exhaust gases to provide the thrust that gives the jet its motive power (Fig. 5.14). Incidentally, the smaller propeller-driven aircraft used in commercial aviation are usually powered by gas turbines and are called *turboprops* to distinguish them from even smaller planes with propellers driven by piston-and-cylinder gasoline engines. To complicate things further, some gasoline and diesel engines employ devices called *turbochargers,* which are essentially gas turbines placed in the exhaust flow. The turbine extracts some of the energy from waste heat that would normally have gone out the tailpipe and uses it to pump air into the engine at a greater rate than would otherwise be possible. However, a turbocharged engine has to work harder to push exhaust through the turbine. Turbochargers generally enhance engine performance without contributing excess weight. Although

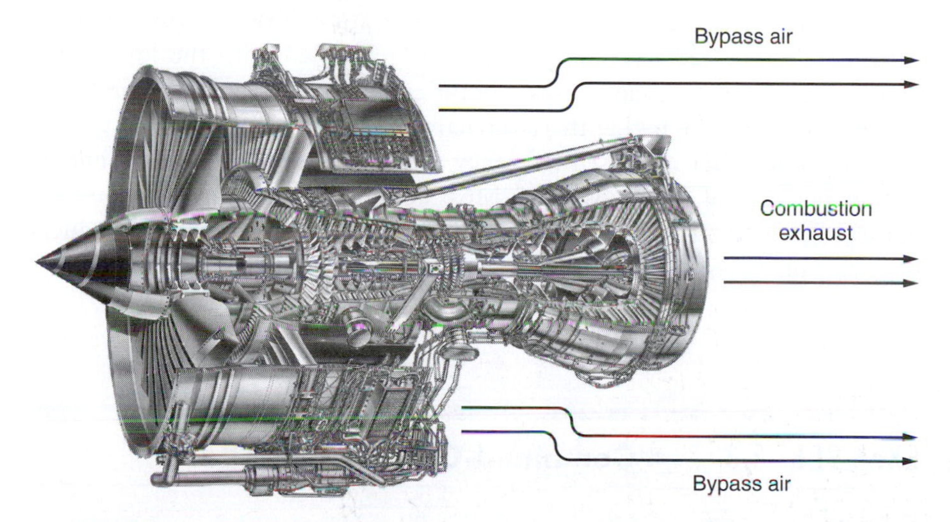

FIGURE 5.14

A jet aircraft engine is an example of a continuous combustion gas turbine. At left are a fan and compressor that pressurize incoming air. Some compressed air enters the combustion chamber, providing oxygen for fuel combustion. The resulting hot gases turn turbines that drive the compressors. As they exit at the right, the exhaust gases also provide some of the jet's thrust. But in a modern jet engine, most of the thrust comes from so-called bypass air that's diverted around the combustion chamber.

they can, in principle, increase fuel efficiency, turbochargers are normally used to achieve enhanced performance, greater acceleration, and higher top speed—all of which lead to higher fuel consumption.

Gas turbines produce lots of power for a given weight, so they're ideal for aircraft engines. But because of their high exhaust temperature (T_c in Equation 4.5), they're not very efficient and, in particular, gas turbines don't make a lot of sense in stationary applications such as power plants. Nevertheless, electric utilities often maintain small gas-turbine-driven generators for so-called peaking—that is, supplying extra power during short periods of high demand—because gas turbines, unlike external-combustion steam power plants, can start and stop very quickly as needed.

COMBINED-CYCLE SYSTEMS

Despite their low second-law efficiency limit, gas turbines increasingly play a role in newer and very efficient **combined-cycle power plants**. In these plants, a gas turbine turns an electric generator. But instead of being dumped to the environment, the hot exhaust from the turbine goes through a so-called heat recovery boiler where some of its energy is used to boil water that then runs a steam turbine to power another generator. Modern combined-cycle power plants can achieve overall efficiencies approaching 60%. Ultimately, the second law of thermodynamics limits the efficiency of a combined-cycle plant just as it does any other heat engine. But the second law doesn't care about the details of the mechanical system used to extract energy, so all that really matters, at least in principle, are the highest and lowest temperatures in the entire combined system. The high temperature of the gas turbine (typically 1,000 to 2,000 K; T_h in Equation 4.5) coupled with the lower final temperature from the steam portion of the combined-cycle plant (T_c in Equation 4.5) makes for a higher thermodynamic efficiency limit than for the gas turbine alone. That increased efficiency manifests itself as a greater energy output per unit of fuel consumed. Most combined-cycle power plants use natural gas as their fuel, and as Figure 5.7 suggested, this further reduces their carbon emissions. Figure 5.15 is a diagram of a typical combined-cycle power plant.

EXAMPLE 5.3 | A Combined-Cycle Power Plant

A combined-cycle power plant has a gas turbine operating between maximum and minimum temperatures of 1,450°C and 500°C, respectively. Its 500°C waste heat is used as the high temperature for a conventional steam turbine cooled by river water at 7°C. Find the thermodynamic efficiency limit for the combined cycle, and compare it with the efficiency limits of the individual components if they were operated independently.

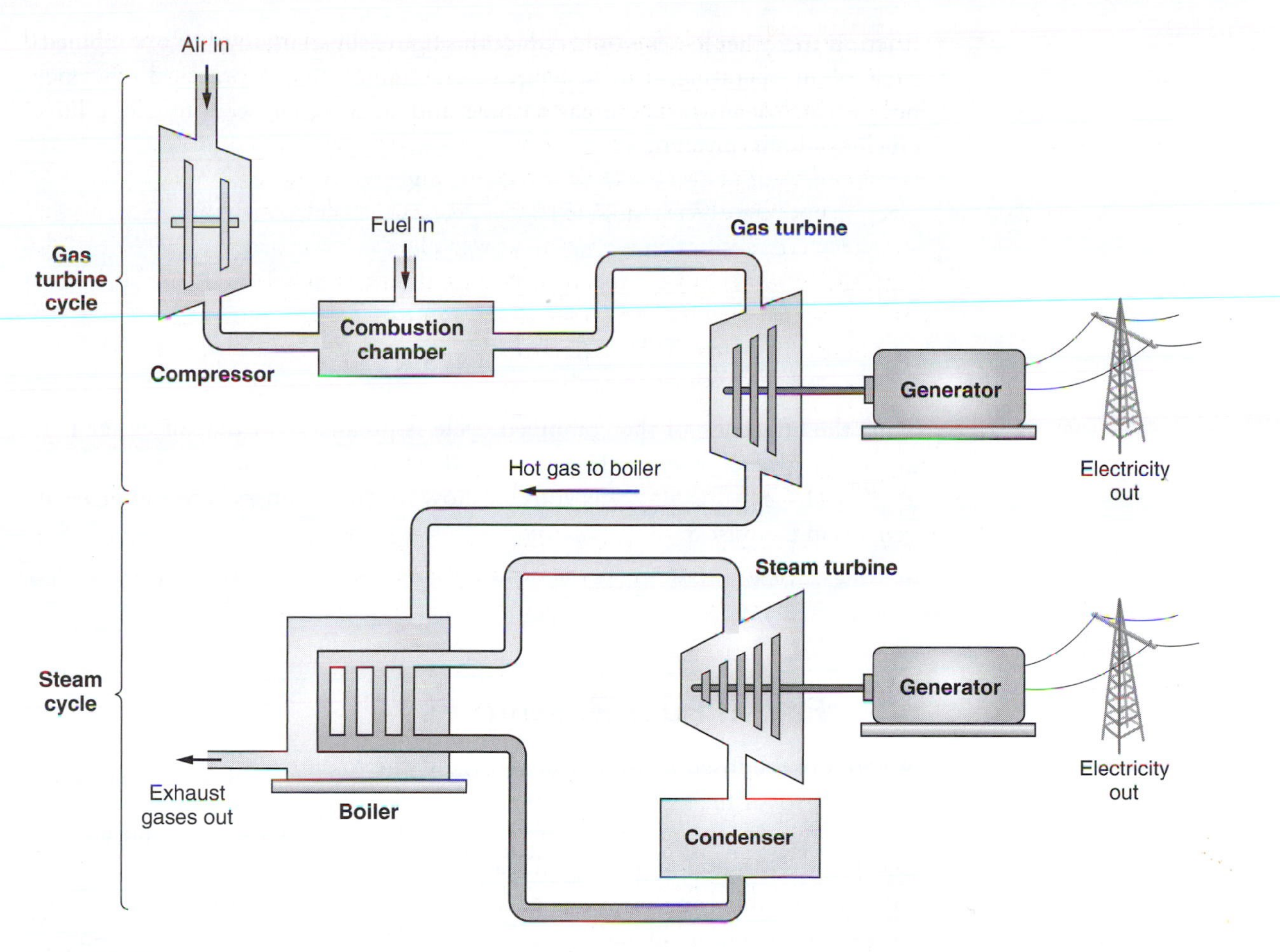

FIGURE 5.15

Diagram of a combined-cycle power plant. The steam section is similar to the one illustrated in Figure 5.9, although details of the cooling and exhaust systems aren't shown. Hot gas from the gas turbine replaces burning fuel in the steam boiler.

SOLUTION

This is a straightforward calculation of thermodynamic efficiency limits from Equation 4.5. First, though, we need to convert all temperatures to absolute units—that is, kelvins. Adding the 273 K difference between degrees Celsius and kelvins, we have a highest temperature of 1,450 + 273 = 1,723 K, an intermediate temperature of 500 + 273 = 773 K, and a lowest temperature of 7 + 273 = 280 K.

The thermodynamic efficiency limit is determined by the second law alone, through Equation 4.5; the details of the heat engine don't matter. So the thermodynamic limit—the maximum possible efficiency for the combined-cycle plant—is

$$e_{\text{max combined}} = 1 - \frac{T_c}{T_h} = 1 - \frac{280\text{ K}}{1{,}723\text{ K}} = 0.84 = 84\%$$

Friction and other losses would reduce this figure substantially, but a combined-cycle plant operating at these temperatures might have a practical efficiency near 60%. Meanwhile, the gas turbine and steam-cycle sections alone have efficiency limits given by

$$e_{\text{max gas}} = 1 - \frac{T_c}{T_h} = 1 = \frac{773\text{ K}}{1{,}723\text{ K}} = 0.55 = 55\%$$

and

$$e_{\text{max steam}} = 1 - \frac{T_c}{T_h} = 1 = \frac{280\text{ K}}{773\text{ K}} = 0.64 = 64\%$$

Thus the efficiency of the combined cycle is greater than that of either part alone. In general, the combined-cycle efficiency can be written $e_{\text{max combined}} = e_{\text{max gas}} + (1 - e_{\text{max gas}})e_{\text{max steam}}$, as you can show for the numbers here and prove in general in Exercise 5.

5.5 Fossil Fuel Resources

Where are the fossil fuels, and how large is the fossil energy resource? We've seen that fossil fuels began forming hundreds of millions of years ago under fairly common circumstances, primarily acidic swamps for coal and ancient seas for petroleum. But geological conditions resulting in significant, economically recoverable concentrations of fossil fuel are less common. Consequently, the distribution of fossil fuel resources around the globe is quite uneven—a fact that has blatantly obvious economic and geopolitical consequences.

Before we proceed, I need to clarify some vocabulary: The term **resource** refers to the total amount of a fuel in the ground—discovered, undiscovered, economically recoverable, or not economically recoverable. Obviously, estimates of fossil fuel resources are necessarily uncertain, because they include fuel we haven't yet discovered. **Reserves**, in contrast, describe fossil fuels that we're reasonably certain exist, based on geological and engineering studies, and that we can recover economically with existing technology. As prices rise and technology improves, reserves may grow even without new discoveries, because it becomes feasible to exploit what were previously uneconomical sources of fossil fuels.

COAL RESOURCES

Coal is the most widely distributed and most abundant of the fossil fuels. Nevertheless, some countries have much more coal than others, as Figure 5.16 shows. The United States is the leader, with more than one-fourth of the world's known coal reserves. Other major coal reserves are in Russia, China, Australia, and India. Total world reserves amount to some 10^{12} tonnes, about half of which is the higher-quality bituminous and anthracite coal. Energy content varies substantially with coal quality, but a rough conversion suggests that the world's 10^{12}-tonne reserve is equivalent to about 25,000 exajoules (EJ) of energy or, again roughly, 24,000 quads (Q) (see Exercise 6).

Much of the world's coal reserves lie deep underground, and this coal is extracted using traditional mining methods. Where coal is closer to the surface, strip mining removes the surface layers to expose coal seams. Both deep mining and strip mining have significant environmental impacts, which I'll discuss in Chapter 6.

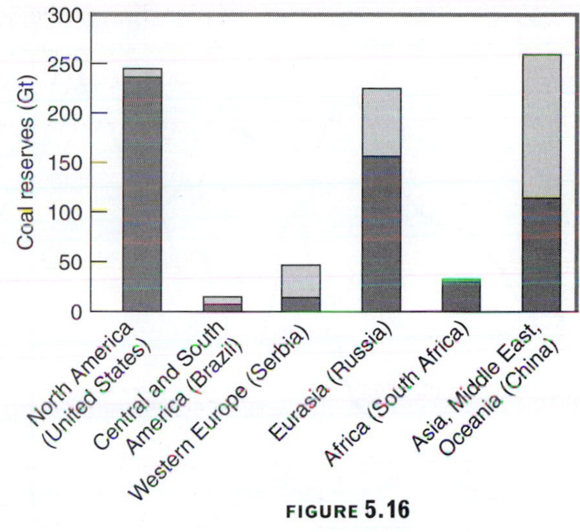

FIGURE 5.16
World coal reserves in gigatonnes (billions of metric tons). The full height of each bar gives the total coal reserves for the indicated continent, while the lower part shows reserves for the listed country, which has the most coal in that continent.

OIL RESOURCES

Citizens of today's world, and especially of the United States, cannot help but be acutely aware of the uneven global distribution of petroleum reserves. Figure 5.17 shows that the politically volatile Middle East holds some 64% of the world's crude oil; Saudi Arabia alone accounts for almost a quarter of the world's known conventional oil reserves. The United States, by far the largest oil consumer with more than one-fifth of total world oil consumption, has less than 2% of the global reserves. That discrepancy, along with ever-increasing demand, explains why U.S. dependence on foreign oil sources has grown dramatically in recent decades, from less than 10% of total petroleum consumption in 1950 to nearly 70% today (Fig. 5.18).

World oil reserves most likely total somewhat over a trillion barrels, with authoritative estimates in the range of 1.2 to 1.4 trillion barrels. You might contemplate that figure in light of a world oil consumption rate of about 30 billion barrels per year, rising at more than 1% per year. As you can show in

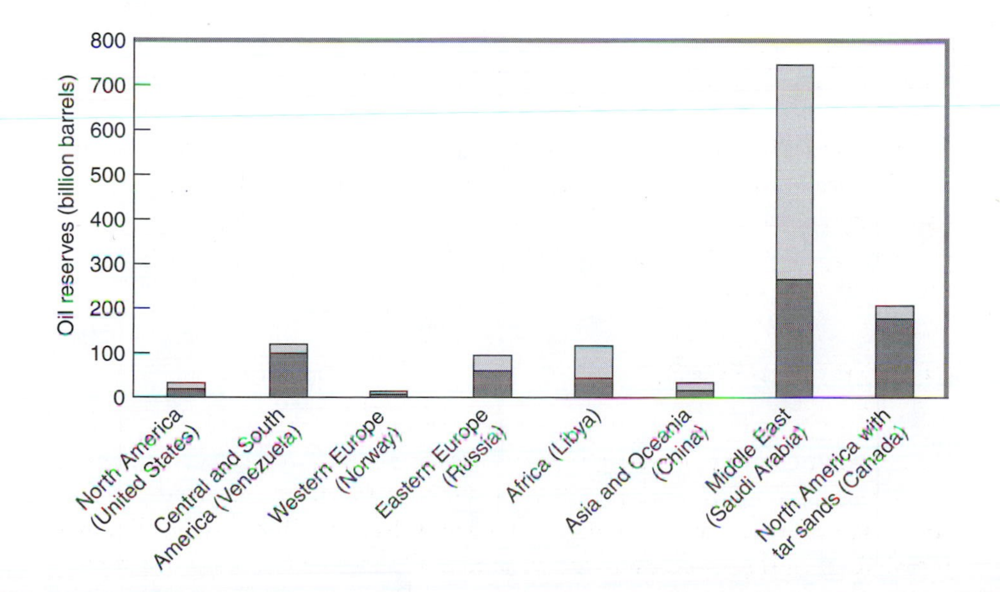

FIGURE 5.17
World oil reserves, presented as for coal in Figure 5.16, with the Middle East broken out separately. North America is shown twice: The United States has the largest conventional reserves in North America (left), but if the Canadian tar sands are included, Canada has almost as much oil as Saudi Arabia (right).

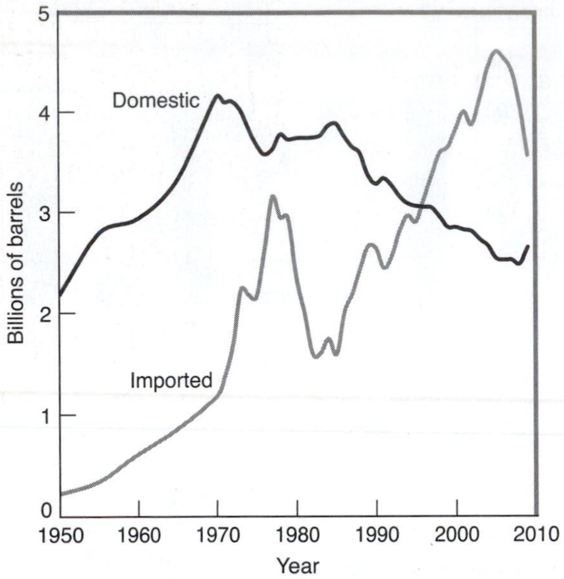

FIGURE 5.18

U.S. oil imports have risen substantially to compensate for declining domestic production. The two large drops are the result of oil supply disruptions, price increases, and economic recessions, including the Great Recession of 2008–2010.

Exercise 7, a trillion-barrel reserve translates into a little over 6,000 EJ, or a little under 6,000 Q. It's instructive to consider this figure in light of the world's total energy consumption rate of, very roughly, 10^{13} W (see Exercise 8).

Oil extraction technology has evolved substantially since Drake struck Pennsylvania oil at a mere 69 feet in 1859. By the 1930s, the deepest wells went straight down some 3 km (about 2 miles), and today some wells exceed 10 km (about 7 miles). An equally significant advance is the development of so-called directional drilling, which allows drills to penetrate rock vertically, horizontally, and diagonally to tap multiple oil deposits from a single wellhead (Fig. 5.19). By the middle of the twentieth century, oil drilling had begun to move offshore. Today more than 30% of the world's oil comes from offshore drilling stations, including platforms mounted on the ocean floor as well as on floating structures. Today's offshore rigs can work in waters approaching 2 miles deep, below which their drill strings may penetrate several additional miles (Fig. 5.20).

NATURAL GAS RESOURCES

Most natural gas occurs in conjunction with crude oil, so its distribution is similar to that of oil. However, quantities of oil and gas don't necessarily correlate; thus Russia has far and away the world's largest gas reserves, followed

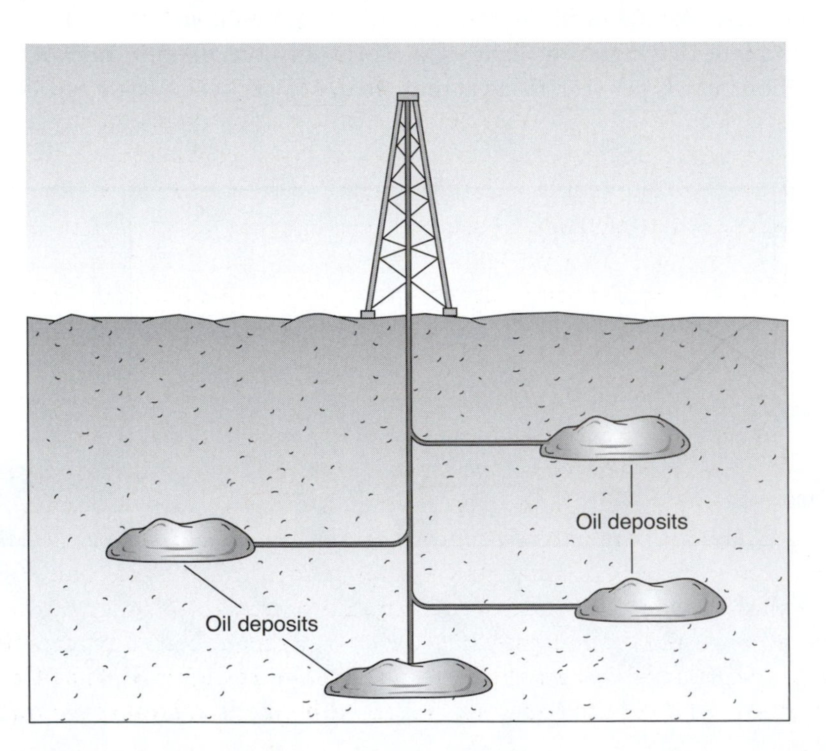

FIGURE 5.19

Directional drilling taps multiple oil deposits from a single wellhead. This technology not only produces more oil, but also reduces the environmental impact at the surface.

by Iran and Qatar. Saudi Arabia, the giant of oil reserves, is comparable to the United States in gas reserves.

Being gaseous, natural gas is harder to handle and transport than liquid oil. But for environmental reasons and because of its availability, gas is playing an ever-larger role in the world's energy supply. In the United States, natural gas is commonly measured in cubic feet; in this unit, the world's reserves are estimated to be around 6 quadrillion cubic feet, which amounts to 6,000 EJ, or 6,000 Q—roughly the same energy content as the world's oil reserves (see Exercise 9).

FIGURE 5.20
Offshore drilling rigs produce some 30% of the world's oil.

UNCONVENTIONAL FOSSIL RESOURCES

There's a lot more fossil fuel than what's included in the proven reserves of coal, oil, and natural gas; most of it is oil and gas locked in geological structures where extraction has not been technologically or economically feasible. But as conventional supplies run short and thus become more expensive, and as technology advances, the world will inevitably turn to these unconventional fossil resources.

I began this chapter with a description of fossil fuel formation through a variety of physical and chemical processes, and I pointed out that today's economically recoverable oil occurs when liquid petroleum migrates and concentrates in geological "traps." But much of the fuel remains more dispersed, sometimes at an early stage in the formation process, in the very rocks where it originated. **Oil shale** is rock that contains a significant amount of kerogen, the waxy substance that represents a step on the way to liquid crude oil. Shales that are particularly rich in kerogen can burn directly and have seen limited use as fuels for centuries. To produce true oil from oil shale involves pulverizing the rock and applying heat and pressure to complete what nature has left undone. The process is expensive and, because huge amounts of rock are involved, it has major environmental impacts. However, the oil shale resource is extensive and not heavily concentrated in one part of the world. The United States alone may have over 1 trillion barrels of oil locked in oil shale, nearly equal to the proven world reserves of liquid crude oil. However, the energy content isn't equivalent, because large amounts of energy go into processing the oil shale. Oil shale might become a significant contributor to our fossil fuel supply, but only if we can surmount numerous technological, economic, and environmental issues.

Tar sands are another unconventional fossil fuel resource. These are sand deposits into which liquid oil has leaked. Over time, the lighter components have evaporated, leaving a heavy tar. The Canadian province of Alberta is home to a large portion of the world's known tar sands, equivalent to some 1.6 trillion barrels of oil, more than the world's reserves of liquid crude. With some 10% or more of this resource considered economically extractable, mining operations in Alberta are already underway (Fig. 5.21). Including Alberta's extractable tar sands dramatically changes the oil-reserve picture in Figure 5.17, as Canada becomes second only to Saudi Arabia. As with oil shale, however, extracting

(a)

(b)

FIGURE 5.21
(a) Tar sands from Alberta, Canada. (b) It takes 2 tons of sand to make a barrel of oil. This shovel suggests the enormous scale of tar-sand extraction.

liquid oil from tar sands is an expensive, energy-intensive, time-consuming, and environmentally damaging process.

Natural gas, too, has unconventional sources. The process of *hydrofracking* uses high-pressure fluids to fracture deep rock formations, permitting extraction of otherwise unaccessible natural gas. Expansion of hydrofracking operations is currently generating controversy in the United States. Gas also occurs in coal seams, where it's responsible for explosive mining accidents. And under extreme pressures in the deep oceans, methane molecules become trapped in icelike crystals to form **methane clathrates**. The amount of this trapped methane could be 100 times conventional gas reserves, but at present we lack the technology to extract or even locate this fuel resource. Methane clathrates can also provoke sudden climate change by releasing large amounts of the greenhouse gas methane into the atmosphere.

5.6 When Will We Run Out?

Fossil fuels take hundreds of millions of years to form and we're using them at a vastly greater rate than they're forming. It's clear we'll run out sometime, but when?

The simplest way to estimate that time is to divide a fossil reserve by the rate at which we're using it, yielding a time known as the **reserve/production (R/P) ratio**. But this approach is inaccurate for several reasons. First, known reserves don't account for as-yet undiscovered resources. Second, even if there were no undiscovered resources, economic and technological factors would continue to expand the reserve as the cost of the increasingly scarce fossil fuel rises and, for example, marginal oil fields or unconventional sources such as oil shale and tar sands become economically viable. On the other hand, our fossil fuel consumption rate is growing, so using today's rate overestimates the remaining supply. Finally, the real question isn't when we'll run out completely (when we pump the last drop of oil from the ground) but when and whether production of fossil fuels begins to drop even as demand continues to rise.

EXAMPLE 5.4 | Running Out: A Rough Estimate

Use Figure 5.6 to estimate the current rate of world oil consumption, expressed in power units. Use that result in conjunction with known oil reserves to estimate the R/P ratio, which provides a very rough estimate of the time remaining until those reserves run out.

SOLUTION

Figure 5.6 shows a global oil consumption rate close to 4 Gt oil equivalent per year; with Table 3.1's conversion factor of 41.9 GJ per tonne oil equivalent, that works out to be about 170 EJ per year. With world oil reserves at some 6,000 EJ, this means we have about

$$\frac{6{,}000\,\text{EJ}}{170\,\text{EJ/year}} = 35 \text{ years}$$

until we run out.

Example 5.4 neglects future growth in oil consumption. Of course, no one can predict the future, but we can extrapolate from past experience, which shows world oil consumption increasing since the mid-1990s at somewhat over 1% per year. This increase means that our 35-year result in Example 5.4 may be an overestimate of how long it will take to consume all known oil reserves. When we account for the mathematics of exponential growth (see Box 5.2 and Exercise 11), we find that we only have about 27 years of oil left, if the current percentage growth rate holds.

BOX 5.2 | Exponential Growth

One day, a gardener noticed a lily pad growing in her garden pond. The next day there were two lily pads, the following day four, and so forth, with the number doubling each day. At first the lily pads covered only a small portion of the pond, but as their incessant doubling continued, the gardener grew alarmed. When the pond was half covered, she decided to take action. At that point, how much time did she have before the pond would be covered?

The answer, of course, is one day.

This story illustrates the power of exponential growth—growth not by a fixed amount but by a fixed percentage of what's already there. In the case of the lily pond, the rate is 100% per day, meaning that each day the total population of lily pads increases by 100% of what there was before.

Many quantities increase exponentially. Money in the bank earns interest at a fixed percentage of what's already there. As your money grows, so does the actual amount of interest you receive. Put a

(continued)

BOX 5.2 | Exponential Growth (Continued)

bacterium in a petri dish with all the food it needs, and soon you'll have 2, then 4, 8, 16, 32, 64, and so on. For centuries the growth of the human population could be described as exponential—although only approximately, because the percentage rate actually increased until the 1960s, after which it has fallen. And the growth in world energy consumption, or in the consumption of a given fossil fuel, has often been approximately exponential.

Unchecked, exponential growth always leads to huge values (Fig. 5.22), but it can't go on forever. Shortages of resources—money, food, oil, whatever—eventually halt the exponential growth and result in a transition to slower growth, decline, or even collapse of the numbers that were once growing so rapidly. Because it starts slowly and then increases dramatically, exponential growth can "sneak up" on us and leave little time for an orderly transition to the postexponential phase. That's our lily-pond gardener's plight, and it's one of the big concerns whenever resource use grows exponentially.

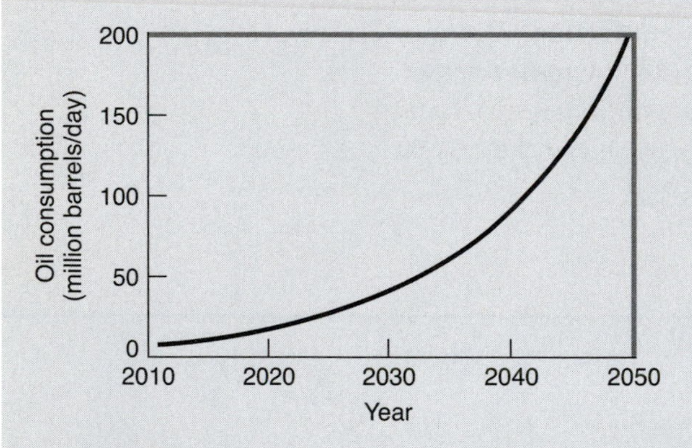

FIGURE 5.22
China's 8% annual increase in oil consumption is an example of exponential growth. If this rate were to hold steady through 2050, China's current consumption of roughly 9 million barrels per day would grow nearly 30 times, to more than 200 million barrels per day—more than double the current *global* daily consumption rate.

Mathematically, exponential growth is characterized by a function with time in the exponent. If at time $t = 0$, we start with a quantity N_0, whether it be dollars or people or barrels of oil per year, and if that quantity grows at some rate r, specified as the fraction by which the quantity increases in whatever time unit we're using, then at a later time t we'll have a quantity N given by

$$N = N_0 e^{rt} \tag{5.1}$$

Here e is the base of natural logarithms, with an approximate value of 2.71828 . . . , but this value isn't important here because your scientific calculator has the exponential function built in.

When exponential growth describes resource consumption, and therefore N represents the consumption rate, then the calculus-savvy can show that the total amount of the resource used up in some time t—the cumulative total of the increasing yearly consumption—is given by

$$\text{Cumulative consumption} = \frac{N_0}{r}(e^{rt} - 1) \tag{5.2}$$

where N_0 is the initial consumption rate. I used Equation 5.2 in making that 27-year estimate for the time remaining until oil might run out. I set the annual growth rate to 2%—that is, $r = 0.02$, typical of the recent past (see Research Problem 7). I used $N_0 = 170$ EJ per year from Example 5.4 as the initial oil-consumption rate, and I solved for the time t when the cumulative oil consumption reaches the 6,000 EJ contained in the total reserves. Exercises 11 through 16 explore exponential growth in the context of fossil fuel reserves and include the derivations of Equation 5.2 and of an explicit expression for the time.

But again, that's not the whole story. As the supplies dwindle, oil will get very expensive and we'll surely reduce our consumption. Reserves may last a bit longer, or the actual reserve may be larger, making that 27-year figure an underestimate. But near the end, whenever it comes, we'll be in a real energy crisis if we haven't found an alternative to oil. "By then" is not an abstract concept; it means something on the order of a few decades from now, and if you're a typical college student, it means the prime of midlife for you. This is your problem, not something for later generations to worry about.

HUBBERT'S PEAK

So the issue isn't when the last drop of oil disappears, but when we begin to feel the pinch of dwindling reserves. In 1956 Shell Oil Company geophysicist M. King Hubbert made a remarkable prediction: Oil production in the United States, then the world's leading oil producer, would peak in the late 1960s or early 1970s. Oil companies, government agencies, and economists dismissed Hubbert's prediction, but he was right. Oil production in the United States peaked in 1970 and has been declining ever since.

Hubbert based his estimate on a simple idea—that the extraction of a finite resource such as oil starts out slowly, rises with increasing demand, eventually levels off as the finite resource becomes harder to extract, and then declines. In short, production should follow, roughly, a bell-shaped curve (Fig. 5.23). This curve depicts the rate of oil extraction in, say, billions of barrels per year. The total amount of oil extracted up to a given time is the area under the curve from the starting point to the time in question (if you've had calculus, you'll recognize this as a statement about the definite integral). Given a perfectly symmetric bell-shaped curve, we reach the peak—**Hubbert's peak**—when we've used up half of the oil reserve.

Now here comes the crunch: Oil consumption has been rising more or less steadily for decades, with no obvious end in sight. Production has been rising, too, to keep up with demand. Although some oil is stored for future use (as in the U.S. Strategic Petroleum Reserve; see Research Problem 1), the amount going into and out of storage is small, so in practice we use oil at essentially the same rate at which it's pumped from the ground. But after Hubbert's peak, production necessarily declines. In the case of U.S. production, which peaked in 1970, we've made up the increasing difference between production and consumption with ever-rising oil imports (recall Fig. 5.18). But globally, the world's oil reserves are all we have. Unless we can begin to slow our global oil consumption well before the world reaches its Hubbert's peak, we may be in big trouble.

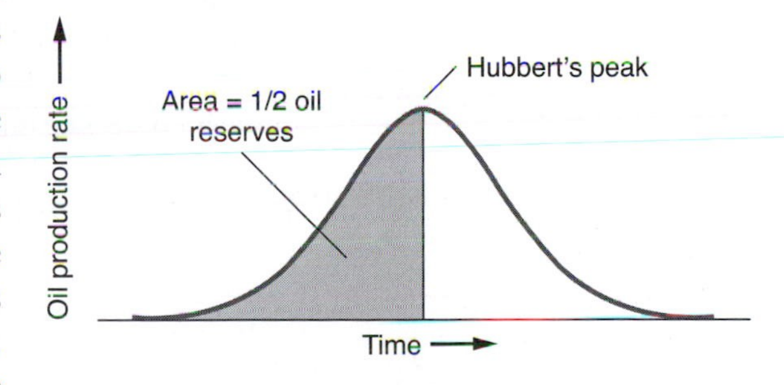

FIGURE 5.23
An idealized bell-shaped curve for oil production known as Hubbert's peak. Production peaks when half the resource has been exhausted; thereafter, the production rate declines as the remaining oil reserves become more difficult and expensive to extract.

So when do we reach the global Hubbert's peak for oil? Serious estimates range from a few years hence to about 2040 (Fig. 5.24). A few pessimists believe we're already at or even past the peak now, and some optimists think that technological advances and new oil discoveries will push Hubbert's peak

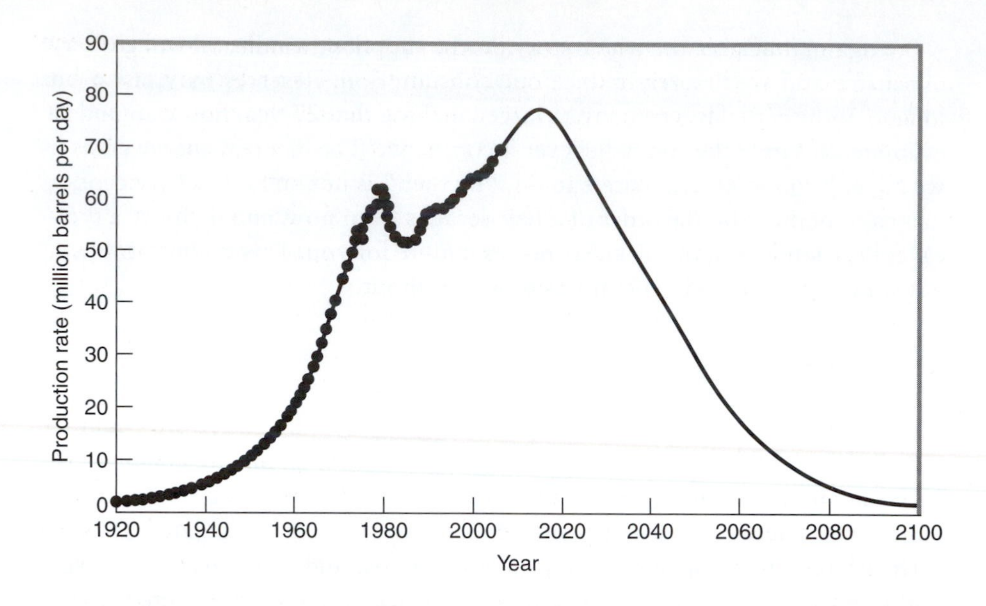

FIGURE 5.24
One of the more sophisticated projections of future oil production is a 2010 study from Kuwait University's College of Engineering and Petroleum; it has world production peaking around 2014. The study analyzed 47 oil-producing countries and incorporated multiple Hubbert-like models to reflect changes in oil-extraction technology, government regulations, and economic changes. Black dots represent actual data; solid curve is the model projection.

beyond the turn of the next century. Hubbert's symmetric bell-shaped curve is only an approximation, and peak estimates are sensitive to the exact form one assumes. There's also the question of the ultimate oil reserve; the U.S. Geological Survey, for example, estimates that actual reserves may be two to four times that trillion-barrel or 6,000 EJ figure I've been using. But others point to an ongoing decline in new oil discoveries as evidence that the currently known reserves may be all we have. Those variations make less difference than you might think, as you can show in Exercise 12. In any event, it's fair to say that a broad consensus has the crunch coming in a matter of a few decades.

The world's dependence on oil has created a huge economic and technological infrastructure dedicated to oil production, supply, and consumption. Think of all the gasoline stations, all the oil-heated homes, all the jet airplanes, all the lubricating oils used in the machinery of the world's factories, all the tankers plying the seas to deliver imported oil to all the refineries, and you get the picture of a vast oil enterprise. There's enormous inertia associated with that enterprise, and it's going to be very difficult to turn it around on a timescale of mere decades.

THE CORNUCOPIAN VIEW

Not everyone subscribes to the impending crunch implicit in Hubbert's idea. So-called **cornucopians** represent the extreme opposite in a wide range of opinions about the future of oil and other finite resources. Cornucopians believe that human ingenuity will remain always a step ahead of Hubbert's peak, continuing to find new reserves, advancing technologies to make currently uneconomical resources viable, or sliding smoothly to an economy based on alternative fuels or altogether new energy sources. In this view, the finite nature of a given resource is simply irrelevant.

Cornucopians point to past energy predictions that have been wildly off the mark. The 35-year R/P ratio we found in Example 5.4 sounds scary, but a cornucopian would note, correctly, that if we had calculated this ratio in 1980, the result would have been just over 25 years, meaning that we would now be out of oil. In 1945 it looked like we had just 20 years of oil left!

What happened? How did we go through decades of oil consumption and end up with what looks like a longer oil future? We discovered more oil, expanding the known reserves. We improved extraction technologies, making new sources accessible and enhancing the yield from older wells. And despite our SUV craze and other energy follies, we became more efficient in our energy use. Energy intensity in the United States, for example, dropped by nearly half between 1980 and 2010, as I showed in Figure 2.8. Although our total annual energy consumption increased substantially in that time, the increase was a lot less than it would have been without efficiency improvements.

So is Hubbert's peak an alarmist notion? Not in the context of Hubbert's correct prediction that U.S. oil production would peak in the early 1970s. And there's good evidence that the rate of new oil discovery is dropping significantly. The issue may not be so much whether the idea of Hubbert's peak is valid or not, but the time until we reach that peak if we stay our present course. If that time is long, then there's a reasonable chance of making a smooth transition to a different energy regime and thus avoiding the peak altogether. But if the time is short—and a few decades is probably short on the timescale needed to change our vast energy infrastructure—then Hubbert's peak really does represent a looming energy crunch.

COAL AND NATURAL GAS SUPPLIES

How about coal and natural gas? As with any finite resource, their production, too, might experience a Hubbert's peak. For gas, this will probably happen a decade or two after oil's peak. Exactly when depends in part on how vigorously the world tries to curtail its carbon emissions by substituting natural gas for coal.

Coal is a different story. A simple calculation like the one we did for oil in Example 5.4 yields several hundred years of remaining supply (see Exercise 10). But growth in coal consumption, which we might expect as oil and gas supplies falter, could reduce that number. Also recall that Hubbert's peak comes when we've extracted only half the world's coal, not when we're down to the last chunk. But still, with coal we don't face the urgency we do with oil; the timescale to peak is a century or more, compared with a few decades. Furthermore, resource limitations may turn out to have little relevance in the case of coal, because coal's deleterious environmental impacts—especially on climate—may prove to be the dominant factor limiting our use of this abundant fossil fuel. In that case, Hubbert's peak would be irrelevant for coal, much of which may remain forever in the ground.

5.7 Policy Issue: Carbon Tax or Cap-and-Trade?

Clearly, it's in our best interest to wean ourselves gradually from fossil fuels. This conclusion follows obviously from the preceding section, and it will be reinforced when we look at the environmental impacts of fossil fuels in Chapter 6 and at climate change in Chapters 12 to 16. But how are we to reduce fossil fuel consumption in a world where energy use is closely tied to our standard of living (recall Fig. 2.7), and where the developing world seeks to emulate the developed world's material well-being?

One approach is to use economic incentives to reduce fossil fuel consumption by encouraging greater energy efficiency and a switch to alternative energy sources. Some argue that this represents tampering with "free markets," but today's price for fossil fuels might not qualify as "free market" because it doesn't account for the considerable damage these fuels do to our environment, health, and even economy. An example of an economic incentive is the **carbon tax,** which would be levied on fuels based on the amount of carbon each fuel puts into the atmosphere. The carbon tax is often proposed as a way of reducing the climate impact of fossil fuel combustion, because it would encourage shifts from coal and oil to natural gas, as well as away from fossil fuels altogether. But unless it's designed carefully, a carbon tax could be regressive, having a greater impact on low-income consumers than on the well-to-do. Proponents of a carbon tax have suggested redistributing some of the tax revenue based on income, but that goal may conflict with another: using carbon-tax revenues to fund research into nonfossil energy sources.

Although carbon-based taxation is a new and relatively untried idea, taxes on fuel itself have long been with us. European countries, in particular, use the gasoline tax as an instrument of environmental policy, explicitly encouraging lower fuel consumption, minimizing environmental impacts, and guiding consumers to fuel-efficient vehicles. The gasoline tax is one big reason why, as I noted earlier, half the new cars sold in Europe are fuel-efficient diesels. European gas taxes are considerable, amounting to some 60% to 80% of the total fuel cost. For the United States, that number is only about 30%. Figure 5.25 compares gasoline prices throughout the world. An increase in the low U.S. gasoline taxes, as sensible as it would seem, has been politically unachievable for many years. The last increase, in 1993, amounted to a mere 4.3 cents—less than the typical variations in gasoline prices from week to week.

An alternative to the carbon tax is a so-called **cap-and-trade** scheme, in which a governing body sets limits on carbon emissions, then issues allowances to individual emitters that permit a certain amount of carbon emissions. These allowances can either be granted free at the outset or auctioned to raise money to be used, for example, for research into energy alternatives. If a company produces less carbon than it's allowed, it can sell its allowances on an open market. Thus the cap-and-trade scheme provides a market-based incentive to reduce emissions. The European Union has operated a mandatory carbon market since 2005, and today you can track the price of carbon allowances just like any

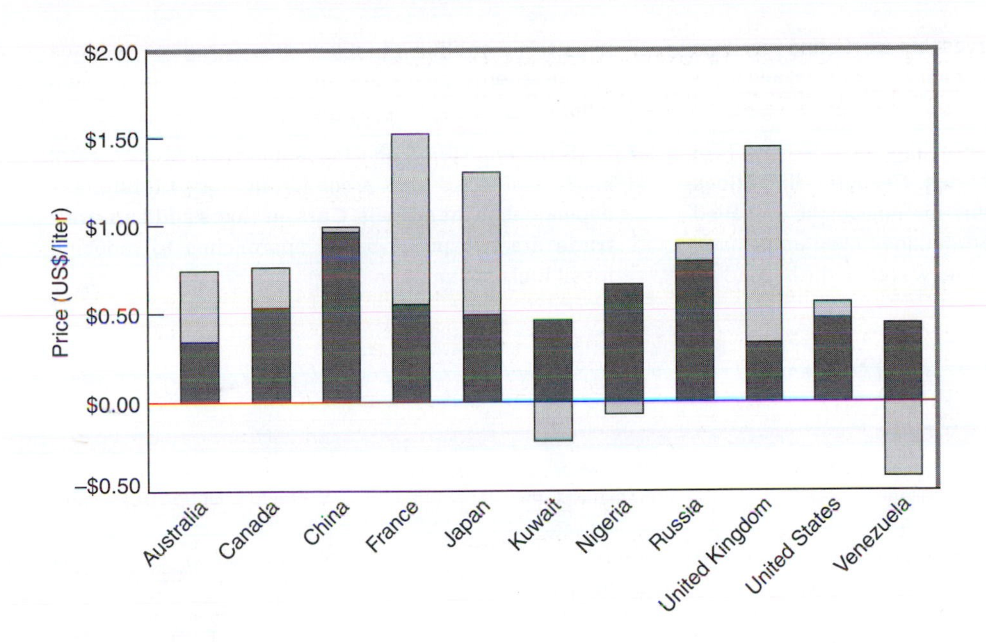

FIGURE 5.25
Gasoline prices vary dramatically as a result of different tax policies. European countries have the highest prices because of taxes intended to encourage energy efficiency. The dark portion of each bar shows the basic gasoline cost; the lighter area indicates taxes. Some major oil producers subsidize gasoline rather than taxing it, hence the negative taxes for Kuwait, Nigeria, and Venezuela.

other commodity or stock. In the United States, the Chicago Climate Exchange operates a similar market, but participation is voluntary. Research Problem 8 explores the cost of carbon allowances on the European market and the Chicago Climate Exchange.

Should we increase the gasoline tax? Should we adopt a carbon tax? Should we require mandatory participation in a cap-and-trade market? Those are questions of policy, not science. But an intelligent answer requires understanding the science behind fossil fuels—their origin, their uses, their availability, and their environmental impact. This chapter discussed the first three of these; Chapter 6 focuses on the last.

CHAPTER REVIEW

BIG IDEAS

5.1 **Coal**, oil, and natural gas are fossil fuels that contain stored solar energy captured by plants hundreds of millions of years ago.

5.2 Coal is mostly carbon, while oil and gas are hydrocarbons, containing a mix of carbon and hydrogen. We **refine** fossil fuels into products ranging from heavy oils to jet fuel to gasoline to the feedstocks for making plastics.

5.3 Burning fossil fuels produces largely CO_2 and water vapor. For a given amount of energy, coal produces the most CO_2 and natural gas the least.

5.4 We extract energy from fossil fuels with **external combustion engines** such as electric power plants and with **internal combustion engines** for transportation. Their efficiencies are limited by the second law of thermodynamics and by design factors. Modern technologies such as **gas–electric hybrid** vehicles and **combined-cycle power plants** provide substantial efficiency gains.

5.5 Fossil fuel **resources** are limited. **Reserves** describe the amount of fuel we're reasonably confident is in the ground. Unconventional sources such as **tar sands**

might increase fossil reserves, but extracting fuel from these sources is expensive, energy intensive, and environmentally disruptive.

5.6 Petroleum reserves will probably last a matter of decades, coal a few centuries. Resource limitations become severe when production peaks (the so-called **Hubbert's peak**) and then declines even as demand continues to increase. Whether we reach the impending peak crisis or slide into a new energy regime depends on our energy policy and the prospect of increasing reserves.

5.7 Both the impending shortages and the environmental impacts of fossil fuels argue for an energy future less dependent on these fuels. **Carbon taxes** and **cap-and-trade** are two policy-based approaches to reducing fossil fuel consumption.

TERMS TO KNOW

cap-and-trade (p. 122)
carbon tax (p. 122)
coal (p. 95)
combined-cycle power plant (p. 110)
compression-ignition engine (p. 106)
condenser (p. 103)
continuous combustion engine (p. 105)
cooling tower (p. 100)
cornucopian (p. 120)
crankshaft (p. 105)
crude oil (p. 97)
electric vehicle (p. 108)
fractional distillation (p. 97)
gas–electric hybrid (p. 107)
gas turbine (p. 109)
Hubbert's peak (p. 119)
intermittent combustion engine (p. 105)
kerogen (p. 96)
methane clathrate (p. 116)
natural gas (p. 97)
oil shale (p. 115)
parallel hybrid (p. 107)
peat (p. 95)
petroleum (p. 96)
plankton (p. 96)
plug-in hybrid (p. 108)
refine (p. 97)
regenerative braking (p. 107)
reserve (p. 112)
reserve/production (R/P) ratio (p. 116)
resource (p. 112)
series hybrid (p. 107)
spark-ignition engine (p. 105)
tar sands (p. 115)
thermal power plant (p. 104)
turbine (p. 103)

GETTING QUANTITATIVE

Coal energy content: averages 29 MJ/kg, ~10,000 kWh per ton

Energy content of petroleum products: ~45 MJ/kg, ~40 kWh/gallon

CO_2 emissions from 1-GW coal-fired power plant: ~260 kg/s, ~500 pounds per second

Mass ratio, CO_2 to carbon: 44/12, or 3.67

Efficiency of typical gasoline engine: 20%, fuel energy to mechanical energy

Efficiency of typical coal-fired power plant: 35%, fuel energy to electricity

Efficiency of typical combined-cycle power plant: 60%, fuel energy to electricity

Coal reserves: 10^{12} tonnes, ~25,000 EJ or ~25,000 Q

Oil reserves: ~10^{12} barrels, ~6,000 EJ or 6,000 Q

Natural gas reserves: ~6,000 EJ or 6,000 Q

World energy consumption rate: ~10^{13} W

$\text{R/P ratio} = \dfrac{\text{reserves}}{\text{consumption rate}}$ (units of time)

Exponential growth: $N = N_0 e^{rt}$ (Equation 5.1; p. 118)

Cumulative consumption: $\dfrac{N_0}{r}(e^{rt} - 1)$ (Equation 5.2; p. 118)

QUESTIONS

1. Explain, in terms of basic chemistry, why natural gas produces less CO_2 per unit of energy generated than do the other fossil fuels.
2. Fossil fuels are continuously forming, even today. So should we consider coal, oil, and natural gas to be renewable energy sources? Explain.
3. Explain how combined-cycle power plants manage to achieve a high thermodynamic efficiency.
4. Why can't a clever engineer design a heat engine with 100% efficiency?
5. In the context of fossil fuels, what's the difference between reserves and resources?
6. How is it possible for oil reserves to increase without the discovery of any new oil fields?
7. In the cornucopian view, the finite nature of a natural resource need not be an issue. How is this possible? Back up your answer with the example of coal.
8. How does the occurrence of Hubbert's peak for U.S. oil production in 1970 relate to the increasing U.S. dependence on imported oil?

EXERCISES

1. Complete combustion of a single octane molecule results in how many molecules of CO_2 and how many of H_2O? How many oxygen (O_2) molecules are needed for this combustion?
2. Here's the analog of Example 5.2 for a natural gas power plant: Consider a gas-fired power plant rated at 600 MWe and having 44% efficiency. Natural gas is supplied via pipeline. What rate of gas flow, in cubic meters per day, is required?
3. A combined-cycle power plant operates with a maximum temperature of 1,080°C in its gas turbine, then dumps waste heat at 10°C from its steam turbine. What is the thermodynamic efficiency limit for this plant?
4. A hypothetical (and impossibly perfect) combined-cycle power plant operates at its thermodynamic efficiency limit of 81%. It dumps waste heat from its steam cycle at 5°C. (a) What is the maximum temperature in its gas turbine? If the gas turbine stage alone has 63% thermodynamic efficiency, what are (b) the intermediate temperature and (c) the thermodynamic efficiency of the steam cycle alone?
5. Show that the efficiency limit for a combined-cycle power plant is given by $e_{\text{max combined}} = e_{\text{max gas}} + (1 - e_{\text{max gas}})e_{\text{max steam}}$, where $e_{\text{max gas}}$ and $e_{\text{max steam}}$ are the efficiency limits for the gas turbine and steam cycles alone.
6. World coal reserves amount to about 520 Gt (gigatonnes) of anthracite and bituminous coal with an average energy content of about 32 MJ/kg, and about 470 Gt of lower-grade coal with an average energy content of about 18 MJ/kg. Verify the text's figure of about 25,000 EJ for the energy equivalent of the world's coal reserves.
7. Use Table 3.1 to show that a world oil reserve of 1 trillion barrels contains roughly 6,000 EJ or 6,000 Q of energy.
8. World oil reserves amount to about 6,000 EJ, and humankind uses energy at the rate of roughly 10^{13} W. If all our energy came from oil and assuming there was no growth in energy consumption, estimate the time left until we would exhaust the 6,000-EJ reserve.
9. Verify the statement made in the text that the world's natural gas reserve of some 6 quadrillion cubic feet has an energy content of about 6,000 EJ.
10. Use this chapter's estimate of coal reserves, along with current world coal consumption of some 6 Gt per year, to make a simple estimate of the time remaining until coal runs out.
11. (a) Use Equation 5.2 to verify my 27-year estimate for the time that proven oil reserves will last with a 2% annual growth in world oil consumption. How would the time estimate change if the growth rate (b) increased to 3%, (c) dropped to 1%, or (d) became negative at −0.5%, indicating a decline in consumption? (Equation 5.2 still applies in this case.)
12. A rough figure for proven world oil reserves is the 1 trillion barrels (10^{12} barrels or 6,000 EJ) that I gave in the text, which led to Example 5.4's simple estimate of

35 years' remaining oil supply if there's no change in the world oil consumption rate. However, the U.S. Geological Survey estimates that recoverable oil reserves may actually be somewhat more than double that figure. Approximating the figure as 2 trillion barrels or 12,000 EJ would obviously double the simple estimate of 35 years. But what would it do to my 27-year estimate, which assumes a 2% annual growth in oil consumption? Study Box 5.2 for help with this exercise.

13 China consumes oil at the rate of about 9 million barrels per day, increasing by 8% annually. The United States consumes about 20 million barrels per day, currently with little change from year to year. If these trends hold, how long before China overtakes the United States in oil consumption?

14 Show that Equation 5.2 implies that the time to reach a cumulative consumption C is

$$t = \frac{\ln (rC/N_0 + 1)}{r}$$

15 If you've had calculus, derive Equation 5.2. Remember that N here is the annual rate of consumption (which is a function of time) and that you can get the total consumption from time 0 to time t by integrating the consumption rate over this time interval.

16 Show that Figure 5.22's assumptions about exponential growth in China's oil consumption would result in China's having consumed roughly the world's entire 1-trillion-barrel oil reserve by 2050.

RESEARCH PROBLEMS

1 How much oil is in the U.S. Strategic Petroleum Reserve? For how much time could the reserve supply the oil we're now importing from other countries?

2 Research advanced combined-cycle power plant designs and find an example of a contemporary high-efficiency design. Give the manufacturer, the total power output, and the overall efficiency. From the highest and lowest temperatures, calculate the thermodynamic efficiency limit—the theoretical maximum for any heat engine operating between these temperatures. Compare this result with the actual overall efficiency.

3 Find the composition of a typical gasoline, listing the major chemical components and the percentage of each.

4 Find the latest edition of the U.S. Department of Energy's *Annual Energy Review* (available online). Under the "International Energy" section, find data you can use to produce a graph of world oil production versus time. Is Hubbert's peak evident yet?

5 Find as many different estimates as you can of current world oil reserves. List each, along with its source.

6 Use a spreadsheet or other software to fit a curve to the data found in Research Problem 4. If your software permits, use a Gaussian curve (the standard bell curve); if not, use a quadratic (a second-order polynomial). From either the graph or its equation (which your software should be able to provide), determine when the curve you've produced reaches its peak. This is a simple way to estimate Hubbert's peak for oil.

7 Consult the latest edition of the U.S. Department of Energy's *Annual Energy Review* (available online), and under the "International Energy" section, find data on world oil consumption. Plot the data from the early 1990s to the present, and fit an exponential function (which will involve the term e^{rt}, where t is the time in years). Compare the growth rate r of your exponential fit with the 2% figure used in Box 5.2.

8 Find the current prices of allowances to emit a ton (or tonne) of carbon on the European Carbon Exchange and the Chicago Climate Exchange.

ARGUE YOUR CASE

1 Your state's sole nuclear power plant is shutting down, and you're attending a hearing on sources of replacement power. Options on the table include building a coal-fired plant or a combined-cycle gas plant. Make your choice and formulate an argument to defend it.

2 A friend dismisses either a carbon tax or cap-and-trade scheme as just another example of government taking the people's hard-earned money. How do you respond?

Chapter 6

ENVIRONMENTAL IMPACTS OF FOSSIL FUELS

Fossil fuels dominate today's global energy supply, so it's important to understand the environmental impact of these fuels. Furthermore, fossil fuels are among the most environmentally deleterious ways we have to produce energy, so their impact on the environment is doubly significant.

In this chapter I interpret the term *environmental impact* broadly, to include not only such obvious factors as air pollution and landscape desecration, but also direct and indirect effects on human health, safety, and quality of life. However, I leave the details of the most significant global impact of fossil fuel consumption—climate change—for later chapters focusing on that big subject.

Environmental impacts result, obviously, from the combustion of fossil fuels and the associated release of combustion products to the atmosphere. Extraction, transportation, and refining of fossil fuels carry additional environmental and health implications. Even the thermodynamically mandated waste heat can affect local ecosystems.

6.1 What's Pollution?

Pollution refers to any substance introduced into the environment that has harmful effects on human health and well-being, or on natural ecosystems. With fossil fuels, pollutants can result during extraction; during refining, transport, and handling; in the release of thermodynamically mandated waste heat; and, most obviously, as a result of combustion.

Fossil fuels contain carbon and hydrogen, so burning them to release stored energy produces water and carbon dioxide. However, fossil fuels may contain other elements as well, and combustion often converts these into pollutants. For example, sulfur in coal and oil burns to form sulfur dioxide and related compounds that damage human health and the environment. Even pure hydrocarbons don't burn completely, and the result is a host of products of incomplete combustion that are often carcinogenic or otherwise harmful. Pollutants resulting from non-hydrocarbon contaminants or from incomplete combustion aren't essential to the energy-releasing process; in fact, with incomplete combustion, their presence indicates an operation that is less than fully efficient.

But what about carbon dioxide? It's different, because it's an *essential* product of fossil fuel combustion. If we're going to burn fossil fuels, then we're going to make CO_2. To use these fuels efficiently, in fact, we *want* to turn as much of their carbon as possible into CO_2. Because nearly 90% of humankind's energy comes from fossil fuels, we want to make a lot of CO_2! So, is CO_2 a pollutant?

For centuries carbon dioxide was considered a benign product of fossil fuel combustion; after all, the gas is nontoxic and its concentration in the atmosphere is normally too low to have any impact on air's breathability. But the reality of CO_2-induced climate change has altered that view. The deleterious effects of climate change—both present and future—are myriad and well documented, so from a climate-change standpoint CO_2 is indeed a substance that's harmful to human health and well-being, and to natural ecosystems. In that sense it's a pollutant. Indeed, a 2007 decision by the U.S. Supreme Court ruled that CO_2 and other climate-changing greenhouse gases qualify as pollutants, and in 2009 the U.S. Environmental Protection Agency issued an "endangerment finding" declaring that CO_2 and five other greenhouse gases endanger public health and welfare. The EPA's finding stopped short of grouping greenhouse gases with traditional pollutants, but it did set the stage for the development of rules to regulate these gases, including CO_2.

Carbon dioxide really is different from traditional pollutants. It's not a toxic substance that causes immediate health effects in humans or other species. It's a natural component of the atmosphere, and it's essential to the growth of plants. Natural CO_2 even plays an important role in establishing the relatively benign climate in which human civilization developed. The deleterious impact of anthropogenic CO_2—namely, climate change—occurs indirectly through CO_2's alteration of natural energy flows. And, as I mentioned earlier, CO_2 is different from other pollutants in that it's an *essential* product of fossil fuel combustion.

Both in its legal status and in the public's mind, CO_2 has become more closely associated with the term "pollution." But it's still important to distinguish CO_2

from traditional pollutants. That's not because it's less harmful, but because of its essential connection to fossil fuels and the indirect, global nature of its impact on Earth's climate. So when I talk about a "low emission" vehicle, for example, I mean one that produces low levels of traditional pollutants; it may or may not have low CO_2 emissions (see Box 6.1 for a practical example of this distinction). When I talk about a "dirty" fuel, I mean one that produces lots of traditional pollutants; it may or may not produce lots of CO_2. You'll often find terms like *low emission*, *dirty*, or *clean coal* used sloppily, sometimes referring to traditional pollutants and sometimes to CO_2. Better to maintain a clear distinction. I'll treat traditional pollutants in this chapter, and save carbon dioxide and its climatic effects for the chapters on climate change.

6.2 Air Pollution

Air pollutants associated with fossil fuels result from several distinct sources. Some, such as sulfur compounds and mercury, arise from substances other than carbon and hydrogen that occur in fossil fuels. Others are compounds resulting from incomplete combustion. A third category includes nitrogen oxides, which form when usually inert atmospheric nitrogen combines with oxygen at the high temperatures present in fossil fuel combustion. Even before combustion, leakage of fuel vapors can also contribute to air pollution. Finally, derivative pollutants, including the complex mix called *smog*, form as a result of atmospheric chemical reactions among primary pollutants.

BOX 6.1 | Gas or Diesel?

Although diesel cars are common in Europe, they're rare in the United States. Among the diesels available in the U.S. market are the Volkswagen TDI (turbo direct injection) models. The TDI engine is designed to minimize common diesel problems, including odor, smoke, and hard starting. Diesel cars offer substantially greater fuel efficiency than comparable gasoline models; for example, the manual-transmission gasoline and diesel models of the New Beetle shown in Figure 6.1 average, respectively, 28 miles per gallon and 42 miles per gallon. (Vehicles with automatic transmissions get lower mileage; see Exercise 2.)

Which New Beetle should you buy? Fewer gallons burned in the diesel translate into substantially

FIGURE 6.1
Which New Beetle should you buy?

(continued)

BOX 6.1 | Gas or Diesel? (Continued)

lower CO_2 emissions, so from a climate perspective the diesel is preferable. Indeed, the U.S. Environmental Protection Agency gives the New Beetle's diesel model a greenhouse gas score of 9 out of a possible 10. But diesels produce more particulate pollution, so the diesel has an air pollution score of only 2. That difference—very good in one index, very poor in the other—illustrates directly the distinction I'm making between CO_2 and traditional pollutants. (The gasoline model rates 6 on both scales.)

In California, and other states that have adopted California's strict air-quality standards, you don't have the choice suggested in Figure 6.1. You can buy a gasoline-powered Beetle engineered specifically for the California market that does better on pollution (9 out of 10) without sacrificing fuel efficiency. But you can't buy the diesel model, because its emissions of traditional pollutants don't meet state standards.

PARTICULATE MATTER

Solid material that doesn't burn comes through the combustion process in the form of tiny particles carried into the air along with the stream of gaseous combustion products. Other particles result from incomplete combustion; soot from a candle flame is a familiar example. Still others form in the atmosphere from reactions among gaseous pollutants. All these constitute **particulate pollution.** Coal and heavier oils, including diesel oil, are the most significant sources of par-

FIGURE 6.2
This satellite image shows a blanket of polluted air—a giant "brown cloud"—over eastern China.

ticulate pollution; a common example is the cloud of black smoke you see when a large truck starts up. Less obvious are tiny sulfur-based particles emitted from coal-burning power plants; these produce atmospheric haze that, among other things, is responsible for significant loss of visibility in scenic areas of the American West. Heavier particulate matter soon falls to the ground or is washed out of the air by rain, but lighter particulates can remain airborne for long periods.

Particles less than 10 microns (10 μm, or one-millionth of a meter) in size reach deep into the lungs and are therefore especially harmful to human health. Breathing particulate-laden air may cause or exacerbate respiratory diseases such as asthma, emphysema, and chronic bronchitis. Indeed, a careful study done for the Clean Air Task Force suggested that some 24,000 premature deaths per year in the United States may be the result of particulate air pollution from fossil-fueled power plants—more than the number of homicides! This toll should be reduced substantially with new power-plant emissions standards taking effect in 2012. Globally, the World Health Organization estimates that air pollution causes some 2 million premature deaths annually.

Once associated primarily with localized industrialized areas, particulate air pollution has now become a more global phenomenon (Fig. 6.2). An example is the so-called *Asian brown cloud*, first reported in 1999 and consisting of particulate and gaseous air pollution that extends over much of South Asia and the Indian Ocean during the months of December through April. In addition to air-pollution health effects described in this chapter, the brown cloud is large enough to exert significant impacts on climate, primarily a cooling that helps counter greenhouse-gas warming in the affected region.

Technologies to reduce particulate pollution are readily available and in widespread use. Long gone, at least in wealthier countries, is the pall of black smoke that once hung over industrialized areas (Fig. 6.3). The very tall smokestacks of fossil-fueled power plants are themselves crude pollution control devices; by injecting emissions into the atmosphere at great heights where winds are strong,

FIGURE 6.3
Air pollution hangs over New York City in this photo from the 1950s.

they ensure that pollutants, including particulates, are dispersed throughout a larger volume of air and are carried away from the immediate area.

Several techniques remove particulate pollution from the gases produced in fossil fuel consumption. **Filters** are perhaps the most obvious particulate-reduction devices. They use fine fabric or other meshes to trap particulate matter while permitting gases to pass through. In large-scale industrial and power-plant applications, fabric filters often take the form of tube-shaped bags, with hundreds or thousands of bags in an enclosure called a **baghouse**. Filtration can be highly effective, removing some 99.5% of all particulate pollution. However, frequent mechanical shaking of the bags is necessary to remove the particulate material that would otherwise block the filter. Bag filters are also expensive, require frequent maintenance, and don't work at the high temperatures of postcombustion **flue gases**.

Cyclones use the principle of inertia—that an object tends to remain in a state of uniform motion unless a force acts on it. In a cyclone, flue gas is forced to travel in a spiral vortex similar to the inside of a tornado. The lighter gas molecules have low inertia, so they readily "go with the flow." But the more massive particulates have high inertia, so they have trouble following the tight, high-speed spiral motion of the gas. These particulates strike the inside walls of the cyclone, then drop into a collection hopper (Fig. 6.4a). You may have noticed the characteristic upside-down cone shape of cyclones on industrial buildings (Fig. 6.4b). Cyclones typically remove between 50% and 99% of particulate matter, depending on particle size. Given their inertial operation, you can understand why they're more effective with larger particles. Among the least expensive particulate-control devices, cyclones are used on their own

FIGURE 6.4
Cyclone separator for reduction of particulate air pollution. (a) A cutaway diagram shows the spiral air path within the device. (b) Cyclones atop an industrial facility.

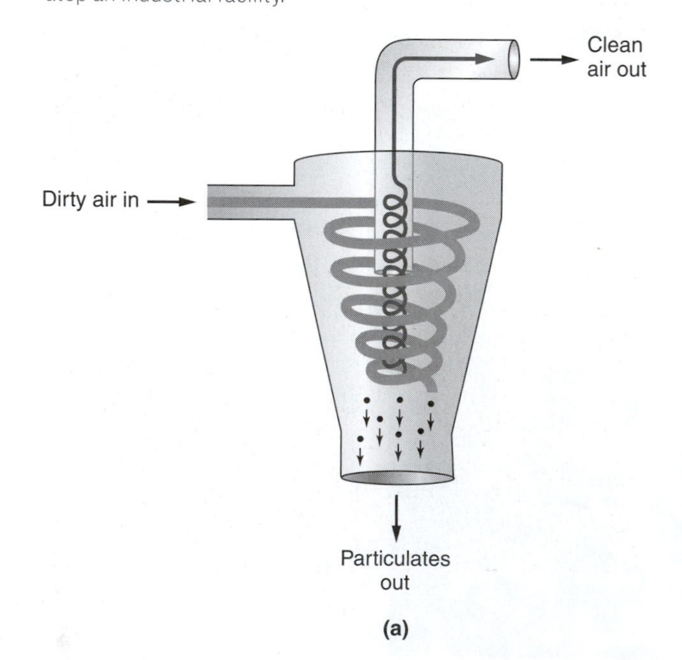

in dust-producing industrial applications, and often as pretreatment for flue gas before it enters more efficient pollution-control devices.

Electrostatic precipitators offer a more sophisticated approach. In these devices, a high voltage is applied between a thin, negatively charged wire and positively charged metal plates. The strong electric field in the vicinity of the wire ionizes gas molecules, resulting in free electrons that attach to particles in the gas stream. The negatively charged particles are driven onto the positively charged metal plates (Fig. 6.5). Periodically, the plates are mechanically shaken, and the particulate matter falls into a hopper where it collects and is eventually trucked away. This so-called **fly ash** collected from coal-burning power plants actually contains metals that may be economically worth extracting; alternatively, fly ash is used in the manufacture of cement, ceramics, and even some plastics.

Electrostatic precipitators can remove more than 99% of particulate matter. Precipitators installed in U.S. factories and power plants since 1940 have reduced sub-10-μm particulate emissions by a factor of 5, even as energy and industrial output grew substantially. But this clean air comes at a cost: Electrostatic precipitators and other particulate removal systems typically consume some 2% to 4% of a power plant's electrical energy output, which is one of the many reasons why actual power-plant efficiencies are well below the theoretical limits imposed by the second law of thermodynamics.

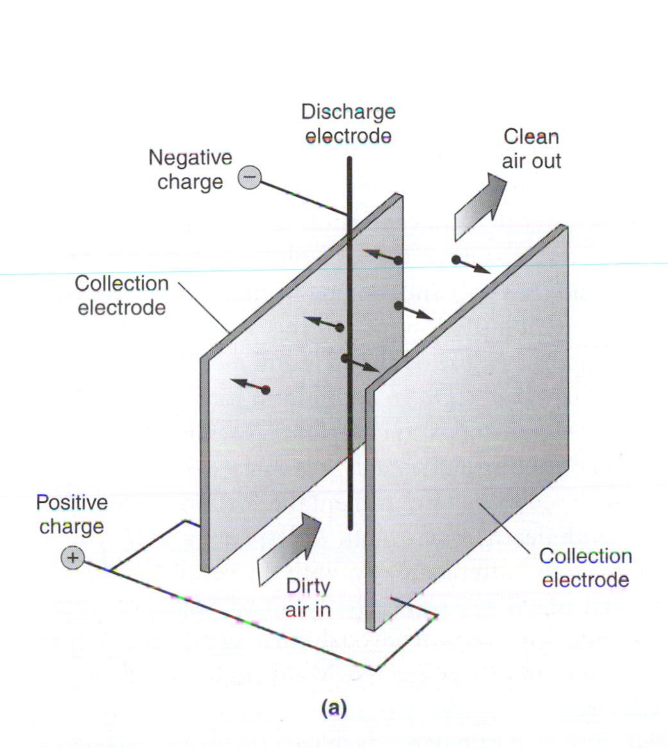

FIGURE 6.5
Electrostatic precipitators are among the most effective devices for particulate pollution control. (a) The discharge electrode is charged to a high negative voltage, which negatively charges particles in the gas stream. The particles are then attracted to the positively charged collection electrodes, from which they are removed by mechanical shaking. (b) Electrostatic precipitators are large devices that consume a lot of energy.

SULFUR EMISSIONS

Sulfur is present in coal and, to a lesser extent, in oil. Unless it's removed before combustion—a technologically feasible but expensive process—the sulfur burns to produce sulfur dioxide (SO_2). The native sulfur content of fuels varies substantially, so sulfur reduction can also be accomplished by changing the fuel source. For example, coal from the western United States typically contains less than 1% sulfur, whereas coals from the Midwest may contain 4% to 5% sulfur.

Sulfur dioxide itself is detrimental to human health. As usual, the health burden falls most heavily on the very young, the very old, and those already suffering from respiratory diseases. Acute episodes of high atmospheric SO_2 concentration have resulted in outright mortality; in a 1952 air pollution incident in London, some 4,000 people perished when atmospheric SO_2 rose sevenfold.

Once in the atmosphere, SO_2 undergoes chemical reactions that produce additional harmful substances. Oxidation converts SO_2 to sulfur trioxide (SO_3), which then reacts with water vapor to make sulfuric acid (H_2SO_4):

$$2SO_2 + O_2 \rightarrow 2SO_3$$
$$SO_3 + H_2O \rightarrow H_2SO_4$$

The second reaction can take place either with water vapor or in liquid water droplets. In liquid water, sulfuric acid dissolves into positive hydrogen ions and negative sulfate ions (SO_4^-). Sulfate ions can join with other chemical species to make substances that, if the water evaporates, remain airborne as tiny particles known as **sulfate aerosols**. These particles, typically well under 1 μm in size, are very efficient scatterers of light. As a result, they reflect sunlight and thus lower the overall energy reaching Earth. Consequently, as you'll see in Chapter 13, sulfate aerosols have a cooling effect on climate.

When water droplets laden with sulfuric acid fall to Earth, the result is **acid rain**, which has significantly greater acidity than normal rain. Acidity is measured on the **pH scale**; a pH of 7 is neutral and anything less than 7 is acidic. Technically, pH is the negative of the logarithm of the concentration of hydrogen ions in a solution; in pure water that concentration is 1 part in 10 million, or 10^{-7}, for a pH of 7. Natural atmospheric CO_2 dissolves to some extent in rainwater, forming carbonic acid (H_2CO_3) and making normal rain somewhat acidic, with a pH of 5.6. Rain with a pH lower than this value is considered acid rain. By comparison, vinegar has a pH of about 3, and most sodas have a pH around 4 because of their dissolved CO_2. In addition to forming dissolved sulfuric acid in rainwater, sulfate aerosols also contribute directly to environmental acidity when they land on the ground or fall into surface waters.

Acidity at a pH level below about 5 has a detrimental effect on aquatic life. Reproduction in fish falters, with death or deformity widespread among young fish. Amphibians and invertebrates suffer similarly, with the result that highly acidic lakes have little animal life. The northeastern United States has been particularly affected by acid rain, much of it from coal-burning power plants in the Midwest, and so has much of northern Europe. Many high-altitude lakes, even those in protected wilderness areas, have pH levels well below 5 and are considered "dead." The effect of acid rain depends in part on geology; lakes or

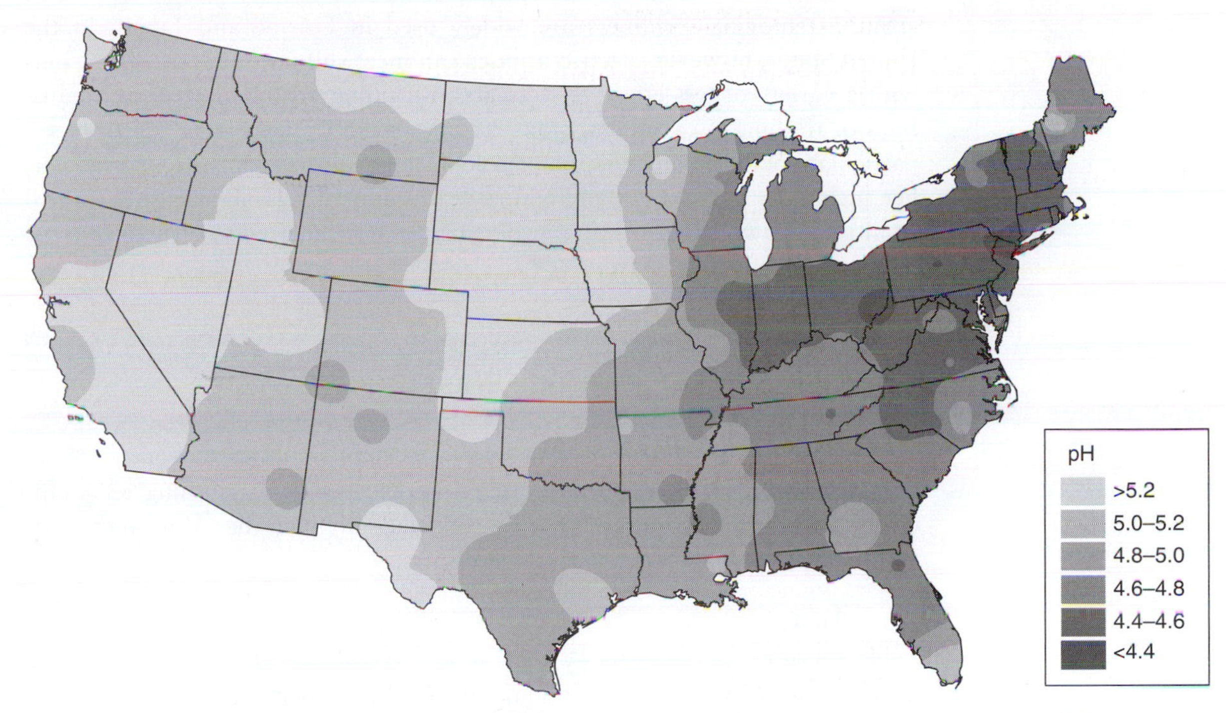

FIGURE 6.6
Values of pH in precipitation falling on the United States in the early twenty-first century. Lower pH corresponds to greater acidity. Excess acidity in the Northeast is largely the result of sulfur emissions from coal-burning power plants in the Midwest.

soils on limestone rock suffer less harm because the limestone can neutralize the excess acidity, but granite and quartz-based rocks lack this buffering effect. Figure 6.6 maps the acidity of precipitation in the United States.

Acid precipitation affects not only water quality; it also damages terrestrial vegetation, especially high-altitude trees, and causes increased corrosion and weathering of buildings. Limestone and marble are particularly susceptible to damage from acid precipitation, as are some paints.

Because sulfur pollution forms initially as the gas SO_2, it's not removed by particulate pollution controls. The most effective sulfur-removal technique now in widespread use is **flue gas desulfurization**, also known as **scrubbing**. In a so-called wet scrubber, flue gases pass through a spray of water containing chemicals, usually calcium carbonate ($CaCO_3$) or magnesium carbonate. In a dry scrubber, the flue gas contacts pulverized limestone, which is largely calcium carbonate. Either way, chemical reactions yield solid calcium sulfite ($CaSO_3$), which drops out of the gas stream or is removed along with other particulates:

$$SO_2 + CaCO_3 \rightarrow CaSO_3 + CO_2$$

Scrubbers remove some 98% of sulfur from the flue gases, but again there's a price. The monetary cost of scrubbers can be 10% to 15% of the total cost of a coal-burning power plant, and the energy cost may be as much as 8% of the plant's total electric power output. This is in addition to the 2% to 4% for particulate removal. Much of that energy goes into reheating the flue gas after its contact with water in a wet scrubber, which is necessary to restore buoyancy so the flue gas can rise up the smokestack and disperse in the atmosphere.

Sulfur-removing scrubbers are widely used in Europe and Japan. In the United States, however, electric utilities can meet sulfur emissions regulations with a variety of less-effective strategies, including switching to lower sulfur fuels or trading so-called emissions credits. In the United States, sulfur-based particulate pollutants play a major role in the health impacts I discussed earlier, including those 24,000 premature deaths annually. Finally, sulfur removal results in its own environmental problems, namely the need to dispose of large amounts of sulfurous waste.

EXAMPLE 6.1 | Sulfurous Waste

Consider a 1-GWe power plant burning coal with a sulfur content of 3% by weight. Estimate (a) the amount of coal burned per day, assuming 33% efficiency, and (b) the amount of $CaSO_3$ produced daily if all the sulfur in the coal is converted to $CaSO_3$ in the plant's scrubbers.

SOLUTION

Table 3.3 shows that the energy content of coal is about 30 MJ/kg. Our 1-GWe power plant is 33% efficient, so it goes through coal energy at the rate of 3 GW to produce its 1 GW of electric power (recall the distinction between GWe—the electric power output of a plant—and the total thermal power extracted from fuel and measured in GWth). That 3 GW is 3,000 MW, or 3,000 MJ/s. At 30 MJ/kg, our plant therefore uses coal at the rate of

$$\frac{3{,}000 \text{ MJ/s}}{30 \text{ MJ/kg}} = 100 \text{ kg/s}$$

That's a tonne of coal every 10 seconds! Since the coal contains 3% sulfur by weight, this amounts to 3 kg/s of sulfur. The scrubbers combine each sulfur atom with one calcium atom and three oxygen atoms to make $CaSO_3$. Refer to the periodic table in the Appendix, and you'll find that sulfur's atomic weight is 32, whereas calcium's is 40 and oxygen's is 16. This means that, for every 32 units of sulfur by weight, we get an additional 40 units from the calcium and $3 \times 16 = 48$ units from the oxygen; equivalently, for every 1 unit of sulfur we get 40/32 units of calcium and 48/32 units of oxygen. So our 3 kg/s of sulfur translates into

$$(3 \text{ kg/s})\left(1(\text{S}) + \frac{40}{32}(\text{Ca}) + \frac{48}{32}(\text{O})\right) = 11 \text{ kg/s}$$

of $CaSO_3$. There are 86,400 seconds per day (figure that out!), so the power plant's scrubbers produce

$$(11 \text{ kg/s})(86{,}400 \text{ s/day}) = 950{,}000 \text{ kg/day}$$

That's about 1,000 tonnes per day of $CaSO_3$! Actually, the $CaSO_3$ generally ends up in a water-based slurry totaling perhaps four times this mass.

CARBON MONOXIDE

Carbon monoxide results from the incomplete combustion of fossil fuels or other carbon-containing materials, such as tobacco, charcoal, or wood. Carbon monoxide formation is enhanced when combustion occurs with insufficient oxygen, as in indoor burning or the smoldering of a cigarette. But all fossil fuel combustion produces some CO. Although CO occurs naturally in the atmosphere at low levels, its concentration in industrialized areas and traffic-congested cities can be 10 to 1,000 times greater.

Carbon monoxide's dominant health effect is a result of the gas's affinity for hemoglobin, the molecule in red blood cells that carries oxygen to the body's tissues. Carbon monoxide binds 240 times more readily to hemoglobin than does oxygen, and it binds more tightly than oxygen does, making it hard to displace. A person breathing CO-contaminated air therefore carries a substantial amount of so-called carboxyhemoglobin (COHb), which reduces the blood's oxygen-carrying ability. At very high concentrations, CO causes death from lack of oxygen in body tissues. At lower concentrations, oxygen deprivation damages tissues and affects a variety of physiological functions. Carboxyhemoglobin levels of 6% can cause irregular heart rhythm in cardiovascular patients, increasing the risk of mortality. Nonsmokers normally carry COHb levels of 1% or lower; for smokers, the normal level averages 4% and can be twice that. Concentrations in the 5% range can also affect the brain, increasing reaction time and decreasing ability to accomplish tasks requiring high coordination. Some studies have suggested that automobile accidents occur more frequently among drivers with high COHb levels resulting from prolonged exposure to traffic pollution.

Control of CO emissions is relatively easy with stationary sources such as power plants, where combustion takes place under steady conditions—in particular, precisely controlled and optimized fuel/air ratios. Careful engineering with attention to adequate mixing of fuel and air greatly reduces the incomplete combustion that causes CO formation. Mobile sources are another story: With the intermittent combustion that occurs in gasoline and diesel engines, and with engines operating under a wide range of speeds and loads, CO production is inevitable and difficult to control. In the United States, cars and trucks produce about half of all CO emissions, while other mobile sources, such as planes, boats, and farm equipment, are responsible for roughly another 20% (Fig. 6.7).

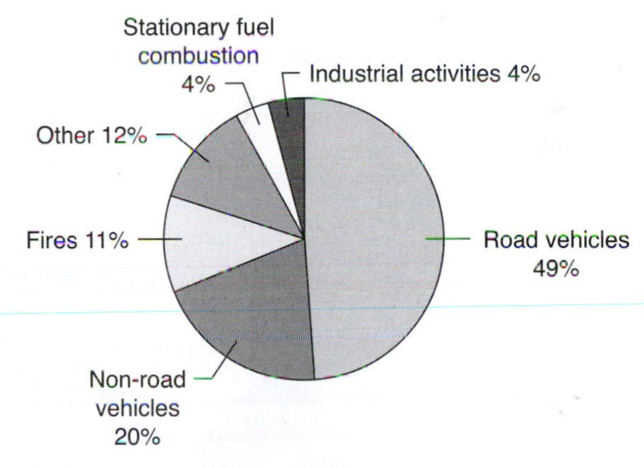

FIGURE 6.7
More than two-thirds of all CO emissions in the United States come from vehicles of all types. Non-road vehicles include recreational vehicles, boats, trains, aircraft, farm equipment, and lawnmowers. Nearly half the emissions in this category are from lawn and garden equipment.

Carbon monoxide emissions would be higher if it weren't for the widespread use of pollution control measures. **Catalytic converters** in automobile exhaust systems essentially "burn" CO to CO_2 with the aid of a chemical catalyst that allows the reaction to proceed at lower temperatures than in combustion. Although it facilitates the reaction, the catalyst itself is not consumed. Another approach to lowering CO emissions is to reformulate gasoline by adding oxygen-containing chemicals that help to ensure complete conversion of all carbon to CO_2. Such additives include alcohols, both methanol and ethanol, and methyl tertiary-butyl ether (MBTE), a substance that reduces CO production but has been linked to water pollution when fuel leaks into the ground.

Because CO represents a higher energy state than CO_2 (the latter being the product of complete combustion), CO in the atmosphere is naturally oxidized to CO_2 on a fairly short timescale of about a month, which means CO never has time to become widely distributed and remains essentially a local pollutant. However, the oxidation of atmospheric CO to CO_2 involves a complex sequence of reactions that affect other atmospheric chemicals and may have implications for climate. Also, microorganisms in the soil rapidly absorb CO, helping to remove it from the air in all but the most heavily paved urban areas. In any event, CO emissions don't stay around for long, and the atmospheric concentration of CO fluctuates substantially according to the time of day, traffic conditions, and specific location.

NITROGEN OXIDES

Earth's atmosphere is predominantly nitrogen (N_2). In that form, nitrogen is relatively inert, meaning it doesn't participate readily in chemical reactions. But at the high temperatures typical of fossil fuel combustion (especially over 1,100°C), nitrogen combines with oxygen to form a variety of compounds, collectively designated as the nitrogen oxides (NO_x, often pronounced "nox"). Among these are nitric oxide (NO), also known as nitrogen monoxide; nitrogen dioxide (NO_2); dinitrogen pentoxide (N_2O_5); and nitrous oxide (N_2O), also known as "laughing gas" and commonly used as a dental anesthetic. Figure 6.8 shows that fossil-fueled electric power plants, industry, and transportation are all significant sources of NO_x emissions.

Most of the nitrogen that forms NO_x comes not from fuel, but from the air itself. So there's no way to prevent NO_x emissions by pretreating fuel. Nor is it practical to remove nitrogen from the air supplied for combustion (after all, nitrogen constitutes some 80% of the atmosphere). In motor vehicles, the same catalytic converters that oxidize CO to CO_2 also convert much of the NO_x into the normal, harmless atmospheric components N_2 and O_2. A variety of additional techniques can help reduce nitrogen oxide production in stationary sources such as power plants and industrial boilers, including modifications to the combustion chamber and flame geometry, adjustments in the fuel and air flows, and recycling of flue gases through the combustion chamber. Recycling flue gases reduces combustion temperature and with it the amount of NO_x produced. Unfortunately, reducing the temperature reduces efficiency, lowering the useful energy available from a power plant.

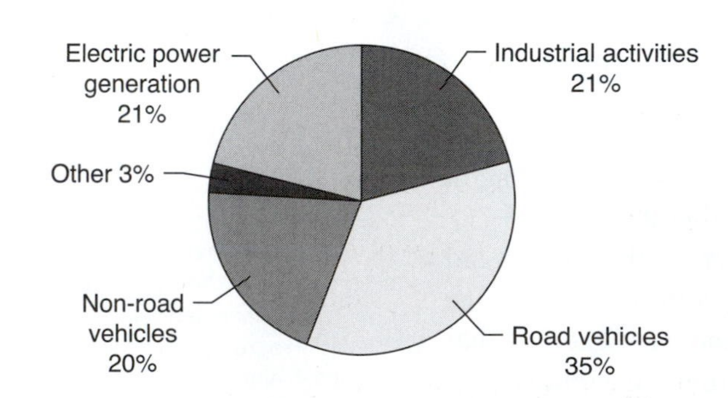

FIGURE 6.8
Vehicles, industry, and power generation all contribute significantly to NO_x emissions.

Nitric oxide (NO), the dominant component of NO_x, is not particularly harmful in itself, but NO_2 is toxic, and long-term exposure may result in lung damage. Oxidation converts NO to NO_2, whose reddish-brown color is responsible for

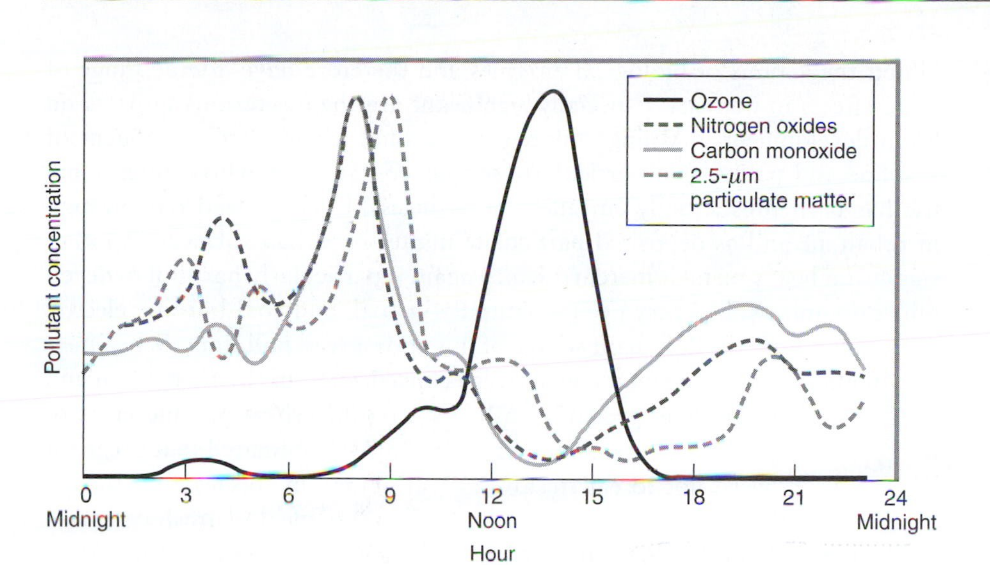

FIGURE 6.11
Hourly concentrations of several pollutants for Los Angeles on a midweek day in January. Notice how ozone peaks after the others since it's a secondary pollutant formed by photochemical processes. Can you identify the effects of the morning and evening rush hours? The concentrations are scaled so they all peak at the same height on the graph.

cause eye irritation as well as respiratory problems. PANs and other reactive molecules can also injure plants and damage materials such as paints, fabrics, and rubber.

The photochemical nature of smog is evident in a plot showing the levels of smog precursors and ozone in an urban area over a 24-hour period. At night, levels of all these substances are low. Morning rush hour quickly increases nitrogen oxides and hydrocarbons. As sunlight strengthens through the morning, NO and hydrocarbons decline and ozone increases. Finally, by midday all these precursor molecules have declined through the formation of more complex molecules, including PANs. Figure 6.11 shows pollutant concentrations measured throughout a typical 24-hour period in Los Angeles.

The best way to minimize photochemical smog is to reduce emissions of smog precursors, especially nitrogen oxides and hydrocarbons. Catalytic converters and other techniques discussed earlier reduce NO_x resulting from combustion. Volatile organic compounds (VOCs) from evaporation of fuels and other products also contribute to smog, and careful handling to avoid spills, leakage, and vapor escape can reduce these emissions. Cars, for example, have vapor recovery systems that capture gasoline vapors in a charcoal canister and recycle them back into the fuel system rather than letting them escape to the atmosphere. And increasingly, vapor release in the fueling process itself is reduced through the use of special gasoline nozzles (Fig. 6.12). Paints, paint thinners, and other solvents can be reformulated to reduce their volatility (tendency to vaporize). Widespread use of water-based paints further reduces VOC emissions.

FIGURE 6.12
The rubber bellows on this gasoline nozzle seals the gasoline tank opening to reduce leakage of hydrocarbon vapors that are precursors of photochemical smog.

HEAVY METALS

Fossil fuels contain a variety of heavy metals that are released as pollutants when fuels burn. Other metals enter the environment through wear in engines, brakes, turbines, and other machinery. In their elemental form (i.e., not combined with other elements to make chemical compounds), the heavier metals inhibit the actions of biological enzymes and therefore have a wide range of toxic effects in humans. Especially significant are the deleterious impacts on brain development in young children. Lead, once a common component of gasoline and paints, is a particularly serious contaminant whose intentional use has been substantially curtailed; nevertheless, levels of lead remain high in urban air and as deposited particulate matter on urban surfaces. Another significant heavy metal is mercury, which again is particularly harmful to developing organisms. Mercury occurs naturally in coal, and coal-burning electric power plants are the dominant source of this widespread pollutant. When mercury pollution lands in surface waters, it's concentrated in the food chain and ends up at high levels in predatory fish. As a result, fish in pristine lakes of the central and eastern United States are so highly contaminated that pregnant women are advised not to eat freshwater fish of certain species, and for the general populace, warnings to eat no more than one meal of freshwater fish per month are common. A particularly disturbing study shows that one-fifth of all Americans may have mercury levels exceeding EPA recommendations of no more than 1 part per million. No other pollutant even comes close to mercury for violating federal standards so widely. The dominant source of this mercury contamination is coal-burning power plants. Figure 6.13 maps the extent of mercury pollution in the United States.

FIGURE 6.13

Average mercury concentration in U.S. fish, in parts per million (ppm), plotted by watershed region. The EPA maximum permissible concentration is 0.3 ppm, and advisories start at 0.15 ppm.

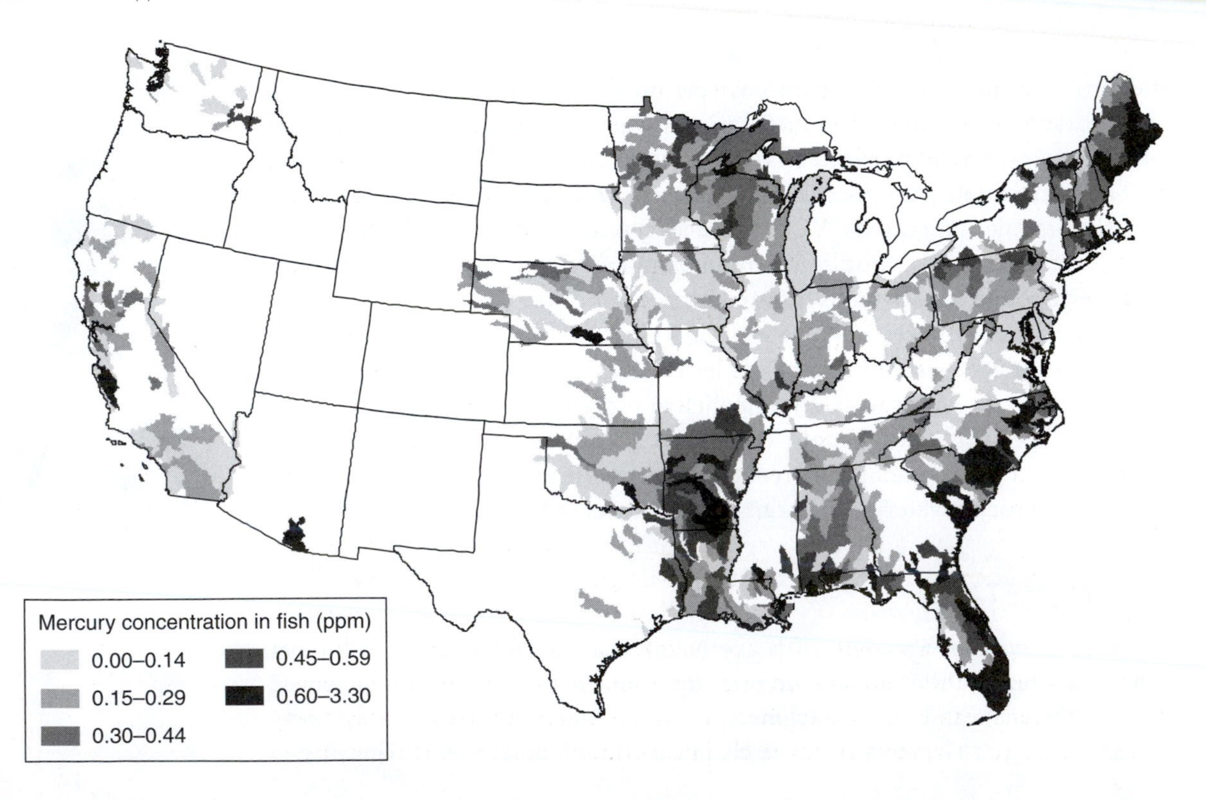

RADIATION

It might surprise you to learn that fossil fuel combustion is a source of radioactive materials. Coal, especially, contains small quantities of uranium and thorium, as well as their radioactive decay products, such as radium. Although most coal contains about 1 part per million (ppm) of uranium, this figure varies widely; one sample of North Dakota lignite coal measured 1,800 ppm of uranium. Radioactive constituents of coal end up in the air or in fly ash recovered from particulate-removal devices. Although radiation from fossil fuels is not a major concern, it's interesting to note that the level of radiation emitted in coal burning is often higher than that emitted in the normal operation of a nuclear power plant. As a result, people living near some coal-burning power plants may be subjected to as much as five times more radiation than those living near nuclear plants.

6.3 Other Environmental Impacts of Fossil Fuels

The environmental impacts of fossil fuels begin long before combustion. Since fossil fuels form underground, we generally drill, dig, mine, or otherwise penetrate the Earth to recover them. After we extract them from the ground, we have to transport and process these fuels before they're ready for combustion.

COAL EXTRACTION

When you think of coal extraction, you probably picture miners toiling away deep underground. That's the traditional method, but today some 60% of all coal comes from **strip mining**, in which surface layers are removed to reach the coal. In the United States, the proportion of strip-mined coal is likely to increase because this coal, which is largely from the American West, has a significantly lower sulfur content than coal from the East and Midwest. In strip mining, a relatively thin layer (typically tens of feet) of rock and soil is removed to expose coal seams that are typically 50 to 100 feet thick. A related method, used in the Appalachian mountains of the eastern United States, is so-called **mountaintop removal**. In this technique, the top 1,000 feet or so of a coal-containing mountain is removed by blasting, the coal is extracted, and excess rubble is deposited in nearby valleys. Figure 6.14 shows the aftermath of mountaintop removal; in recent decades, hundreds of peaks have met similar fates.

Obviously, surface mining by stripping or mountaintop removal results in major environmental alteration. In the United States, the Surface Mining Control and Reclamation Act of 1977 requires that strip-mined land be restored to its original biotic productivity. Whether that is realistically possible in the arid West is unclear. Serious erosion damage can occur before plants have a chance to reestablish themselves. Complete restoration of mountainous topography after mountaintop removal is obviously impractical. The removal process often

FIGURE 6.14
A mountaintop-removal coal operation encroaches on the West Virginia community at the bottom of the photo.

buries streams, altering natural water flows and resulting in water pollution and additional erosion. With all types of coal mining, acidic runoff from mines adds further insult to surface waters; such **acid mine drainage** has rendered many Appalachian streams lifeless.

Even old-fashioned underground mining has serious implications for the environment and for human health. Despite tightened health and safety regulations, mining remains among the most dangerous occupations in the United States. China, where rapid industrialization drives a growing appetite for coal, averages more than a dozen deaths each day from coal-mining accidents. Miners around the world continue to suffer from black lung disease and other infirmities brought on by exposure to dust-laden air in underground mines. Environmental dangers persist long after mining ends; for example, fires in abandoned coal mines can burn for years or even decades, threatening natural and human communities that have the misfortune to be located above the mines. Coal fires are also a significant source of air pollution, and they produce enough CO_2 to worry scientists concerned about climate change.

OIL AND NATURAL GAS EXTRACTION

Conventional extraction of oil and natural gas is, at least in principle, less messy than coal mining. These fluids come from deep in the ground and are "mined" through narrow, precision-drilled wells. However, geologically trapped petroleum is often under high pressure, and therefore requires careful handling. Although modern drilling techniques and safety systems have reduced "gushers" and "blowouts," disasters still occur. The 2010 *Deepwa-*

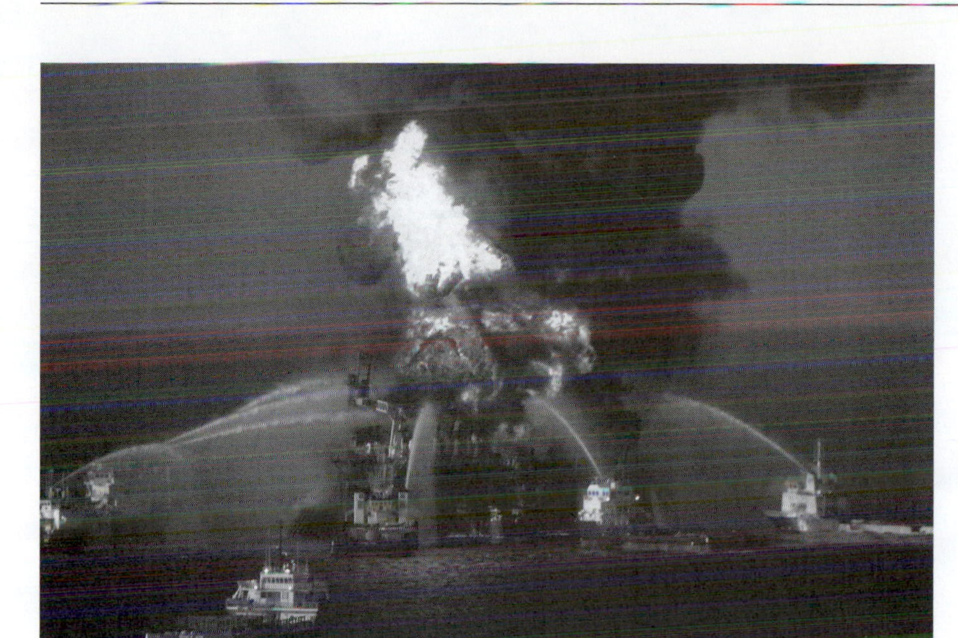

FIGURE 6.15
An oil fire burns fiercely on the *Deepwater Horizon* drilling platform following the disastrous blowout in 2010.

ter Horizon blowout in the Gulf of Mexico is a case in point; here, failure of a blowout prevention device resulted in oil spewing into the ocean for 3 months (Fig. 6.15). By the time the wellhead was finally sealed, some 5 million barrels (about 200 million gallons) of oil had contaminated the ocean—the largest marine oil spill to date.

The existence of oil reserves in environmentally sensitive areas pits environmentalists against the oil industry and its political allies; witness the debates surrounding construction of the Trans-Alaska oil pipeline in the 1970s, drilling in the Arctic National Wildlife Refuge in the 2000s, and the debates on offshore drilling following the 2010 *Deepwater Horizon* disaster. In many other parts of the world, oil exploration and extraction proceed with little regard for the environment or local human populations. Sabotage also contributes to pollution associated with oil extraction; Iraq's deliberate burning of oil wells during the 1991 Persian Gulf war briefly produced air pollution at 10 times the rate of all U.S. industry and power generation, and pollution from these oil fires was detected halfway around the world (Fig. 6.16).

FIGURE 6.16
Oil wells burning in Kuwait after the 1991 Persian Gulf War.

The potential environmental impacts of extracting unconventional petroleum resources may be orders of magnitude greater than those of conventional drilling. Extracting oil from oil shale requires grinding up masses of rock. Exploiting the coal sands of Alberta, Canada, has turned vast areas into moonscapes (Fig. 6.17). If the world's demand for oil continues to grow, the unconventional sources of petroleum will surely become major environmental battlegrounds.

FIGURE 6.17
Extraction of oil from tar sands entails massive environmental degradation.

TRANSPORTATION

All fossil fuels contain on the order of 30 to 60 MJ of energy per kilogram. The world's population consumes energy at the rate of around 500 EJ per year, most of it from fossil fuels. This means we have to burn a great many kilograms of fossil fuels. But first we have to get all those kilograms from mines and oilfields to the power plants, cars, homes, and factories where we're going to burn them, so there must be a huge mass of fossil fuels in transit at any given time. You can estimate how much in Exercise 3; the answer is something approaching 100 million tons! We can't move that much material around without significant environmental impact.

One immediate consequence is that we have to use even more fuel to run all the tanker ships, trains, trucks, and pipelines that move our fossil fuels. In fact, a small but significant fraction of our fossil fuel consumption goes into transporting the fuels themselves. For Alaskan oil that travels by pipeline and ship to ports on the U.S. West Coast the energy required to move the oil amounts to about 1% of the oil's energy content. A supertanker carrying oil from the Middle East to the United States can use the equivalent of 15% of its cargo's energy (see Exercise 15). This additional energy use compounds the environmental impacts of fossil fuel combustion, but fuel transport also has its own unique impacts on the environment and on human well-being.

Oil spills from supertanker accidents are perhaps the most dramatic examples of environmental damage from fossil fuel transport (Fig. 6.18). As tankers have grown in size, so have the spills; the largest have involved more than a quarter million tons of oil—although tanker accidents don't involve as much oil as marine drilling disasters like the 2010 *Deepwater Horizon* blowout. Some spills are caused when supertankers run aground or collide with other vessels, while others result from fires, hull failures, or accidents during loading and unloading. With 37,000 tonnes spilled, the *Exxon Valdez* accident in 1989 was relatively small, but it occurred in an ecologically sensitive Alaskan bay and caused widespread environmental damage.

Oil spills block sunlight from reaching photosynthetic organisms, reduce dissolved oxygen needed to sustain life, and foul the feathers of birds and the gills of fish (Fig. 6.18b). Economic impacts increase when oil piles up on beaches. Because supertanker accidents involve large oil releases that occur over short periods and affect relatively small areas, their specific impact can be very large. In the past few decades, tightened regulations and technological fixes such as the use of double-hulled oil tankers have greatly reduced both the number and size of spills (Fig. 6.19). Although plenty of smaller spills occur, it's the big headline-grabbing accidents that are responsible for most of the spilled oil; in actuality, the total amount of oil spilled worldwide varies substantially from year to year.

Although oil spills provide the most dramatic examples of environmental degradation resulting from marine oil transport, a lesser-known effect is the spread of exotic species to coastal ecosystems around the globe. This occurs when ballast water, used to provide stability, is pumped from ships when they take on more oil. Ironically, legislation mandating separate ballast and oil chambers to

(a)

(b)

FIGURE 6.18
(a) Oil spills from the stricken *Exxon Valdez*. (b) An oil-soaked bird.

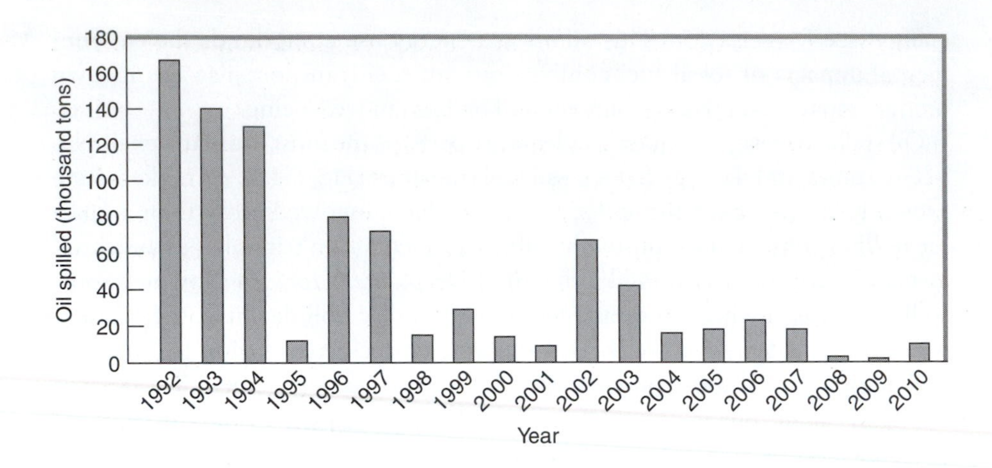

FIGURE 6.19
The total quantity of spilled oil varies from year to year, but there has been a general downward trend despite greater quantities of oil being transported at sea.

prevent oil discharge into the oceans has actually enhanced shipboard conditions for the survival of invasive species. Many coastal ecosystems around the world have been transformed dramatically in recent decades as invasive exotic species outcompete native organisms.

On land, crude oil is transported by pipeline to refineries, where it's processed into a wide variety of fuels, lubricants, and other products. Refineries themselves are major pollution sources, as I'll describe in the next section. Liquid and gaseous fuels leave refineries by pipeline, truck, train, or ship. We hear regularly of small-scale accidents involving tanker trucks or rail cars; both the flammability and environmental toxicity of petroleum fuels make these serious incidents that too often result in loss of life. Pipelines can rupture or leak, causing spills or, worse, explosions. In 2010 a natural gas pipeline exploded in San Bruno, California, killing eight people and devastating an entire neighborhood. In some impoverished countries, pipeline leaks are intentional and can have disastrous consequences. In 1998 in Nigeria, more than 1,000 people died while attempting to scavenge gasoline from a ruptured pipeline, which exploded in an inferno of liquid fire. Similar disasters have killed many hundreds since then.

Transport of natural gas is especially tricky. Unlike oil, it can't be poured into ships, tanker cars, or trucks. Gas is so diffuse that it must be compressed substantially for economical transport. This process takes energy, and it results in dangers associated with high pressure. Pressure exacerbates any tendency to leakage, which is an explosive danger in confined spaces and a climate-change worry when natural gas escapes to the open atmosphere. Once a purely domestic product, natural gas is now transported across the open ocean in special tanker ships that can maintain the gas in liquid form at very low temperatures (about −160°C, or −260°F). This **liquefied natural gas (LNG)** has a volume only 1/600 of its gaseous state. LNG tankers pack a large amount of energy (see Exercise 4), and although LNG transport has an enviable safety record, there's plenty of worry among communities near LNG docking centers about the impact of an explosion or a terrorist attack on LNG tankers or storage facilities.

Coal, being a solid, presents less serious transportation challenges. The dominant mode of coal transport is by so-called unit trains of about 100 cars each (see Example 5.2). A typical unit train carries about 10,000 tonnes (10 kilotons, or 10^7 kg) of coal. Coal-train derailments are not infrequent accidents, although spilled coal, unlike oil, rarely causes significant environmental damage. However, every year several hundred people die in collisions between vehicles and trains at grade crossings; given that coal accounts for nearly one-third of all train car loadings, we can infer that roughly 100 people die each year in the United States from grade-crossing accidents involving coal trains.

I mention some of these seemingly mundane consequences of fossil fuel transport because the outcomes—in terms of environmental damage or loss of human life—are often greater than other things we tend to worry much more about. You may be anxious about terrorism or nuclear power accidents or airplane crashes, but do you worry about being hit by a coal train? Yet the number of people killed in coal-train accidents—just one tiny aspect among many death-dealing consequences of fossil fuel use—likely exceeds all deaths from some more widely publicized dangers, such as nuclear power.

FOSSIL FUEL PROCESSING

Fossil fuels don't come from the mine or well ready to burn. Even coal needs washing, crushing, sizing, and other processing, while crude oil undergoes extensive refining. Among industrial polluters in the United States, refineries emit the greatest amount of volatile organic compounds—chemicals such as the carcinogen benzene and the neurotoxin xylene—which are harmful in their own right and also contribute to smog production. Refineries are the second-largest industrial source of sulfur dioxide emissions and are third in industrial production of nitrogen oxides. A typical refinery produces some 10,000 gallons of waste each day in normal operation; overall, refineries emit some 10% of all pollutants required to be reported to the federal government, despite their constituting less than 1% of the industries required to report emissions. Anyone who has driven through the oil-refining regions of New Jersey or Texas has experienced first-hand the degradation in air quality that accompanies large-scale refining operations.

THERMAL POLLUTION

I've emphasized that the second law of thermodynamics requires that much of the energy released in burning fossil fuels remains as low-grade heat. Somehow this energy must be disposed of, either by using it for heating or by simply dumping it to the environment. The waste heat from large-scale electric power plants presents special problems, since it's enough to alter substantially the temperature of the surrounding environment. All power plants require enormous volumes of water to cool and condense the steam from their turbines; this is where most of the second-law waste heat is extracted. Power plants are almost always built on rivers, lakes, or seashores for a ready supply of cooling water.

Simply dumping the now-heated water back into its natural source usually won't do; this constitutes thermal pollution, whose associated temperature rise can substantially alter the local ecology (although sometimes this effect can be put to advantage, encouraging the growth of commercially desirable aquatic species). Most of the time the water is first run through an air-based cooling system that drops its temperature significantly, effectively transferring the power plant's waste heat to the atmosphere. The huge concrete cooling towers at large power plants accomplish this transfer (recall Fig. 4.14). In so-called **wet cooling towers**, cooling water comes into direct contact with air, resulting in evaporative cooling (ultimately associated with the latent heat discussed in Chapter 4). The evaporated water often recondenses to form clouds of steam rising from the cooling towers. **Dry cooling towers** are used where water is scarce. In these, water remains in sealed pipes while air blows past to extract the heat without evaporating any water. In arid climates, power plants may use closed cooling systems in which water is continuously cycled without the need for a large body of water. Even where water is available, some plants use open-cycle cooling in the winter, when ample cold water is available, and closed-cycle cooling in the drier, hotter summer months. Exercise 14 illustrates the magnitude of the cooling-water problem faced by large power plants.

6.4 Policy Issue: The Clean Air Act

As this chapter clearly suggests, air pollution is (aside from climate change) probably the greatest single environmental impact of the world's voracious appetite for fossil fuels. In the United States, though, despite several decades' growth in fossil fuel consumption, air quality is generally better than it was in the 1970s. Largely responsible for that welcome environmental news is the **Clean Air Act** and especially its amendmants of 1970, 1977, and 1990. These amendments substantially strengthened the 1967 Air Quality Act, which itself was a successor to the 1963 Clean Air Act and the 1965 Motor Vehicle Pollution Act. Still earlier, the 1955 Air Pollution Control Act spelled out the first nationwide air quality policy. Individual cities got into the act much earlier, with Chicago and Cincinnati enacting pollution regulations in the late nineteenth century. And as early as 1306 in London, England's King Edward I banned the burning of so-called sea coal, a particularly dirty-burning coal found in coastal outcroppings.

One caveat here: As I emphasized at the beginning of the chapter, by "clean air" I mean air free of the traditional pollutants, not including CO_2. Although it now has the legal authority to do so under the Clean Air Act, the Environmental Protection Agency, as of 2010, had not yet begun regulating CO_2. In any event the level of that climate-change agent has risen steadily and substantially during the time of modern air-quality policy.

Currently, the Clean Air Act identifies six substances that it calls **criteria pollutants**, and it establishes national standards for their maximum concentrations in ambient air. These **National Ambient Air-Quality Standards (NAAQS)** include **primary standards**, designed to protect human health, and **secondary standards**, associated with more general welfare that includes environmental quality and protection of property. Note the emphasis on standards for air quality: The NAAQS do not directly address emissions of pollutants, but rather their concentrations in the ambient air. This leaves states and municipalities somewhat free to choose policies that ensure their air meets the national standards. In addition, however, the Clean Air Act Amendments require the EPA to establish emissions standards for a number of pollutants and their sources.

Table 6.1 shows the six criteria pollutants and their ambient air-quality standards, both primary and secondary (seven pollutants are listed because the NAAQS distinguishes two size ranges for particulate matter). All are substances I've addressed to some extent in this chapter. Note that each standard has an associated averaging time, or in some cases two times, which means a pollutant level can exceed its standard for short periods as long as the average over the specified time meets the standard.

Table 6.1 specifies the acceptable quality of the ambient air; it says nothing about emissions restrictions to achieve that quality. Table 6.2, in contrast, shows

TABLE 6.1 | **NATIONAL AMBIENT AIR-QUALITY STANDARDS**

Pollutant	Primary standards	Averaging times	Secondary standards
Carbon monoxide	9 ppm (10 mg/m^3)	8 hours	None
	35 ppm (40 mg/m^3)	1 hour	None
Lead	1.5 μg/m^3	Quarterly average	Same as primary
Nitrogen dioxide	0.053 ppm (100 μg/m^3)	Annual	Same as primary
Particulate matter:			
10 μm	150 μg/m^3	24 hours	
2.5 μm	15.0 μg/m^3	Annual	Same as primary
	35 μg/m^3	24 hours	
Ozone	0.075 ppm	8 hours	Same as primary
	0.12 ppm	1 hour	Same as primary
Sulfur dioxide	0.03 ppm	Annual	0.5 ppm
	0.14 ppm	24 hours	(1,300 μg/m^3)
	0.075 ppm	1 hour	—

TABLE 6.2 | EMISSIONS STANDARDS

Source	Pollutant	Emissions standard	Comments
Coal-fired power plant	SO_2	516 grams/gigajoule (g/GJ) fuel energy	71% further reduction in total SO_2 emissions for 31 states under EPA Transport Rule; fully effective in 2014
Oil-fired power plant	SO_2	86 g/GJ	
Gas-fired power plant	NO_x	86 g/GJ	52% further reduction in total NO_x emissions for 31 states under EPA Transport Rule; fully effective in 2014
All power plants	PM	13 g/GJ	PM = particulate matter
Cars	CO	3.4 grams/mile (g/mi)	
	NO_x	0.4 g/mi	Tier 1 1994; must meet standard for first 50,000 miles
		0.07 g/mi	Tier 2 fleet average 2004–2009
	PM	0.08 g/mi	Tier 1
		0.01 g/mi	Tier 2
SUVs, light trucks	CO	4.4 g/mi Tier 1 3.4 g/mi Tier 2	Vehicle category LDT2 (light duty truck 2): medium-weight SUVs and light trucks
	NO_x	0.7 g/mi	Tier 1
		0.07 g/mi	Tier 2 fleet average
	PM	0.08 g/mi	Tier 1
		0.01–0.02 g/mi	Tier 2

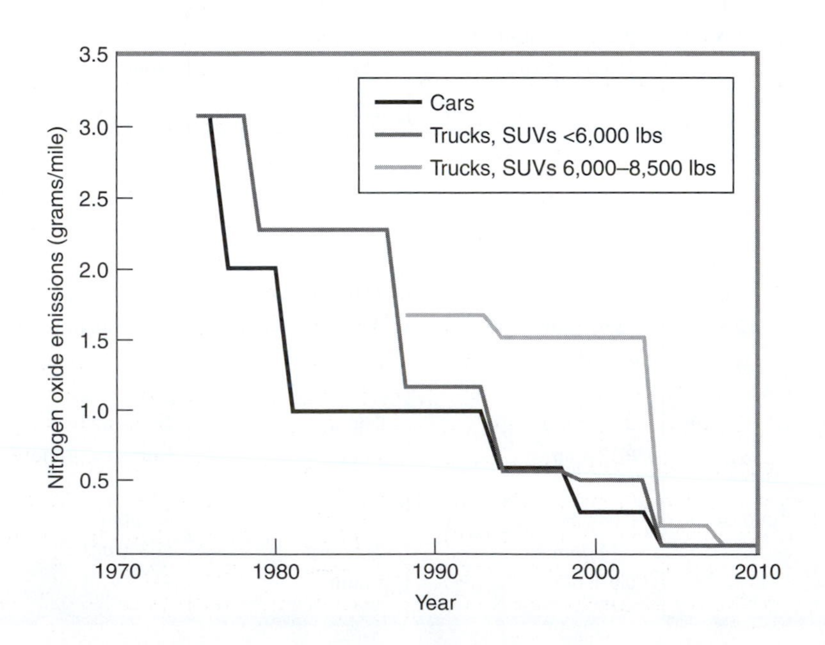

FIGURE 6.20

EPA standards for NO_x emissions from motor vehicles. The tightening of NO_x and other emissions standards over time is largely responsible for cleaner air in the United States.

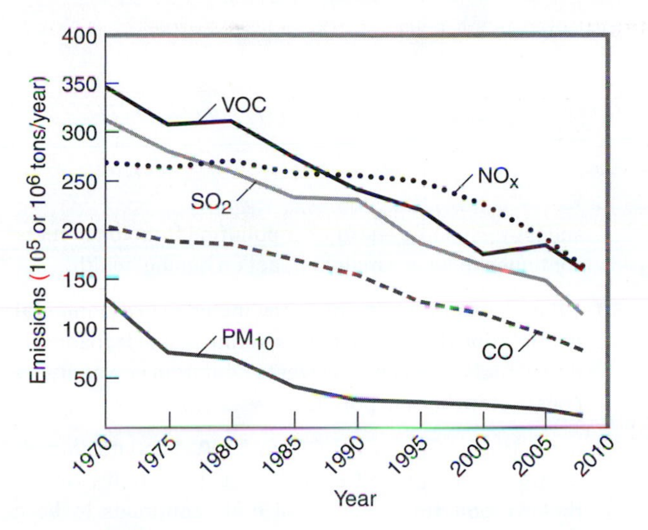

FIGURE 6.21

Pollutant emissions have declined significantly under the U.S. Clean Air Act. The curves show nationwide total emissions, in millions of tons per year for CO and in 100,000-ton units for other pollutants.

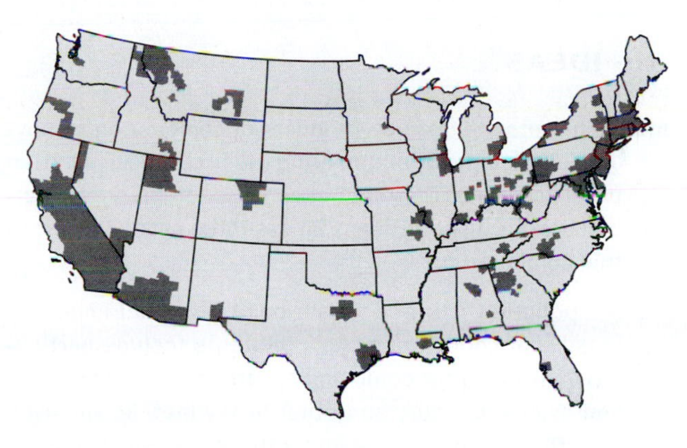

FIGURE 6.22

U.S. counties not meeting EPA air quality standards for one or more of the NAAQS criteria pollutants, as of April, 2011. Although the area covered is small, these counties are home to more than half of the U.S. population.

some individual emissions standards for a variety of pollutants and sources. These are drawn from separate regulations, including New Source Performance Standards for power plants and Vehicle Emissions Standards for motor vehicles. Emissions standards are not set in stone, but evolve over time. Figure 6.20, for example, shows the evolution of vehicular emissions standards for nitrogen oxides from the 1970s. These standards progress through several levels, called *tiers*. The new Tier 2 vehicle emissions standards, phasing in through the first decade of the twenty-first century, are complex and allow vehicle manufacturers some flexibility in emissions among their different vehicles.

So how have we done? Have these complex regulations and standards improved air quality? The answer is an unequivocal yes. Figure 6.21 shows that emissions of the criteria pollutants have declined substantially since the 1970s, despite rising energy use, vehicle miles traveled, and GDP. However, most of this reduction occurred before 1995; wringing further emissions reductions from technological advances in the face of rising energy consumption and vehicle use is proving more difficult. And remember that emissions reductions are only a means to the goal of improving ambient air quality. There, too, we've seen substantial improvement, but as Figure 6.22 suggests, much of the U.S. population still breathes air that fails to meet national standards.

From health effects of air pollution to oil spills, from strip mining to coal-train accidents, the impacts of fossil fuels are vast and often environmentally deleterious. Yet the world will continue to rely on fossil fuels for the major share of its energy for at least several decades to come. And the biggest single emission from fossil fuel consumption—CO_2—has consequences we've hardly begun to discuss. Before we go there, we'll explore alternatives to fossil fuels.

CHAPTER REVIEW

BIG IDEAS

6.1 Pollutants are toxic by-products of combustion, removable in principle by engineering the fuel, the combustion process, or the treatment of combustion products. Carbon dioxide, in contrast, is an essential product of fossil fuel combustion.

6.2 Air pollution is the most serious traditional impact of fossil fuel combustion. Air pollutants include particulate matter, sulfur compounds, carbon monoxide, nitrogen oxides, mercury and other heavy metals, and even radiation. Air pollutants under the influence of sunlight produce **photochemical smog**. Technologies exist to reduce air pollution significantly, but they're expensive and they consume energy. Air pollution from fossil fuels continues to have a major impact on human health.

6.3 Other impacts of fossil fuels include environmental degradation during extraction, refining, and transportation of fuels. Discharge of waste heat from power plants constitutes thermal pollution.

6.4 In the United States, the **Clean Air Act** and its amendments have improved air quality substantially. Nevertheless, pollution from fossil fuels continues to be a major problem both in the United States and globally.

TERMS TO KNOW

acid mine drainage (p. 144)
acid rain (p. 134)
baghouse (p. 132)
catalytic converter (p. 137)
Clean Air Act (p. 150)
criteria pollutant (p. 151)
cyclone (p. 132)
dry cooling tower (p. 150)
electrostatic precipitator (p. 133)
filter (p. 132)
flue gas (p. 132)
flue gas desulfurization (p. 135)
fly ash (p. 133)
lapse rate (p. 139)
liquified natural gas (LNG) (p. 148)
mountaintop removal (p. 143)
National Ambient Air-Quality Standards (NAAQS) (p. 151)
ozone (p. 140)
particulate pollution (p. 130)
pH scale (p. 134)
photochemical reaction (p. 139)
photochemical smog (p. 139)
pollution (p. 128)
primary standards (p. 151)
scrubbing (p. 135)
secondary standards (p. 151)
strip mining (p. 143)
sulfate aerosol (p. 134)
temperature inversion (p. 140)
wet cooling tower (p. 150)

GETTING QUANTITATIVE

Particulate size that has an impact on human health: $<10\ \mu m$

Premature deaths caused by coal pollution in the United States: approximately 24,000 per year

Acid rain formation: $2SO_2 + O_2 \rightarrow 2SO_3$, $SO_3 + H_2O \rightarrow H_2SO_4$

Neutral pH: 7

QUESTIONS

1 A gallon of gasoline weighs about 6 pounds. Yet combustion of that gallon yields nearly 20 pounds of CO_2. How is that possible?

2 Strip mining obviously presents serious environmental challenges. Yet strip-mined coal from the American West is in demand partly because it's less environmentally damaging in one aspect than is Eastern coal. What aspect is that?

3 Why is CO harmful to human health?

4 Carbon dioxide is heavier than air because its triatomic molecules are more massive than the O_2 and N_2 molecules that dominate the atmosphere. Although CO_2 is nontoxic, people have been killed in accidents or natural disasters that release large quantities of CO_2. How is this possible?

5 Explain the difference between primary and secondary air quality standards.

6 Acid rain is mostly a problem in the northeastern United States and northern Europe. Why?

7 What's the purpose of an automobile's catalytic converter?

8 What coal pollutant is generally removed by scrubbing? Why won't filters work on this pollutant?

9 Most constituents of smog aren't emitted directly in vehicle exhaust, and yet vehicle emissions are largely responsible for smog. Explain.

10 Why do coal-burning power plants emit radiation?

11 Switching from coal to natural gas for electrical energy production lowers CO_2 emissions per unit of electrical energy for what two distinct reasons?

12 Figure 6.22 shows that only a small area of the United States fails to meet the National Ambient Air-Quality Standards for ozone. Why is the impact on population more significant than a quick glance at the figure might suggest?

EXERCISES

1 Estimate the amount of CO_2 released in burning 1 gallon of gasoline. Give your answer in both kilograms and pounds. The density of gasoline is about 730 kg/m^3, and gasoline is about 84% carbon by weight.

2 Choose the automatic-transmission gasoline Volkswagen Beetle mentioned in Box 6.1 and your fuel mileage drops to 26 miles per gallon. Suppose you buy the manual-transmission diesel and your friend buys the automatic-transmission gasoline model. If you each drive 10,000 miles per year, how do (a) your annual fuel consumption and (b) your annual CO_2 emissions compare? You'll find the answer to Exercise 1 useful.

3 In the section of this chapter about the transportation of fossil fuels, I pointed out that humankind uses about 500 EJ of energy each year, nearly all of it from fossil fuels, and that the energy content of fossil fuels is around 30 to 60 MJ/kg. (a) Using an approximate figure of 40 MJ/kg, estimate the total weight of fossil fuels we burn each year. (b) Assume that each kilogram of fuel has to be moved 1,000 km from its source to where it's burned (more for imported oil, less for some coal), at an average speed of 20 km per hour. Estimate the total weight of fuel that must be in transit at a given time.

4 A large LNG tanker carries 138,000 m^3 of LNG. (a) Use Table 3.3 and the fact that LNG is 600 times denser than the gaseous state to estimate the total energy content of the LNG tanker's cargo. (b) How long would all that natural gas last if it were used to satisfy the total U.S. energy consumption rate of about 10^{12} W?

5 If all the CO in car exhaust were converted to CO_2, what would be the mass of CO_2 emitted for every kilogram of CO?

6 How much uranium is in each tonne of North Dakota lignite coal, which averages 360 parts per million of uranium?

7 Manual-transmission cars typically average 1 to 2 miles per gallon better than their automatic-transmission counterparts (although the gap has narrowed considerably thanks to computer-controlled transmissions and engines). Assume there are about 150 million cars in the United States, each driven about 12,000 miles per year, and that 90% are automatics. (a) How much gasoline would we save each year if all drivers of automatics switched to comparable manual-transmission cars? Assume an average mileage of 25 miles per gallon for manual transmissions and 23 miles per gallon for automatics. Express your answer in gallons. (b) How does this compare with the 5 million barrels that poured into the Gulf of Mexico during the 2010 *Deepwater Horizon* blowout?

8 Repeat the calculation used in Example 6.1 for a scrubber using magnesium carbonate ($MgCO_3$) and forming magnesium sulfite ($MgSO_3$); your answer should be the daily production of $MgSO_3$.

9 If a 33% efficient coal-burning power plant with electric power output of 1 GWe were to emit the maximum amount of SO_2 allowed as listed in Table 6.2, how much SO_2 would it emit in a year? Note that the energy listed in Table 6.2 refers to the primary energy in the coal, not the electrical output.

10 A 500-MWe gas-fired power plant with 48% efficiency emits 2.4 tonnes of particulate matter each day. Is it in compliance with the particulate-matter standard listed in Table 6.2? Justify your answer quantitatively.

11 If you drive your car 12,000 miles each year, what's the maximum amount of particulate matter you could emit over the year and still meet the standards of Table 6.2? Answer separately for the Tier 1 and Tier 2 standards.

12 By what factor does the concentration of hydrogen ions in rainwater change when the pH drops from 5 to 4.5?

13 A city occupies a circular area 15 km in diameter. If the city's air just barely meets Table 6.1's 24-hour standards for particulate matter of all sizes, what is the total mass of particulate matter suspended in the air above the city to an altitude of 300 m, about the height of the tallest buildings? Assume the particulate matter is uniformly distributed over this volume.

14 During winter, the Vermont Yankee nuclear power plant draws cooling water from the Connecticut River at the rate of 360,000 gallons per minute. The plant produces 650 MW of electric power and has an efficiency of 34%. (a) What is the rate at which the plant discharges waste heat? (b) By how much does the temperature of the cooling water rise as a result of the waste heat it absorbs? You may want to review Section 4.4 on specific heat before tackling this exercise.

15 Ship transportation requires approximately 300 kJ of fuel energy per kilometer for each tonne transported. Consider a supertanker so big it can't fit through the Suez Canal; it carries Middle Eastern oil to the United States on a 21,000-km route around the Cape of Good Hope. Calculate the total fuel energy required for each tonne of oil transported, and compare your result with the energy contained in a tonne of oil.

RESEARCH PROBLEMS

1 Check the EPA web site to find the fuel economy (city and highway), air pollution score, and greenhouse gas score for your or your family's car, SUV, or light truck.

2 Find a source of daily air-quality reports for your community (or the nearest big city for which such reports are available). Keep a week-long record of either overall air quality or the concentration of a particular pollutant, if quantitative data on the latter are available.

3 Look up the most recent large oil spill anywhere in the world. What were its environmental and economic consequences?

4 Determine the five most mercury-contaminated species of fish sold commercially in your country, and rank them in order by degree of contamination.

5 Describe any clean-air legislation currently before the U.S. Congress, your state legislature, or, if you're not from the United States, before your national or regional government.

6 Has the number of air-quality alerts for your community (or the nearest big city) increased or decreased over the past decade? Back up your answer with a graph or table.

ARGUE YOUR CASE

1. Your state's air-quality agency is wrestling with whether or not to permit diesel passenger cars to be sold in your state. Formulate an argument one way or the other, drawing on the latest information on "clean diesel" technology.

2. You're the CEO of an East Coast utility that currently gets coal for your power plants from a mountaintop removal operation in West Virginia. The coal has a high sulfur content, and under new EPA rules your power plants are going to have to emit less sulfur dioxide. Two of your vice presidents are arguing; one wants to install state-of-the-art scrubbers to remove SO_2 from flue gases; the other wants to switch to low-sulfur coal strip-mined in Wyoming. Each VP claims their approach is more sound environmentally. What do you argue?

Chapter 7

NUCLEAR ENERGY

Fossil fuels supply nearly 90% of humankind's energy. A distant second is nuclear energy, used almost exclusively to generate electricity. Today, nuclear energy accounts for some 6% of the world's primary energy and about 14% of its electrical energy. Figure 7.1 shows that those figures vary dramatically from country to country and, in the United States, from state to state. France, for example, gets nearly 80% of its electricity from nuclear plants—the result of a national decision to pursue energy independence. Although the U.S. percentage of nuclear-generated electricity is much lower, about 20%, some regions of the United States depend heavily on nuclear energy. Figure 7.1 shows the fraction of nuclear-generated electricity in a number of countries and U.S. states. Despite its relatively low percentage of nuclear electricity, Figure 7.2 shows that the United States is nevertheless the leader in nuclear generating capability.

Nuclear energy remains controversial. Some encourage it as a clean, carbon-free energy source. Others see nuclear as an unmitigated evil, poisoning the planet with nuclear waste and entangled with the horror of nuclear weapons. Even the environmental community is split in its opinions of nuclear power. Serious concerns follow from the very nature of nuclear energy, but objections to nuclear energy need to be weighed against the risks of fossil and other energy

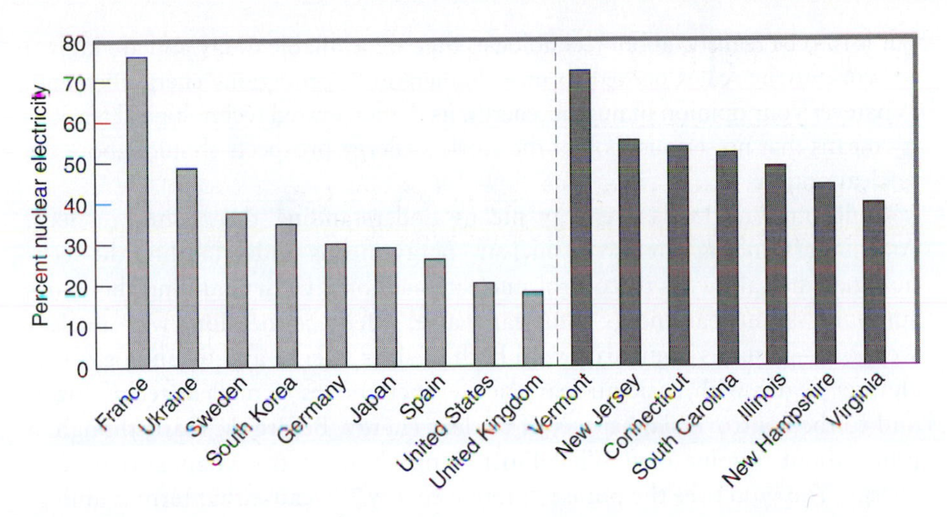

FIGURE 7.1
Percent of electricity generated from nuclear power plants in selected countries and in the most nuclear-dependent U.S. states. Interstate electricity sales mean the U.S. state figures don't necessarily reflect the fraction of nuclear electricity consumed within the state.

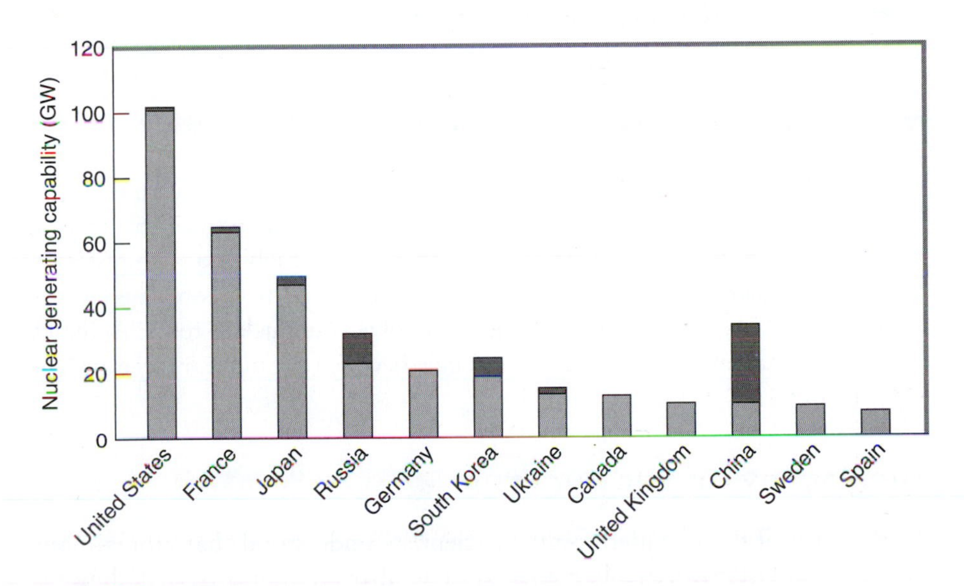

FIGURE 7.2
Countries with the most nuclear generating capability. Darker bars indicate nuclear facilities under construction.

sources. One goal of this chapter is to give a balanced look at the complex questions surrounding nuclear energy, and to help you formulate your own reasoned opinion.

Nuclear energy is one of only two technologically proven nonfossil energy sources capable right now of making a significant contribution to the world's energy supply (the other is hydropower). That is not to denigrate other alternatives, but it is to be realistic about technologies that are available today and are known to work at the scales needed to meet humankind's prodigious energy demand. Whatever your opinion of nuclear energy, its demonstrated technological feasibility means that no serious look at the world's energy prospects should ignore the nuclear option.

Understanding nuclear energy means understanding the atomic nucleus. And understanding our use of nuclear energy means understanding the technologies that allow us to control nuclear reactions. Understanding the issues surrounding nuclear energy—nuclear waste, safety, connections with nuclear weapons—requires familiarity with both nuclear science and technology. This chapter begins with basic nuclear science, then moves to nuclear technologies and to the controversial aspects of nuclear energy. Before we start, though, a point about terminology: What I'm talking about in this chapter is *nuclear* energy. You won't see the phrase "atomic energy," because that term is ambiguous. The chemical energy that comes from rearranging *atoms* to form new molecules—for example, by burning fossil fuels—has just as much claim to the term "atomic energy" as does the energy we get from splitting atomic nuclei. So I'll be unambiguous, and speak of *nuclear* energy, *nuclear* power plants, *nuclear* waste, *nuclear* weapons, *nuclear* policy.

7.1 The Atomic Nucleus

Back in Chapter 3, I introduced the fundamental forces that govern all interactions in the universe—gravity, electromagnetism and its weak-force cousin, and the nuclear force. The chemical energy of fossil fuels is a manifestation of electromagnetism alone, in its incarnation as electric fields and the electric force. As you might guess, nuclear energy involves the nuclear force. Actually, though, nuclear energy depends on a struggle between the nuclear force and the electric force.

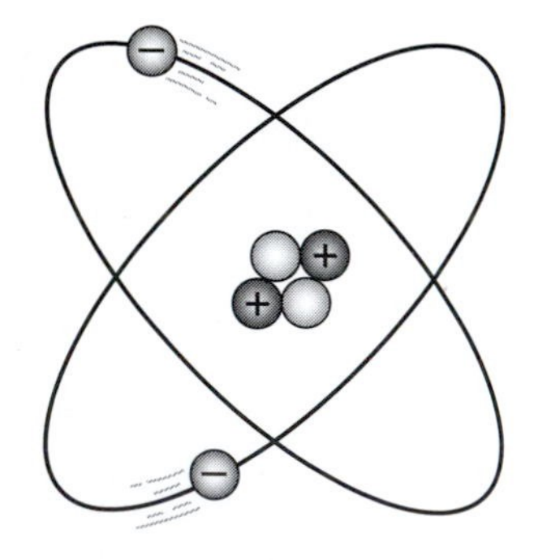

FIGURE 7.3
"Solar system" model of the helium atom, whose nucleus consists of two protons and two neutrons. Figure is not to scale; the electrons are typically some 100,000 times more distant from the nucleus than suggested here.

DISCOVERY OF THE NUCLEUS

By the turn of the twentieth century, scientists understood that atoms—long thought indivisible—actually consisted of smaller components, including the negatively charged electron. Between 1909 and 1911 Ernest Rutherford and colleagues bombarded a thin gold foil with high-energy subatomic particles from a radioactive source. Most passed right through the foil, but on rare occasion one bounced back—evidence that the particle had hit something very massive but very small. Rutherford's group had discovered the **atomic nucleus**, and from their work came the familiar picture of the atom as a tiny but massive, electrically positive nucleus surrounded by orbiting electrons—much like a miniature solar system (Fig. 7.3). Although this picture is not fully consistent with modern quantum physics, it's adequate for our purposes.

By the 1930s, scientists had established that the nucleus itself consists of two kinds of particles: electrically positive **protons** and neutral **neutrons**—

collectively called **nucleons**. Because of their protons, nuclei are positively charged, so they attract electrons to form atoms. Atoms are normally neutral, meaning they have the same number of electrons as protons. It's the outermost electrons that determine the chemical properties of an atom—how it bonds with other atoms to make molecules—and since the number of electrons equals the number of protons, it's the protons that ultimately determine an atom's chemical species. Every chemical element has a distinct number of protons in its nucleus—a quantity called the **atomic number** and given the symbol Z. Hydrogen, for example, has atomic number 1; helium is 2; carbon has $Z=6$, oxygen is 8, iron 26, and uranium has $Z=92$. The atomic number uniquely identifies the element; the element name "nitrogen," for example, is synonymous with atomic number 7.

KEEPING NUCLEI TOGETHER

Every nucleus but hydrogen contains more than one proton and these repel one another because of the electric force. Therefore there must be another—and stronger—force to hold the nucleus together. That's the nuclear force, and it acts between nucleons without regard for electric charge. Thus the nuclear force attracts protons and neutrons, neutrons and neutrons, and even protons and protons. Without it there would be no nuclei, no atoms, no you or me.

Incidentally, if you know something about particle physics, you've probably heard of quarks, the sub-subatomic particles that make up protons and neutrons. What I'm calling the nuclear force is really a residual effect of the so-called strong force that binds quarks together.

The nuclear force is strong, but its range is short. Its strength drops exponentially with increasing distance, so the nuclear force becomes negligible over distances a few times the diameter of a single nucleon—about 1 femtometer (fm, or 10^{-15} meter). The electric force decreases, too, but as the inverse square of the distance—a lot slower decrease than exponential. This means that protons are bound to nearby neutrons and even protons, but repelled by more distant protons. For the nucleus to stick together, nuclear attraction has to overcome electrical repulsion. The neutrons provide the "glue" that makes this possible. Since they offer nuclear attraction without electrical repulsion, neutrons stabilize the nucleus against the tendency of electrical repulsion to tear it apart.

How many neutrons and protons are in a carbon nucleus? There are 6 of each—at least in most carbon. How about oxygen? Eight of each. Helium has 2 protons and 2 neutrons, and most nitrogen has 7 of each. But iron, with 26 protons, has 30 neutrons in most of its nuclei. Iodine has 53 protons and 74 neutrons, while uranium has 92 protons and, usually, 146 neutrons. There's a pattern here: the larger the nucleus, the greater the proportion of neutrons. That's because larger nuclei have protons that are relatively far apart. Distant protons experience electrical repulsion but, because of the short range of the nuclear force, essentially no nuclear attraction. To compensate for this repulsion, larger nuclei need more nuclear "glue"—neutrons—to stick together.

ISOTOPES

Most carbon has 6 protons and 6 neutrons; *most* iron has 26 protons and 30 neutrons; *most* uranium has 92 protons and 146 neutrons. Why *most*? First off, absolutely *all* carbon nuclei have 6 protons; that's what determines the number of electrons and hence the chemistry of the element—in other words, that's what it means to be carbon. To be iron is to have 26 protons in your nucleus, and having 92 protons means you're uranium—period. But the neutrons play no electrical role; add or subtract a neutron and you don't affect the number of electrons in the atom and thus you don't change its chemistry. Carbon with 7 neutrons instead of 6 is still carbon. Oxygen with 10 neutrons is still oxygen (and provides an important diagnostic of global climate change, as you'll see later). Iron with 28 instead of the usual 30 neutrons is still iron. And uranium with 143 neutrons is still uranium. What I'm describing here are different **isotopes** of the same element—versions of the element with different numbers of neutrons. They're *chemically* identical (except for very subtle changes resulting from their slightly different masses), but their nuclear properties can vary dramatically.

We designate isotopes with a symbolic shorthand that includes the symbol for a given element preceded by a subscript indicating the atomic number and a superscript giving the total number of nucleons—also called the **mass number** and approximately equal to the nuclear mass in unified atomic mass units (u, being approximately the mass of the proton, or 1.67×10^{-27} kg). Figure 7.4 shows some isotopes of well-known elements, along with their symbols.

Most elements have several stable isotopes. Carbon-13, also designated C-13 or ${}^{13}_{6}C$, is a stable isotope that comprises 1.11% of natural carbon; the rest is

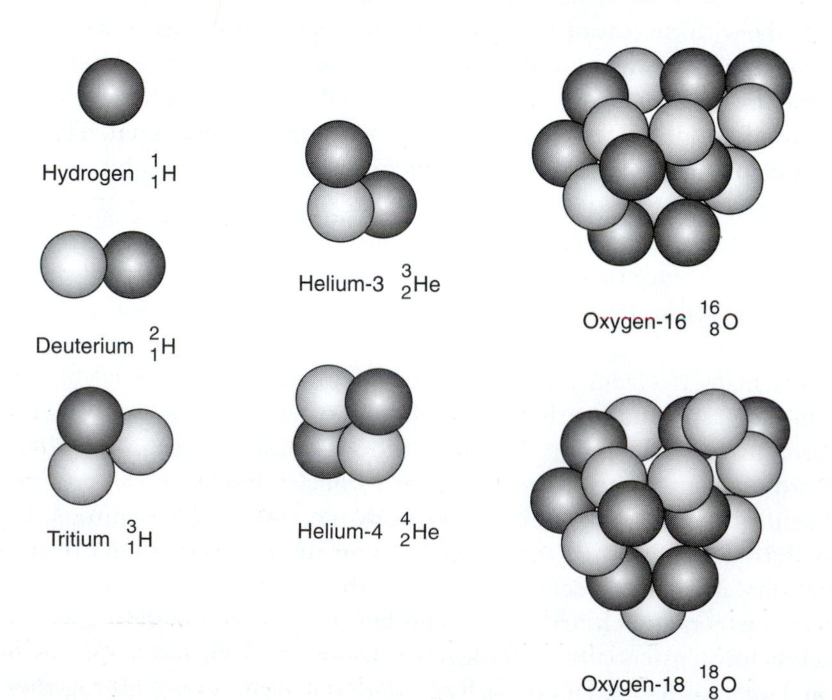

FIGURE 7.4
Isotopes of hydrogen, helium, and oxygen. Each isotope is designated by its element symbol (H, He, O, etc.), a preceding subscript giving the number of protons, and a preceding superscript giving the total number of nucleons. Only the hydrogen isotopes have their own names.

the common carbon-12. Iron-54 ($^{54}_{26}$Fe), with 28 neutrons, comprises about 6% of natural iron. Most of the rest is iron-56 with 30 neutrons, but there's also Fe-57 at about 2% and Fe-58 at 0.3%. Even the hydrogen nucleus isn't always a single proton; one in about every 6,500 hydrogen nuclei is hydrogen-2 (H-2 or $^{2}_{1}$H), also called **deuterium** (and sometimes given its own symbol, D), whose nucleus consists of 1 proton and 1 neutron. And the two uranium isotopes I've mentioned—neither of them completely stable but which last for geological timescales—occur with natural abundances of 99.3% for U-238 and 0.7% for U-235. That particular difference has enormous consequences for our security in the nuclear age.

EXAMPLE 7.1 | Isotopes

The two stable isotopes of chlorine are Cl-35 and Cl-37; they occur naturally with abundances of 75.8% and 24.2%, respectively. Assuming their atomic masses are very nearly equal to their mass numbers, estimate the average atomic mass of natural chlorine.

SOLUTION

We find the average atomic mass by weighting each isotope's mass by its proportion in natural chlorine. Thus the average atomic mass of chlorine is

$$(35)(0.758) + (37)(0.242) = 35.5$$

To three significant figures, this is the value listed for chlorine in the periodic table of the elements. By the way, you'll find many elements in the periodic table whose atomic masses are very close to whole numbers; these typically have one dominant isotope, whose atomic mass is close to but not exactly equal to its mass number. Chlorine is an exception, with two stable isotopes that contribute substantially to the mix of natural chlorine.

7.2 Radioactivity

Neutrons provide the "glue" that holds nuclei together. A nucleus with too few neutrons eventually comes apart as a result of its protons' mutual repulsion. This "coming apart" is an instance of **radioactive decay**—the phenomenon whereby a nucleus spews out particles in order to assume a more stable configuration. The emitted particles constitute **nuclear radiation**. Understanding radioactivity and radiation is crucial in evaluating health and environmental impacts of nuclear energy.

For the largest nuclei—those with atomic number 84 or greater—no amount of neutron "glue" can hold the nucleus together. So all elements with $Z \geq 84$ are radioactive, and they most commonly undergo **alpha decay**—the emission of a helium nucleus ($^4_2$He), also called an **alpha particle**. Alpha decay reduces the number of protons and neutrons each by two. A typical alpha-emitting process is the decay of uranium-238, yielding thorium-234:

$$^{238}_{92}\text{U} \rightarrow {}^{234}_{90}\text{Th} + {}^{4}_{2}\text{He} \qquad \text{(alpha decay)} \qquad (7.1)$$

Note how this equation balances: The number of positive charges (atomic number, at the bottom of each nuclear symbol) is 92 on the left and 90 + 2 = 92 on the right. The mass numbers at the top also agree: 238 on the left and 234 + 4 = 238 on the right.

Although neutrons help stabilize nuclei, too many neutrons make a nucleus unstable. Such a nucleus eventually undergoes **beta decay**, a subtle process enabled by the weak force I mentioned in Section 3.1. In beta decay, a neutron decays into a positive proton, a negative electron, and an elusive, nearly massless *neutrino*. The electron—called a **beta particle** in this context—emerges with high energy, leaving the nucleus with one more proton than before. Thus its atomic number increases while—given the electron's tiny mass—its mass number is unchanged. A typical beta decay is that of carbon-14, exploited by archeologists in radiocarbon dating:

$$^{14}_{6}\text{C} \rightarrow {}^{14}_{7}\text{N} + {}^{0}_{-1}e + \bar{v} \qquad \text{(beta decay)} \qquad (7.2)$$

Here e is the electron, with subscript −1 indicating that it carries 1 unit of negative charge and superscript 0 showing that its mass is negligible. The final symbol, $\bar{v}$, is the neutrino (actually an antineutrino, indicated by the bar). Note again how the numbers add up: the subscripts show +6 units of charge on the left and +7 + (−1) = +6 units on the right. The masses, too, agree: 14 on each side of the equation. The nuclear end-product of this beta decay is ordinary nitrogen-7, the dominant nitrogen isotope.

Following radioactive decay or other interactions, a nucleus may end up with excess energy. It sheds this by **gamma decay**, emitting a high-energy photon called a **gamma ray**. Recall from Chapter 3 that a photon is just a bundle of electromagnetic radiation. The photon has no charge or mass, so, unlike alpha and beta processes, gamma decay doesn't change the type of nucleus.

EXAMPLE 7.2 | Making Plutonium

When uranium-238 absorbs a neutron (^1_0n; no charge but 1 mass unit), the result is uranium-239: $^{238}_{92}\text{U} + {}^{1}_{0}n \rightarrow {}^{239}_{92}\text{U}$. The U-239 then undergoes two successive beta decays, like that of carbon-14 in Equation 7.2. Write equations for these two decays, and identify the final products.

SOLUTION

Each beta decay increases the atomic number by 1, so we're going to end up with elements 93 and 94. A look at the periodic table shows that these are, respectively, neptunium (Np) and plutonium (Pu). So the reactions are:

$$^{239}_{92}U \rightarrow ^{239}_{93}Np + ^{0}_{-1}e + \bar{\nu}$$

and

$$^{239}_{93}Np \rightarrow ^{239}_{94}Pu + ^{0}_{-1}e + \bar{\nu}$$

The end-product is plutonium-239, a potential bomb fuel. The reactions described in this example occur in nuclear reactors—a fact that inexorably links nuclear power and nuclear weapons.

HALF-LIFE

The decay of a single radioactive nucleus is a truly random event, but given a large number of identical radioactive nuclei a statistical pattern emerges: After a certain time just about half of the nuclei will have decayed. This time is the **half-life**. Wait another half-life and half of those remaining nuclei are gone, too—leaving only one-fourth of the original number. After 3 half-lives, only one-eighth of the original nuclei remain. Wait 10 half-lives, and only about one-thousandth remain (actually, $1/2^{10}$, or 1/1024). Twenty half-lives, and you're down to about a millionth of the original sample. Figure 7.5 depicts this exponential decay, whose mathematics is the same as that of exponential growth described in Box 5.2, but with a minus sign.

Half-lives of radioactive isotopes vary dramatically, from fractions of a second to billions of years. Radioactive isotopes found in nature are either those with very long half-lives, shorter-lived nuclei that result from the continual decay of longer-lived isotopes, or isotopes formed by nuclear reactions involving cosmic rays. Table 7.1 lists some important radioactive isotopes, their half-lives, their origin, and their scientific, environmental, or health significance.

The shorter its half-life, the more rapidly a radioactive substance decays—and therefore the more vigorously radioactive it is. Thus the uranium isotopes, with their half-lives in the billion-year range, aren't strongly radioactive. For that reason fresh nuclear fuel doesn't pose a great radiation hazard. But after it's

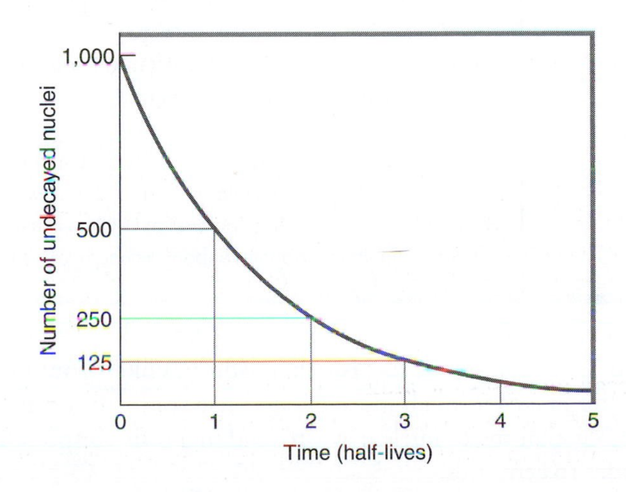

FIGURE 7.5
Decay of a radioactive sample containing initially 1,000 nuclei.

TABLE 7.1 | SOME IMPORTANT RADIOACTIVE ISOTOPES

Isotope	Half-life (approximate)	Significance
Carbon-14	5,730 years	Formed by cosmic rays; used in radiocarbon dating for objects up to 60,000 years old.
Iodine-131	8 days	Product of nuclear fission; released into the environment from nuclear weapons tests and reactor accidents. Lodges in thyroid gland where it can cause cancer.
Potassium-40	1.25 billion years	Isotope comprising 0.012% of natural potassium. The dominant radiation source within the human body. Used for dating ancient rocks and to establish Earth's age.
Plutonium-239	24,000 years	Isotope produced in nuclear reactors from U-238. Can be used in nuclear weapons.
Radon-222	3.8 days	Gas formed by the decay of natural radium in rocks and soils, ultimately from uranium-238. Can be a health hazard when it seeps into buildings.
Cesium-137	30 years	Product of nuclear fission responsible for widespread land contamination at Chernobyl and Fukushima. Water soluble and mimics potassium in the body.
Strontium-90	29 years	Product of nuclear fission that chemically mimics calcium, so it's absorbed into bone. Still at measurable levels in the environment following above-ground nuclear bomb tests of the mid-twentieth century.
Tritium (hydrogen-3)	12 years	Produced in nuclear reactors. Radioactive hydrogen isotope used to "tag" water and other molecules for biological studies. Used to boost the explosive yield of fission weapons.
Uranium-235	704 million years	Fissile isotope comprising 0.7% of natural uranium; fuel for nuclear reactors and some nuclear weapons.
Uranium-238	4.5 billion years	Dominant uranium isotope (99.3%). Cannot sustain a chain reaction, but boosts the yield of thermonuclear weapons. Depleted uranium—after removal of U-235—is used for armor-penetrating conventional weapons because of its high density.

been "burned" in a reactor, the fuel is rich in isotopes like strontium-90, whose 29-year half-life makes it much more dangerously radioactive. It's short-lived isotopes like ^{90}Sr that make nuclear waste so dangerous—especially on timescales of decades to centuries.

EXAMPLE 7.3 | Nuclear Terrorism

A serious concern in this age of terrorism is the so-called "dirty bomb" that would use conventional chemical explosives to disperse radioactive material. Spent nuclear fuel would make a particularly dirty weapon. Suppose such a device were made with strontium-90, and ended up spreading Sr-90 into the environment at levels 100,000 times what's considered safe. If the Sr-90 could not be removed, how long would the contaminated region remain unsafe?

SOLUTION

In one half-life, the level of radioactive material drops to half its original value; in two half-lives to one-fourth, and so on—dropping to $1/2^n$ after n half lives. Here we need the level to drop to 1/100,000 of its original value, so we want n such that $1/2^n = 1/100{,}000$, or, inverting, $2^n = 100{,}000$. We can isolate the n in this equation by taking logarithms of both sides:

$$\log(2^n) = \log(100{,}000)$$

But $\log(2^n) = n \log(2)$, and $\log(100{,}000) = \log(10^5) = 5$, using base 10 logarithms. So we have

$$n \log(2) = 5$$

or

$$n = \frac{5}{\log(2)} = \frac{5}{0.301} = 16.6$$

for the number of half-lives needed. Table 7.1 shows that strontium-90's half-life is 29 years, so this answer amounts to (29 years/half-life)(16.6 half-lives), or just under 500 years. That's a long time!

If you're not fluent with logarithms, you might recall that 20 half-lives drop the level of radioactivity by a factor of a million, which is more than we need here. Working backward, 19 half-lives will drop by a factor of half a million, 18 by a quarter-million, 17 by one-eighth million or 125,000—just over what we need. So the answer must be a little under 17 half-lives, as our more rigorous calculation shows.

7.3 Energy from the Nucleus

THE CURVE OF BINDING ENERGY

It takes energy to tear a nucleus apart, working against the strong nuclear force that binds the nucleons. That energy is called **binding energy**. Conversely, when a nucleus is formed from widely separated neutrons and protons, energy equal to the binding energy is released. In small nuclei, individual nucleons don't have many neighbors to exert attractive nuclear forces, and the binding

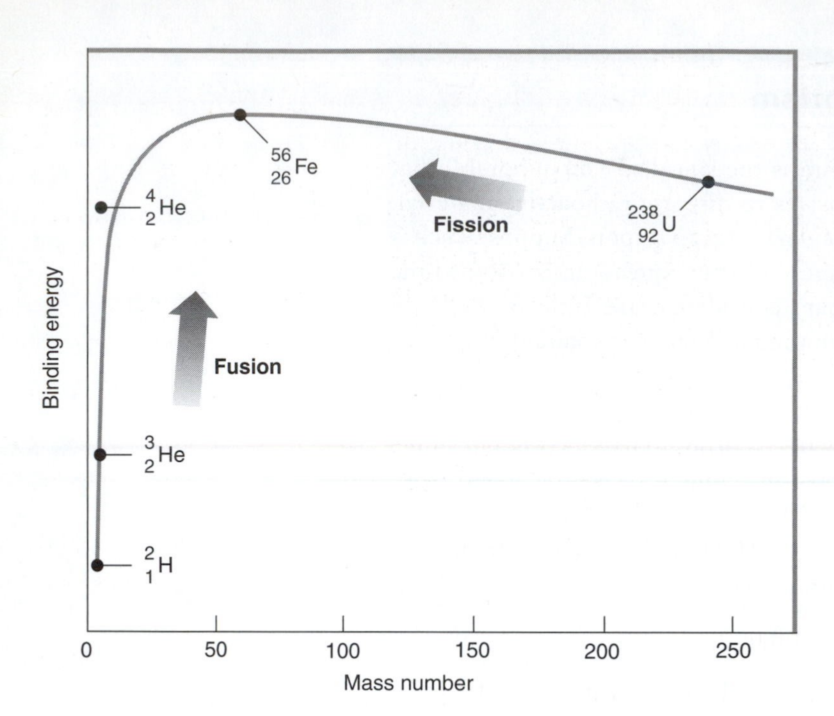

FIGURE 7.6

The curve of binding energy, a plot of binding energy per nucleon versus mass number. Individual isotopes may lie slightly off the general curve, as with ${}^4_2$He. Arrows indicate the energy-releasing paths of nuclear fusion and fission.

energy associated with a single nucleon isn't particularly large. But as nuclei get larger, the force on each nucleon grows, and with it the binding energy per nucleon. Still larger and the binding energy per nucleon begins to level off because additional nucleons are so far away that the short-range nuclear force doesn't contribute much more to the binding energy. Then at some point—at the size of the iron nucleus, in fact—the repulsive effect of the protons begins to tip the balance. As a result, binding energy per nucleon has the greatest value for iron and decreases for heavier elements. This trend is evident in the famous **curve of binding energy**, a plot of the binding energy per nucleon versus atomic number (Fig. 7.6).

The curve of binding energy shows that there are two fundamentally different strategies for extracting energy from atomic nuclei. The process of **fusion** joins two lighter nuclei, increasing the binding energy and thus releasing energy equal to the difference between the total binding energies of the original nuclei and that of the combined nucleus. I've marked the fusion process with an arrow in Figure 7.6. Nuclear fusion is a crucial process in the universe; it's what powers the stars. The Sun, specifically, fuses hydrogen into helium; in more massive stars fusion makes heavier elements as well. Thus the energy that comes to Earth in the steady flow of sunlight, as well as the energy locked away in fossil fuels, is energy that originated from nuclear fusion in the Sun. Fusion is also the dominant energy source in thermonuclear weapons, or "hydrogen bombs." We haven't yet learned to control fusion for the steady generation of energy.

The curve of binding energy heads downward beyond iron, showing that nuclei much heavier than iron have less binding energy per nucleon than those only a little heavier. Therefore splitting a massive nucleus into two medium-weight nuclei is also an energy-releasing process. This is **fission**. Unlike fusion, fission plays almost no role in the natural universe, but it's fission that drives our nuclear power plants. (A natural fission reaction did occur 2 billion years ago in what's now a uranium mine in Africa, and has provided us with valuable information on the migration of nuclear wastes through the environment.) I've marked an arrow symbolizing fission energy release in Figure 7.6.

THE NUCLEAR DIFFERENCE

Chemical reactions involve rearrangement of atoms into new molecular configurations, but the atoms' nuclei remain unchanged. The energy released in such reactions is modest; for example, formation of one carbon dioxide molecule

(a)

(b)

FIGURE 7.7
The nuclear difference. (a) Unloading fuel at a nuclear power plant. Refueling takes place typically once a year, with one or two truckloads of uranium fuel. (b) A 110-car trainload of coal arriving at a fossil-fueled power plant, one of 14 such deliveries each week.

from carbon and oxygen releases about 6.5×10^{-19} joules, or somewhat under 1 attojoule (aJ). In contrast, the energy released in forming a helium nucleus is about 4 million aJ, and splitting a single uranium nucleus gives about 30 million aJ. So nuclear reactions provide, in rough terms, something like 10 million times the energy of chemical reactions. I call this the **nuclear difference**, and it's the reason nuclear fuel is such a concentrated energy source. The nuclear difference has a great many practical, and sometimes ominous, implications. For example, it's why we need many 100-car trainloads of coal each week to fuel a coal-burning power plant (recall Example 5.2), while a nuclear plant refuels maybe once a year or so with a truckload of uranium fuel (Fig. 7.7). It's also the reason why terrorists could fit a city-destroying bomb in a suitcase. The nuclear difference is why major nuclear accidents spread anxiety throughout the whole world. Yet it's also why the volume of waste produced by the nuclear power industry is minuscule compared with the tons and tons of carbon dioxide, fly ash, and sulfurous waste produced in burning coal. That's a comment about quantity, not danger; nuclear waste is, on a pound-for-pound basis, a lot more dangerous than the products of fossil fuel combustion. But there are a lot fewer pounds of it—by, again, that factor of roughly 10 million.

Physically, the nuclear difference reflects the relative strengths of the forces binding nuclei and molecules. Because the nuclear force is so strong, rearranging nuclei necessarily involves a lot of energy. Chemical reactions, in contrast, involve the much weaker electrical forces on the outermost atomic electrons. Consequently the energies involved in atomic—that is, chemical—rearrangements are much smaller.

7.4 Nuclear Fission

As Figure 7.6 showed, splitting any massive nucleus should release energy. But for most nuclei splitting doesn't happen readily. The exceptions, which occur among the heavier nuclei, are designated **fissionable**. Some nuclei undergo spontaneous fission, but this process occurs rarely, making the total energy

BOX 7.1 | $E = mc^2$

Most people think of Einstein's famous equation $E = mc^2$ as the basis of nuclear energy. Actually, Einstein's equation applies to all energy-releasing processes, from the burning of fossil fuels to the metabolism of food to the snapping of a rubber band to the fissioning of uranium. What $E = mc^2$ expresses is a universal equivalence between matter and energy. $E = mc^2$ says that if a process releases energy E, then there's an associated decrease in mass given by E/c^2. The nature of the energy-releasing process doesn't matter. Consider two electric power plants, one nuclear and one coal. Suppose they have exactly the same efficiency, and produce the same amount of electric power. Then they convert matter to energy at exactly the same rate—the coal plant no more and no less than the nuclear plant. Weigh all the uranium that goes into the nuclear plant in a year, and all the waste that comes out; the latter will weigh less. Weigh all the coal and oxygen going into the fossil plant, and all the products of combustion; the latter will weigh less—and the difference will be the same as in the nuclear case. What distinguishes the two is the fraction of the fuel mass that gets converted to energy. Only a minuscule fraction of the coal's mass is converted—which is why, as Figure 7.7 graphically illustrates—the plant consumes coal at a huge rate. To achieve the same energy output, the nuclear plant converts the same amount of matter to energy—but given the nuclear difference, the fraction of nuclear fuel converted is some 10 million times greater (although still well under 1%). In fact, it's only with nuclear reactions that it's practical to measure changes in mass associated with energy release—but nevertheless mass change occurs in all energy-releasing processes.

release insignificant. More productive is **neutron-induced fission**. Here, a nucleus captures a neutron and becomes a new isotope. With most nuclei that's the end of the process (although the new isotope may be unstable and eventually undergo radioactive decay). But with fissionable nuclei, neutron capture can cause the nucleus to split into two middleweight nuclei (Fig. 7.8). These **fission products** fly apart with a lot of kinetic energy; in fact, most of the approximately 30 picojoules (pJ, or 30×10^{-12} J) released in uranium fission goes into kinetic energy of the fission products.

The fission products have nearly the same neutron-to-proton ratio as the uranium they came from—but because they're much lighter nuclei, that gives them too many neutrons for stability. Fission products are therefore highly radioactive, most decaying by beta emission as described in Section 7.2. Fission products constitute the most dangerously radioactive component of nuclear waste, although they have relatively short half-lives.

Many heavier nuclei are fissionable, including the common uranium isotope U-238. In most cases, though, the bombarding neutron must deliver substantial energy, making fission difficult to achieve. A much smaller group of nuclei undergo fission with neutrons of arbitrarily low energy. Principal among these **fissile** isotopes are uranium-233 (U-233), uranium-235 (U-235), and plutonium-239 (Pu-239). Of these, only U-235 occurs in nature, and it comprises just 0.7% of natural uranium (the rest is U-238). Uranium-235 is therefore the important ingredient in nuclear fuel, although nonfissile U-238 and Pu-239 play significant roles in nuclear energy as well.

THE DISCOVERY OF FISSION

The discovery of nuclear fission marks an important moment in human history. In the 1930s, the Italian physicist Enrico Fermi and his colleagues tried bombarding uranium with neutrons, in the process creating a host of new radioactive materials. Most were so-called **transuranic** elements, heavier than the heaviest naturally occurring element, namely uranium. The German chemist Ida Noddack was the first to suggest that such experiments might also lead to what we now call fission—the splitting of uranium into large chunks. Noddack's suggestion went largely unnoticed, while the neutron-uranium experiments continued.

In 1938, German chemists Otto Hahn and Fritz Strassmann analyzed the results of their experiments on neutron bombardment of uranium. To their surprise, they found barium among the products. Barium's atomic number is only 56, just over half that of uranium's 92. So where did the barium come from? Hahn and Strassmann's former colleague, the physicist Lise Meitner, found the answer in December 1938. Now in Sweden after fleeing Nazi Germany, Meitner discussed the uranium experiments with her nephew, Otto Frisch. Meitner drew a sketch similar to Figure 7.8, suggesting how uranium might split into two middleweight fragments, one of them Hahn and Strassmann's barium. Meitner and Frisch calculated that the energy released in the process would be tens of millions of times that of chemical reactions—a recognition of the nuclear difference. In a paper describing their work, Meitner and Frisch first used the word *fission* in the nuclear context.

Fission's discovery came on the eve of World War II and involved scientists on both sides of Europe's great divide. The weapons potential of fission was obvious, and soon both sides had secret fission research programs. In the United States, where a stream of scientists fleeing Fascism contributed to the effort, success came when, in 1942, a group under Fermi achieved the first sustained fission reaction at the University of Chicago. Detonation of the first fission bombs came in 1945, with the test at Trinity Site in New Mexico, followed weeks later by the bombing of the Japanese cities Hiroshima and Nagasaki. Since then the destructive potential of fission bombs and more advanced nuclear weapons has been a guiding force in international affairs and an ongoing threat to humanity's very existence.

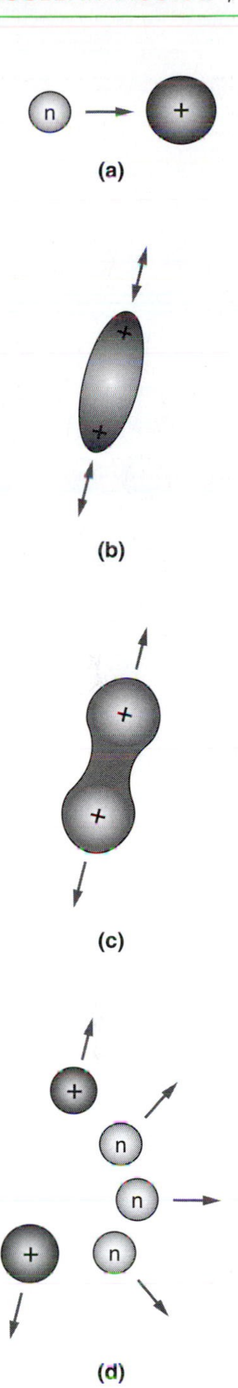

FIGURE 7.8
Neutron-induced fission. (a) A neutron strikes a heavy nucleus and is absorbed. (b) The nucleus begins to oscillate. (c) It takes on a dumbbell shape, and the repulsive electric force begins to dominate. (d) The nucleus fissions into two unequal middleweight nuclei, emitting several neutrons in the process.

THE CHAIN REACTION

Fission of U-235 produces more than fission products and energy. As Figure 7.8 showed, typically two or three neutrons also emerge. And it's neutrons that induce U-235 fission in the first place, so each fission event releases neutrons that can go on to cause more fission. The result is a **chain reaction**, a self-sustaining process that may continue until there's no U-235 left to fission (Fig. 7.9).

Imagine starting with a lump of uranium containing at least some of the fissile isotope U-235. Soon a U-235 nucleus fissions spontaneously, releasing two or three neutrons. If at least one of those neutrons strikes another U-235 nucleus, the reaction continues. But that's a big "if," for several reasons. First of all, most of the nuclei in natural uranium—99.3%—are non-fissile U-238. So a neutron

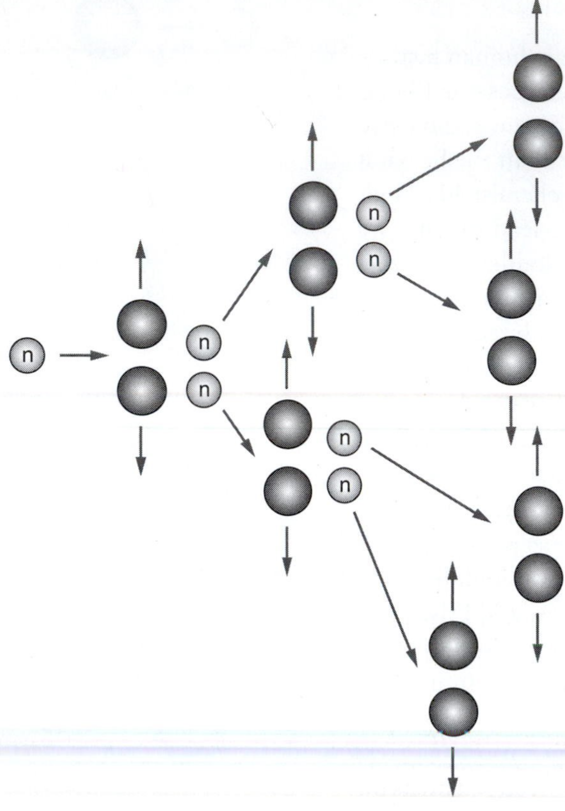

FIGURE 7.9

A nuclear chain reaction. At left, a neutron causes fission in a U-235 nucleus, producing two fission-product nuclei (larger spheres) and, in this case, two neutrons. Each neutron causes an additional fission, and the chain reaction grows exponentially. This is what happens in a bomb; in a reactor the neutrons are carefully controlled to ensure that, on average, each fission results in only one additional fission.

striking U-238 is "lost" to the chain reaction. Other substances in or around the nuclear fuel may also absorb neutrons and thus squelch the chain reaction. And if the volume of nuclear fuel is small, some neutrons will escape altogether rather than encounter a nucleus. For this last reason there's a **critical mass** required to sustain a chain reaction. The critical mass depends on the geometrical shape of the material and on the concentration of fissile isotopes. For pure U-235, the critical mass is about 30 pounds; for plutonium-239 it's 5 pounds or so.

Those numbers are frighteningly small. Consider a chain reaction like the one shown in Figure 7.9, in which, on average, two fission-released neutrons go on to cause additional fission. Soon after the initial fission event, 2 fissions occur; soon after that, 4, then 8, then 16—and the fission rate grows exponentially. In a mass of pure U-235, the time between successive fissions is about 10 nanoseconds (10 billionths of a second). At that rate, a 30-pound mass would fission in a microsecond, releasing about 10^{15} J of energy. That's the energy in 30,000 tonnes of coal (three of those 100-car trains; see Exercise 5), but released in a millionth of a second. What I've described, of course, is a nuclear fission bomb, and you can see why the relatively small size of a critical mass is so frightening.

EXAMPLE 7.4 | Critical Mass

Using an energy of 30 pJ per fission event in U-235, confirm that fissioning of a 30-pound critical mass releases about 10^{15} J.

SOLUTION

A mass of 30 pounds corresponds to (30 lb)/(2.2 kg/lb) = 13.6 kg. The mass of a U-235 nucleus is approximately 235 u, or

$$(235\ \text{u})(1.67 \times 10^{-27}\ \text{kg/u}) = 3.92 \times 10^{-25}\ \text{kg}$$

So our critical mass contains

$$\frac{13.6\ \text{kg}}{3.92 \times 10^{-25}\ \text{kg/nucleus}} = 3.47 \times 10^{25}\ \text{U-235 nuclei}$$

At 30 pJ (30×10^{-12} J) per fission, the total energy released is

$$(3.47 \times 10^{25}\ \text{nuclei})(30 \times 10^{-12}\ \text{J/nucleus}) = 1.04 \times 10^{15}\ \text{J}$$

By the way, that's just about the equivalent of 250,000 tons of chemical explosive, making this a 250-kiloton bomb—some 20 times more powerful than the Hiroshima bomb, which actually fissioned only about 1 kg of its 100 kg of uranium.

CONTROLLING THE CHAIN REACTION

To produce useful energy from fission, we want, on average, exactly one of those neutrons released by each fissioning nucleus to induce another fission. Less than one, and the reaction fizzles. More than one, and the reaction grows explosively, although perhaps not as fast as in a bomb. The all-important quantity here is the average number of neutrons from each fission event that cause additional fission; that's called the **multiplication factor**.

Nuclear reactors for power generation use a variety of approaches to controlling the chain reaction. We'll explore technological details in the next section; here I present the basic principles behind reactor control. In designing a reactor, a nuclear engineer has two goals: first, to ensure that there are enough neutrons to sustain a chain reaction, and second, to ensure that the reaction rate doesn't start growing exponentially. In other words, the goal is to keep the multiplication factor at 1, not more and not less. Because the reaction rate varies with factors like temperature and age of the nuclear fuel, a nuclear reactor needs active systems that can respond to such changes to keep the multiplication factor at exactly 1.

It turns out that low-energy, slow-moving neutrons are far more likely to induce fission in U-235 than are the fast neutrons that emerge from fissioning uranium. That's not an issue in a bomb, where fissile uranium is so tightly packed that fast neutrons can still do the job. But it does mean that, in most nuclear reactor designs, it's necessary to slow the neutrons. The substance that slows neutrons is called the **moderator**. That term can be a bit misleading; you might think it means something that "moderates" or limits the chain reaction, but the opposite is true: the moderator slows the neutrons, helping sustain the chain reaction. Without a moderator, the chain reaction in most nuclear reactors would stop.

What to use for a moderator? The most effective way to slow something is to have it collide with an object of similar mass. (You've learned this if you've had introductory physics or if you've played pool: If one ball hits another head-on, the first stops and transfers all its energy to the second ball.) Something with very nearly the same mass as the neutron is the proton. There are protons galore in water (H_2O), since the nucleus of ordinary hydrogen (^1_1H) is just a proton. So water makes an excellent moderator. However, ordinary water has an undesirable property, namely that hydrogen readily absorbs neutrons, turning into the heavier hydrogen isotope deuterium (^2_1H, or D). A neutron absorbed is a neutron that can't cause fission. That problem, coupled with the absorption of neutrons by the common but nonfissile uranium isotope U-238, makes it impossible to sustain a chain reaction in natural uranium with ordinary water as the moderator. Fuel for such reactors must be enriched in U-235, from that isotope's natural abundance of 0.7% up to about 4%. Enrichment is expensive and technologically challenging. More importantly, there's nothing to prevent a country with enrichment capability from enriching to 80% or more U-235—and that's bomb-grade material.

An alternative to ordinary water is **heavy water** ($^2_1H_2O$, or D_2O) containing the hydrogen isotope ^2_1H, deuterium. Deuterium absorbs fewer neutrons, so the reactor fuel can be unenriched natural uranium. But heavy water is rare, and extracting it from ordinary water is expensive. A third moderator is carbon, in the form of graphite. This has the advantage of being a solid, although carbon's greater mass means it's less effective in slowing neutrons. And graphite is flammable—a significant safety consideration. Overall, there's no obvious choice for the ideal moderator. In the next section I'll show how the choice of moderator figures centrally in the different nuclear-reactor designs in use today.

Precise control over the chain reaction is achieved with neutron-absorbing **control rods** that can move in among bundles of nuclear fuel, absorbing neutrons and decreasing the reaction rate. Pulling the rods out increases the reaction rate. The whole system can be automated, using feedback, to maintain the multiplication factor at exactly 1 and therefore keep the chain reaction going at a steady rate. In an emergency shutdown, the control rods are inserted fully and the chain reaction quickly stops.

All those fission reactions produce heat, which arises as the high-energy fission fragments and neutrons collide with the surrounding particles, giving up their energy in what becomes random thermal motion. If that energy weren't removed, the nuclear fuel would soon melt—a disaster of major proportions. More practically, we want to use that energy to generate electricity. A nuclear power reactor is, after all, just a big device for heating water to make steam. So every reactor needs a **coolant** to carry away the energy generated by nuclear fission.

Moderator and control rods are the essential elements that control the nuclear chain reaction; the coolant extracts the useful energy. Together, the choice of these three materials largely dictates the design of a nuclear reactor.

FIGURE 7.10
Bundles of nuclear fuel rods being lowered into a reactor.

7.5 Nuclear Reactors

A nuclear power reactor is a system designed to sustain a fission chain reaction and extract useful energy. In most of today's reactors, the fuel takes the form of long rods containing uranium oxide (UO_2; Fig. 7.10). Moderator, control rods, and coolant occupy the space between the fuel rods. But there the similarity ends, as the choice of these materials, especially the moderator, leads to vastly different designs.

LIGHT-WATER REACTORS

More than 80% of the world's nuclear power plants use **light-water reactors (LWRs)**, with ordinary water as the moderator. The same water also serves as coolant. A thick-walled **pressure vessel** contains the nuclear fuel, and the moderator/coolant water circulates among the fuel rods to slow neutrons and carry away thermal energy. The simplest LWR is the **boiling-water reactor (BWR)**, in which water boils in the reactor vessel to make steam that drives a turbine connected to an electric generator, just as in a fossil-fueled power plant. About a third of the world's LWRs are BWRs. The rest are **pressurized-water reactors (PWRs)**, in which the water is kept under such high pressure that it can't boil. Superhot water from the reactor flows through a **steam generator**, where it contacts pipes containing lower-pressure water that's free to boil. Steam in this secondary water system then drives a turbine. One advantage of the PWR is that water in the secondary loop never contacts the nuclear fuel, so it doesn't get radioactive. Figure 7.11 shows what's inside the pressure vessel of a typical light-water reactor.

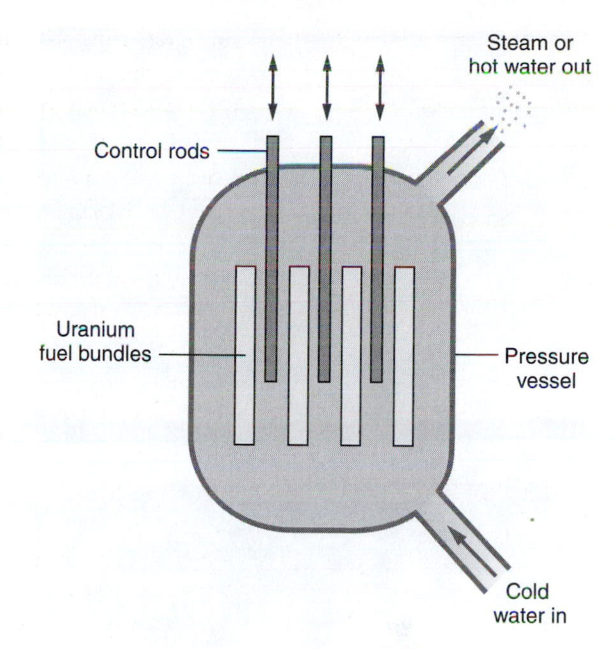

FIGURE 7.11
Inside the pressure vessel of a light-water reactor. Fuel bundles are assemblies of individual fuel rods as shown in Figure 7.10. Control rods, made of neutron-absorbing material, move up and down to control the chain reaction.

Light-water reactors are based on designs first used to power submarines in the 1950s. They're fairly straightforward and have one important safety feature: Loss of coolant—among the most serious of possible reactor accidents—also deprives the reactor of its moderator, which immediately stops the chain reaction. That's not quite as promising as it sounds, though, because radioactive decay continues to produce energy at nearly one-tenth the rate of the full-scale fission reaction. But it does mean that the nuclear chain reaction itself stops automatically in the event of coolant loss.

Refueling a light-water reactor is difficult, because it requires shutting down the entire operation and opening the pressure vessel. So refueling takes place at intervals of a year or more, and usually entails weeks of downtime (Fig. 7.12). There's a security advantage here, though, because it's difficult to remove fuel from an operating reactor and extract plutonium for weapons.

OTHER CONTEMPORARY REACTOR DESIGNS

Canada chose its own route to nuclear power, with the CANDU design (for CANadian-Deuterium-Uranium). CANDU reactors use heavy water for their moderator, meaning they can use natural, unenriched uranium fuel. That's an advantage as far as weapons-proliferation potential is concerned. CANDU reactors are used throughout the world, and comprise around 10% of operating reactors. The CANDU's continuous refueling capability minimizes downtime, and CANDU reactors are generally competitive economically and technologically with light-water reactors.

Still other reactors use solid graphite as moderator, with fuel rods and coolant channels interspersed among graphite blocks. This design was popular in Russia and other states of the former Soviet Union. Like the CANDU, the Russian design can be refueled continually, and some Soviet power reactors served the dual purpose of producing both civilian electric power and plutonium for nuclear weapons. Graphite reactors lack the inherent safety of the light-water design, because coolant loss doesn't stop the chain reaction. In fact,

FIGURE 7.12
Refueling a light-water reactor is a major operation, undertaken once a year or even less frequently. Here workers remove the top of the pressure vessel.

the neutron-absorbing properties of the hydrogen in light water mean that loss of coolant can actually accelerate the chain reaction. That phenomenon played a role in the 1986 Chernobyl accident. So did combustion of the graphite moderator, which burned fiercely and helped spew radioactive materials into the environment.

A variant on the graphite-moderated design is the gas-cooled reactor (GCR). Here, helium gas flows through channels in a solid graphite moderator to carry away the heat from fissioning uranium. Britain and France constructed most of the 18 GCRs in operation today. The only commercial GCR in the United States was not particularly successful and was eventually converted to a natural-gas-fired facility.

BREEDER REACTORS

Example 7.2 showed how neutron absorption in U-238 leads to plutonium-239. Pu-239 is fissile, like U-235, so it, too, can participate in a chain reaction. Pu-239 has a lower critical mass than U-235, and other properties that make it more desirable and dangerous as a bomb fuel—although one that's more difficult than uranium to engineer into a bomb.

Conversion of nonfissile U-238 into fissile Pu-239 occurs gradually in a normal nuclear reactor. Plutonium joins in the fission chain reaction, and, as the nuclear fuel ages, Pu-239 contributes significantly to the reactor's energy production. In a typical light-water reactor, roughly two Pu-239 nuclei are produced for every three U-235 nuclei that fission. Since reactor fuel contains relatively little U-235 compared with U-238, that means only a small fraction of the U-238 gets converted to plutonium.

A **breeder reactor** is designed to optimize plutonium production, actually making more plutonium than the uranium-235 it consumes. Breeder reactors could, in principle, utilize much of the 99.3% of natural uranium that's nonfissile U-238. Breeding works best with fast neutrons and hence no moderator. That means breeders need to be more compact, sustain a more vigorous chain reaction, and operate at higher temperatures than ordinary slow-neutron reactors. In other words, a breeder is a little more like a bomb. No reactor could ever blow up like a bomb, but the breeder design probably risks more serious accidents than light-water reactors. And the breeder is chemically more dangerous, because its coolant—chosen to minimize neutron absorption—is liquid sodium, a substance that burns spontaneously on contact with air.

More significantly, an energy economy dependent on breeder reactors is an economy that necessarily traffics in plutonium. That's a serious threat to world security, and for that reason the United States in the 1970s decided to forgo work on breeder reactors. Other nations went forward, especially France, although the results were not entirely satisfactory. France once had two breeders in commercial operation, but shut down its 1,200-MW Superphénix reactor in 1998, and followed by closing the smaller Phénix reactor in 2009.

EXAMPLE 7.5 | Dry Clothes and Bombs

You can either toss your clothes into a 5-kW electric clothes dryer for an hour, or you can hang them on the clothesline. If your electricity comes from a 33%-efficient nuclear plant, how much plutonium-239 are you responsible for making if you choose the electric dryer?

SOLUTION

That 5-kW dryer uses energy at the rate 5 kJ/s, so in an hour the amount of energy it uses is (5 kJ/s)(3600 s) = 18 MJ. This energy comes from a 33%-efficient power plant, so back at the plant a total of 3×18 MJ = 54 MJ of energy is released from fissioning uranium. As I noted earlier, each uranium fission releases 30 pJ, so the total number of U-235 nuclei that fission to dry your clothes is

$$\frac{54 \times 10^{6}\ \text{J}}{30 \times 10^{-12}\ \text{J/U-235}} = 1.8 \times 10^{18}\ \text{U-235 nuclei}$$

Because two Pu-239 nuclei are produced for every three U-235 fission events, your choice to use the electric clothes dryer results in the production of some 1.2×10^{18} Pu-239 nuclei. That's a million trillion plutonium nuclei, and, as you can show in Exercise 4, it amounts to about 0.5 milligram of Pu-239. Plutonium's critical mass is about 5 kg, so it would take 10 million households running their dryers for an hour to produce one bomb's worth of Pu-239. In the United States, plutonium ends up as nuclear waste, not bombs, but this example illustrates quantitatively a potential link between nuclear power and nuclear weapons.

ADVANCED REACTOR DESIGNS

The first nuclear power reactors, built in the 1950s, were prototypes called **generation-I reactors**. The worldwide boom in nuclear construction during the 1960s through 1980s produced **generation-II reactors**, which comprise the majority of reactors in service today. These incorporated evolutionary advances based on experience with the early gen-I prototypes. Reactor evolution continued with **generation-III** designs in the late 1990s and early 2000s. Gen-III reactors should be inherently safer, relying more on passive safety systems—for example, gravity-fed emergency cooling water—as opposed to active systems requiring powered components. That's especially significant in light of the 2011 disaster at Japan's Fukushima Daiichi nuclear plant, which resulted when an earthquake cut off power to cooling systems in Fukushima's 1970s vintage gen-II reactors, and a subsequent tsunami disabled backup power generators. Reactors being built today are **generation-III+** designs, which incorporate further safety and efficiency advances. Gen-III+ reactors are standardized for economies of construction and maintenance as well as efficient regulatory approval. But these new reactors are still only modifications of the basic light-water designs that trace their roots to nuclear submarine reactors of the 1950s. Can we do better?

On the drawing boards are a plethora of **generation-IV reactors**. Diverse in design, they nevertheless share common goals: more efficient use of uranium resources, greater safety, resistance to weapons proliferation, reduction in the amount and lifetime of nuclear waste, and better economics. Some gen-IV proposals are for moderated slow-neutron reactors; others are fast-neutron designs employing gas or liquid-metal coolants. But commercial gen-IV reactors are decades away, so they aren't going to help us replace fossil fuels any time soon.

Even further off are proposals to combine conventional nuclear fission with high-energy particle accelerators to "burn" a variety of fissionable but not necessarily fissile isotopes. Such reactors could operate with less than a critical mass, since the accelerator provides the particles that induce fission. That's a big safety plus, since it means that turning off the accelerator immediately stops the chain reaction. But accelerator-based designs have a long way to go before they prove themselves either technologically or economically.

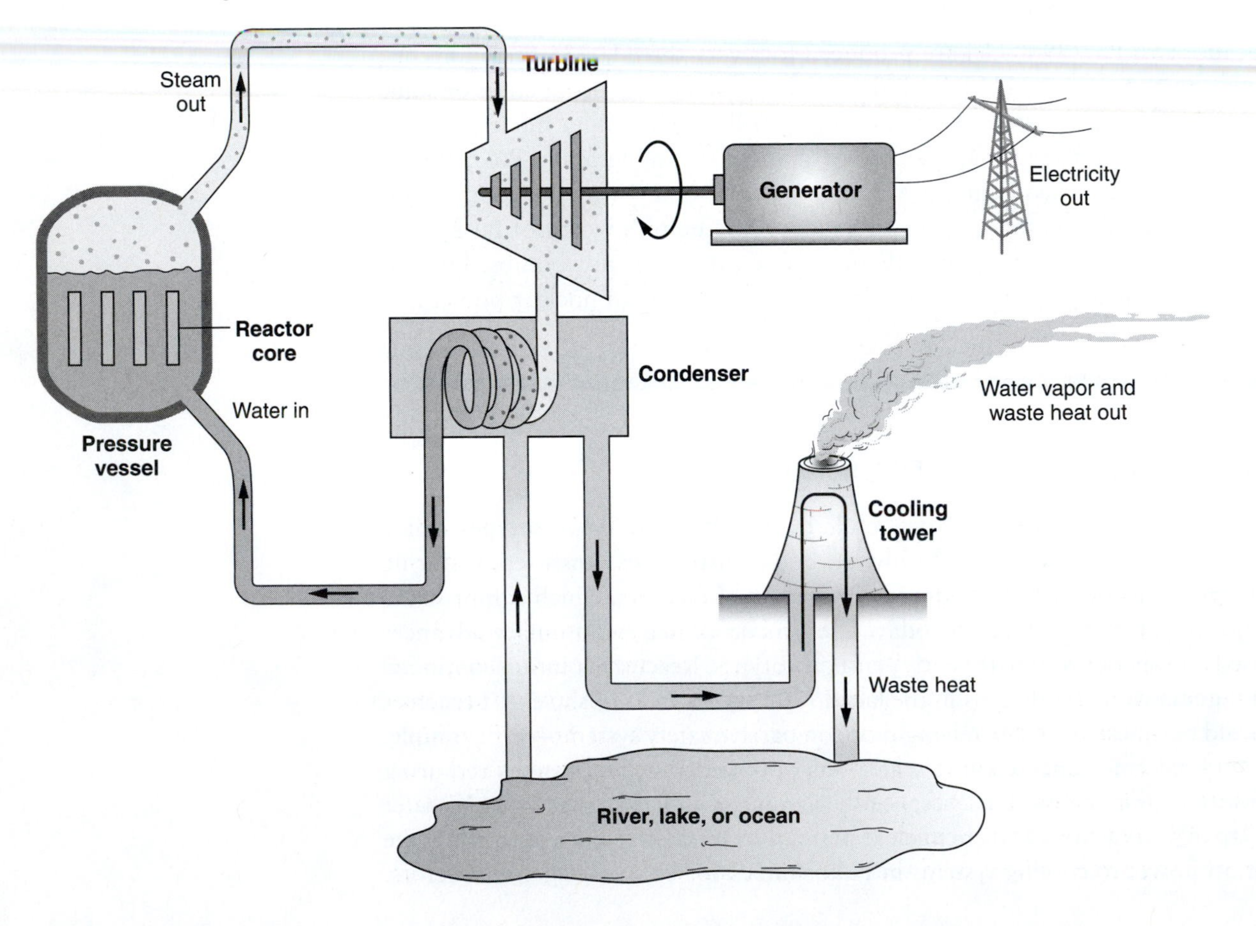

FIGURE 7.13
A nuclear power plant using a boiling-water reactor. Compare with the fossil-fueled plant shown in Figure 5.9; the reactor replaces the boiler, but otherwise they're essentially the same. The reactor core comprises the nuclear fuel and other components immersed in the cooling water.

NUCLEAR POWER PLANTS

The reactor is the heart of any nuclear power plant. Ultimately, the reactor is just a big heater, producing thermal energy from fissioning uranium. From there the power plant is similar to a fossil-fueled plant: Thermal energy boils water, and the steam drives a turbine connected to an electric generator. Spent steam is condensed, and the cycle continues. Cooling towers usually dump some of the waste heat to the atmosphere to avoid thermal pollution of surface waters. As always, it's the second law of thermodynamics that ultimately limits the overall efficiency. The practical efficiency limit for nuclear plants is similar to that for coal plants, although with light-water reactors the need for a large, thick-walled pressure vessel limits maximum steam temperature and thus reduces efficiency somewhat from that of a comparable coal plant. But the similarities between nuclear and fossil plants are striking—so much so that some nuclear plants have been converted to fossil-fueled facilities. Figure 7.13 diagrams a typical nuclear power plant.

7.6 The Nuclear Fuel Cycle and Uranium Reserves

Coal and oil come out of the ground, are processed and refined, and go into power plants where they're burned. Out come electrical energy and waste heat, CO_2, fly ash, sulfurous waste, and a host of toxic and environmentally degrading air pollutants. A similar sequence happens with nuclear energy: Uranium is mined, processed into nuclear fuel, and fissioned in a reactor. Out come electrical energy, waste heat, and radioactive waste. Small amounts of waste are released, intentionally or inadvertently, to the environment. But the majority of radioactive waste stays locked physically in the fuel rods. Eventually the rods' fission energy declines, and they're replaced with new uranium fuel. Those spent fuel rods constitute the major waste product of nuclear power plants, and they have to be disposed of somehow.

URANIUM MINING

What I've just described is a "once-through" **nuclear fuel cycle**—the sequence from uranium mining through processing, then fissionning in a reactor, and finally removal and disposal of nuclear waste (Fig. 7.14). As with fossil fuels, each step in the fuel cycle has environmental consequences. Although uranium itself is only mildly radioactive, its natural decay products are more so and include radon gas. Early in the nuclear age, radon-induced lung cancer was a serious health problem for uranium miners; however, advances in mining technology and radiation protection have dropped exposure levels by factors of 100 to 1,000. Another mining issue is the disposal of *tailings*, the rock that's been pulverized to extract uranium. Uranium mine tailings are mildly radioactive, and wind can easily spread the radioactive material. In the past, uranium

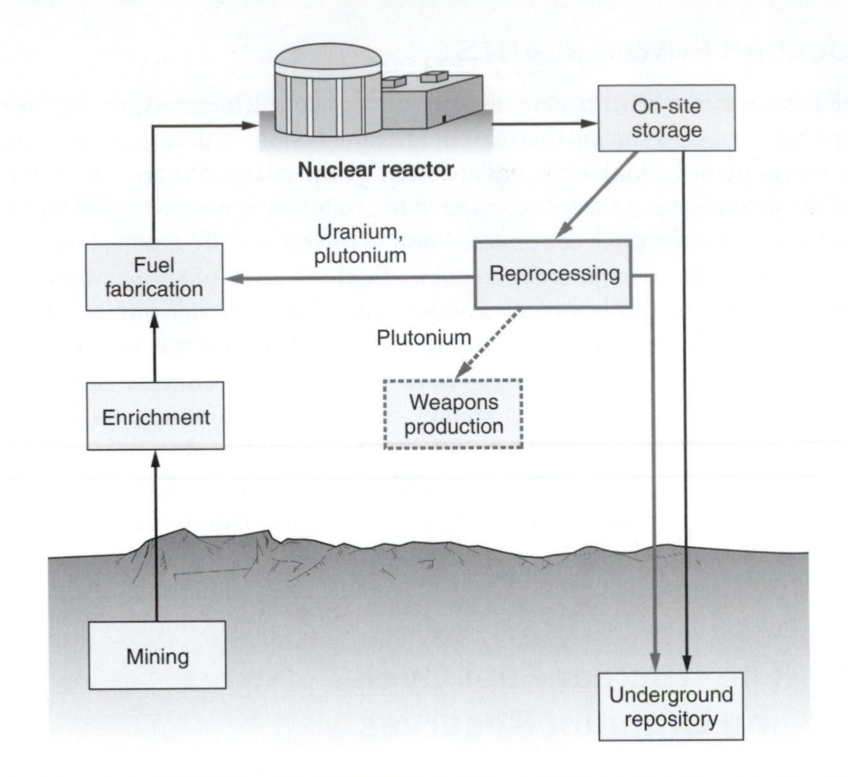

FIGURE 7.14
Nuclear fuel cycles. A once-through cycle is in black; additional steps in a reprocessing cycle are in gray. On-site storage involves pools of water for short-term storage of fresh, highly radioactive waste, followed by longer-term dry cask storage.

tailings were actually incorporated into construction materials, and as a result thousands of buildings have had to be abandoned. Today, tailings piles in the United States must be covered, and this particular source of radioactive pollution has been substantially reduced.

Remember the nuclear difference: It means that we would need only about one ten-millionth as much U-235 as we would coal even if uranium and coal supplied equal percentages of our energy—which they don't. Of course U-235 comprises less than 1% of natural uranium, and uranium itself is only a fraction of the material in its ore—but nevertheless the nuclear difference means that uranium mining involves far less material than coal mining. Consequently there are in the United States fewer than 500 uranium miners, compared with more than 80,000 coal miners—the latter subject to their own health hazards, especially mining accidents and black lung disease.

URANIUM ENRICHMENT

Uranium comes from its ore in the form of "yellowcake," the oxide U_3O_8. But only 0.7% of that uranium is fissile U-235. For all but the CANDU reactor, fuel must be enriched in U-235. Because uranium isotopes are chemically similar, **enrichment** schemes that increase the percentage of U-235 make use of the small mass difference between the two isotopes. In the most common approach today, uranium first reacts chemically to produce uranium hexafluoride gas (UF_6). This is injected into centrifuges (spinning cylinders), where its slightly greater inertia causes $^{238}UF_6$ to concentrate more toward the outside of the cylinders, leaving UF_6 enriched in U-235 toward the interior. The enriched gas is fed to additional centrifuges, eventually producing reactor fuel at around 4% U-235 or weapons-grade uranium at 90% or more U-235.

Uranium enrichment itself has little environmental impact, although it is energy intensive. But its waste product, so-called depleted uranium (DU) with less U-235, is still mildly radioactive. Because of its high density, depleted uranium is made into armor-piercing ammunition—leaving some modern-day battlefields strewn with this weakly radioactive material. Soldiers injured with DU munitions have detectable radiation levels in their urine. This environmental impact pales when compared with any single effect of fossil fuels—but it shows once again that our energy use can affect the environment in surprising ways.

Although uranium enrichment itself is not of great environmental significance, enrichment is one of the two main routes to nuclear weapons (the other being the extraction of plutonium from spent reactor fuel). The environmental impact of nuclear war would be horrendous, and if the spread of enrichment technology enabled such a war, then my statement here about the minimal environmental impact of uranium enrichment would prove horribly false.

NUCLEAR WASTE

Following enrichment, uranium is fabricated into the fuel rods that go into nuclear reactors. After about 3 years, most of the U-235 has fissioned. Plutonium-239, the fissile isotope produced when nonfissile U-238 absorbs a neutron, has reached a level where it's being fissioned as fast as it's created. And a host of fission products "poison" the fuel by absorbing neutrons and reducing the efficiency of the chain reaction. It's time to refuel (Fig. 7.15).

In a typical light-water reactor, about one-third of the fuel is replaced each year. The incoming uranium is so mildly radioactive that you could handle it without harm. But spent fuel is nasty, because of those highly radioactive fission products. A few minutes in the neighborhood of a spent fuel bundle would kill you. Consequently, removal of spent fuel takes place under water, using remote-controlled equipment. Energy released from radioactive decay also makes the spent fuel so hot—in the old-fashioned thermal sense—that for this reason, too, it has to be kept under water.

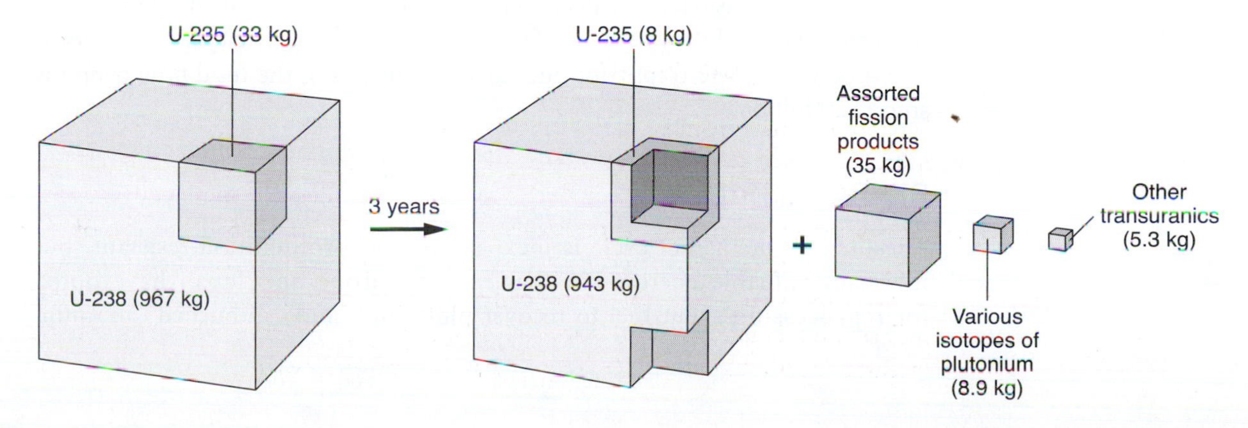

FIGURE 7.15
Evolution of 1,000 kg of 3.3%-enriched uranium in a nuclear power reactor. After 3 years the fission products interfere with the chain reaction, and it's time to refuel.

After a few years, the most intensely radioactive isotopes have decayed and both radioactivity and thermal energy have decreased enough that spent fuel can be removed from the reactor site for disposal. But where? The longest-lived isotopes, the transuranics, have half-lives as long as tens of thousands of years. Any disposal site must therefore keep radioactive waste out of the environment for 100,000 years or more—far longer than the history of human civilization. We're more than half a century into the nuclear age, and there's still no operational long-term waste disposal facility anywhere in the world. At many nuclear plants, spent fuel sits in temporary cooling pools or dry-cask storage because there's no other place for it to go. The shortcomings of short-term storage in pools became dramatically obvious in the 2011 Fukushima accident, where spent-fuel pools posed nearly as great a danger as reactor meltdowns.

Table 7.2 shows some solutions that have been proposed for nuclear waste. Today most experts believe that underground storage is the best long-term solution, and that well-engineered repositories in geologically appropriate settings can keep radioactive materials out of the environment for the tens of thousands of years it would take the longest-lived isotopes to decay. Others argue that long-term geological stability cannot be guaranteed, and that radiation-generated heat can lead to the migration of water to the waste site, or fracturing of rock, or other processes that might provide pathways for the escape of radioactive waste. Some would have us bury nuclear waste in holes miles deep, where it could never be recovered. Others claim that nuclear waste may someday be valuable or that we'll develop technologies to render it harmless, making burial a half-mile down or so more appropriate. As these arguments proceed, nuclear waste continues to pile up. Again, the nuclear difference means that the sheer volume of waste is small compared with that from fossil energy; the high-level waste from one year's operation of a nuclear plant could fit under your dining room table. But this is nasty stuff, and with more than 400 commercial reactors operating globally, some for more than 40 years, the world faces a substantial waste disposal problem.

In the United States, hope for a permanent nuclear waste solution focused for years on a single site, inside Yucca Mountain in arid Nevada. Tunnels and storage areas were constructed, but technical, geological, and political issues keep postponing the repository's opening. Finally, in 2009, the government abandoned Yucca Mountain. The United States, with its 100,000-tonne inventory of nuclear waste, was back to square one. Meanwhile, an assessment of one-tenth of a cent on every kilowatt-hour of nuclear-generated electricity has been accumulating since 1982 to pay for nuclear waste disposal; the fund now contains some $20 billion.

REPROCESSING

The once-through fuel cycle is inexpensive and proliferation-resistant, but it wastes valuable energy. For that reason Europe and Japan have opted for **reprocessing** spent fuel to recover plutonium and "unburned" uranium

TABLE 7.2 | SOME NUCLEAR WASTE OPTIONS

Option	Advantages	Disadvantages
Dry-cask storage	Available short-term option Waste kept on site; no need to transport	Most nuclear plants are near population centers and waterways
Shallow burial (~1 km)	Relatively easy construction and access Waste is recoverable and easily monitored	Proximity to groundwater poses contamination risk Subject to geologic disturbance
Sub-seabed burial	Keeps waste far from population centers	Probably not recoverable Requires international regulatory structure Currently banned by treaty
Deep-hole burial (~10 km)	Keeps waste well below groundwater	Not recoverable Behavior of waste at high temperature and pressure not well understood Subject to geologic disturbance
Space disposal (dropped into Sun or stored on Moon)	Permanent removal from Earth environment	Impractical and economically prohibitive Risk of launch accidents
Ice sheet disposal	Keeps waste far from population centers	Expensive due to remoteness and weather Recovery difficult Global climate change is diminishing ice sheets Treaty bans radioactive waste from Antarctica
Island disposal	Burial under remote islands keeps waste away from population centers	Ocean transport poses safety issues Possible seawater leakage into waste repository Seismic and volcanic activity common at island sites
Liquid-waste injection	Waste locked in porous rock below impermeable rock	Requires processing waste to liquid form Movement of liquid waste might result in radiation release
Transmutation	High-energy particle accelerators or "fusion torches" induce nuclear reactions that render waste non-radioactive or very short-lived	Technology not proven nor available Requires shorter-term recoverable storage option until technology is operational

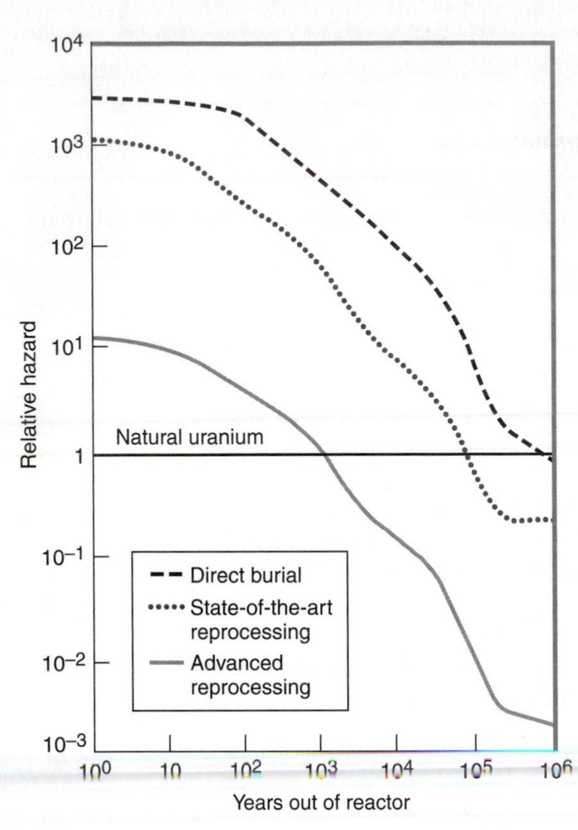

FIGURE 7.16

Decline in hazardousness of nuclear waste with direct burial, today's state-of-the-art reprocessing, and future advanced reprocessing that would remove 99.9% of transuranic isotopes. Hazard is measured relative to natural uranium. Even without reprocessing, the waste hazard drops by an order of magnitude in the first 1,000 years. With advanced reprocessing it's no more hazardous than natural uranium after that time.

(see Fig. 7.14). Reprocessing is expensive and technically difficult. It also entails dangers associated with increased transport of radioactive materials, and raises the danger that terrorists could steal plutonium to use in nuclear weapons. Plutonium extracted in reprocessing is mixed with uranium to produce **mixed-oxide fuel (MOX)**, which is then "burned" in reactors. This reduces the need for raw uranium and thus extends uranium reserves. "Burning" reprocessed plutonium also reduces both the level and duration of the radiation danger from nuclear waste (see Fig. 7.16). But in the event of a nuclear accident, a MOX-fueled reactor can spread dangerous plutonium—an issue with one of the reactors involved in the 2011 Fukushima nuclear accident.

THE URANIUM RESOURCE

Nuclear fission looks like a serious alternative to fossil fuels, at least for the generation of electric power. But is there enough uranium?

A superficial check of the proven uranium resource—some 3 million tons, containing some 1,500 EJ of energy—suggests that the time until we face uranium shortages could, as for oil, be only a matter of decades. One solution to this apparent uranium shortage is the breeder reactor, discussed earlier, which would convert the uranium-238 that comprises 99.3% of natural uranium into fissile plutonium-239. Right there our nuclear fuel reserve would increase by a factor of 99.3/0.7 or nearly 150-fold. Even reprocessing of fuel from ordinary reactors—as is done in Europe and Japan—stretches the uranium supply somewhat.

But nuclear energy hasn't seen the growth once expected, and as a result the world is currently awash in cheap uranium. At present there's little incentive to explore for more—although such exploration would surely increase the known reserves substantially. And there's plenty of lower-grade ore—probably 100 to 300 million tons—that could be exploited at higher cost. However, fuel plays a much lower part in the cost of operating a nuclear plant than it does for a fossil plant (again, the nuclear difference), so the cost of extracting uranium from low-grade ore wouldn't have much impact on the cost of nuclear-generated electricity.

Uranium is not particularly rare. It's dispersed throughout Earth's crust at an average concentration of about 4 parts per million. And it's soluble in water to the extent that uranium is more abundant in the oceans than some common metals such as iron. In fact, the world's oceans contain some 4.5 billion tons of uranium—enough to last for thousands of years. At present it's much too expensive to extract, but this uranium energy resource is certainly available and could become economically viable as energy costs rise and technological advances bring down extraction costs.

The bottom line here is that high-grade uranium ore could become scarce in a fission-powered world, but that the overall uranium resource is sufficient that fission represents a serious alternative to fossil fuels, at least for stationary power sources like electric power plants.

7.7 Environmental and Health Impacts of Nuclear Energy

The major concern with nuclear power is the effect of nuclear radiation on human health and the environment. But there's a looming issue that could overshadow even those concerns: the connection between nuclear power and nuclear weapons, and the impact of nuclear war. Here we'll examine both issues.

QUANTIFYING RADIATION

The basic measure of radioactivity is the **becquerel** (Bq), named after Henri Becquerel, who discovered natural radioactivity in 1896. The becquerel measures the rate at which a given sample of radioactive material decays, without regard to the type or energy of the radiation. One becquerel is 1 decay per second. An older unit, the **curie**, is defined as 37 billion decays per second, and is approximately the radioactivity of 1 gram of radium-226. Activity in becquerels or curies is inversely related to half-life. Given equal numbers of atoms of two different radioactive materials, the one with the shorter half-life has the greater activity because it decays faster.

We're most concerned with radiation's effect on materials, especially biological tissue. Two additional units address these effects. The **gray** (Gy) is a purely physical unit that describes the energy absorbed by an object exposed to radiation. One gray corresponds to the absorption of 1 J of radiation energy per kilogram of the exposed object. An older unit still in widespread use is the **rad**, equal to 0.01 Gy.

The biological effects of radiation depend not only on the total absorbed energy but also on the type and energy of the particles comprising the radiation. Yet another radiation unit, the **sievert** (Sv), also measures absorbed energy per kilogram, but it's adjusted to account for the biological impacts of different radiation types and energies. Again, you'll often encounter an older unit, the **rem**, equal to 0.01 Sv.

The biological impact of radiation depends in part on its penetrating power. Alpha radiation tends to have the lowest energy of the three types of radiation, and is typically stopped by a layer of clothing or a few centimeters of air. Still, alpha-emitting material that's ingested or inhaled can wreak havoc because of its immediate proximity to internal organs and tissues. Beta radiation, consisting of high-energy electrons, is more penetrating; it's capable of traversing several feet of air or a fraction of an inch in solids, liquids, and body tissues. Again the greatest concern is internal contamination with beta-emitting radioisotopes. Gamma radiation, consisting of high-energy photons, can penetrate several feet of concrete or even lead.

BIOLOGICAL EFFECTS OF RADIATION

The high-energy particles that comprise nuclear radiation affect biological systems by breaking, ionizing, or otherwise disrupting the biochemical molecules within living cells. The results include mutations, cancer, and outright cell death. The likelihood of these outcomes, and the overall impact on a living organism, depends on the total radiation dose.

A handful of nuclear accidents and the bombings of Hiroshima and Nagasaki show the effects of large radiation doses on human beings. A dose of 4 Sv delivered in a relatively short period will kill approximately half the people exposed. Death rates go up with higher exposures, and severe radiation sickness persists after somewhat lower doses. However, very few people are exposed to such high radiation levels. For example, neighbors of Pennsylvania's Three Mile Island reactor experienced average individual exposures of about 10 microsieverts (μSv), or 1 millirem during TMI's partial meltdown in 1979—the worst commercial nuclear accident in the United States. In contrast, the average American is exposed to about 500 μSv each year through X rays and other radiation-based medical procedures. At such low doses it's impossible to ascertain with certainty a direct relation between radiation dose and health effects. Most radiation standards are based on the idea that the effects of high radiation exposure extrapolate linearly to low doses. Others estimate the effect of low radiation doses as even smaller, arguing that so-called **repair mechanisms** in cells correct damage caused by the occasional high-energy particle that strikes a cell during low-dose radiation exposure. On the other hand, radiation preferentially damages fast-growing cells, which means that children and the unborn are especially susceptible. This preferential effect on fast-growing cells also means that radiation—itself a carcinogen—is also useful in fighting cancer.

The U.S. National Academy of Sciences, in its study *Biological Effects of Ionizing Radiation* (BEIR VII), concludes that exposure to 100 mSv will give an individual a 1-in-100 chance of developing radiation-induced cancer in their lifetime and roughly half that chance of fatal cancer. That same individual has a 42-in-100 chance of developing cancer from other causes. By comparison, a typical chest X-ray gives a radiation dose of 0.1 mSv, while a full CAT scan gives 10 mSv. Cancer risk is assumed to scale linearly with dose, meaning, for example, that a 10-mSv CAT scan results in a 1-in-1000 chance of cancer. The same statistics apply to entire populations; for example, 10 Sv (100 times the 100 mSv that gives a 1-in-100 chance of cancer) spread over a population should result in roughly one case of cancer. But it's impossible to ascribe any given cancer to low-level radiation; the most we can do is to look for statistically significant increases in cancer among radiation-exposed populations. Given that cancer is common and a leading cause of death, it takes a really large radiation dose for radiation-induced cancers to stand out against the background cancer rate.

EXAMPLE 7.6 | Fukushima Cancers

During the 2011 Fukushima nuclear accident in Japan, the government's Ministry of Health, Labor, and Welfare raised the maximum radiation dose for nuclear plant workers to 250 mSv. How would such a dose increase one's risk of cancer? If all 400 emergency workers experienced this dose during the accident (most experienced far less), how many radiation-induced cancers would result?

SOLUTION

At 0.01 cancers per 100 mSv, a dose of 250 mSv should induce 0.025 cases of cancer. So an individual exposed to 250 mSv would incur a 2.5% risk of radiation-induced cancer. If all 400 workers had received that dose, the result would be

$$(400 \text{ doses})(0.025 \text{ cancers/dose}) = 10 \text{ cases of cancer}$$

This increase might be just detectable statistically against the background of about 200 cancers from all causes expected among 400 Japanese.

BACKGROUND RADIATION

Radiation from nuclear power should be evaluated in the context of our normal exposure to other radiation from artificial and natural sources, collectively constituting **background radiation**. On average, individual radiation exposure averages about 2.4 mSv per year worldwide; in the United States it's about 3.6 mSv per year. The figure varies substantially with such factors as where you live, how much you travel, and even your house construction. Move to high-altitude Denver, or fly across the Atlantic, and your exposure to cosmic rays goes up. Build a house with a cinder-block basement over bedrock that's high in natural uranium, and you'll be exposed to significant levels of radioactive radon gas right in your own home. In fact, radon is the dominant source of radiation exposure for most people, at just over half the total radiation dose. Rocks and soils contribute just under 10%, as do cosmic rays. Your own body provides over 10% of your radiation dose, mostly from natural potassium-40. Medical procedures amount to another 15% of the average American's yearly radiation dose, and consumer products like smoke detectors, exit signs, and even tobacco contribute some 3%. Radiation from nuclear power and nuclear weapons activity weighs in at less than 1% of the average American's radiation exposure (Fig. 7.17).

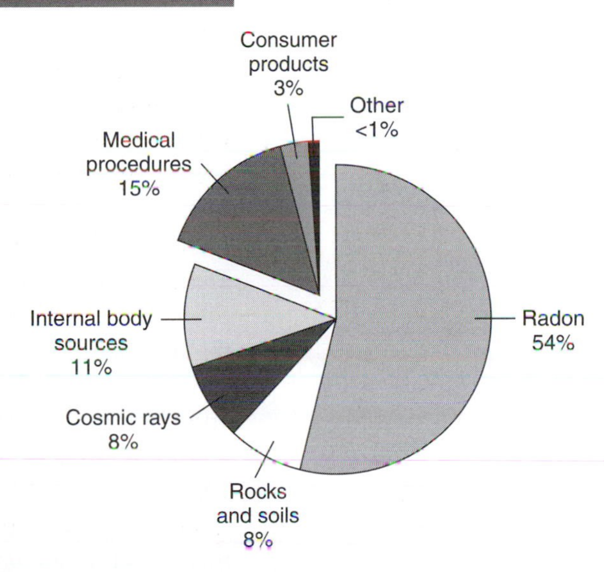

FIGURE 7.17
Sources of background radiation for the average U.S. resident. The "other" category includes nuclear power, radioactive waste, and fallout from weapons tests. Most background radiation is natural; artificial sources are in the separate wedge at top left.

RADIATION FROM NUCLEAR POWER

In normal operation, nuclear power plants emit very small amounts of radioactive materials, mostly to the atmosphere. These include the chemically inert gases krypton and xenon, and small amounts of iodine-131. Of these, only iodine is biologically significant. The U.S. Nuclear Regulatory Commission limits the exposure of the general public to a maximum yearly dose of 0.08 mSv from nuclear power plant emissions—about 2% of the normal background. Actual exposures are far less; for example, immediate neighbors of nuclear plants typically receive about 0.01 mSv per year. If the entire U.S. population received this dose, the result would be about 150 fatal cancers (see Exercise 10). Even if you work in the U.S. nuclear power industry, you're allowed a maximum yearly exposure of only 0.05 Sv, about 15 times the average background dose. Airline flight crews routinely experience higher doses from cosmic radiation.

NUCLEAR ACCIDENTS

The inventory of radioactive material in a nuclear power reactor is huge—enough to cause large-scale environmental contamination and widespread health effects in the event of an accidental or intentional release. There's still more radioactivity in the spent fuel stored onsite at many nuclear power facilities. Although there have been plenty of incidents involving radiation release at commercial nuclear power plants, three stand out as major events with long-term impacts on the nuclear power industry and, in two cases, on human and environmental health.

The 1979 Three Mile Island (TMI) accident in the United States was a partial meltdown resulting in a minor radiation release that probably caused at most a few fatal cancers in the exposed population. TMI was triggered by a pump failure and exacerbated by a stuck valve and instrumentation failures that kept the plant's operators from realizing they faced a serious loss-of-coolant accident. The unfolding accident was marked by confusion on the part of operators and officials, and by the unanticipated danger of a hydrogen gas explosion that, in this accident, never materialized.

In contrast, the 1986 Chernobyl accident in Ukraine involved a massive radiation release that affected much of Europe and that leaves hundreds of square miles contaminated to this day. Chernobyl resulted, ironically, from a runaway nuclear reaction that occurred during a test of the reactor's safety systems. A fire in the reactor's graphite moderator compounded the radiation release; helped by the lack of a containment structure, the fire sent radioactive material high into the atmosphere. An authoritative study nearly 20 years after the accident suggests that some 4,000 deaths worldwide will have resulted from Chernobyl in the 50 years following the accident. That's a lot less than the 24,000 premature deaths that probably occur in the United States alone each year as a result of pollution from coal-burning power plants (Section 6.2), but it's a horrific testament to what can go wrong at a single nuclear plant.

Intermediate in severity between TMI and Chernobyl was the 2011 disaster at the Fukushima Daiichi nuclear plant in Japan. Here, three of the plant's six reactors were operating at the time of a massive earthquake. Safety systems immediately shut down the reactors, but the earthquake cut power to the plant's cooling pumps. A subsequent tsunami disabled backup generators, leaving the plant with no way to cool either the three reactor cores—still generating substantial heat through radioactive decay—or the spent-fuel pools for all six reactors. Partial meltdowns occurred before operators took the extreme and not entirely effective action of pumping seawater into the reactors. Hydrogen generated by the reaction of hot fuel rods with water exploded in three of the reactor buildings, blowing off the buildings' outer shells. Spent fuel caught fire at an inactive reactor, spewing radioactive materials into the air. The crisis dragged on for months, with low-level radiation releases punctuated by more serious incidents. Levels off-site were high enough to necessitate evacuation of some 200,000 people, and during radiation spikes it became impossible for workers to approach the reactors to continue emergency cooling operations. Although it involved multiple reactors, the Fukushima accident probably released only about one-tenth the radiation of Chernobyl—a difference attributable to the safer design of Fukushima's 40-year-old boiling-water reactors. It's far too early to estimate the death toll from Fukushima, but it will likely number in the hundreds.

Nuclear accidents at the scale of Chernobyl, Fukushima, and TMI are rare, but clearly not impossible. Each of these accidents was characterized by chains of events that were either totally unanticipated or deemed so unlikely as to be effectively impossible. Confusion about what was going on during each accident highlighted the sense of a dangerous technology spinning out of control. Each accident had a major effect on the nuclear power industry; in the United States, for example, TMI marked the beginning of a 30-year stretch that saw no new nuclear plant orders (although economic factors played a role as well). The Fukushima accident is likely to prove devastating to an industry that had gone a quarter century without a major accident; in particular, it will surely reduce nuclear power's attractiveness as an alternative to fossil fuels just at the time when the world urgently needs such alternatives. Indeed, in the wake of Fukushima, several European countries have announced plans to phase out their nuclear plants, although both Japan and the United States have reaffirmed their commitments to nuclear power.

It's difficult to assess the overall impacts of large-scale nuclear power accidents because these are extremely rare events, but with potentially huge consequences. In this sense, nuclear power's impacts are very different from those of fossil fuels, where a steady output of pollution results in a significant but politically acceptable rate of death, illness, and environmental degradation. A nuclear incident with even a small fraction of the impact we experience from routine fossil-fuel pollution would be totally unacceptable.

A number of studies have attempted to assess the probability of severe nuclear accidents. A comprehensive report by the U.S. Nuclear Regulatory Commission (NUREG-1150) looked at five generation-II nuclear plants, and concluded that a large-scale radiation release to the environment beyond a plant's boundaries might be expected to occur, on average, about once every 250,000 **reactor-years.** A reactor-year means one nuclear reactor operating for one year. With approximately 100 reactors operating in the United States, that means we could expect a large-scale radiation release once every 2,500 years. Over the next 25 years—which takes us beyond the operating lifetime of nearly all U.S. nuclear plants, there's a 1% chance of such an accident. Generation-III+ reactors are supposed to be much safer; for the Westinghouse AP-1000, now under construction in several countries, the probability of a large-scale radiation release is estimated at less than 1 in 10 million reactor-years. It's fair to say, however, that those numbers need more scrutiny in light of the Fukushima accident.

TERRORISM

The terrorist attacks of September 11, 2001 raise the frightening the prospect of an attack on a nuclear power plant. Thick, concrete containment structures surround commercial light-water reactors, and license hearings for U.S. nuclear plants included discussion of accidental aircraft impacts. But given the relatively small size of a nuclear facility, the probability of an accidental airplane crash into a reactor containment structure is minuscule. Clearly that's not the case with intentional terrorism. The U.S. Nuclear Regulatory Commission admits that existing plants are not designed explicitly to withstand impacts by large airliners. However, a 2002 study done for an industry group, the Nuclear Energy Institute, used computer modeling to suggest that an intentional crash of a Boeing 767 would not breach a reactor containment structure. The study also claimed that on-site nuclear waste storage facilities would survive an aircraft attack. A National Academy of Sciences report was less optimistic, suggesting particularly that spent fuel stored in water-filled pools at nuclear power plants is vulnerable to terrorist attack, a possibility made more real by spent-fuel radiation releases during the Fukushima accident. In any event, the vulnerability of nuclear plants to terrorist attack remains uncertain and controversial.

COMPARATIVE RISKS

Nuclear energy and its perceived dangers stir worries and passions far beyond the emotions associated with fossil fuels. The link—part real, part imagined—between nuclear energy and the horrific destructive power of nuclear weapons is one reason for public concern with all things nuclear. The invisible, odorless, tasteless nature of nuclear radiation is another. The sheer concentration of nuclear energy—that 10-million-fold nuclear difference—adds to the sense of nuclear danger, as does the fact that each nuclear accident has involved new and unexpected calamities. And the long-term hazard of nuclear waste makes nuclear energy look like a burden that humanity might not want to take on.

No energy source is without risk, however, and a serious examination of energy alternatives should compare those risks. As I've suggested here and in Chapter 6, objective analyses show that the risks of nuclear energy are likely to be well below those associated with fossil fuels. Recall that estimate of 24,000 premature deaths each year from air pollution due to coal burning in the United States alone. There's simply nothing close to this with nuclear power; if there were, the nuclear industry would be shut down immediately. Table 7.3 com-

TABLE 7.3 | **IMPACTS OF TYPICAL 1 GWE COAL AND NUCLEAR PLANTS**

Impact	Coal	Nuclear
Fuel consumption	360 tons coal per hour	30 tons uranium per year
Air pollutants	400,000 tons per year	6,000 tons per year
Carbon dioxide	1,000 tons per hour	0
Solid waste	30 tons ash per hour	20 tons high-level radioactive waste per year
Land use (includes mining)	17,000 acres	1,900 acres
Radiation release	1 MBq per minute from uranium, thorium, and their decay products, but varies and can exceed nuclear plant radiation	50 MBq per minute from H-3, C-14, inert gases, I-131
Mining deaths	1.5 per year from accidents; 4 per year from black lung disease	0.1 per year from radon-induced lung cancer
Deaths among general public	100 premature deaths per year from air pollution	0.1–10 deaths per year from radiation-induced cancer

pares impacts from coal and nuclear power plants. Estimates of total deaths from a nuclear plant range from the most conservative to those of the most virulently antinuclear groups; even the latter are an order of magnitude below the death rate from coal. The impacts in Table 7.3 represent those from power-plant operation only, and don't include such things as fuel processing and transportation. It takes considerable fossil energy to prepare nuclear fuel, and that's responsible for some traditional pollutants and greenhouse gases. A German study on this issue concluded that, for example, greenhouse emissions associated with nuclear power are a few percent those of coal.

THE WEAPONS CONNECTION

There is a scenario where globally catastrophic environmental and health impacts result indirectly from the use of nuclear power. That is the development of nuclear weapons as an offshoot of a nuclear power program, and the subsequent use of those weapons. Although nuclear weapons and power reactors are very different things, they share some connections. Enrichment of uranium to reactor-fuel levels is harmless enough, but the same enrichment capability can be taken further, to make weapons-grade uranium. Nuclear power plants produce plutonium-239, a potent weapons fuel. Extracting and purifying plutonium is difficult and hazardous, but it can be done. More generally, the technical personnel and infrastructure needed to maintain a civilian nuclear power program can easily be turned to weapons development. Nuclear weapons, especially in the hands of rogue nations or terrorist groups, are a potentially grave threat to world security. Some of today's thorniest international dilemmas involve countries that exhibit erratic behavior and links to terrorism—and that are seeking or have acquired nuclear weapons. It's debatable just how tightly linked the nuclear weapons threat is to nuclear power, but it would be irresponsible to deny the existence of such a linkage.

It should be evident from this chapter that I regard nuclear fission as a viable part of the world's energy mix, fraught—like other energy sources—with dangers, but not to be dismissed out of hand. And in stable, responsible countries I believe that nuclear power may be preferable to coal as means of generating electricity (although I'm happier with other alternatives, such as photovoltaics). But the weapons connection gives me pause. Who's to tell one country's citizens they can't be trusted with nuclear power, while another's can? If nuclear fission is to remain part of the mix, then it seems to me a long-term solution must involve international control over enrichment and processing of nuclear fuel. That would deny individual countries the ambiguous technologies that can produce both reactor fuel and weapons-grade fissile material. Given today's politically complex world, complete international control of nuclear fuel is probably a distant goal. But we have seen progress; for example, multilateral programs to convert weapons-grade uranium and plutonium from the former Soviet Union into reactor fuel. And in 2010 a generous gift from a private billionaire helped the United Nations establish an international nuclear fuel bank.

7.8 Policy: A Nuclear Renaissance?

The number of nuclear power plants—some 450 worldwide today—has grown substantially in the decades since the first commercial reactors went online in the 1950s. But growth has been far less than nuclear advocates had anticipated in the early years, when nuclear electricity was thought "too cheap to meter" and peaceful uses of nuclear energy were to include not only electric power but nuclear explosives for excavation and even nuclear-powered spaceflight. And projections for the future suggest that nuclear energy's share of world electrical energy production could well shrink in the coming decades despite increases in nuclear power generation (Fig. 7.18). That's because world electricity consumption is projected to grow so fast that even optimistic estimates of nuclear power growth barely change the nuclear share of world electricity.

Today, nuclear power is growing slowly, but that growth is hardly uniform across the planet. The most rapid growth is in Asia, where burgeoning economies fuel increasing demand for energy (recall Fig. 7.2, showing reactors under construction). European nuclear energy is largely stagnant, with some modest

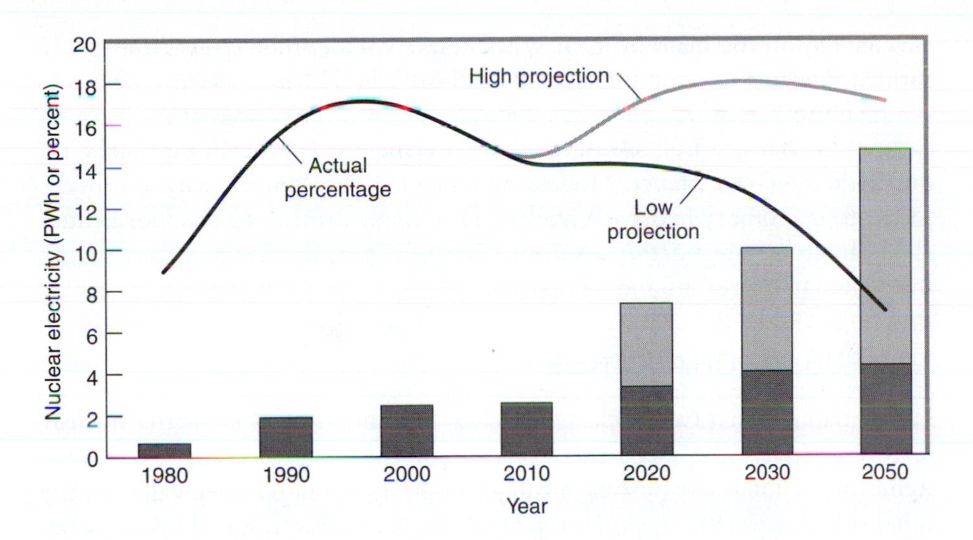

FIGURE 7.18
Historical trends and projections of nuclear energy production in petawatt hours (columns) and percent of electricity from nuclear sources (line graphs). Dark line represents historical data and low estimates; gray represents high estimates.

growth in Eastern Europe. In the United States the number of operational reactors is slowly declining following a three-decade hiatus in reactor construction. However, new government incentives have resulted in more than 20 applications for new reactors in the United States. It's too early to say what effect the 2011 Fukushima accident will have on the worldwide growth of nuclear power, but it will surely be negative.

A FUTURE FOR NUCLEAR POWER?

Few technologies inspire as much fear and controversy as nuclear power. Yet the nuclear option is arguably less harmful to the environment than continued use of fossil fuels. On the other hand, nuclear energy is an enormously concentrated source whose potential for catastrophic accidents exceeds anything that could happen with fossil fuels. And there's a connection, subtle or not, between nuclear power and the nuclear weapons capable of destroying civilization. Are the environmental benefits of nuclear power real, or do uncertainties about nuclear waste, the weapons connection, and nuclear accidents cloud the picture? Should we embrace nuclear power to help satisfy, temporarily or permanently, our prodigious and growing energy appetite? Even if your answer is "yes," will logistical and economic realities allow nuclear energy's contribution to grow significantly?

7.9 Nuclear Fusion

Nuclear fission dominates this chapter because well-established fission energy technology is a reality that already contributes significantly to the world's energy supply. But, as Figure 7.6 showed, nuclear energy can also come from fusion of light nuclei. Fusion powers the stars, but we humans have managed to harness it only in our most destructive weapons. Yet fusion, if it could be controlled, offers nearly limitless energy.

In principle, any nuclei lighter than iron can fuse and release energy. Fusion reactions in massive stars join successively heavier nuclei, building all the elements up to the mass of iron. When stars explode, they spew these newly formed elements into the interstellar medium, where they're eventually incorporated into new stars and planets. Stellar fusion is the source of the oxygen, carbon, nitrogen, sulfur, silicon, and other elements up to iron that make up our bodies and our planet. Modest stars like the Sun "burn" only hydrogen, fusing four ordinary hydrogen nuclei ($^1_1$H, a single proton) to produce helium ($^4_2$He). Intermediate reactions convert two of the protons into neutrons, giving the helium its 2 protons and 2 neutrons.

THE FUSION CHALLENGE

Fusion requires that two nuclei get so close that the strongly attractive nuclear force overcomes the electrical repulsion of the positively charged nuclei. Achieving fusion is like pushing a ball up a hill to get it into a deep valley on the other side (Fig. 7.19). The hill is the repulsive electric force, and it takes energy to get up it. But you get even more energy back once the ball drops into the valley; that's the fusion process. Actually, quantum physics lets the approaching nuclei "tunnel" through the hill, reducing the energy required, but it's still substantial. So how do we get nuclei close enough to fuse? One approach is a temperature so high that the average thermal speed is great enough for nuclei to approach closely despite their mutual repulsion. Once we've got that temperature, we need to confine the hot material long enough for a significant energy release from fusion reactions. That's the fusion challenge: High temperature and confinement. The stars solve both problems with their immense gravities. Gravitational attraction compresses the stellar core, heating it to million-kelvin temperatures. Once fusion "turns on" in a star, the hot, fusing gas remains gravitationally confined. But gravitational heating and confinement aren't options here on Earth.

The most achievable reaction for controlled fusion is **D–T fusion**, depicted in Figure 7.20, which joins the hydrogen isotopes deuterium and tritium to make helium; a neutron and energy are also released:

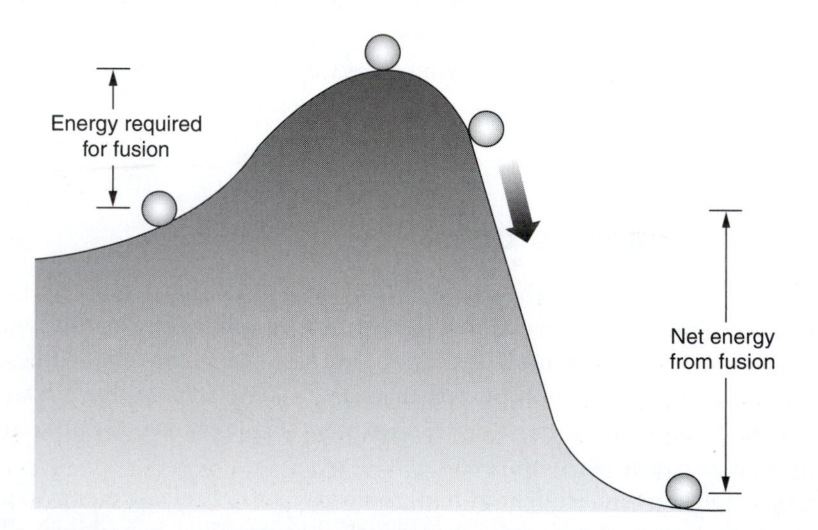

FIGURE 7.19
An analogy for nuclear fusion. It takes energy to get the ball up the hill, but a lot more is released when it drops into the valley.

$$^2_1H + ^3_1H \rightarrow ^4_2He + ^1_0n + 2.8\ pJ \quad \text{(D–T fusion)} \qquad (7.3)$$

Note that the energy here is in picojoules (pJ; 10^{-12} J), as opposed to the attojoules (aJ; 10^{-18} J) used to describe chemical reactions. That's yet another reminder of the nuclear difference, and shows that fusion reactions release about a million times as much energy as chemical reactions.

D–T fusion requires a so-called **ignition temperature** of about 50 million kelvins (50 MK). At that temperature atoms are completely stripped of their electrons, forming an ionized gas called **plasma**. The denser the plasma, the more frequent are deuterium–tritium collisions that result in fusion. Thus we could get the same energy by confining dense plasma for a short time, or more diffuse plasma for a longer time. Quantitatively, what counts is the **Lawson criterion**, the product of particle density and confinement time; for D–T fusion, it's 10^{20} s/m^3. The two fusion schemes being explored today take diametrically opposite approaches to satisfying the Lawson criterion.

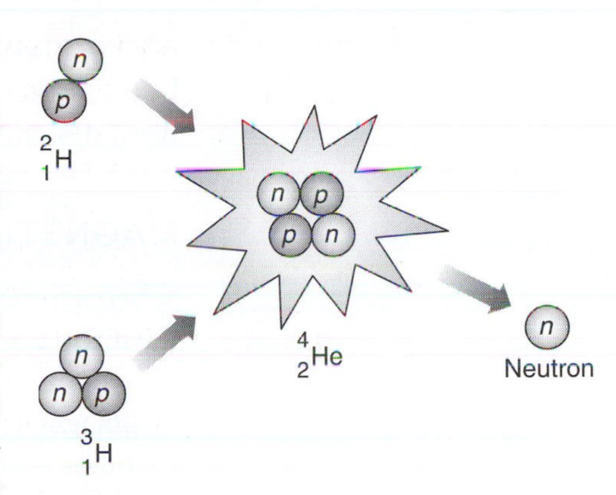

FIGURE 7.20
The deuterium–tritium fusion reaction of Equation 7.3 produces a helium nucleus (^4_2He), a neutron, and energy.

INERTIAL CONFINEMENT

At one extreme is **inertial confinement**, which uses high densities and confinement times so short that the fusing particles' own inertia—their resistance to change in motion—ensures that they stick around long enough. Inertial confinement works in thermonuclear weapons, where energy from a nuclear fission explosion is focused on a deuterium–tritium mix that's consumed in fusion reactions before it has time to blow apart. Inertially confined controlled fusion anticipates a similar approach, focusing colossal laser systems on tiny D–T targets and inducing what are effectively thermonuclear explosions in miniature. The most ambitious laser fusion experiment is the National Ignition Facility (NIF) at California's Lawrence Livermore National Laboratory. The size of a football stadium, this device focuses 192 laser beams on a millimeter-size D–T pellet, delivering laser energy at the rate of 500 TW (Fig. 7.21). That's right—500 TW, some 30 times humankind's total energy-consumption rate. But the lasers are only on for a few nanoseconds, discharging energy that was stored, much more gradually, in advance. By 2011 NIF's lasers had compressed simulated fuel pellets to the temperatures and densities required for fusion. Experiments with actual deuterium–tritium fuel pellets should soon follow, and are expected to yield 10 times as much fusion energy as it takes to power the lasers. NIF and similar laser-fusion experiments explore the basic physics of inertial confinement, but they're a long way from a fusion power plant. NIF, for example, fires its lasers only every 4 to 8 hours; a working power plant would

FIGURE 7.21
Target chamber of the National Ignition Facility during installation. The chamber is 11 m in diameter and weighs 130 tonnes, yet at its center sits a millimeter-size fuel pellet that's the focus of 192 laser beams converging through the many holes in the chamber.

need to blast new fuel pellets around 10 times each second. And solutions to a host of practical engineering problems, from energy transfer to overall power-plant design, are at this point no more than ideas.

MAGNETIC CONFINEMENT

Whereas inertial confinement exploits extreme densities and minuscule confinement times, **magnetic confinement** takes the opposite approach. Under development since the 1950s, magnetic confinement holds diffuse hot plasma in "magnetic bottles"—configurations of magnetic fields that keep charged particles away from the physical walls of the device. Early optimism for magnetic fusion was tempered by plasmas going unstable and escaping from their magnetic bottles. Instability diminishes as the size of the plasma increases, so scaling up the magnetic bottles makes fusion plasmas more manageable. The most promising magnetic configuration is the **tokamak**, a Russian design that features a donut-shaped plasma chamber whose magnetic field spirals around without ever touching the chamber walls (Fig. 7.22).

By the mid-1990s, tokamaks in the United States, Japan, and Europe had approached so-called scientific breakeven, where the energy produced in fusion exceeds the energy required to heat the plasma. However, the overall fusion process is still far from breakeven, because it takes a lot more energy to run a fusion device, including energy to sustain the magnetic fields. Achieving a more practical breakeven is one goal of ITER, a fusion device being constructed in France by a large international consortium. Scheduled for operation around 2019, ITER should produce fusion power of 500 MW for tens of minutes. If ITER works, its successor may be a prototype fusion power plant.

FIGURE 7.22
Interior of Japan's JT-60 tokamak shows clearly the machine's toroidal shape.

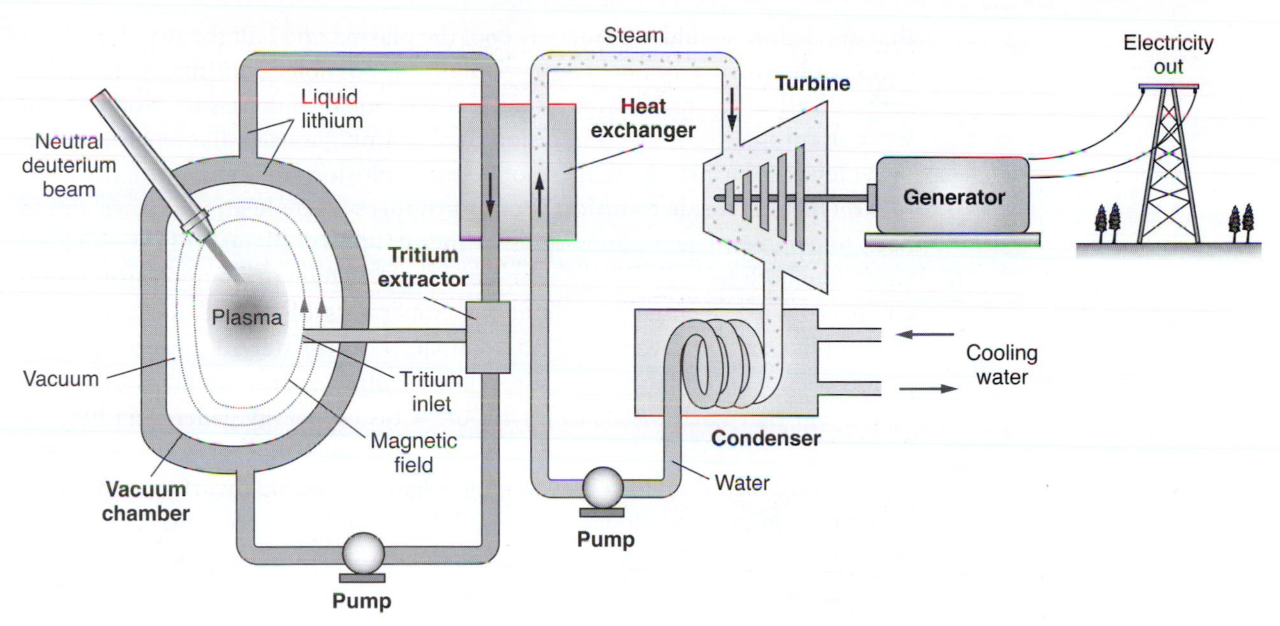

FIGURE 7.23
Possible design for a D–T fusion power plant with a tokamak fusion reactor. D-shaped structure at left is a cross section through the tokamak. The neutral deuterium beam supplies both fuel and energy to heat the plasma.

Early fusion plants will be thermal systems, like fossil and nuclear fission plants. A coolant, possibly liquid lithium metal, will surround the reaction chamber and absorb fusion-produced energy (Fig. 7.23); this design can also "breed" tritium for D–T fusion from neutron bombardment of lithium. In the longer term, we may learn to harness the more difficult deuterium–deuterium (D–D) fusion. This produces high-energy charged particles, which could generate electricity directly without the need for a steam cycle—hence giving much greater thermal efficiencies.

ENVIRONMENTAL IMPACTS OF FUSION

Nuclear fusion, if successful, would be cleaner than fossil fuels or nuclear fission, would have zero carbon emissions, and would have less visual and land-use impact than renewables such as wind, solar, and biomass. The main product of fusion reactions is helium, a benign inert gas and a useful commodity. Some tritium is also produced, and any that isn't "burned" needs to be managed as a nuclear waste; however, given tritium's 12-year half-life, waste management is nowhere near as daunting as with the long-lived wastes from nuclear fission.

The neutrons produced in fusion bombard the reaction chamber walls, inducing radioactivity and structural weakness. Therefore the fusion reactor itself becomes mildly radioactive, presenting hazards during maintenance and requiring eventual disposal as radioactive waste. Again, though, the half-lives involved are far shorter than for fission products. More important, the total amount of radioactivity produced in a fusion power plant would be far less than with fission.

Fusion has the additional safety advantage that there's no chain reaction to go out of control. It's so hard to sustain the high temperature needed for fusion that any failure would immediately cool the plasma and halt the reaction. Thus catastrophic fusion accidents are virtually impossible. And magnetic fusion isn't related to the production of nuclear weapons, so there's no worry about weapons proliferation. Inertial fusion is more ambiguous in that regard, since it provides a tool for studying thermonuclear explosions in miniature.

Is there a downside to fusion? Yes: It produces waste heat, which we would need to dispose of as we do with other thermal power plants. If fusion proved successful both technically and economically, then human energy consumption could grow exponentially, without the constraints of limited fuel resources, cost, pollution, or greenhouse gas emissions. Ultimately, however, all that fusion energy and its thermal waste—like all other energy we generate—ends up as heat that Earth needs to get rid of. If fusion energy generation became even 1% or so of the solar input (our energy production is now 0.01% of the solar input), then we'd have a heating problem that would dwarf today's worry about carbon-induced global warming. I don't think that scenario is likely, but it's a limitation we need to keep in mind.

THE FUSION RESOURCE

Throughout this book, I frame every energy source with a quantitative look at the associated resource. Is a given energy flow sufficient to make a significant contribution to our energy supply? Is there enough of a given fuel to last more than a few decades? For fusion, the answer to the fuel resource question is a resounding "YES!" The fusion fuel deuterium comprises one in 6,500 hydrogen atoms, and there's a lot of hydrogen in the water of Earth's oceans. Example 7.7 shows that each gallon of seawater contains, in deuterium, the energy equivalent of 350 gallons of gasoline. At our current energy-consumption rate, that amounts to some 25 billion years of fuel—five times longer than the Sun will continue to shine. So deuterium–deuterium fusion really does offer the prospect of nearly limitless energy. Deuterium–tritium fusion is a bit more constrained; with its dozen-year half-life, tritium doesn't occur naturally. But, as I suggested earlier, it can be "bred" by neutron bombardment of lithium. And, at 20 parts per million by weight, lithium is modestly abundant in Earth's crust.

EXAMPLE 7.7 | Fusion Resources: Energy from Water

Deuterium fusion releases 330 TJ of energy per kilogram of deuterium. Considering that one of every 6,500 hydrogen nuclei is deuterium, estimate the fusion energy content of 1 gallon of water, and compare it with gasoline.

SOLUTION

A gallon of water weighs a little over 8 pounds, for a mass around 4 kg. Because water is H_2O with nearly all the hydrogen in the isotope ${}^{1}_{1}H$ and nearly all the oxygen as ${}^{16}_{8}O$, water's average molecular weight is $(2 \times 1) + 16 = 18$, of which

2/18 or 1/9 is hydrogen. One out of every 6,500 hydrogen nuclei is deuterium; since deuterium ($^2_1$H) has twice the atomic weight of ordinary hydrogen, this means deuterium comprises 1/3,250 of all the hydrogen by mass. So our 4-kg gallon of water contains (4 kg)/9/3,250 = 0.14 gram of deuterium. At 330 TJ/kg, the corresponding energy content is (330 TJ/kg)(0.00014 kg) = 0.046 TJ, or 46 GJ.

Table 3.3 lists the energy content of gasoline at 36 kWh/gallon; since 1 kWh = 3.6 MJ, this means 1 gallon of gasoline contains (36 kWh/gallon)(3.6 MJ/kWh) = 130 MJ, or 0.13 GJ. With our 46 GJ of deuterium energy per gallon of water, this gives 1 gallon of water the energy equivalent of (46 GJ/gal water)/(0.13 GJ/gal gasoline), or about 350 gallons of gasoline!

PROSPECTS FOR FUSION ENERGY

Since the 1950s, fusion enthusiasts have declared that fusion energy is only a few decades away. Half a century of experience with the challenges of fusion has tempered that optimism. Some think fusion will never be economically viable, even if we overcome all its scientific and engineering problems. And national commitments to expensive fusion research have wavered; the ITER collaboration nearly fell apart in squabbles over where to site the project, and the United States withdrew from ITER for several years because of the project's cost. Nevertheless, the fusion community has made steady, if slow, progress toward its goals. Still, a commercial fusion reactor is unlikely before 2050. At this time we can't know whether fusion will be a long-term solution to our energy needs. Even if it is, it won't be in time to reduce our immediate dependence on fossil fuels.

CHAPTER REVIEW

BIG IDEAS

Nuclear energy is a distant second to fossil fuels in its share of the world's energy supply.

7.1 The **atomic nucleus** consists of **neutrons** and **protons** bound by the strong nuclear force. The **atomic number** is the number of protons, which determines the chemical element. Different **isotopes** of an element have different numbers of neutrons.

7.2 Unstable nuclei undergo **radioactive decay**. A radioactive isotope's **half-life** is the time it takes half the nuclei in a large sample to decay. Common decay schemes include **alpha decay** (emission of a helium-4 nucleus, also called an **alpha particle**), **beta decay** (emission of a high-energy electron, or **beta particle**), and **gamma decay** (emission of a high-energy photon, or **gamma ray**).

7.3 **Binding energy** is the energy released when a nucleus forms; the **curve of binding energy** per nucleon peaks at iron, with 26 protons. It's possible to extract nuclear energy by **fission** of elements heavier than iron or by **fusion** of elements lighter than iron. Nuclear reactions release some 10 million times more energy than chemical reactions do.

7.4 **Fissile** isotopes fission when struck by low-energy neutrons; they include U-235 and Pu-239. Fission produces intermediate-size nuclei called **fission products**; it also releases energy and typically 2 to 3 neutrons. The neutrons may cause additional fission, resulting in a **chain reaction**. To generate energy safely from nuclear fission, it's necessary to control the chain reaction. In most cases it's also necessary to slow the neutrons with a **moderator** so they're effective in inducing fission.

7.5 Nuclear power reactors come in different designs. Most common are **light-water reactors** that use ordinary water as moderator and coolant. Other designs include heavy-water reactors and gas-cooled reactors. **Breeder reactors** convert nonfissile U-235 into fissile Pu-239. New reactor designs promise greater intrinsic safety. Whatever the reactor type, nuclear reactors are used to boil water and drive a steam cycle as in fossil-fueled power plants.

7.6 The **nuclear fuel cycle** includes uranium mining, **enrichment** to increase U-235 content, fuel fabrication, "burning" in a reactor, and waste storage or reprocessing. Nuclear waste includes isotopes with half-lives as long as tens of thousands of years, and isolation of these wastes from the environment is one of nuclear power's thornier issues. **Reprocessing** removes fissile plutonium from spent nuclear fuel rods for use in new reactor fuel, and renders the remaining waste less radioactive. But reprocessing is difficult, and it makes plutonium a commercial substance that could be diverted for use in nuclear weapons. Reprocessing, especially when combined with breeder reactors, could substantially extend the life of Earth's uranium reserves.

7.7 The dominant environmental impact of nuclear power is the health effects of radiation, through mutations, cancer, or cell death. The effects of high radiation doses are well known, but it's hard to quantify the effects of low doses against the background of mutations and cancers from other causes. Radiation from nuclear power constitutes a tiny fraction of the radiation humans receive from natural and anthropogenic sources. Realistic estimates of nuclear energy's health and environmental impacts suggest they're far lower than that of fossil fuels. However, nuclear power has the potential for large-scale accidents and is vulnerable to terrorist attacks, and the technology and expertise that enable nuclear power programs can be used to produce nuclear weapons.

7.8 Nuclear power has grown more slowly than anticipated because of safety and economic issues. The need for carbon-free energy coupled with a new generation of simpler, less expensive, and intrinsically safer nuclear reactors could bring about a renaissance of nuclear power. But such a renaissance is unlikely to increase significantly the nuclear contribution to the world's energy mix, and a nuclear renaissance must confront the link between nuclear power and nuclear weapons.

7.9 Nuclear fusion offers the prospect of nearly limitless energy based on the hydrogen isotope deuterium. Despite decades of research, we are still a long way from a fusion power plant. Approaches to fusion include **inertial confinement** and **magnetic confinement**.

TERMS TO KNOW

alpha decay (p. 163)
alpha particle (p. 164)
atomic nucleus (p. 160)
atomic number (p. 161)
background radiation (p. 187)
becquerel (p. 185)
beta decay (p. 164)
beta particle (p. 164)
binding energy (p. 167)
boiling-water reactor (BWR) (p. 174)
breeder reactor (p. 176)
chain reaction (p. 171)
control rods (p. 174)
coolant (p. 174)
critical mass (p. 172)
curie (p. 185)
curve of binding energy (p. 168)
deuterium (p. 163)
D–T fusion (p. 194)
enrichment (p. 180)
fissile (p. 170)
fission (p. 168)
fissionable (p. 169)
fission product (p. 170)
fusion (p. 168)
gamma decay (p. 164)
gamma ray (p. 164)
generation-I reactor (p. 177)
generation-II reactor (p. 177)
generation-III reactor (p. 177)
generation-III+ reactor (p. 178)
generation-IV reactor (p. 178)
gray (p. 185)
half-life (p. 165)
heavy water (p. 173)
ignition temperature (p. 195)
inertial confinement (p. 195)
isotope (p. 162)
Lawson criterion (p. 195)
light-water reactor (LWR) (p. 174)
magnetic confinement (p. 196)
mass number (p. 162)
mixed-oxide fuel (MOX) (p. 184)
moderator (p. 173)
multiplication factor (p. 173)
neutron (p. 160)
neutron-induced fission (p. 170)
nuclear difference (p. 169)

GETTING QUANTITATIVE

Isotope symbol example: $^{4}_{2}\mathrm{He}$

U-235 as fraction of natural uranium: 0.7%

Mass-energy equivalence (Box 7.1): $E = mc^2$

Nuclear difference: $\dfrac{\text{Energy released in typical nuclear reaction}}{\text{Energy released in typical chemical reaction}} \sim 10^7$

Alpha decay example: $^{238}_{92}\mathrm{U} \rightarrow {}^{234}_{90}\mathrm{Th} + {}^{4}_{2}\mathrm{He}$ (Equation 7.1; p. 164)

Beta decay example: $^{14}_{6}\mathrm{C} \rightarrow {}^{14}_{7}\mathrm{N} + {}^{0}_{-1}e + \bar{\nu}$ (Equation 7.2; p. 164)

Cancer incidence from radiation: 1% lifetime chance of radiation-induced cancer per 100 mSv radiation dose; ~ 50% of cancers fatal

Background radiation dose: world average, 2.4 mSv/year; United States, 3.6 mSv/year

Waste from 1-GWe nuclear power plant: 20 tons high-level radioactive waste per year

Nuclear power reactors worldwide, 2010: ~ 450

D–T fusion: $^{2}_{1}\mathrm{H} + {}^{3}_{1}\mathrm{H} \rightarrow {}^{4}_{2}\mathrm{He} + {}^{1}_{0}n + 2.8\ \mathrm{pJ}$ (Equation 7.3; p. 195)

Deuterium energy content: 330 TJ/kg

D–T fusion ignition temperature: 50 MK

QUESTIONS

1. Explain the role of the moderator in a nuclear reactor, and describe why the use of the same water as moderator and coolant is an inherent safety feature of light-water reactors.
2. In what sense does a breeder reactor produce more fuel than is put into it? Do breeders violate the principle of energy conservation?
3. What are some of the possible connections between nuclear power and nuclear weapons?
4. Describe and contrast the two approaches to controlled fusion.
5. Describe the nuclear difference, and explain how it affects the relative quantities of waste from fossil fuels and nuclear fission.

EXERCISES

1 Copper's atomic mass is 63.6 u. Natural copper consists of just two stable isotopes, with 69.1% of copper being Cu-63. What's the other isotope?

2 The Sun generates energy at the rate of about 3.9×10^{26} W. Study Box 7.1, and use Einstein's equation to verify that the Sun loses mass at the rate of about 4 million tons per second.

3 Fission of a uranium-235 nucleus releases about 30 pJ (30×10^{-12} J). If you weighed all the particles resulting from a U-235 fission, by what percentage would their total mass differ from that of the original uranium nucleus?

4 Find the mass of the plutonium-239 produced by your decision to use the electric clothes dryer in Example 7.5.

5 Confirm the assertion in Section 7.3 that the 10^{15} joules released in fissioning 30 pounds of U-235 is approximately equal to that contained in three 100-car trainloads of coal, with each car carrying 100 tonnes.

6 For every three U-235 nuclei that fission in a reactor, approximately two nuclei of Pu-239 are produced. Suppose you walk out of your room and leave your 100-W light on for 1 hour. If your electricity comes from a 33% efficient nuclear plant, how many plutonium-239 nuclei are produced as a result of your leaving the light on for that hour?

7 Radium results from a sequence of reactions, beginning with U-238 undergoing the alpha emission described in Equation 7.1. Two beta decays follow, and then two more alpha decays. Write equations for these four additional reactions, correctly identifying the intermediate isotopes.

8 After Japan's 2011 Fukushima nuclear accident, levels of iodine-131 radioactivity in some milk samples reached 17 times the government's safety limit. Given I-131's half-life of 8.04 days, how long would the Japanese have to wait for that milk to become safe? Compare with the typical 2-week shelf life of fresh milk.

9 (a) In the U.S. population of approximately 310 million, how many cancer cases should result each year from the average individual background radiation dose of 3.6 mSv per year? (b) Consult Figure 7.17 to determine the number of these deaths attributable to natural and artificial radiation sources.

10 Confirm the text's estimate of 150 fatal cancers per year if everyone in the United States was exposed to radiation at the 0.01-mSv per person dose received in the neighborhood of nuclear power plants (recall that roughly half of radiation-induced cancers are fatal).

11 Assuming the ocean's average depth is 3 km, use the result of Example 7.7 to estimate the length of time D–D fusion using deuterium from ocean water could supply humankind's current energy consumption rate of 16 TW.

12 The masses of deuterium, tritium, helium-4, and the neutron are, respectively, 3.3434, 5.0073, 6.6447, and 1.6749 in units of 10^{-27} kg. Use these figures to compute the mass difference between the two sides of the reaction in Equation 7.3, and use $E = mc^2$ to find the corresponding energy. Compare with the energy yield in Equation 7.3.

13 Japan's JT-60 tokamak has a confinement time of 1 second. What density is required for the JT-60 to meet the Lawson criterion for D–T fusion?

14 How much pure deuterium would you need in order to have the same energy content as 1 tonne of coal?

RESEARCH PROBLEMS

1 Determine the fraction of electricity generated by nuclear power plants in your state, country, or local electric utility's region.

2 Locate the nearest nuclear plant to your home and determine as much of the following information as you can: (a) type of reactor; (b) maximum rated power output in MWe or GWe; (c) year of initial operation; (d) capacity factor (actual annual energy production as a fraction of the production if the plant produced energy at its maximum rate all the time); (e) the status of its high-level nuclear waste (spent fuel).

3 If you're a resident of the United States, find your average annual radiation dose using the Environmental Protection Agency's calculator at www.epa.gov/rpdweb00/understand/calculate.html.

ARGUE YOUR CASE

1 Your state needs new sources of electricity, and the local utility is considering either a coal-burning power plant or a nuclear plant. You live downwind of the proposed plant site. Decide which type of plant would you rather have, and formulate an argument explaining why.

2 Nuclear power advocates have claimed that more people died as a result of the Three Mile Accident from extra coal that had to be burned when the damaged reactor went offline than died from nuclear effects of the accident. Do you agree? Formulate a wide-reaching and quantitative argument in support of your view.

Chapter 8

ENERGY FROM EARTH AND MOON

Having covered fossil and nuclear energy sources, we're done with the Earth-stored fuels that we currently utilize for most of our energy. But, as we saw in Chapter 1, there are also energy flows that we can tap into. This chapter deals with two of those flows, geothermal and tidal energy, while the next two chapters cover the much more significant solar energy flow.

8.1 The Geothermal Resource

Earth's interior is hot, a result of naturally occurring radioactive materials and primordial energy released during our planet's formation. The temperature difference between Earth's interior and its surface drives an upward heat flow. That flow sets the ultimate size of the sustainable geothermal energy resource—although there's nothing to keep us from using that resource, at least temporarily, at an unsustainable rate.

Figure 1.8 shows that the geothermal energy flow supplies only about 0.025% of the energy coming to Earth's surface; nearly all the rest is from the Sun. Averaged over the entire planet, that 0.025% amounts to about 0.087 watts per square meter—or 87 mW/m^2. That's pretty feeble compared with the average solar flow, in direct sunlight, of 1,000 watts per square meter—1 kW/m^2. As you can show in Exercise 1, the total geothermal flow is some 40 TW, about three times humankind's current energy consumption rate. But the low temperature associated with near-surface geothermal energy means this is energy of very low quality. That, along with the diffuse nature of the geothermal flow, says that we'll never be able to use any but a tiny fraction of the geothermal energy flow for high-quality energy needs (see Exercise 2 for a quantitative look at this point). So sustained use of geothermal energy will never go very far toward meeting humankind's total energy demand. Nevertheless, geothermal energy today is a significant source of renewable energy, behind wind but ahead of solar. And unlike solar and wind, geothermal provides a steady, uninterruptible energy flow that can be used for baseload power generation.

On average, the temperature in Earth's crust increases by about 25°C for every kilometer of depth—a figure known as the **geothermal gradient.** That temperature gradient drives the geothermal energy flow through heat conduction, as described in Chapter 4. Using Table 4.1's value for the thermal conductivity of rock, you can show in Exercise 3 that that the geothermal energy flow and geothermal gradient are quantitatively consistent with thermal conduction.

There's nothing to keep us from extracting geothermal energy at a greater rate than it's being replenished by the geothermal flow—but such energy use shouldn't count as renewable or sustainable. With a geothermal gradient of 25°C/km, the average temperature in the top 5 kilometers of Earth's crust is about 60°C (half the 125°C excess at 5 km) above the surface temperature. As Example 8.1 shows, that amounts to a thermal energy around 100 million EJ, far more than the known reserves of fossil fuels. But, as Exercise 4 shows, large-scale extraction of geothermal energy from the near-surface crust is neither efficient nor practicable. So even nonsustainable extraction of geothermal energy isn't likely to make a significant contribution to humankind's total energy demand.

EXAMPLE 8.1 | The Geothermal Energy Resource

Estimate the thermal energy extractable from the top 5 km of Earth's crust, assuming it's granite (density 2,700 kg/m^3) and that the average temperature is 60°C above the surface temperature. Here "extractable" means the energy that would become available in cooling this entire layer to the surface temperature. Compare with Earth's total known reserves of fossil energy.

SOLUTION

The total thermal energy content is the energy that would be released in reducing temperature of the 5-km layer by 60°C. That energy, Q, depends on the mass m of material being heated and on its specific heat c, as described in Equation 4.3: $Q = mc\,\Delta T$. We can calculate the mass of rock involved from the given density and the volume, which is the Earth's surface area $4\pi R_E^2$ multiplied by the 5-km layer thickness:

$$m = density \times volume = (2{,}700\ \text{kg/m}^3)(4\pi)(6.37 \times 10^6\ \text{m})^2(5{,}000\ \text{m})$$
$$= 6.88 \times 10^{21}\ \text{kg}$$

Here we found Earth's radius in the physical constants list inside the back cover. For the next step we'll need the specific heat of granite, given in Table 4.3 as 840 J/kg·K and equally well written 840 J/kg·°C because temperature *changes* in kelvins and Celsius degrees are equivalent. Then Equation 4.3 gives

$$Q = mc\Delta T = (1.72 \times 10^{21}\ \text{kg})(840\ \text{J/kg·°C})(60\text{°C})$$
$$= 3 \times 10^{26}\ \text{J}$$

rounded to one significant figure. That's 300 million EJ. Adding the known reserves of oil (about 6,000 EJ), gas (another 6,000 EJ), and coal (roughly 25,000 EJ) as given in Chapter 5 shows that this geothermal energy resource is nearly 10,000 times the energy content of the known fossil fuels. But again, low energy quality and the diffuse nature of the resource—spread over the entire globe, with some 70% beneath the oceans—means that this comparison is not very meaningful.

Although geothermal energy underlies every place on Earth, today's technology limits its use to specific regions with larger-than-normal geothermal flows. In such regions, hot magma lies closer to Earth's surface, resulting in a more rapid increase in temperature with depth—that is, a larger geothermal gradient. That helps in two ways: first, it means the sustained flow of geothermal energy is greater. Second, it puts high temperatures and hence higher quality energy closer to the surface, making energy extraction technically and economically more viable.

Geothermal sites with the highest energy-producing potential generally lie near the boundaries of Earth's tectonic plates, where earthquakes and volcanic eruptions are most likely (Fig. 8.1). Other sites, like the Hawaiian Islands, are

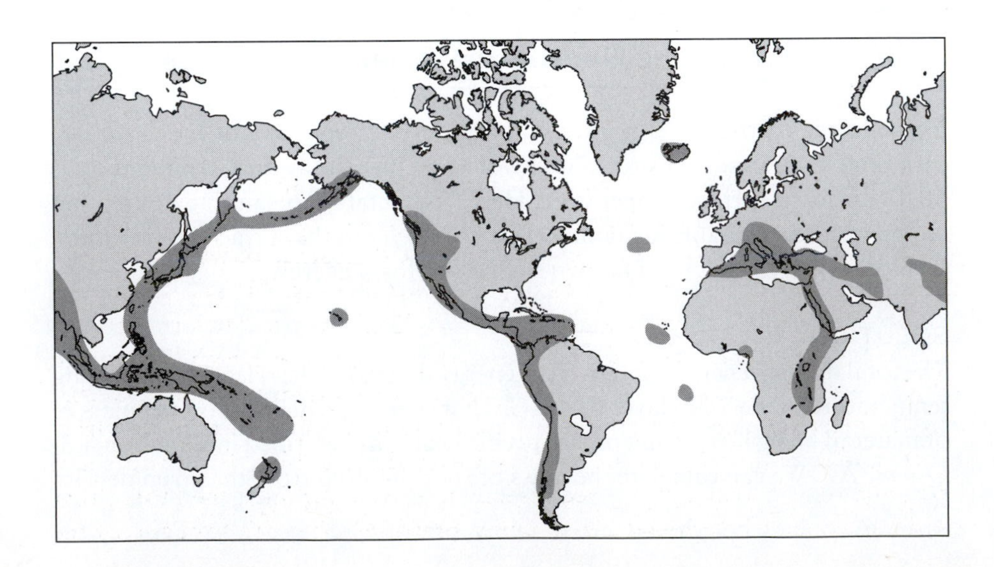

FIGURE 8.1
Regions with the greatest geothermal energy potential, indicated in darker gray, tend to be located along geological plate boundaries.

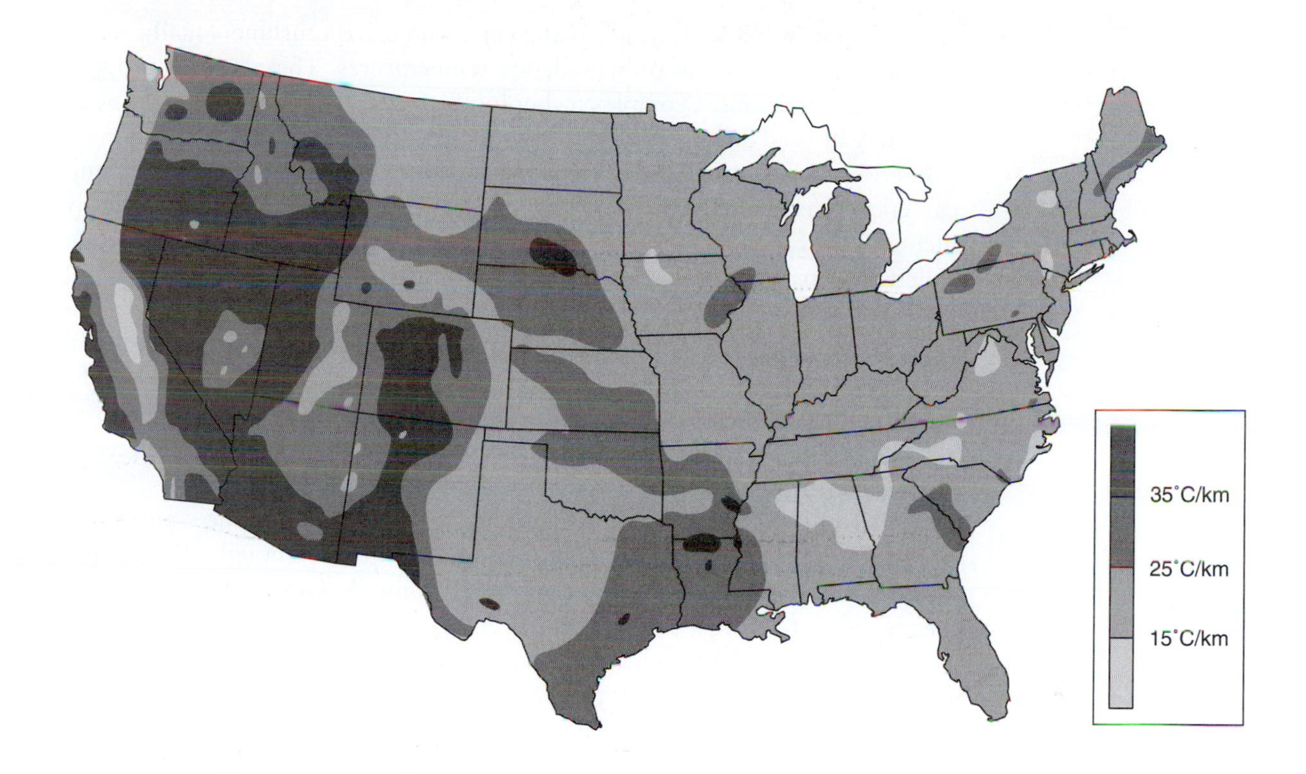

FIGURE 8.2
Geothermal gradient in the continental United States. Note the preponderance of geothermal resources in the western states.

located over local "hot spots" associated with individual plumes of magma that extend close to the surface. The so-called "ring of fire" surrounding the Pacific Ocean is home to some of Earth's most active volcanoes, some of the most damaging earthquakes, and some of the most promising sites for geothermal energy. A look at geothermal resources in the United States (Fig. 8.2) shows that sites with the greatest potential are located, not surprisingly, in the western states. Geothermal gradients in the continental United States range from less than 10°C/km to more than 60°C/km.

WET OR DRY? TYPES OF GEOTHERMAL ENVIRONMENTS

Temperature and quantity of thermal energy aren't the only factors that determine the potential of a geothermal energy resource. Also important are the associated geological structure and the presence or absence of water. Geothermal resources come in two broad categories, wet and dry. Wet systems are the most valuable for energy production. In these systems, fractures in the rock allow water to penetrate to depths where the temperature is high. Systems in which water or steam can circulate freely are called **hydrothermal systems.** In the most valuable hydrothermal systems, termed **vapor-dominated**, fractures and pores in the rock are saturated with steam at high temperature and pressure—steam that can be used directly to drive a turbine-generator. Other systems contain hot, liquid water under pressure; when it's brought to the surface, some water boils to steam and can again drive a turbine. Finally, many systems contain water at only moderate temperatures. These cannot produce steam and require more complex technologies to generate electricity—although they may be used directly to supply low-quality energy for heating.

In addition to thermal energy, some hydrothermal systems contain mechanical energy associated with high water pressure. Water pressure naturally increases with depth due to the weight of the overlying material, but in so-called **geopressured systems** the water-saturated zone is effectively cut off from the surface, usually by layers of sedimentary rock. The result is a reservoir of hot water at high pressure. In principle, both thermal and mechanical energy could be extracted simultaneously from such systems, although the technology to accomplish this extraction has not proved economical. Geopressured systems often contain dissolved methane, making them also sources of chemical energy.

Dry geothermal environments lack either water or the fractures and pores necessary for groundwater to contact hot rock. Drill deep enough, though, and you'll find hot, dry rock anywhere on Earth. Pump down water and up comes steam, ready to drive a turbine. However, dry-rock energy extraction has not yet proven either technologically or economically viable—or even safe. A pilot project in Basel, Switzerland, penetrated 3 miles into hot rock, but was suspended in 2006 when it triggered earthquakes.

8.2 Geothermal Energy Technology

Humanity's use of geothermal energy goes back thousands of years. Surface geothermal activity such as geysers, steam vents, and hot springs would have been obvious to our prehistoric ancestors, and it's likely that gentler geothermal features were used for bathing and food preparation. Ancient Japanese, Greeks, and Romans constructed spas at geothermal sites, and the use of geothermal hot springs for bathing, relaxation, and physical therapy continues today (Fig. 8.3).

FIGURE 8.3
Bathers enjoy geothermally heated waters in Iceland's Blue Lagoon.

GEOTHERMAL HEATING

Geothermally heated water at modest temperatures can be used directly for purposes requiring low-quality thermal energy. Temperatures in such direct-use geothermal applications range from as low as about 40°C to about 150°C. Geothermal resource temperatures below 90°C are classified as low-temperature, and are generally suitable only for direct-use applications. Temperatures between 90°C and 150°C are considered moderate, and may be used directly as thermal energy or, with some difficulty and low efficiency, to produce electricity. Resources above 150°C are most valuable for producing electricity.

In heating applications, hot water from surface features or shallow wells circulates through individual buildings or community-wide hot-water systems. Reykjavik, Iceland; Copenhagen, Denmark; Budapest, Hungary; Paris, France; and Boise, Idaho, are heated in part with geothermal energy. A novel community use of direct geothermal heat is to melt snow on sidewalks (Fig. 8.4).

Many industrial and agricultural processes require only low-quality energy, which can be supplied with modestly heated water. Direct use of geothermal energy can substitute for fossil fuels in such applications. Geothermally heated greenhouses provide fresh produce year-round in cold regions. Dairies, fish farms, mushroom growers, and light industries also benefit from low-grade geothermal heat. Figure 8.5 shows the distribution of direct-use geothermal energy throughout the world. Globally, systems making direct use of geothermal energy have a capacity of nearly 30 GWth, but their average output is only 31% of capacity.

FIGURE 8.4
Geothermal hot water melts snow from walkways in Klamath Falls, Oregon. An antifreeze solution heated with water from geothermal wells first passes through buildings in the city's central district, then through sub-walkway tubing.

ELECTRIC POWER GENERATION

Hotter geothermal sources can generate electricity, albeit with only modest thermodynamic efficiency. The most valuable geothermal resources for electric power generation are the vapor-dominated systems where a hydrothermal environment is saturated with steam. Not only is the steam at high temperature and pressure, but also the absence of liquid water simplifies power-plant design. Drill a well into such a system, and up comes the steam, ready to use. Such a geothermal power plant is like the fossil and nuclear plants we saw in earlier chapters, with the geothermal steam source replacing a fossil-fueled boiler or fission reactor. The rest is the same: High-pressure steam turns a turbine connected to an electric generator. The spent steam goes through a condenser, which extracts the waste heat required by the second law of thermodynamics and turns the steam back to liquid water. The water is re-injected into the geothermal heat source in order to replace the extracted steam; were this not done, the plant's power output would decline, due not to the exhaustion of energy but of water. Additional makeup water is sometimes needed, as steam is inevitably lost to the atmosphere. Figure 8.6 shows the essential features of a steam-based geothermal power plant, while Box 8.1 describes California's Geysers geothermal power systems.

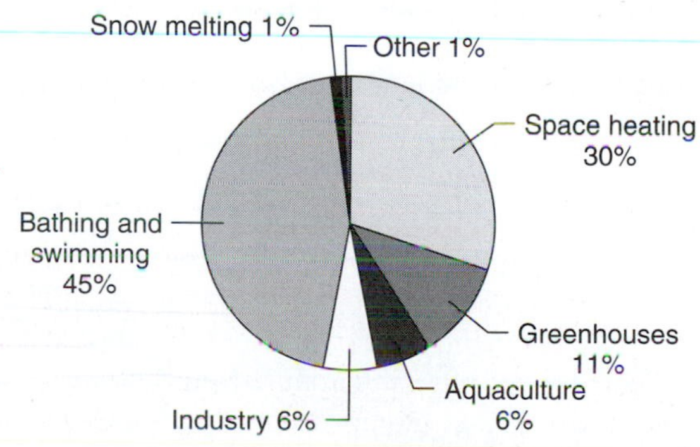

FIGURE 8.5
Direct uses of geothermal energy, worldwide. Geothermal heat pumps are not included, for reasons discussed at the end of Section 8.4.

As with fossil and nuclear plants, the waste heat from a geothermal steam cycle can itself be used to supply low-quality thermal energy. Iceland has pioneered such geothermal cogeneration plants, providing both electricity and community-wide heating. Although some geothermal sites are too remote for cogeneration to be practical, California is exploring community heating and snow-melting using waste heat from geothermal plants at Mammoth Lakes.

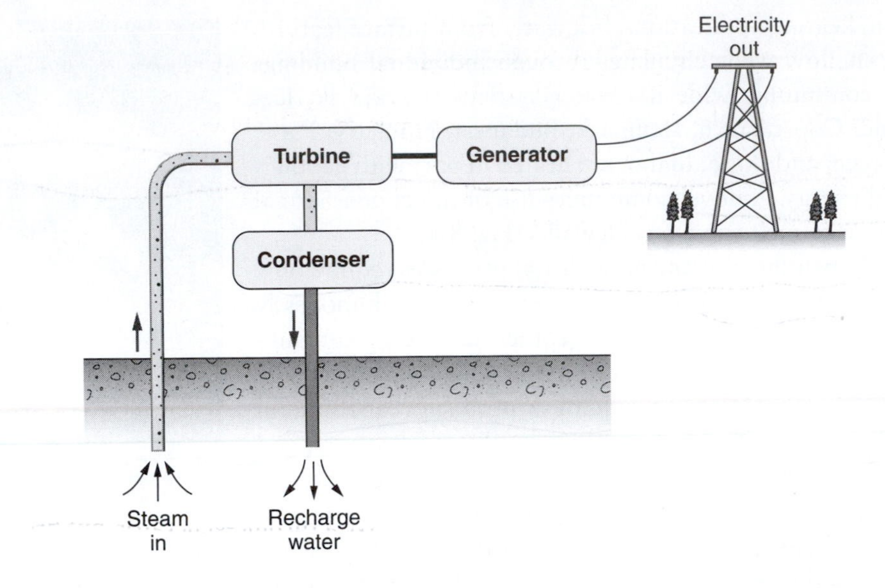

FIGURE 8.6
Essential features of a geothermal power plant using steam from a vapor-dominated geothermal resource.

High-quality vapor-dominated systems like Geysers are rare. More common, and still valuable for power generation, are liquid-water systems at temperatures above water's atmospheric-pressure boiling point of 100°C. High below-ground pressure keeps water liquid even at such high temperatures, but when the water reaches the surface in a geothermal well some of it flashes—boils

BOX 8.1 | The Geysers Geothermal Power Plants

The Geysers, a region in northern California about 120 km north of San Francisco, has long been known for its geothermal features, including steam vents, spouting hot springs, and mudpots—but no true geysers. Adjacent to a volcanic area that erupted as recently as a few thousand years ago, Geysers sits atop a large vapor-dominated hydrothermal system containing steam at high temperature and pressure.

Wells first drilled in 1924 used this geothermal resource to drive turbine-generators capable of a few kilowatts. By the 1950s, wells several hundred meters deep tapped into the main hydrothermal system, and by 1960 a single power plant at the Geysers was producing 12 MW of electric power. Development at the Geysers continued to accelerate until about 1990, at which time the total generating capacity of its 26 individual plants was over 2 GWe—the equivalent of two large coal or nuclear plants (Fig. 8.7).

FIGURE 8.7
A geothermal power plant at Geysers, California.

Given its high rates of energy extraction, however, the Geysers definitely does not qualify as a sustainable or renewable energy source, and in fact power production declined after 1988. Much of this decline was due to loss of steam, which was not fully compensated by the injection of condensate water into the ground. In 1997, a 50-km pipeline began supplying treated municipal wastewater from nearby communities for ground injection; the resulting increase in geothermal steam immediately raised the Geysers' power output by 75 MWe, while helping to dispose of wastewater in an environmentally sound way. Expansion of the wastewater system to the city of Santa Rosa in 2003 resulted in an additional 85-MWe increase in generating capacity. Nevertheless, the Geysers' capacity of some 1.5 GW remains well below its peak in the late 1980s, and average power output is under 1 GW. Some Geysers plants have been retired, and one has been designated a National Historical Mechanical Engineering Landmark. Yet geothermal development continues at Geysers, with a new 25-MW plant scheduled to go online in 2013.

The Geysers geothermal plants are a major contributor to the 4.5% of California's electrical energy that comes from geothermal sources. Geysers power sells at about 3.5 cents per kilowatt-hour, comparable to fossil-generated power.

abruptly—to produce steam. That steam can drive a turbine-generator, but first it's necessary to remove liquid water. Thus a hot-water-based geothermal power plant is more complicated than one using geothermal steam, because it needs a device to separate steam and water. That drives up cost, although hot-water geothermal systems can still compete with conventional sources of electrical energy. Figure 8.8a shows a typical water-based system; note that both condensate and water separated from steam are injected back into the ground.

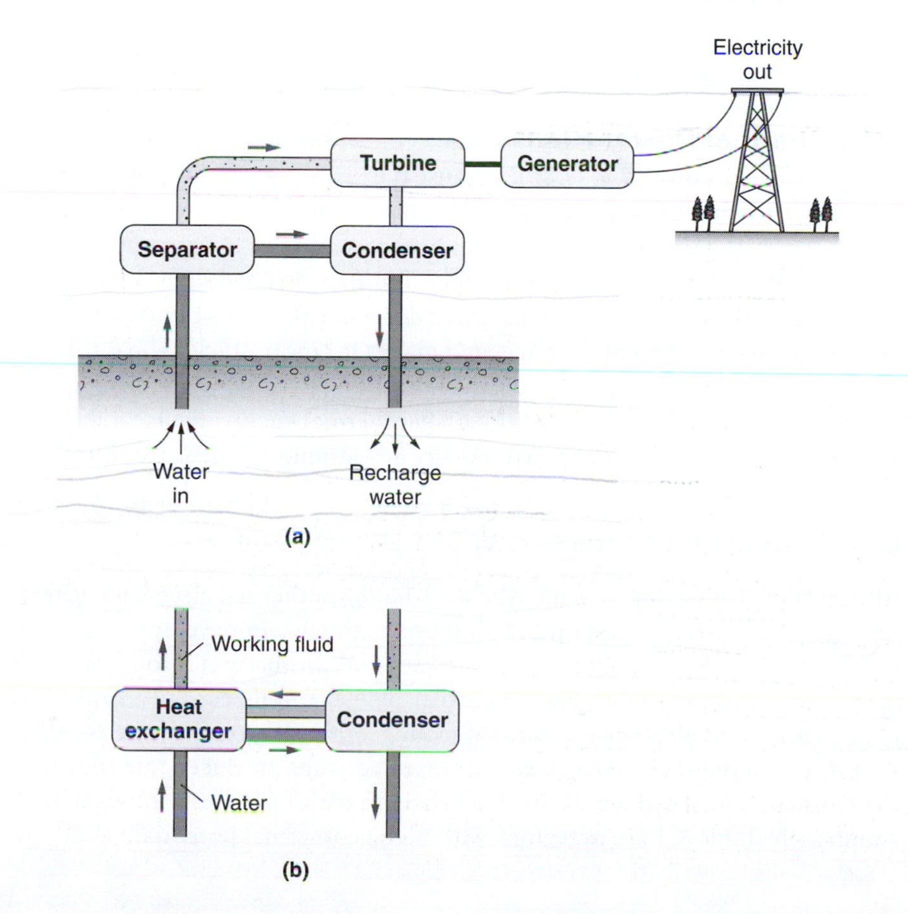

FIGURE 8.8
(a) In a hot-water geothermal power plant, water boils in the well pipe, and a separator extracts the steam to drive a turbine. Separated water and condensate return to the geothermal source. (b) A binary geothermal plant uses a closed loop with a separate working fluid, whose boiling point is lower than water's. The heat exchanger transfers energy from geothermal water to the working fluid. Heat exchanger and associated piping replace the steam separator shown in (a).

FIGURE 8.9
Binary-cycle geothermal power plant at Mammoth Lakes, California. The plant transfers heat from geothermal water at 170°C to its working fluid, isopentane, which boils at 28°C. Four separate units produce a total of 40 MW of electric power. The plant was designed to blend with its scenic surroundings, and the closed-loop binary cycle minimizes emissions.

Finally, even moderate-temperature hydrothermal systems—from a little below to somewhat over 100°C—can produce electrical energy, but not with steam turbines. Instead, geothermal hot water extracted from the ground passes through a **heat exchanger**, transferring energy to a *working fluid* with a lower boiling point than water. The working fluid boils to vapor, which drives a turbine-generator. It then passes through a condenser, liquefies, and is recycled into the heat exchanger (Fig. 8.8b). Because it involves two fluids, such a power plant is called a **binary system**. Binary systems are more complex and expensive than steam-based geothermal plants, and because of their lower temperatures, they have lower thermodynamic efficiency. But they do have some advantages. The geothermal water doesn't boil, which decreases the loss of both water and pressure in the geothermal resource, and makes for more efficient heat transfer from the surrounding rocks. And binary technology minimizes some of the environmental problems associated with geothermal energy. Figure 8.9 shows a 40-MW binary geothermal power facility at Mammoth Lakes, California.

GEOTHERMAL ENERGY: A WORLDWIDE LOOK

Although the United States is the world leader in geothermal electricity generation, other countries have been much more aggressive in developing geothermal energy, usually because they have substantial geothermal resources. Table 8.1 lists countries with the greatest geothermal generating capacity, ranked by the percentage of electricity generated from geothermal sources. I've included California among the "countries" because the state produces one-fourth of the world's geothermal electricity. Comparison with Figure 8.1 shows that all countries in Table 8.1 are in regions with high geothermal potential.

The data in Table 8.1 are just for electrical energy. The geothermal energy extracted worldwide for direct heating is more difficult to quantify, but it's roughly comparable to what's used for electrical energy. And some countries get substantially more of their *total* energy from geothermal sources than Table 8.1 might suggest. Iceland, for instance, uses geothermal energy extensively for community heating, with some 87% of Iceland's homes relying on geothermal heat. Coupled with its significant geothermal electricity production (nearly all the rest is hydropower), this makes Iceland the world leader with more than 50% of its total primary energy coming from geothermal sources.

TABLE 8.1 | GEOTHERMAL ELECTRICITY: TOP PRODUCING COUNTRIES

Country	Percent of electricity generation capacity from geothermal sources	Total geothermal generation capacity (MWe)
El Salvador	26	204
Iceland	25	575
Philippines	18	1,904
Kenya	14	167
Costa Rica	13	166
New Zealand	10	628
California	4.5	2,566
Indonesia	3.7	1,197
Mexico	1.6	958
Italy	1.2	843
United States	0.4	3,086
Japan	0.2	536
World	0.23	10,715

8.3 Environmental Impacts of Geothermal Energy

Geothermal energy is unquestionably "greener" than fossil fuels. But no energy source is completely benign. We've already seen that geothermal energy isn't always renewable, in which case we need to worry about depleting the local geothermal resource. And geothermal energy has significant environmental impacts—although they pale in comparison with the environmental consequences of fossil fuel combustion.

Geothermal fluids contain dissolved gases, which can escape to the atmosphere when fluids are brought to the surface. Typically, geothermal power plants emit about 13% as much carbon dioxide per unit of electricity generated as do coal plants, and more than one-third the CO_2 of gas-fired plants. Geothermal water also contains dissolved sulfur and nitrogen compounds, some of which escape to the atmosphere. The quantity of sulfur dioxide from a typical geothermal plant is only about 1% of what a comparable coal-fired plant emits, and NO_x emissions are even lower. Another gas found in geothermal fluids is hydrogen sulfide (H_2S). Although toxic in high concentrations, H_2S typically occurs at low concentrations, where it's a nuisance because of its characteristic "rotten egg" smell.

Early geothermal plants dumped condensed steam and geothermal water to nearby rivers or ponds. But geothermal fluids are rich in salts and other contaminants—as much as 30% dissolved solids in water-based systems—and the result was significant water pollution. Today, geothermal fluid is re-injected into the ground in order to sustain the hydrothermal system and maintain geothermal energy output. This process also minimizes water pollution, but fallout from steam vented to the atmosphere can still contaminate land and surface waters.

Binary geothermal systems (recall Figs. 8.8 and 8.9) greatly reduce the potential for land and water pollution. Geothermal fluids in these systems never boil, and flow in a strictly closed loop from the ground, through the heat exchanger, and back into the ground. Despite their greater complexity and lower efficiency, this environmental benefit is one reason these plants are becoming increasingly popular.

Geothermal energy presents some unique environmental issues. Many geothermal sites are in scenic natural areas or in regions that are valued as attractions because of geothermal features themselves. Building industrial-scale power plants in such areas can be an obvious aesthetic affront, although it's possible to reduce the visual impact (recall Fig. 8.9). Geothermal plants can also be noisy, particularly with the shrill sound of escaping steam.

More seriously, removal of geothermal fluids from the ground can cause land subsidence—slumping of the ground level as the land drops to fill the void left by the departed fluid and decreased pressure. Modern plants, with water re-injection, have substantially reduced this problem. However, injection of geothermal fluid or additional wastewater poses another problem: The process lubricates seismic faults and increases the frequency of small earthquakes. At the Geysers, for example, where water is re-injected to compensate for steam extraction, the rate of "micro-earthquakes" correlates directly with steam extraction. That rate increased dramatically with the additional injection of Santa Rosa wastewater in 2003. Whether these quakes pose a hazard to geothermal power facilities or even to the public is an open question.

8.4 Geothermal Heat Pumps

I introduced heat pumps in Chapter 4, as refrigerator-like devices used for heating and/or cooling buildings. Your kitchen refrigerator extracts thermal energy from the food you put inside it, and transfers that energy to the refrigerator's surroundings. In consequence, the kitchen gets a bit warmer. A heat pump in cooling mode does exactly the same thing, with the building replacing the refrigerator's interior. In heating mode, a building's surrounding environment plays the role of the refrigerator's interior, as the pump extracts energy from the environment and pumps it into the building. In warmer climates, so-called *air source heat pumps* work both ways: In summer they cool by transferring energy from a building to the outside air; in the winter they pump energy back in, cooling the outside air in the process. In cooler climates, **geothermal heat pumps** are used primarily for heating. Also called **ground-source heat pumps** or **geoexchange heat pumps**, these devices extract energy from the ground at depths of a meter or more, where the temperature stays nearly constant year-round. In the northern United States, that temperature is typically 5°C to 10°C (40°F to 50°F).

Ground-source heat pump systems come in several flavors. **Closed-loop systems** circulate a heat-exchange fluid through a loop of buried pipes (Fig. 8.10). **Open-loop systems** extract heat from water in wells or ponds, chilling the water in the process, then returning it to its source.

Section 4.9 introduced heat pumps as being essentially heat engines run in reverse. A heat engine extracts energy from a hot source, delivers some

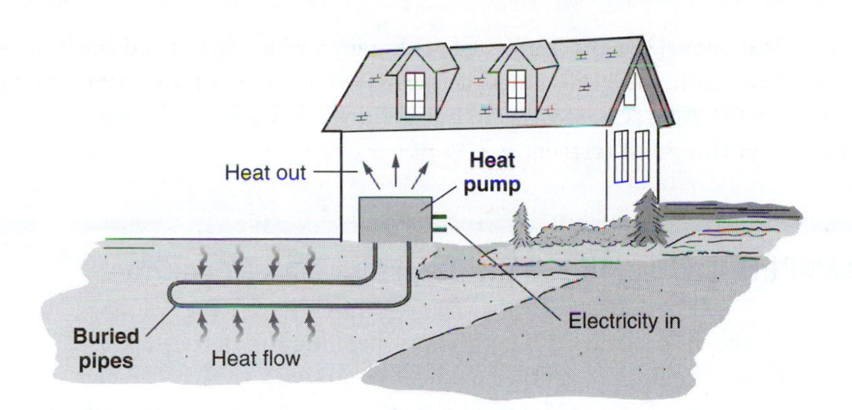

FIGURE 8.10
Closed-loop ground-source heat-pump system using buried pipes.

high-quality mechanical energy, and transfers additional thermal energy to a lower-temperature environment. Turn the engine around and you have a refrigerator or heat pump. Thermal energy moves from a cool region to a hotter one—something that doesn't happen spontaneously but that requires an additional input of high-quality energy. Figure 8.11 combines Figures 4.13 and 4.16 in summarizing this relation between heat engines and refrigerators or heat pumps. In practice, refrigerators and heat pumps aren't actually built like heat engines, but the conceptual similarity remains.

Heat pumps operate on electricity, and are valuable because they transfer more thermal energy than the electrical energy they consume; you can see this in Fig. 8.11b, where the arrow indicating thermal energy delivered to the hot region includes both the thermal energy from the cool region and the electrical energy supplied to the device. A geothermal heat pump moves thermal energy from the ground—energy that's available for free—into the building. But the whole process isn't free, because we have to supply high-quality electrical energy to effect the unnatural energy transfer from cool to hot.

We defined the efficiency of an engine in Equation 4.4 as the ratio of mechanical work delivered to the heat extracted. For a heat pump, the analogous measure of efficiency is the coefficient of performance (COP), defined as the ratio of heat delivered to the electrical energy required to operate the pump:

$$\text{COP} = \frac{\text{heat delivered}}{\text{electrical energy required}} \tag{8.1}$$

A COP of 4, for example, means for every one unit of electrical energy used, the pump delivers four units of energy into the building being heated. Where do the other three units come from? As Figure 8.11 shows, they're extracted from the environment, in this case the ground. So the heat pump gives us a good deal: four units of energy for the price of one! And those four units of heat energy delivered carry the environmental impact associated with only one unit of electrical energy.

One caveat, though: If that electricity comes from a thermal power plant—fossil or nuclear—with a typical efficiency of around 33%, then that 1 unit

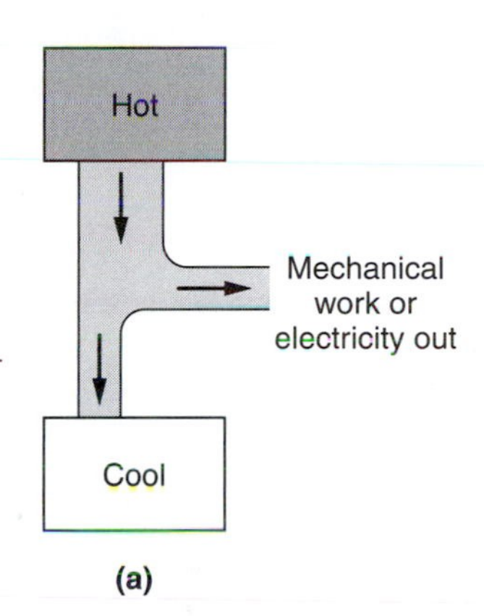

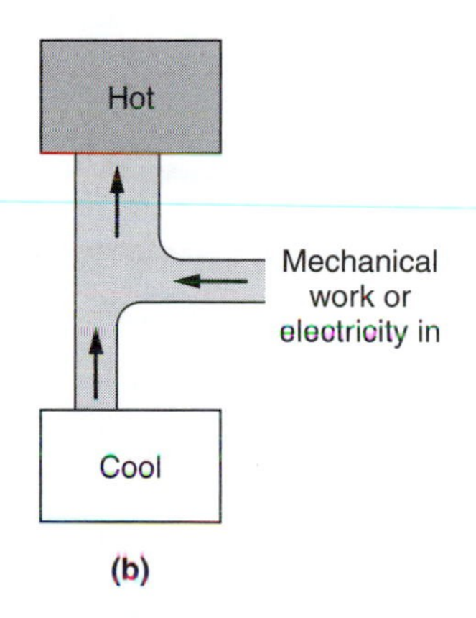

FIGURE 8.11
Conceptual diagrams of (a) a heat engine and (b) a refrigerator or heat pump. In principle, the two are similar devices, operated in reverse.

of electrical energy means 3 units of fuel energy were consumed back at the power plant. Still, we get 4 units of thermal energy for those 3 units of fuel energy—better than if the fuel were burned directly to produce heat. Example 8.2 explores this point further.

EXAMPLE 8.2 | Heat Pump versus Gas Furnace

A gas furnace operates at 85% efficiency, meaning it converts 85% of its fuel energy into useful heat. A geothermal heat pump with COP of 3.6 gets its electricity from thermal power plants whose efficiency averages 33%. Which heating system uses the least fuel?

SOLUTION

With 85% efficiency, the gas furnace uses 1/0.85 = 1.2 units of fuel energy to supply one unit of thermal energy to the house. The heat pump's COP of 3.6 means it delivers 3.6 units of heat for every unit of electrical energy. Equivalently, for every one unit of heat delivered to the house, the pump uses 1/3.6 = 0.28 units of electrical energy. But the electricity comes from power plants that average 33% efficiency, so to produce that 0.28 units of electrical energy requires 0.28/0.33 = 0.85 units of fuel energy.

So here's the comparison between the two approaches to heating our house: burning gas directly in the home furnace requires 1.2 units of fuel energy, while the heat pump takes 0.85 units of fuel energy. Both, by the way, are vastly more efficient than using electricity directly to heat the house. Although that approach is essentially 100% efficient in converting high-quality electrical energy to heat, the power-plant inefficiency means that we would still need to burn 1/0.33 = 3 units of fuel energy to supply the house with 1 unit of heat.

Clearly, the heat pump of this example is energetically superior to the furnace. Whether it's economically superior depends on the relative costs of gas and electricity, tempered by the substantially greater installation costs for the heat pump. Exercise 11 and Research Problem 3 explore heat pumps in comparison with conventional heating systems, including economic issues.

THERMODYNAMIC LIMITS ON HEAT PUMPS

A heat pump, being a heat engine in reverse, is subject to a thermodynamic limit related to that of Equation 4.5 for an engine. The same analysis that leads to Equation 4.5 gives the theoretical maximum possible COP of a heat pump:

$$\text{COP}_{\text{max}} = \frac{T_h}{T_h - T_c} \qquad \text{(maximum possible COP)} \qquad (8.2)$$

Here T_h is the temperature inside the building and T_c that of the environment from which heat is extracted. A look at the denominator of Equation 8.2

shows that the maximum possible COP increases as the temperature difference between interior and outside decreases. As with Equation 4.5, temperatures here must be given in absolute units—kelvins in the SI system.

Remember that Equation 8.2, like its counterpart Equation 4.5, gives a theoretical *maximum* value. The COPs of real heat pumps are significantly lower than Equation 8.2's limit. For example, a typical geothermal heat pump might extract heat from the ground at 10°C (50°F) and deliver heated water at 50°C (about 120°F) to circulate within a building. Converting to kelvin temperatures and using Equation 8.2 then gives the maximum possible COP as

$$\mathrm{COP}_{\max} = \frac{T_h}{T_h - T_c} = \frac{323\ \mathrm{K}}{323\ \mathrm{K} - 283\ \mathrm{K}} = 8.1$$

Real geothermal heat pumps operating at these temperatures have actual COPs in the range of 3 to 5.

HEAT PUMPS: THE ENERGY SOURCE

I've been using the term *geothermal heat pump* because it's become established terminology for heat pumps that extract energy from the ground or from groundwater. And heat pumps are commonly cited among applications of geothermal energy; that's why they're included in this chapter. In fact, heat pumps are often listed as the dominant use of geothermal energy for direct heating. That's because they use the modest subsurface temperatures available anywhere, not the higher temperature resources available only in geologically active regions.

But do ground-source heat pumps really qualify as geothermal energy? In one sense, yes, because they extract energy from the ground. But the top layers of the ground are strongly influenced by surface conditions, and in particular much of the thermal energy stored in those layers is actually solar in origin. One clue to that fact is that the near-surface geothermal gradient reverses in the summer, when the surface temperature is greater than the temperature a meter or so down. One has to go to significant depth, on the order of 100 meters or more, before the surface influence disappears and the temperature starts to rise in response to Earth's high interior temperature. So although I'm including heat pumps in this chapter on geothermal energy, be aware that their energy source isn't unambiguously geothermal. That's also why the term *ground-source heat pump* is preferable to *geothermal heat pump*.

This energy-source question is further complicated when a ground-source heat pump is used for both heating and cooling. Summertime cooling transfers energy from the building to warm the ground; given that the ground has relatively low thermal conductivity, some of that extra energy sticks around and is available to be pumped back into the building in the winter. Inevitable heat loss and the fundamental limitations of the second law of thermodynamics mean this process isn't perfect. Nevertheless, here's a rare case of "energy recycling." The larger the heat-pump system, the larger this thermal recycling effect, so it's most significant with commercial buildings. On the other hand, limited thermal conductivity can hinder heat flow into the source region during the winter, resulting in falling temperatures and thus declining COP.

8.5 Tidal and Ocean Energy

Tidal energy is unique in that it's the only one of Earth's energy flows that originates as mechanical energy. You're certainly aware that the Moon's gravity is in some way responsible for the tides. Actually, it's not so much gravity itself but the *variation* in gravity with position that causes tides. The Sun's *direct* gravitational effect on Earth is some 200 times stronger than the Moon's, but the Moon's proximity makes the variation in its gravity more noticeable. As a result, the Moon's *tidal* effect is greater—although only about double the Sun's.

ORIGIN OF THE TIDES

Figure 8.12 shows how tides would occur in a simplified Earth covered entirely with water. The Moon's gravitational attraction decreases with distance, so the part of the ocean closest to the Moon experiences a stronger attractive force than Earth itself. This results in a tidal bulge in the ocean facing the Moon. The solid Earth, in turn, experiences a stronger pull than the ocean water on the far side of the planet, and the effect is to leave the water bulged outward on the far side as well. As Earth rotates beneath the Moon, the tidal bulges sweep by a given point twice a day. That's why there are (usually) two high tides each day. Because the Moon revolves around Earth every 27.3 days, at a given time it appears at a slightly different position in the sky on successive days—which is why the time of high tide shifts from day to day. The Sun's tidal contribution adds directly to the Moon's, making for especially high tides when Sun, Moon, and Earth are all in a line—namely, at new Moon and full Moon.

But things aren't as simple as Figure 8.12 suggests. The tilt of the Moon's orbit introduces tidal variations with latitude. Tides affect the entire ocean from surface to bottom, and therefore water depth strongly affects tidal ampli-

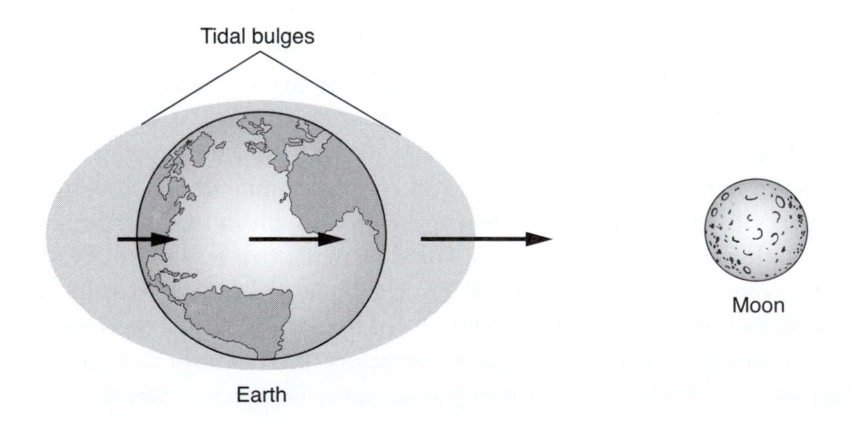

FIGURE 8.12
Tides result from the variation in the Moon's gravity (arrows). In an ideal ocean-covered Earth, there would be two tidal bulges on opposite sides of the planet. Bulges shown are greatly exaggerated.

FIGURE 8.13
Canada's Bay of Fundy has the world's highest tides. Shown here are Hopewell Rocks, photographed at low and high tide.

tude. In the open ocean the tidal bulge is just a fraction of a meter in height, but in shallow coastal water it can be much higher. Bays and inlets, particularly, affect tidal amplitude through their shape and bottom structure, and through resonance effects involving wave motion of tidal flows. The world's highest tides, at Nova Scotia's Bay of Fundy, range up to 17 meters (Fig. 8.13).

THE TIDAL ENERGY RESOURCE

The energy associated with the tides comes ultimately from the motions of Earth and Moon. The result is a minute slowing of Earth's rotation and an increase in the Moon's orbital radius. Estimates of the tidal energy input to Earth come from sensitive measurements of the decrease in Earth's rotation. These yield a tidal energy rate of about 3 TW—about 20% of humankind's total energy consumption rate. About two-thirds of this tidal energy, however, is dissipated in the open ocean. The remaining third—1 TW—appears in coastal waters. This energy is available to us to harness, before it's lost to frictional heating.

Harvesting tidal energy is practical only where the tidal range is several meters or more, and where the tidal flow is concentrated in bays, estuaries, and narrow inlets (Fig. 8.14). As a result, probably only about 10% of that 1 TW—about 100 GW, or the equivalent of 100 large fossil or nuclear power plants—is actually available for human energy supply. And that may not all be economically harvestable. So, although tidal energy may play a role in a handful of favorable locations, it's not going to make a significant dent in humankind's energy supply.

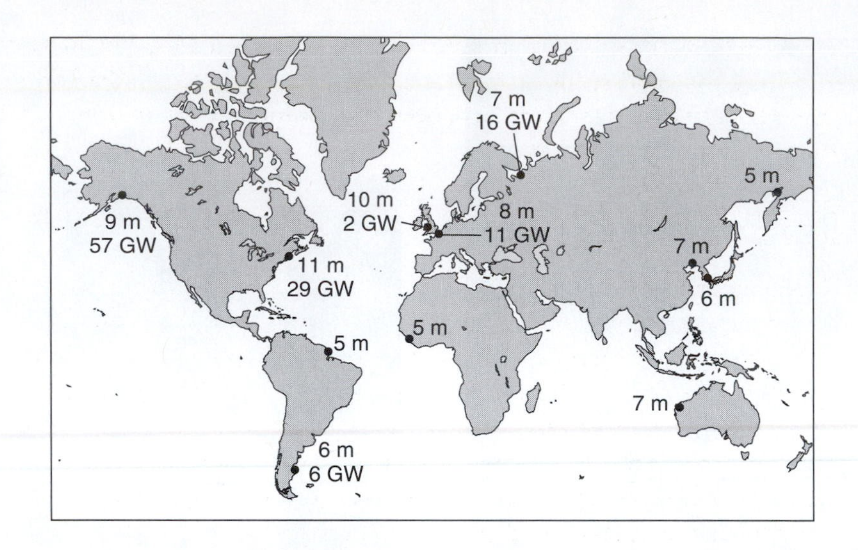

FIGURE 8.14
Locations of some of the world's highest tides, with estimated tidal power generation capacity listed for some sites. Capacity depends not only on tidal height, but also details of coastline and potential power-plant sites.

HARNESSING TIDAL ENERGY

Tidal power has been used for centuries to grind grain and for other tasks traditionally powered by running water. Impounding water at high tide and then letting it turn a water wheel as it flows out provides a reliable but intermittent power source. In the latter part of the twentieth century, this same idea saw large-scale application with the construction of a handful of tidal electric generating plants. The first and largest, near the outlet of La Rance River in France, was completed in 1967. This plant has a peak power output of 240 MW—although it only operates near times of high and low tides, so its average output is one-fourth of its peak.

The twenty-first century has seen a resurgence of interest in tidal power, now using what are essentially underwater versions of wind turbines (Fig. 8.15). These devices capture the energy in tidal currents without the need for dams. Like their wind-driven cousins, tidal turbines rotate to harness energy, and they work on both inflow and outflow. Individual tidal turbines have power outputs around 300 kW, but plans call for large-scale turbine farms generating as much as 100 MW—significant, but still small compared with full-scale fossil and nuclear plants.

FIGURE 8.15
An undersea turbine for tidal power generation.

ENVIRONMENTAL IMPACTS OF TIDAL ENERGY

The traditional approach to tidal energy involves large dams at the inlets to bays or estuaries. Such dams reduce the overall tidal flow and the water's salinity; they also impede the movement of larger marine organisms. Shifting sediments and changes in water clarity may have additional adverse impacts. These changes are particularly significant because estuaries—nutrient-rich regions where fresh and salt water mix—harbor a rich diversity of marine life and are the "nurseries" for many commercially important fish species.

The newer tidal turbines are more environmentally friendly, but in large numbers they, too, will reduce tidal flows and therefore alter the balance of fresh and salt water. To date there have been few comprehensive reviews of the environmental impact of tidal power.

WAVES AND CURRENTS

Waves and currents, like the shallow "geothermal" heat pumps we discussed earlier, don't quite belong in this chapter. That's because solar-induced temperature differences between tropical and higher latitude waters drive the ocean currents, which are therefore mainly indirect manifestations of solar energy. Earth's rotational energy, however, plays a modest role in ocean circulation. Similarly, winds arise from solar-induced atmospheric temperature differences, and they in turn drive ocean waves. But because waves and currents are considered "ocean energy" and thus related to tidal power, we'll deal with them here.

Trials of wave-energy converters with peak outputs around 1 MW began in the early 2000s (Fig. 8.16), and these devices are making very modest contributions to electrical energy supplies, especially in coastal Europe. The most ambitious wave project to date is Wave Hub, deployed in 2010 some 16 km off the English coast. With a seafloor connection to the power grid, Wave Hub serves as a test bed for different wave-energy technologies; when fully configured, it could produce 20 MW of electric power. But wave energy is limited: The total wave power delivered to all the world's coasts is some 3 TW, and clearly we could harness only a minuscule portion of that. Estimates for the power economically extractable with current wave-conversion technologies is estimated at between 0.01 and 0.1 TW—a tiny fraction of the world's 16-TW primary energy consumption.

Ocean currents are even less developed for energy. Some estimates suggest that we may be able to capture as much as 450 GW of power worldwide from these vast "rivers" that flow within the oceans. Although this is significant—about a quarter of the world's electric power consumption—the high capital costs of producing equipment that can operate reliably for years on the ocean floor mean that it will be a long time, if ever, before ocean currents make even a modest contribution to our energy supply.

FIGURE 8.16
This snakelike device generates 750 kW of electrical power as it flexes in the ocean waves. "Wave farms" consisting of several of these units are linked to the shore with a single cable laid on the sea floor.

Tides, waves, and ocean currents all present intriguing possibilities for converting the mechanical energy of moving water into electricity. But the overall energy resources of all three are limited, and their technologies have a long way to go. As with geothermal energy, there are local regions in which ocean energy may become significant. But realistically, it's unlikely to make a substantive contribution to the global energy supply.

CHAPTER REVIEW

BIG IDEAS

8.1 Geothermal heat supplies only about 0.02% of the energy reaching Earth's surface, but in a few localized regions the geothermal resource is significant.

8.2 Technologies for extracting geothermal energy include direct use of geothermal heat as well as electric power generation.

8.3 Environmental impacts of geothermal energy include air and water pollution from geothermal fluids, CO_2 emissions, and land subsidence.

8.4 **Geothermal heat pumps** use electrical energy to transfer larger amounts of energy from the ground to heat buildings. Typically the amount of energy transferred as heat is several times the amount of electricity used.

8.5 Tidal energy originates from the Earth–Moon system and is a limited resource that can provide a practical energy supply only in a few localities. The energy available from ocean waves and currents is greater but technologically difficult to harness.

TERMS TO KNOW

binary system (p. 212)
closed loop system (p. 214)
geoexchange heat pump (p. 214)
geopressured system (p. 208)
geothermal gradient (p. 206)
geothermal heat pump (p. 214)
ground-source heat pump (p. 214)
heat exchanger (p. 212)
hydrothermal system (p. 207)
open-loop system (p. 214)
vapor-dominated system (p. 207)

GETTING QUANTITATIVE

Geothermal gradient, average: ~25°C/km

Geothermal electric capacity, worldwide: ~11 GW, 0.23% of total

Coefficient of performance: $COP = \frac{\text{heat delivered}}{\text{electrical energy required}}$ (Equation 8.1; p. 215)

Thermodynamic COP limit: $COP_{max} = \frac{T_h}{T_h - T_c}$ (Equation 8.2; p. 216)

QUESTIONS

1 What are the two fundamental sources of geothermal energy?

2 Why does a binary geothermal plant require a working fluid other than water?

3 In what sense are most "geothermal" heat pumps not actually a geothermal energy sources?

4 Name some environmental impacts of geothermal electric power generation.

5 What are two factors that can lead to depletion of a geothermal resource?

6 Why does the confluence of tidal energy resources and estuaries increase the environmental impact of dam-based tidal power plants?

7 What is the ultimate source of the energy that powers a lightbulb whose electricity comes from a tidal power plant?

8 Are any of the energy sources discussed in this chapter likely to make a major contribution to humankind's energy needs? Discuss.

EXERCISES

1 Calculate the total global geothermal energy flow rate, assuming an average of 87 mW/m^2.

2 Consider a heat engine operating between Earth's surface, at a typical temperature of 295 K, and the bottom of a geothermal well 1 km deep. (a) Given a geothermal gradient of 25°C/km, what would be the maximum possible efficiency of such a heat engine? (b) If we could tap the geothermal flow over 10% of Earth's land area (exclude the oceans) with such heat engines, roughly what percentage of humankind's 16-TW energy usage could we supply?

3 Use granite's thermal conductivity from Table 4.1 to find the energy flow rate per square meter through granite when the geothermal gradient is 25°C/km.

4 A city occupies a circular area 15 km in diameter and uses energy at the rate of about 10 GW. Make the approximation that the top 4 kilometers of Earth's crust consists of granite (density 2,700 kg/m^3, specific heat 840 J/kg·K, and thermal conductivity 3.37 W/m·K), and that the average temperature in this layer is 50°C higher than the surface temperature. (a) Estimate the total energy that could be extracted in cooling the 4-km-thick layer of crust beneath the city to the temperature of the surface. (b) If the efficiency in converting this low-quality thermal energy to useful energy is 5%, how long would the 4-km geothermal resource below the city last? (c) Show that the initial rate at which the geothermal energy is replenished by conduction from below is far less than the rate of extraction, implying that this use of geothermal energy is probably not renewable.

5 Compare the thermodynamic efficiency limits for heat engines operating from (a) a 150°C geothermal source (considered about the minimum for electric power generation) and (b) magma at 1,000°C. Assume the lowest temperature accessible to the engine is the ambient environment at 300 K.

6 The average U.S. home consumes energy at the rate of about 3.5 kW. If the "footprint" of a typical house is 150 m^2, and if the average geothermal heat flow is 87 mW/m^2, what percentage of the home's energy needs could the geothermal flow through the floor supply?

7 (a) Find the maximum possible COP of a heat pump that transfers energy from the ground at 10°C to water at 60°C. (b) At what rate would such a pump consume electricity if it were used to heat a house requiring 15 kW of heating power?

8 The actual efficiency of a geothermal power plant using a 200°C geothermal resource is only about 7%. Take this value as typical of California's Geysers geothermal field, which produces 900 MW of electric power. (a) Find the actual rate of geothermal heat extraction at Geysers. (b) It's estimated that 35 EJ of energy could be extracted from the Geysers before the temperature dropped from 250°C to the 150°C minimum for a vapor-dominated geothermal plant. Given your answer in (a), how long will the Geysers geothermal resource last before the temperature reaches 150°C?

9 At the beginning of the heating season, your geothermal heat pump operates with COP = 4.0. But at the end of the season, the COP has dropped to 2.5 because the ground has chilled as a result of the heat pump's energy extraction. Assuming identical average temperatures in the first and last months of the heating season, how much greater is your energy use in the last month than in the first?

10 The binary geothermal power plants at Casa Diablo, California, operate from a geothermal resource at 170°C. If the discharge temperature is 90°C, what is the maximum thermodynamic efficiency of these power plants?

11 Your home's 85%-efficient furnace burns natural gas costing $3.85 per million Btu (MBtu) of energy content.

You could replace it with a COP-3.5 heat pump that runs on electricity costing 11 cents per kWh. (a) Compare the fuel energy used by each system to deliver 1 MBtu of heat, assuming the efficiency of electricity generation is 31%. (b) Compare the costs of each MBtu of heat.

12 Consider an ocean current flowing at 2.5 m/s. (a) How much kinetic energy is contained in cubical block of water 1 m on a side? (The density of seawater is about 1,030 kg/m^3.) (b) If the flow is perpendicular to one of the cube faces, what is the rate at which the current carries kinetic energy across each square meter? Your answer gives an upper limit for the power that could be extracted from the flow—although an unrealistic limit because you'd have to stop the entire flow.

RESEARCH PROBLEMS

1 Consult the latest "Existing Electric Generating Units in the United States," available from the U.S. Department of Energy's Energy Information Administration (www.eia.doe.gov), to name five geothermal power plants along with their locations and rated power outputs.

2 Nearly half the states in the United States have at least one geothermal energy facility. Does yours? Research the answer, and give the capacity, in megawatts, of both direct-use facilities and geothermal electric generating plants in your state.

3 Find the initial cost of a heat-pump system in your area, and compare with that of a conventional heating system. Then compare fuel and electricity costs, and make a rough estimate of how long, if ever, you would have to operate the heat-pump system before its total cost was less than that of the conventional system.

ARGUE YOUR CASE

1 A friend is extolling the virtues of his new geothermal heat pump. It has a COP of 4, so he argues that it uses only one-fourth the energy of a conventional heating system. Formulate a more realistic assessment of the heat pump's energy consumption.

2 A member of your national legislature claims that "the answer to all our energy problems lies right beneath our feet," and proposes a crash program of intensive geothermal energy development. Develop an argument either supporting or opposing this program.

Chapter 9

DIRECT FROM THE SUN: SOLAR ENERGY

The first thing I want to emphasize in this chapter is that there's *plenty* of solar energy. Don't let anyone try to convince you that solar energy isn't practical because there isn't enough of it. In round numbers, the rate at which solar energy reaches Earth is some ten thousand times humankind's energy-consumption rate. And it's entirely renewable; solar energy keeps on coming regardless of how much we might divert to human uses. That's not true of the fossil fuels, which will last at most a few decades, or perhaps a little longer for coal. It's not true of nuclear fission fuels, because recoverable resources of U-235 are also measured in decades, and breeder technologies that convert U-238 to plutonium might buy us a few centuries to a millennium or so. Finally, as I made clear in Chapter 8, flows of geothermal and tidal energy aren't sufficient to meet global energy demand. But the Sun will shine for another 5 billion years, bathing Earth in a nearly steady stream of energy into the unimaginable future.

Direct use of solar energy isn't going to displace fossil fuels as our dominant energy source in the next few decades, but there are plenty of near-term opportunities to develop significant solar contributions to our energy mix. For the long term, however, solar energy is the one proven source that can clearly and sustainably meet our energy needs. With the possible exception of technologically elusive nuclear fusion, which would essentially build miniature Suns here on Earth, there's really no other viable long-term energy source.

9.1 The Solar Resource

Deep in the Sun's core, fusion of hydrogen into helium releases energy at the prodigious rate of 3.84×10^{26} W. At the core temperature of 15 million K, most of this power takes the form of high-energy electromagnetic radiation—gamma rays and X rays. As it travels outward through cooler layers, the electromagnetic energy scatters off the dense solar material and its wavelength increases until, at the surface, the Sun radiates that 3.84×10^{26} W in the visible and near-visible portions of the electromagnetic spectrum. This energy radiates outward in all directions. The total remains essentially constant, but the intensity drops as the solar radiation spreads over ever-larger areas.

THE SOLAR CONSTANT

At Earth's average distance from the Sun (which is the average radius of the planet's orbit, $r = 150$ Gm), solar energy is spread over a sphere of area $4\pi r^2$; the intensity, or power per unit area, is then about 1,364 W/m^2 (Exercise 1 explores the calculation leading to this figure). There's some uncertainty in the exact value at the level of a few watts per square meter, but I'll stick with 1,364 W/m^2. This number is the **solar constant**, *S*, and it measures the rate at which solar energy impinges on each Sun-facing square meter in space at Earth's orbital distance. The solar constant isn't quite constant; it varies by about 0.1%—around 1.4 W/m^2—over the 22-year solar cycle. There are also longer-term variations that may be climatologically significant, as we'll see in Chapters 13 and 14. But overall, the Sun is a steady and reliable star, and for the purposes of this chapter we can consider its energy output to be constant.

INSOLATION

If Earth were a flat disk of radius R_E facing the Sun, its area πR_E^2 would intercept a total solar power given by $P = \pi R_E^2 S$. With $R_E = 6.37$ Mm and $S = 1{,}364$ W/m^2, this amounts to 1.74×10^{17} W, or 174 PW (Exercise 2). In principle, this is the total rate of solar energy input that's available to us on Earth. Currently, humankind uses energy at the rate of about 16 TW (16×10^{12} W), so the total solar input is just over 10,000 times our energy needs. This is the basis of my assertion that there's plenty of solar energy. Put another way, every 40 minutes the Sun delivers to Earth the amount of energy that humankind uses in a year (see Exercise 3). Earth isn't a flat disk, of course, but its sunward side nevertheless intercepts this same 174 PW (Fig. 9.1). However, our planet's curvature means that **insolation**—incoming solar energy per square meter of Earth's surface—isn't the same everywhere. Insolation is greater in the tropics, where the surface is nearly perpendicular to the incoming sunlight, than at high latitudes, where the surface is inclined away from the Sun. That's why it's generally cooler at high latitudes.

Figure 9.1 shows that the incident sunlight is spread over the Sun-facing half of Earth's spherical surface (total area $4\pi R_E^2$)—that is, over $2\pi R_E^2$ rather than the flat-disk area of πR_E^2. If all the power quantified by the solar constant *S* reached the surface, this would mean an average insolation of $S/2$ on the sunward-facing or daytime side. Averaging this $S/2$ energy over the area of the entire globe (both the nighttime and daytime sides) introduces another factor-of-2 decrease in average insolation, which would then be $S/4$ or 341 W/m^2. Reflection by clouds and other light surfaces returns some of the incident solar energy to space, leaving an average of about 239 W/m^2 actually available to the Earth–atmosphere system; of this, some is absorbed in the atmosphere before reaching Earth's surface. We'll take a closer look at these figures when we consider climate in Chapter 12. For now we'll consider that we have a global average of a little over 200 W/m^2 of available solar energy, with more getting through to lower latitudes and regions with clearer-than-average skies. Another rough figure to keep in mind is that midday sunlight on a clear day carries about 1,000 W/m^2, or 1 kW/m^2, and that this energy is available on any surface oriented perpendicular to the incoming sunlight.

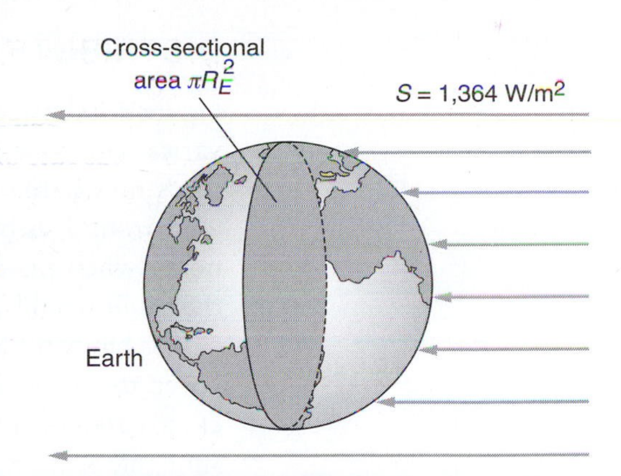

FIGURE 9.1
The round Earth intersects the same amount of solar energy as would a flat disk with an area equal to Earth's cross-sectional area (πR_E^2). The quantity S measures the power per unit area incident on an area oriented perpendicular to the incoming sunlight.

EXAMPLE 9.1 | Solar Houses versus Solar Cars

The roof area for a typical house is around 150 m^2 and that of a car is around 7 m^2. (a) Find the midday solar power incident on a typical house roof, and compare it with the average U.S. household energy-consumption rate of 3.5 kW. (b) Find the midday solar power incident on a car roof, and compare it with the 300-horsepower (hp) rating of a typical engine. Assess the practicality of solar-powered houses and solar-powered cars.

SOLUTION

(a) At 1 kW/m^2, the house roof is exposed to 150 kW in bright sunlight—far more than the household's average energy-consumption rate of 3.5 kW. Nonoptimal roof angles, conversion inefficiencies, clouds, seasonal variations, and nighttime periods all decrease the lopsidedness of this comparison, but nevertheless it's clearly not unrealistic to consider meeting all or most of a home's energy needs through the solar energy falling on its roof.

(b) The car is a different story. One kilowatt per square meter on its 7-m^2 area gives only 7 kW. Since 1 hp is about 0.75 kW, our 7 kW is about 10 hp, which conversion inefficiencies would reduce to perhaps 2 hp. So we'll not see cars powered by roof-mounted solar energy collectors, at least as long as we insist on having hundreds of horses under our hoods. Pure solar cars have been built and raced across entire continents (Fig. 9.2), but they'll probably never make for practical everyday transportation.

FIGURE 9.2
Stanford University's *Solstice*, a solar-powered car that won the stock class competition in the 2005 North American Solar Challenge.

DISTRIBUTION OF SOLAR ENERGY

Figure 9.1 shows that insolation varies significantly with latitude; clearly it also varies with time of day, season, and local weather conditions. At sunrise the Sun's rays barely skim across the horizon, making for minimal insolation on a horizontal surface. As the day progresses, insolation increases through local noon, when the Sun is highest, and then decreases. On a perfectly clear day, the result would be a smooth curve of insolation versus time, but when passing clouds block the Sun, insolation fluctuates (Fig. 9.3). Note that even with cloud cover, the insolation in Figure 9.3 doesn't drop to zero. That's partly because clouds aren't completely opaque, and also because Earth's atmosphere scatters sunlight. Scattering means that solar energy comes not only straight from the direction of the Sun—so-called **direct insolation**—but also from throughout the sky—called **diffuse insolation**. (Incidentally, atmospheric scattering is the reason why the sky appears blue.) Some solar energy collectors can use both direct and diffuse sunlight, so they continue to function, albeit less effectively, on cloudy days.

The total daily solar energy input on a horizontal surface depends on latitude, season, and length of day. As you move toward the equator, the peak insolation on a horizontal surface increases, but in summer the length of day shortens as one nears the equator. The combined effect of the Sun's peak angle in the sky along with the length of day gives a surprising result for the maximum possible average daily summer insolation on a horizontal surface: It's actually lowest at the equator and greater toward the poles (Fig. 9.4). But the long polar night of the winter months overcompensates, giving lower latitudes a substantial edge in overall yearly insolation. Also significant is cloud cover, which tends to be greater in the equatorial zones and at higher latitudes; the result is that the actual average daily summer insolation peaks at around 30° latitude, in a zone containing most of the world's major deserts. Figure 9.4 shows actual insolation for selected cities in the Northern Hemisphere, with an approximate curve depicting this peaking effect at about 30° north latitude. Variations in climatic

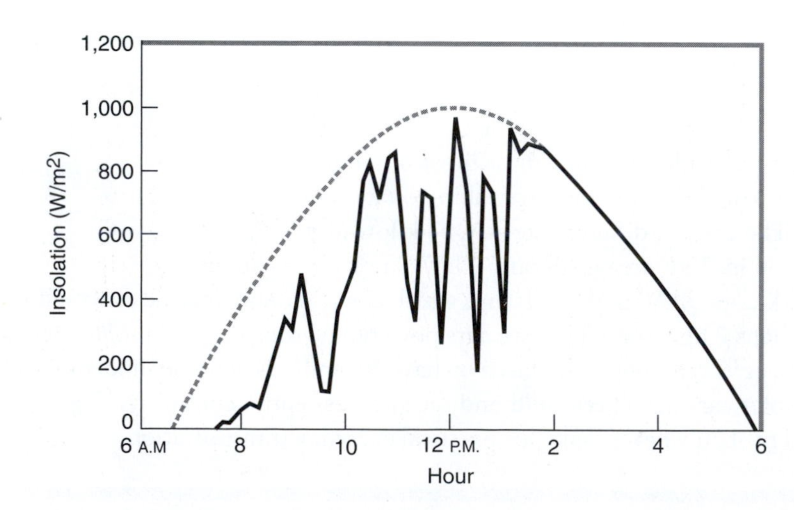

FIGURE 9.3
Insolation measured on a mid-October day in Middlebury, Vermont. The dashed curve represents an ideal cloudless day. Can you tell from the numbers whether the insolation was measured on a horizontal or a tilted surface?

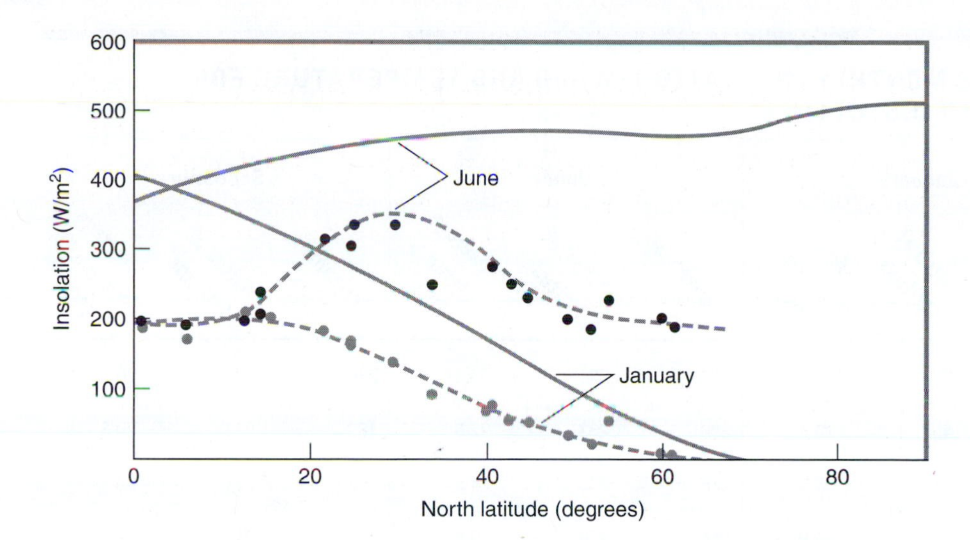

FIGURE 9.4
Theoretical maximum horizontal-surface insolation (solid curves) versus latitude for January and June. Also shown are actual data points for selected world cities in January (gray circles) and June (black circles). Dashed curves are trends in the insolation data, indicating the summer insolation peak at around 30° north latitude, above which the actual insolation drops because of increasing cloudiness. Cities, in order of latitude, are Singapore, Jakarta, Bangalore, Dakar, Manila, Honolulu, Abu Dhabi, Riyadh, Cairo, Atlanta, Ankara, Madrid, Detroit, Bucharest, Paris, London, Edmonton, Oslo, Anchorage.

conditions with longitude also greatly influence availability of solar energy, as shown for world cities in Figure 9.5 and, in more detail, for the United States in Figure 9.6 and Table 9.1. Of course it doesn't make sense to put a solar collector flat on the ground, so Table 9.1 also lists insolation for surfaces tilted at a location's latitude—a common compromise orientation for simple solar collectors. A collector that tracks the Sun can maximize solar energy collection, and Figure 9.6 shows insolation for Sun-tracking surfaces. However, tracking mechanisms greatly increase the expense of solar collector systems.

The solar resource varies substantially but not overwhelmingly with geographic location. Sunny Albuquerque, for example, gets only about 50% more year-round average than northern cities such as Chicago, Detroit, or Burlington, Vermont. So although some places are better endowed with solar energy, even the worst sites aren't orders of magnitude worse. It probably doesn't make sense to put expensive large-scale solar power stations in places like Seattle or

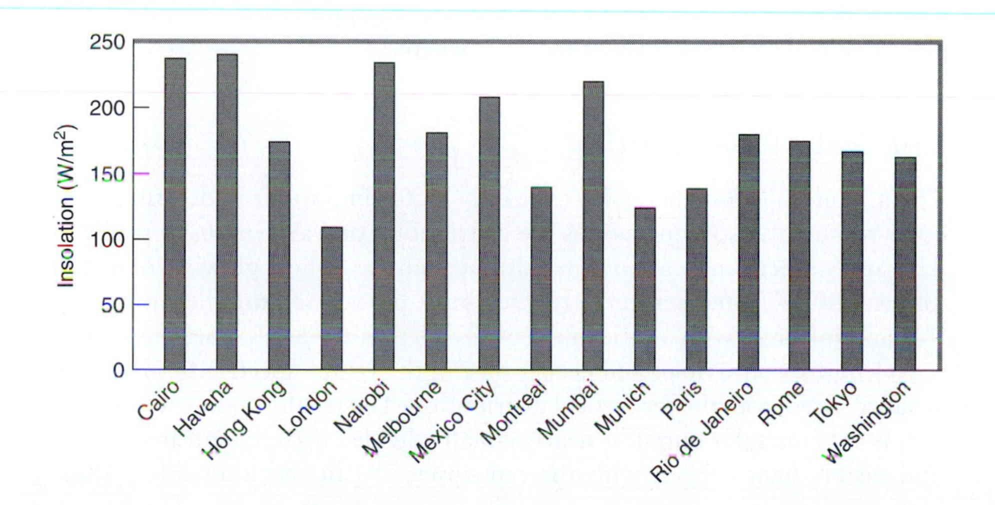

FIGURE 9.5
Annual average insolation on a horizontal surface for selected world cities.

TABLE 9.1 | **AVERAGE MONTHLY INSOLATION (W/m²) AND TEMPERATURE FOR SELECTED U.S. CITIES**

	January			June			September		
City	Horizontal	Tilted at latitude	Average temp. (°C)	Horizontal	Tilted at latitude	Average temp. (°C)	Horizontal	Tilted at latitude	Average temp. (°C)
Albuquerque, NM	133	221	1.2	338	296	23.4	246	283	20.3
Baltimore, MD	88	146	−0.1	258	233	22.5	183	213	20.3
Birmingham, AL	104	154	5.3	258	233	24.6	200	221	23
Boston, MA	79	142	−1.9	254	229	19.8	179	213	18.2
Burlington, VT	67	121	−8.7	250	225	18.4	167	200	14.9
Chicago, IL	75	129	−6.2	263	238	20.3	175	204	18
Denver, CO	100	183	−1.3	288	254	19.4	208	250	16.8
Detroit, MI	67	113	−5.1	258	233	19.8	171	200	17.3
Fairbanks, AK	4	29	−23.4	233	196	15.4	96	142	7.5
Miami, FL	146	196	19.6	233	213	27.4	204	213	27.7
Minneapolis, MN	75	146	−11.2	263	233	20.1	171	208	15.8
San Francisco, CA	92	146	9.3	300	271	16.4	225	267	18.1

Detroit, but some of the more individualized solar applications I discuss in this chapter could work effectively almost anywhere.

THE SOLAR SPECTRUM

The amount of solar energy isn't the only factor important in describing the solar resource. Also significant is the distribution of wavelengths in sunlight. The Sun's spectrum is approximately the same as a hot, glowing object at about 5,800 K, a temperature at which most of the radiation lies in the visible and infrared, with a little in the ultraviolet. In the Sun's case, the shorter wavelengths—ultraviolet and deeper blue in the visible spectrum—are somewhat reduced from the theoretical distribution. The result, as shown in Figure 9.7, is that the solar output is nearly equally divided between infrared (47%) and visible light (46%), with the remaining 7% in the ultraviolet. These proportions are found at the top of the atmosphere, but most of the ultraviolet and some longer-wavelength infrared are absorbed in the atmosphere and don't reach Earth's surface.

You've probably seen a magnifying glass used to concentrate sunlight and set paper on fire. In principle, this approach can achieve temperatures up to that of the 5,800-K solar surface (Question 2 asks you to consider why we can't go higher). The high solar temperature makes sunlight a high-quality energy source. As you can show in Exercise 12, a heat engine operated between 5,800 K and Earth's average ambient temperature of around 300 K would have a thermodynamic efficiency limit of 95%. Practical constraints on solar energy concentration and heat engine design reduce this figure considerably, but nevertheless concentrated sunlight remains a high-quality energy source.

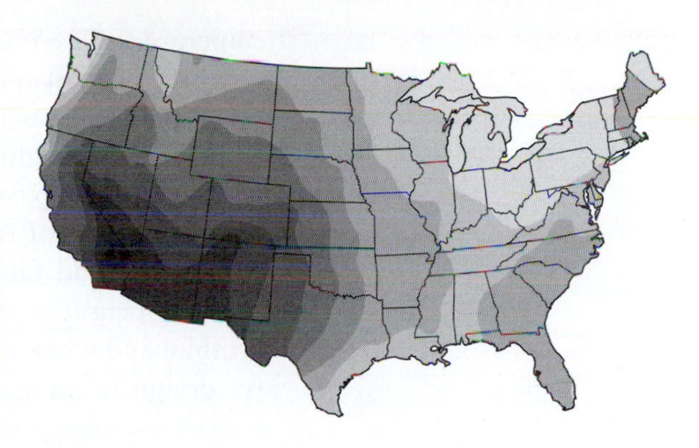

FIGURE 9.6
Annual average insolation in the continental United States, for surfaces oriented perpendicular to the incident sunlight. Values range from about 150 W/m² for the lightest shading to over 500 W/m² for the darkest.

At the same time, the fact that just over half the solar energy comes to us as infrared radiation means that most solar photons have lower energy than those of visible light (recall Equation 3.1, which describes the quantization of electromagnetic wave energy). Later in this chapter we'll see how this limits the efficiency with which we can convert sunlight directly into electricity.

These energy-quality and photon-energy issues aren't relevant to the simplest of solar applications, namely solar heating, because solar heating systems absorb nearly all the incident solar energy and convert it into low-quality heat. We'll consider solar heating in the next two sections; after that we'll turn to the use of solar energy to generate electricity, an application for which energy quality and photon energy are important considerations.

9.2 Passive Solar Heating

A simple, straightforward, and effective use of solar energy is to heat buildings. In **passive solar heating** schemes, the building design itself maximizes the solar energy input and ensures that energy is distributed evenly throughout the building and over time. What distinguishes a passive system is the absence of active devices such as pumps and fans to move energy about.

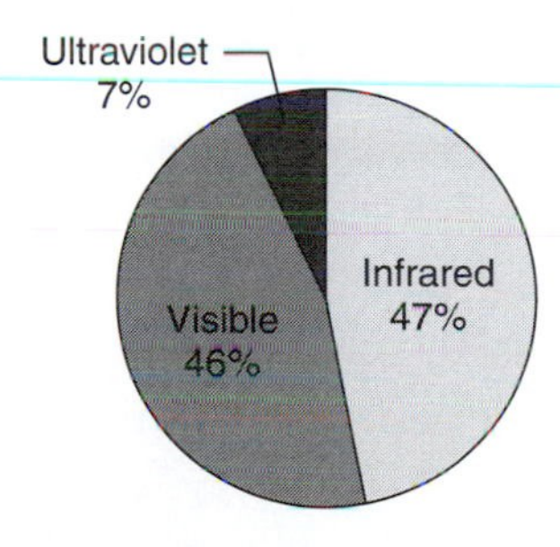

FIGURE 9.7
Distribution of solar energy among infrared, visible, and ultraviolet portions of the spectrum. Wavelengths between 380 nm and 760 nm make up the visible region.

Passive heating begins with building orientation. Siting a conventional house with the majority of its windows facing south (in the Northern Hemisphere) can provide considerable solar gain on clear winter days, as we found in Example 4.3. This, in turn, reduces the fuel consumption of a conventional heating system. But passive solar designs can do much better, with the best of them eliminating the need for conventional heating systems in all but the coldest, cloudiest climates.

Passive solar design entails four distinct elements. First is the collection of solar energy through large areas of south-facing glass. Sunlight passes through the glass, and its energy is absorbed in the building's interior, raising the temperature. There's no point in collecting solar energy only to have the solar-generated heat escape through walls and windows, so the second essential element is high-R insulation (recall Chapter 4's discussion of building insulation and *R* values). High-R insulation is important in *any* building, but it's especially crucial in a passive solar structure. Third, the solar design must promote the distribution of energy throughout the building—something that's done with pumps and fans in conventional heating. Finally, passive solar design must compensate for times when the Sun isn't shining, such as at night (predictable) and when it's cloudy (less predictable). So the fourth element in passive design is energy storage, which allows the building to accumulate and store excess energy when the Sun shines, then gradually release that energy as needed.

Figure 9.8 shows a simple house incorporating all these design elements. The south wall is largely glass, to admit sunlight. The north-facing wall is windowless and heavily insulated. There's more insulation in the roof and under the floor. A thick, dark-colored concrete floor slab absorbs the incoming solar energy and warms. Concrete's relatively large thermal conductivity (1 W/m·K) ensures an even temperature throughout the concrete slab and therefore throughout the house. Concrete's substantial heat capacity (880 J/kg·K) means that it takes a lot of energy to change the concrete's temperature. Once heated, the concrete slab is therefore a repository of stored thermal energy, and it takes a long time to cool down. An added bonus is that this property of concrete helps to prevent

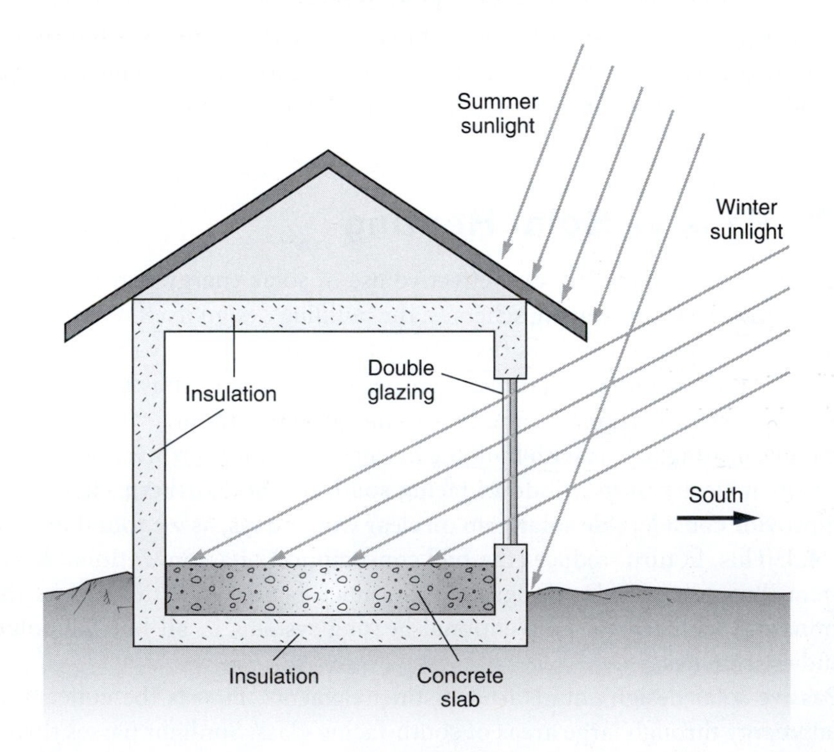

FIGURE 9.8
A passive solar house using direct gain and a concrete slab for thermal mass.

overheating on sunny, warm days. Thus the concrete slab in Figure 9.8 functions as a **thermal mass**, which, because it's slow to heat and cool, helps to promote an even temperature over both time and space. Although the house in Figure 9.8 is designed for winter heating, it has one feature to combat summer heat: a roof overhang that reduces solar gain by blocking the more vertical summertime solar radiation.

A black concrete floor may not seem aesthetically appealing, but black isn't essential. A dark color helps, but once sunlight gets into the building it reflects around and is eventually absorbed, whatever the color. A darker color absorber is more effective because it ensures immediate absorption before light has a chance to bounce right back out of the windows. And the thermal mass doesn't have to be concrete. As Table 4.3 shows, water's high specific heat makes it a particularly effective heat-storage medium. Although it's hard to use water as a building material, there are plenty of passive solar homes and greenhouses that use stacks of black-painted 55-gallon drums filled with water as their thermal mass. Even wood can function as a thermal mass; Table 4.3 shows that wood's specific heat is higher than that of concrete, but wood's much lower density means a much larger volume of wood is required for the same thermal storage.

Another approach to thermal energy storage is to use materials that change phase at appropriate temperatures. In Chapter 4 we discussed the considerable energy required to change ice to water, and water to steam. While such a phase change is occurring, the ice/water or water/steam mixture maintains a constant temperature of 0°C or 100°C, respectively. Changing back to the lower-energy phase then releases the energy that went into melting or boiling. Choosing a substance that melts at a temperature just a bit above what's comfortable in a house thus allows one to store large quantities of energy when the material is in the liquid phase. One such material is Glauber salt ($Na_2SO_4 \cdot 10H_2O$), which melts at 32°C (90°F). Glauber salt's heat of fusion is 241 kJ/kg, so a tonne of Glauber salt (1,000 kg) takes 241 MJ to melt. In a solar house this energy would come from sunlight, and it would then be released slowly during sunless times as the salt solidifies, all the while maintaining its 32°C melting-point temperature. Exercise 4 explores a Glauber-salt storage system.

The glass used in passive solar construction presents a dilemma. Ideally, it should admit the most solar energy while minimizing heat loss. Double or triple glazing reduces convective and conductive losses, while low-emissivity coatings reduce radiation loss. But these measures also decrease the solar energy coming into the house. So there's a tradeoff between maximizing solar energy gain and minimizing heat loss. Furthermore, there's no solar gain when the Sun isn't shining, but there's still heat loss. So it's important to provide extra insulation for the glass, especially at night. Insulation schemes include heavy curtains or slabs of foam that swing downward to cover the glass, preferably with a vapor barrier and edge seal to prevent air leakage and condensation.

EXAMPLE 9.2 | A Passive Solar House

A passive solar house like the one shown in Figure 9.8 has a level of insulation that gives a heat loss of 120 W/°C through the walls and roof. Its south-facing glass measures 8 m wide by 3 m high, passes 82% of the incident sunlight, and has an R value of 4. On average, the admitted sunlight completely illuminates 48 m^2 of the black concrete floor. Use data from Table 9.1 to find the average interior temperature of the house in January if it was located in (a) Albuquerque and (b) Detroit. (Ignore the heat loss through the floor.)

SOLUTION

We need to balance heat loss with solar gain to find the interior temperature. We're given a heat-loss rate of 120 W/°C through the walls and roof, which means the total loss through these surfaces is (120 W/°C)ΔT, where ΔT is the temperature difference between inside and outside. For the glass we're given the R value and the area (3 m × 8 m = 24 m^2), so we need to calculate the loss rate. Recall from Chapter 4 that R values are given in English units, namely ft^2·°F·h/Btu. Here we're given all other values in SI units, so we need to convert the R value. Exercise 17 of Chapter 4 handles this calculation; the result is that an R value of 1 ft^2·°F·h/Btu is equivalent to 0.176 m^2·°C/W. The glass has an R value of 4, or 0.176 × 4 = 0.70 m^2·°C/W, meaning that each square meter of glass loses 1/0.70 W for each degree Celsius of temperature difference across the glass. With 24 m^2 of glazing, the loss rate is then 24/0.70 = 34 W/°C. Thus the glass loses energy at the rate (34 W/°C)ΔT. Adding the wall and roof loss gives a total loss of 120 + 34 = (154 W/°C)ΔT.

Now for the solar gain. The absorber here is the 48 m^2 of illuminated floor, so we use Table 9.1's entries for horizontal surfaces: 133 W/m^2 for Albuquerque and 67 W/m^2 for Detroit. Multiplying by the illuminated floor area and by 0.82 to compensate for the windows passing only 82% of the incident sunlight, we get average solar gains of (133 W/m^2)(48 m^2)(0.82) = 5.23 kW for Albuquerque and, similarly, 2.64 kW for Detroit. On average the house is in energy balance, so we can equate loss and gain to get

$$(154 \text{ W/°C})\Delta T = 5.23 \text{ kW}$$

or

$$\Delta T = \frac{5.23 \text{ kW}}{0.154 \text{ kW/°C}} = 34\text{°C}$$

for Albuquerque. Table 9.1 lists Albuquerque's average January temperature as 1.2°C, so the Albuquerque house is going to be uncomfortably warm at 35°C, or 95°F. No problem: Drawing some shades across the glass will reduce the solar input. Clearly, though, the Albuquerque house is over-designed and is easily capable of supplying 100 percent of its heat from solar energy.

A similar calculation for the Detroit house, with only 2.64 kW of solar input, gives ΔT = 17°C. But Detroit averages −5.1°C in January, so the interior temperature will be only 12°C, or 54°F. A backup heat source, more insulation, or a different solar design is necessary to keep the Detroit house comfortable.

The numbers in this example are averages; insolation and outside temperature fluctuate of course, and so therefore would the interior temperature. But a large thermal mass could minimize such fluctuations.

An alternative to the direct-gain passive solar house of Figure 9.8 and Example 9.2 is to use **indirect gain**, heating a Sun-facing thermal mass placed so that natural convection then carries energy throughout the house. One such indirect-gain design is the vented **Trombe wall** shown in Figure 9.9. Here sunlight strikes a massive vertical wall that serves as both absorber and thermal mass. Air confined between the wall and glass warms and rises, resulting in natural convection that circulates warmed air throughout the house and returns cold air to be heated at the wall. Radiation from the back of the wall provides additional heat transfer into the house.

9.3 Active Solar Heating

Active solar heating systems use pumps or fans to move solar-heated fluids from **solar collectors** to an energy storage medium, and from there to where it's needed within a building. Active systems are more compact and allow for more flexible placement of components. As a result, they often fit better with

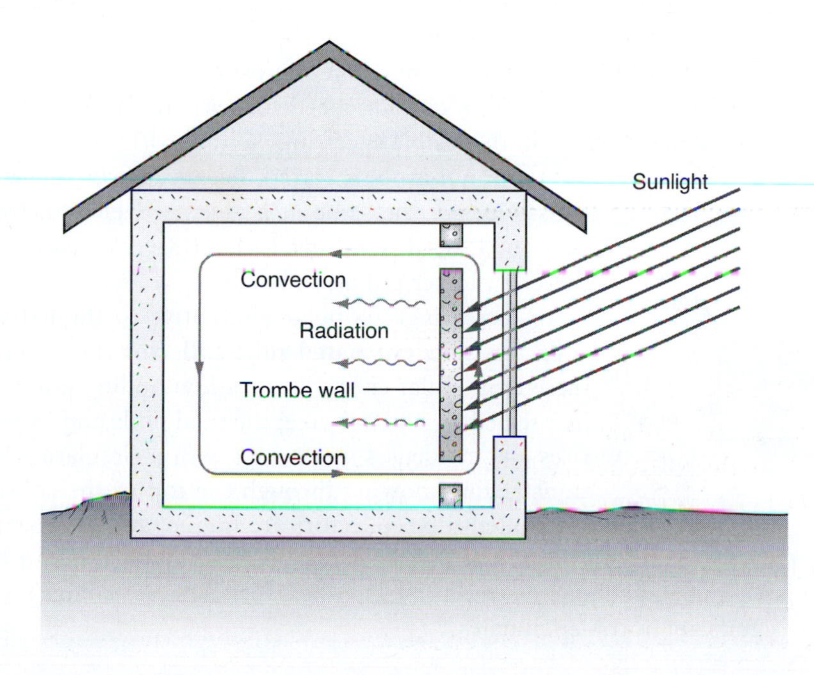

FIGURE 9.9
Indirect-gain solar house using a Trombe wall.

conventional architecture. Because active solar collectors are just solar energy–catching devices that don't have to function as walls or floors, they're often more efficient than passive designs. Finally, active systems can achieve higher temperatures, making them particularly useful for domestic hot-water heating—usually the second-largest home energy use after space heating. The disadvantages of active systems are that they're more complex, with a host of moving parts from pumps to controls, and most require electric power. They also take more energy to manufacture, and cost more than most passive measures.

FLAT-PLATE SOLAR COLLECTORS

The simplest and most economical solar collector is the **flat-plate collector**. Its essential parts are a black **absorber plate**, usually metal, and a glass cover. The back of the absorber plate is heavily insulated to prevent heat loss. A heat-transfer fluid flows through the collector, where it warms on contact with the absorber plate. With air-based systems, air simply passes over the absorber plate. With liquid-based systems, water or antifreeze solution flows through pipes either bonded to or built into the absorber plate. In liquid systems the absorber plate is usually copper or aluminum, chosen for their high thermal conductivity. Figure 9.10 shows a cross section of a typical liquid-based flat-plate collector.

Flat-plate collectors lose energy by infrared radiation and by conduction. As the absorber plate gets hotter, the losses increase and the collection efficiency goes down. At the so-called **stagnation temperature**, the loss equals the solar gain, and there's no net energy collected. The stagnation temperature depends on the ambient temperature and sky conditions, the insolation, and the characteristics of the collector. Again there are tradeoffs: Collectors with multiple glass covers have lower losses, but they admit less sunlight. They tend to perform better at high absorber-plate temperatures or low outdoor temperatures but not as well under the opposite conditions. A collector's efficiency can be increased by coating the absorber plate with a **selective surface**—one that is black in the visible spectrum, so it absorbs incident sunlight, but not in the infrared. Because absorptivity and emissivity are the same at a given wavelength (recall Section 4.3), this means less infrared emission and therefore lower radiation losses.

An increasingly popular alternative to the flat-plate collector is the **evacuated-tube collector** (Fig. 9.11). In these units, solar energy heats a low-boiling-point fluid in a sealed, evacuated tube; the fluid boils and its vapor rises and condenses on contact with a circulating heat-transfer fluid flowing through the top of the collector. Evacuated-tube collectors are fragile, and more expensive than flat-plate collectors. But their efficiency is much greater, because the vacuum eliminates conductive and convective heat loss. This advantage is especially significant in cooler climates.

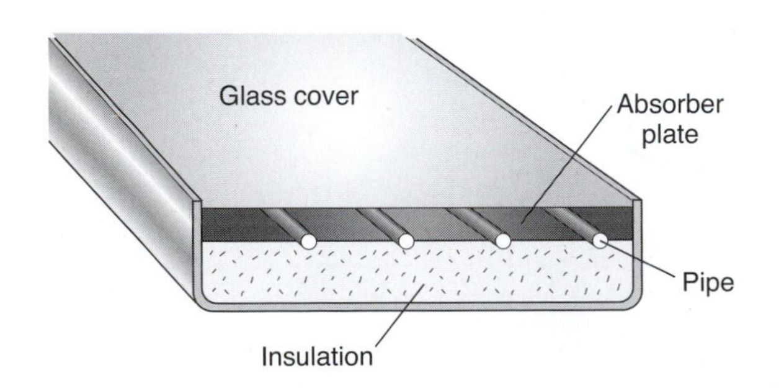

FIGURE 9.10
Cross section of a flat-plate solar collector using a liquid heat-transfer fluid.

SPACE HEATING

FIGURE 9.11
An evacuated-tube solar collector. Inside the evacuated glass tubes are smaller tubes containing fluid that boils and transfers heat to circulating liquid at the top of the collector. The vacuum in the tubes eliminates conductive and convective heat loss.

Active systems for space heating can be air or water based; the two are analogous to fossil-fueled heating systems using forced hot air and circulating water, respectively. In air-based systems, a fan delivers heated air from flat-plate collectors either directly to the building's interior or through an insulated bin of loose rocks or other material that warms and serves as an energy-storage medium. Water-based systems circulate either water or an antifreeze mix through the solar collectors. In the former case, the heated water can go directly into a storage tank. Antifreeze, in contrast, circulates in a separate loop that includes a heat exchanger immersed in the storage tank, where hot antifreeze transfers its energy to the water. Either way, the heated water is circulated through the house for heating, and it may also be withdrawn for domestic hot water. In addition, an auxiliary heat source can be used to boost the solar-heated water to the necessary temperatures when sunlight alone isn't sufficient to supply the building's heating needs.

Despite its potential, the use of solar energy as the primary source for space heating is not widespread. In 2009, the U.S. Census Bureau listed only 11,000 solar-heated homes, or roughly 1 in 10,000 homes and down considerably from a decade earlier. Considerably more significant is the use of solar energy for domestic water heating.

SOLAR DOMESTIC HOT WATER

Hot-water heating accounts for some 20% of home energy use, and solar energy is well suited for water heating. An advantage of solar water heating is that it's easy to retrofit into existing homes; solar space-heating systems, in contrast, are usually incorporated into the initial design and construction. Worldwide, solar hot-water heating is a booming industry, driven by its relatively short payback period and governments bent on encouraging renewable energy. China is the leader in solar hot water, with nearly two-thirds of the world's installed capacity. In Dezhou, home to the world's largest solar panel manufacturing plant, 90% of the city's residences have their water heated by sunlight. More than 7% of Australia's homes have solar hot water, a number that's doubled since 2005. In 2006, Spain mandated solar hot-water systems in all new and renovated buildings, and sunny Israel has long required solar hot-water heating for all residences. In the United States, in contrast, only about 1 in 1,000 homes has solar-heated water—although the solar business is picking up as a result of energy tax credits that run through 2016.

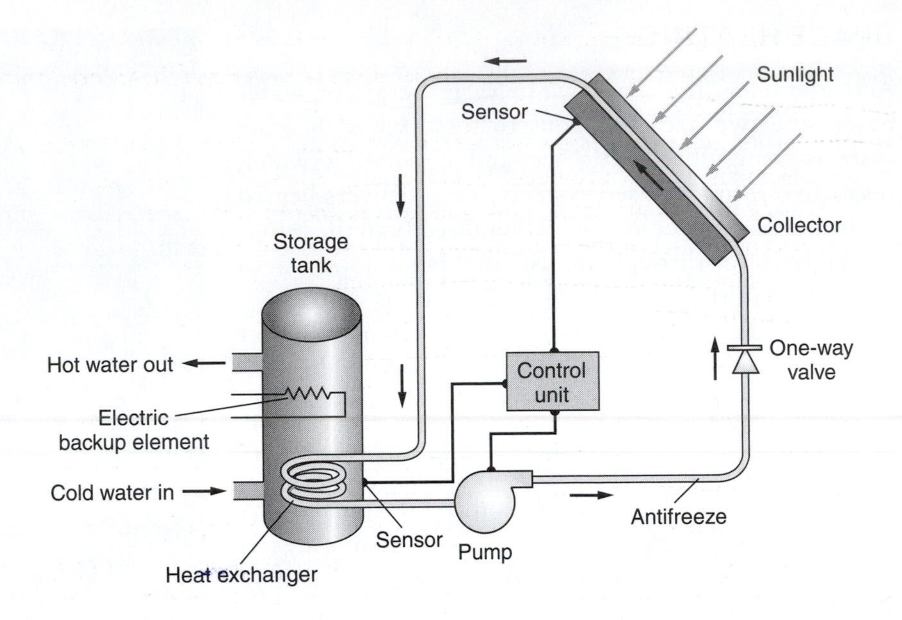

FIGURE 9.12
A solar domestic hot-water system.

In warm climates, solar hot-water systems circulate water directly through solar collectors. Where freezing temperatures occur, nontoxic antifreeze circulates through collectors and through a heat exchanger at the bottom of the water tank (Fig. 9.12). An alternative is the drain-back system, which pumps water through the collectors when solar energy is available, but drains back into the storage tank to protect the collectors from freezing when the Sun isn't shining. Increasingly, newer hot-water systems use evacuated-tube collectors because of their greater efficiency.

The system in Figure 9.12 incorporates a number of other essential details. The control unit compares temperatures from sensors at the collector and at the bottom of the storage tank; when the collector temperature exceeds the storage by some set amount—typically around 10°C—the control unit turns on the pump. Pumping continues as long as the temperature differential stays above a few degrees. In bright sunlight this strategy has the pump running continuously; under cloudy or intermittent sunlight it gives the collectors time to warm up before circulating the cool heat-transfer fluid for a short while until they cool down again.

It's crucial that the heat exchanger and tank sensor be at the bottom of the tank where the cold water enters. This is because hot water, being less dense, rises in the storage tank. The result is **stratification**, in which the tank temperature typically varies from cool at the bottom to hot at the top. Locating the heat exchanger near the bottom of the tank keeps the heat-transfer fluid cooler, and thus the collectors stay cooler. This, in turn, minimizes heat loss from the collectors and therefore increases the energy collection efficiency. On a very hot day the entire tank may reach a uniform temperature, but as soon as hot water is drawn off the top it's replaced by cold water at the bottom, and again the tank stratifies. Energy collection ceases if the heat-transfer fluid reaches the collectors' stagnation temperature, which varies with ambient temperature and insolation. In winter the stagnation temperature is lower and heat loss limits the collection of solar energy.

Another feature of the solar hot-water system in Figure 9.12 is the electric heating element, which provides backup energy for long cloudy periods. It's placed in the upper part of the tank and thus heats only the uppermost water; this leaves cool water at the bottom and allows solar collection at lower collector temperatures and therefore with less than full sunlight.

Finally, there's the one-way valve in the heat-transfer loop. Without that, heat flowing from the water into the heat-transfer fluid at night could result in the warmed and therefore less dense heat-transfer fluid rising from the heat exchanger to the top of the collectors, against the normal flow direction shown in Figure 9.12, and then flowing downward through the collector, which would radiate hard-earned energy back to the cool sky. However, if the tank were located *above* the collector, the density-driven flow would be in the right direction to collect energy rather than lose it. Such **thermosiphon** systems avoid the need for a pump. They're therefore simple, inexpensive, self-contained, and widely used in developing countries. However, they present architectural difficulties because of the need to locate the storage tank above the collectors.

Solar domestic hot-water systems are viable in almost any climate. Their technology is proven, simple, and straightforward, and they probably represent one of the easiest ways to displace a significant part of our fossil fuel use in homes and other buildings. My own house includes a solar hot-water system that's built into the roof and that supplies some 90% of the home's hot water from May to October in cloudy Vermont (Fig. 9.13).

FIGURE 9.13
The solar domestic hot-water system in your author's home, with collectors built into the roof structure. Inset photos show the basement storage tank (right) and control unit (left). Insulated pipes at the bottom of the tank carry heat-transfer antifreeze to and from the heat exchanger at the bottom of the tank. The control unit displays the collector temperature in degrees Fahrenheit, in this case on a hot, sunny summer day.

EXAMPLE 9.3 | Solar Water Heating

The solar collectors in Figure 9.13 each measure 2 feet by 8 feet, and there are six of them (the central glass is a skylight). Each has a single-pane glass cover that admits 84% of the incident sunlight. The loss rate for these collectors is 10 W/m²·°C, meaning that every square meter of collector loses energy at the rate of 10 W for every degree Celsius (or kelvin) of temperature difference (ΔT) between the absorber plate and the ambient air temperature. (Because radiation plays a significant role in the loss of heat from solar collectors, the correct measure of ambient temperature is a bit subtle, and depends on the clarity of the atmosphere. Here we'll stick with air temperature.)

(a) Find the maximum collector temperature on (i) a bright summer day with insolation S = 950 W/m² and ambient temperature of 30°C (86°F), (ii) a cloudy summer day with S = 200 W/m² and T = 27°C, and (iii) a sunny winter day with S = 540 W/m² and T = −10°C. (b) Find the rate of energy collection when the insolation is 900 W/m², the ambient temperature is 22°C, and the collector temperature—kept down by cool heat-transfer fluid—is 43°C.

SOLUTION

(a) The maximum collector temperature—the stagnation temperature—occurs when collector loss equals the solar gain, or $0.84S = (10\ \text{W/m}^2\cdot{}^\circ\text{C})(T_c - T_a)$, with T_c and T_a the collector and ambient temperatures, respectively, and where the factor 0.84 accounts for transmission through the glass. Solving for T_c gives

$$T_c = \frac{0.84S + (10\ \text{W/m}^2\cdot{}^\circ\text{C})T_a}{10\ \text{W/m}^2\cdot{}^\circ\text{C}}$$

We have values of S and T_a for three cases; substituting and working the arithmetic gives (i) 110°C, (ii) 44°C, and (iii) 35°C. The first of these is higher than water's boiling point, and a well-designed solar heating system will have safety cutoffs to prevent the tank water from actually reaching this temperature. The other two temperatures, while well above ambient, are low for domestic hot water (44°C is 111°F, whereas typical water heaters would be set at around 120°F and a hot shower might be mixed with enough cold water to bring it down to around 105°F). So in both of these cases a backup heat source would be needed. Once the collectors reach these maximum temperatures, of course, they're no longer collecting energy, since at that point they lose as much energy as they gain. Note that these collectors do better on the cloudy summer day than on the sunny winter day; on the latter, the low Sun angle limits insolation and the low ambient temperature increases heat loss.

(b) In this scenario, the cool heat-transfer fluid keeps the collectors below their stagnation temperature, and there's a net energy gain equal to the difference between the solar input and the collector loss. We have six 2 ft × 8 ft collectors, for a total area of 96 ft², or 8.9 m². So the solar energy input is

$$(0.84)(900\ \text{W/m}^2)(8.9\ \text{m}^2) = 6.73\ \text{kW}$$

Meanwhile the loss is

$$(10\ \mathrm{W/m^2 \cdot ^\circ C})(43^\circ\mathrm{C} - 22^\circ\mathrm{C})(8.9\ \mathrm{m^2}) = 1.87\ \mathrm{kW}$$

This leaves a net gain of 4.86 kW, close to the power of a typical 5-kW electric water heater. Incidentally, the collector efficiency under these conditions is 4.86 kW/6.73 kW = 72%.

9.4 Solar Thermal Power Systems

In describing the solar resource, I pointed out that the Sun's high temperature makes solar energy a high-quality energy source. By concentrating sunlight to achieve high temperatures, we have the potential to run heat engines with substantial thermodynamic efficiency. Such a heat engine can then drive an electric generator, supplying electricity to the power grid. The combination of concentrator, heat engine, and generator constitutes a **solar thermal power system**. Quantitatively, such systems are characterized by their **concentration ratio**—the ratio of concentrated sunlight to direct sunlight intensity—and the temperature they achieve. For the flat-plate collectors discussed in Section 9.3, the concentration ratio is 1, and the highest temperatures achievable are somewhat over 100°C.

CONCENTRATING SOLAR ENERGY

Both lenses and mirrors can serve as light-concentrating devices. But large-scale lenses are expensive, fragile, and hard to fabricate, so mirrors are the concentrators of choice in solar power systems. The most effective concentrator is a **parabolic reflector** (Fig. 9.14), which has the property that parallel rays reflect to a common point called the **focus**. The Sun is so distant that its rays at Earth are very nearly parallel, so a three-dimensional **parabolic dish** (Fig. 9.14b) brings sunlight very nearly to a point (although not quite, otherwise we could achieve

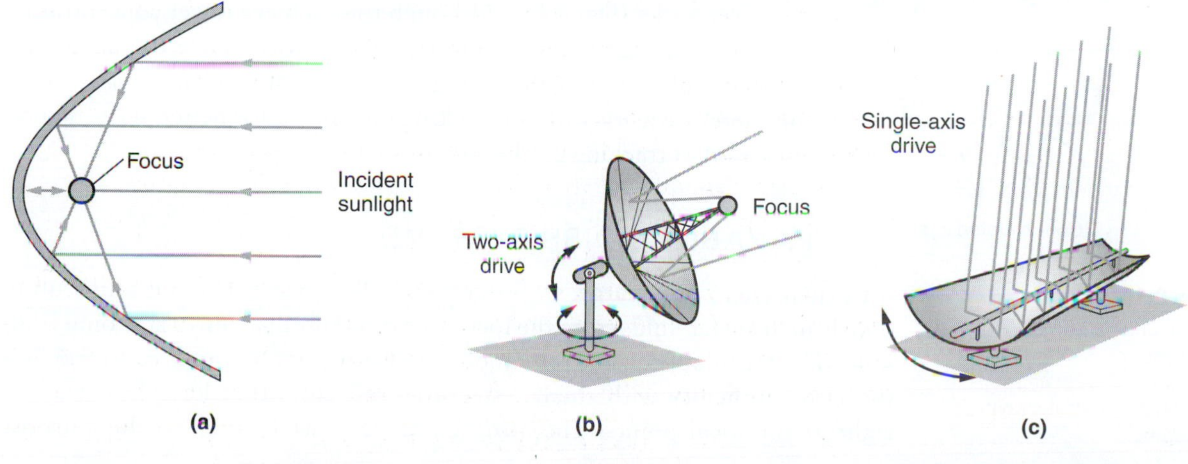

FIGURE 9.14 Solar concentrators. (a) A parabola reflects parallel rays of sunlight to a common focus. (b) A parabolic dish focuses sunlight to a point and requires two-axis tracking. (c) A parabolic trough concentrator focuses sunlight to a line and requires only one-axis tracking.

an infinite temperature, in violation of the second law of thermodynamics; see Question 2). You're familiar with parabolic dish reflectors in another application: They're the satellite dishes used to focus TV signals from communications satellites. And parabolic reflectors—or their spherical approximations—are widely used as the light-gathering elements in astronomical telescopes. Parabolic dish concentrators can have very high concentration ratios, and they reach temperatures measured in thousands of kelvins. A "solar furnace" at Odeillo, France, has a concentration ratio around 10,000 and focuses 1 MW of solar power to achieve maximum temperatures of 3,800 K. That's too hot for a practical heat engine, and this concentrator is not for energy production but high-temperature materials research.

For maximum energy collection, a parabolic dish concentrator needs to point toward the Sun, which requires a complex and expensive mechanism to steer the entire dish and the energy-absorbing component at its focus in two independent directions, as illustrated in Figure 9.14b. This makes the parabolic dish a **two-axis concentrator**. A simpler and lower-cost alternative is the **parabolic trough concentrator** shown in Figure. 9.14c, which focuses sunlight to a line rather than a point. Trough concentrators track the Sun in only one direction, so they're **single-axis** devices, with a correspondingly less complex steering mechanism. The disadvantage is that the concentrated sunlight is spread out along a line, rather than being concentrated into a point, and therefore trough concentrators achieve lower temperatures than dishes.

Another concentration strategy is to use multiple flat mirrors to approximate a parabolic shape. Individual mirrors can be steered to keep sunlight concentrated on a central absorber that itself doesn't need to move. This approach—using individual steerable mirror segments—is also used in some of the largest astronomical telescopes.

The concentrating schemes I've just introduced are all capable of forming actual images. In fact, nighttime astronomers have actually used a mirror-field solar power system to image astronomical objects. But solar power systems don't need imaging, just energy concentration. **Compound parabolic concentrators** achieve such concentration using two parabolic surfaces that funnel light incident on a broad opening to a narrower outlet (Fig. 9.15). Although they achieve lower concentration ratios than imaging concentrators, these devices have the advantage that they concentrate light from a wide range of incidence angles. They can therefore work without tracking the Sun or, for better performance, with modest tracking mechanisms.

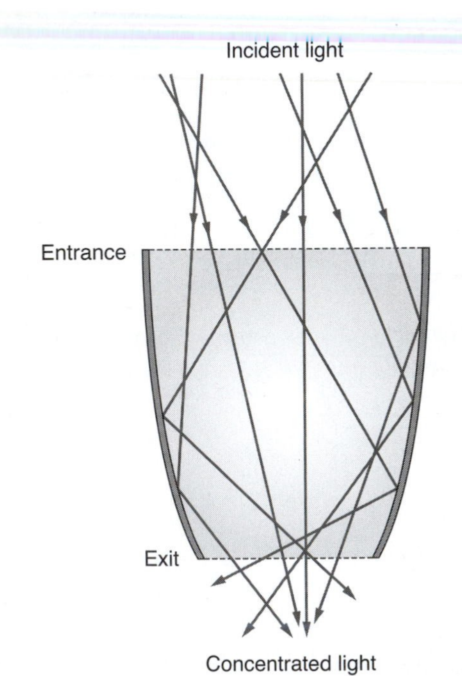

FIGURE 9.15
A compound parabolic concentrator funnels incident light from a range of angles, providing concentration without imaging.

PARABOLIC DISH SYSTEMS

It's expensive to build large, steerable parabolic reflectors, and it's difficult to pipe heat-transfer fluids from the focus of a movable dish into a stationary turbine. The latest approach to parabolic concentrators for solar energy avoids the latter difficulty with smaller steerable reflectors that have heat engines right at the focal points. The *Stirling engine* is well suited to this purpose because it's an external combustion engine that can use concentrated solar energy instead of heat released in combustion. Invented in 1816 by a Scottish minister but never widely used, the Stirling engine's working fluid is a gas sealed permanently into a piston–cylinder system. The American company Stirling Energy Systems, with support from the U.S. Department of Energy, has developed a 25-kW parabolic dish and Stirling engine combination dubbed SunCatcher (Fig. 9.16). This system boasts the efficiency record for conversion of solar energy to grid-supplied electricity: over 30%. The first commercial SunCatcher installation, a 1.5-MW pilot plant using 60 SunCatcher units, opened in 2010 near Phoenix, Arizona.

FIGURE 9.16
Stirling-engine solar concentrators at the 1.5-MW solar power plant in Maricopa, Arozona. Each unit produces 25 kW of electric power.

POWER TOWERS

Power tower solar thermal power plants use a field of **heliostats**—flat, Sun-tracking mirrors—to focus sunlight on an absorber at the top of a central tower. Steam is generated either directly at the tower or indirectly via a heat-transfer fluid. Demonstration units operated in the California desert in the 1980s and 1990s. Since then, Spain has become the world leader in power-tower development (Fig. 9.17). The most advanced of the Spanish plants, the 19-MW Gemasolar project, uses molten salt as both a heat-transfer fluid and energy storage medium. By extracting energy from the solar-heated molten salt, Gemasolar can produce electricity for 15 hours in

FIGURE 9.17
Spain's 19-MW Gemasolar power-tower system uses a molten-salt heat-transfer fluid, and can generate power for 15 hours in the absence of sunlight.

the absence of sunlight, and is expected to achieve a long-term capacity factor (percent of time the plant is operating) of 74%.

A very different solar technology sometimes described using the word "tower" is the **solar chimney**. This design would use sunlight to heat air under a vast glass roof surrounding a central chimney-like tower. Heated air rushing up the chimney would drive turbine-generators for electric power production. The system would use no water—an advantage in arid climates like Australia's, where a large solar chimney was once proposed. But its very low thermodynamic efficiency and exorbitant cost make the solar chimney a rather improbable technology.

TROUGH CONCENTRATORS

Parabolic troughs are simpler than dish concentrators or heliostat mirrors because they rotate around only one axis. The downside, a lower operating temperature and therefore lower efficiency, is offset by lower costs. Trough concentrators' long axes run north–south, so they rotate east to west as the Sun crosses the sky. At the focus of each trough is a pipe containing heat-transfer fluid; the pipe is surrounded by an evacuated glass tube to limit heat loss by conduction and convection. Some parabolic-trough plants incorporate heat storage using their heat-transfer fluid or molten-salt systems, so they can continue to produce electricity after the Sun goes down. Others are fossil–solar hybrids, burning natural gas to drive their turbine-generators when the Sun isn't shining.

In the 1980s and 1990s, Luz International built nine parabolic-trough plants in the California desert, totaling 354 MW of solar-generated electricity. The plants continue to operate, and have provided valuable technical and economic experience with the trough-concentrator design. Subsequent parabolic-trough installations include the 64-MW Nevada Solar One (Fig. 9.18), online in 2007, and a 250-MW Spanish solar plant consisting of five 50-MW parabolic-trough arrays that became operational in 2010–2011.

9.5 Photovoltaic Solar Energy

Imagine an energy technology that has no moving parts, is constructed from an abundant natural resource, and is capable of converting sunlight directly into electricity with no gaseous, liquid, or solid emissions whatsoever. Too good to be true? Not at all: This is a description of photovoltaic (PV) technology.

The basis of PV devices is the same semiconductor technology that fuels the ongoing revolution in microelectronics, bringing us computers, smartphones, high-definition TV, digital photography, GPS receivers, iPods, robotic vacuum cleaners, and the host of other electronic devices that seem essential to modern life. The significant difference is that the semiconductor "chips" in our electronic devices are small—typically a few millimeters on a side—whereas PV devices must expose large areas to sunlight to capture significant solar power.

FIGURE 9.18
Parabolic trough concentrators at the 64-MW Nevada Solar One plant near Boulder City, Nevada. The plant uses molten salt for energy storage, so it continues to produce electricity when the Sun isn't shining.

SILICON AND SEMICONDUCTORS

Silicon is the essential element in semiconductor electronics, and it's widely used in solar PV cells. Silicon is the second-most abundant element on Earth (after oxygen); ordinary sand is largely silicon dioxide (SiO_2). So silicon is not something we're going to run out of, or need to obtain from despotic and unstable foreign countries, or go to war over.

Silicon is a **semiconductor**—a material whose electrical properties place it between insulators, such as glass and plastics, and good conductors, such as metals. Whereas a metal conducts electricity because it contains abundant free electrons, a semiconductor conducts electricity because a few electrons are freed from participation in interatomic bonds. Such a free electron leaves behind a **hole**—the absence of an electron, which acts like a positive charge and is able to move through the material and carry electric current. Thus a semiconductor, unlike a metal, has *two* types of charge carriers: negative electrons and positive holes. Electron–hole pairs can form in silicon and other semiconductors when random thermal energy manages to free an electron or, more important in photovoltaics, when a photon of light energy dislodges an electron. Furthermore, we can manipulate the structure of silicon to give it abundant free electrons or holes, thus controlling its electrical properties.

The silicon atom has four outermost electrons that participate in bonding to other silicon atoms, and in pure silicon only a tiny fraction of those electrons manage to escape their bonds. But by "doping" the silicon with small amounts of impurities—typically 1 part in 10 million or less—we can radically alter its electrical properties. Adding an impurity such as phosphorus, which has five outermost electrons, leaves one electron from each phosphorus atom with no way to participate in the interatomic bonding. These electrons are therefore

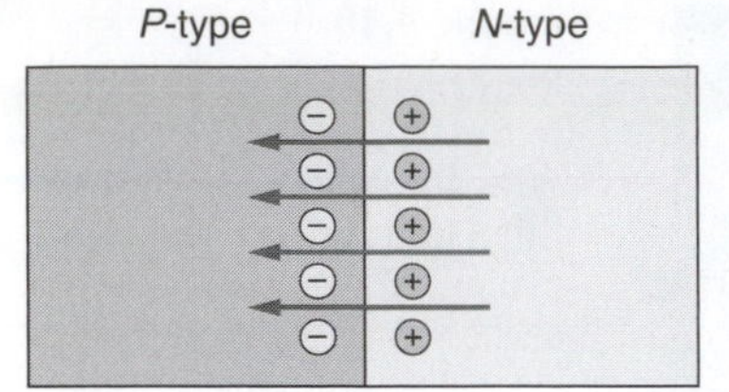

FIGURE 9.19
A *PN* junction. When the junction is first formed, electrons (minus signs) migrate from the *N*-type side to the *P*-type side, and holes (plus signs) migrate from *P* to *N*. This sets up a strong electric field (arrows) pointing from *N* to *P*.

free to carry electric current. A semiconductor with excess free electrons is called an ***N*-type semiconductor**, because its dominant charge carriers are negative electrons. Dope silicon with boron, on the other hand, and you create a ***P*-type semiconductor**, whose dominant charge carriers are positive holes. That's because boron has only three outer electrons, so when it fits into the silicon structure, there's a missing electron—a positive hole.

The key to nearly all semiconductor electronics, including PV cells, is the ***PN* junction**. Joining pieces of *P*- and *N*-type silicon causes electrons from the *N* side to diffuse across the junction into the *P*-type material, while holes from the *P*-type side diffuse across into the *N*-type material. This sets up a strong electric field across the junction, pointing from the *N* side to the *P* side (Fig. 9.19). In silicon, the corresponding voltage difference is about 0.6 volt. The *PN* junction has some remarkable properties; for example, it conducts electricity in only one direction—from *P* to *N*, but not from *N* to *P*. These properties are exploited in transistors and the other "building blocks" of modern electronics.

PHOTOVOLTAIC CELLS

Figure 9.20 shows a photovoltaic cell, which consists of a *PN* junction whose upper *N* layer is thin enough—typically 1 μm—for light to penetrate to the junction. Photons of sufficient energy can eject individual electrons from the silicon crystal structure, creating electron–hole pairs. If that occurs near the junction, the electric field pushes the positive holes downward, into the *P*-type material. Electrons, being negative, are pushed in the opposite direction, into the *N*-type material. Metallic contacts on the bottom and top of the cell thus become positively and negatively charged, respectively—just like the positive and negative terminals of a battery. In fact, the PV cell is essentially a solar-powered battery, and connecting the cell's contacts through an external circuit causes a current of electrons to flow and provide energy to the circuit.

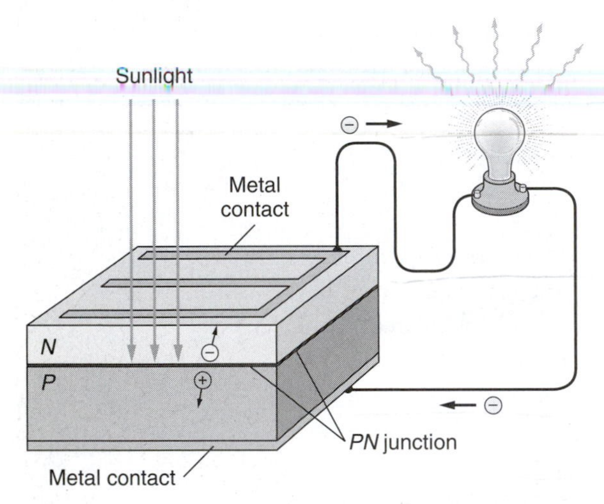

FIGURE 9.20
A photovoltaic cell. The energy of sunlight photons creates electron–hole pairs at the *PN* junction, and the electric field at the junction sends electrons into the *N*-type semiconductor and holes into the *P*-type semiconductor. Metal contacts at the top and bottom become negatively and positively charged, respectively, and as a result a current flows through an external circuit.

Several factors determine a PV cell's efficiency. The most significant relates to the distribution of wavelengths in the solar spectrum described in Figure 9.7. Semiconductor physics shows that there's a minimum energy—called the *band-gap energy*—that's required to eject an electron from the crystal structure, creating an electron–hole pair. In silicon that energy is 1.12 electron volts (eV), or 0.18 aJ. So it takes a photon with at least this much energy to create an electron–hole pair. Equation 3.1 relates photon energy and wavelength, and it shows that an energy equal to silicon's 0.18-aJ band gap corresponds to a wavelength of 1.1 μm, which is in the infrared but not far from the boundary of the visible region. As a result, roughly one-fourth of the total solar radiation lies at wavelengths too long to create electron–hole pairs in silicon. Furthermore, photons with energies above the band gap give up their excess energy as heat. And some photon-created electron–hole pairs recombine before the junction's electric field has a chance to separate them; their energy, too, is lost as heat. The result, for silicon, is a theoretical maximum PV cell efficiency of around 33%.

Other losses reduce the efficiency of commercially available PV cells to around half the maximum value. Some light is reflected off the top surface, although this can be reduced but not eliminated with special antireflection coatings or textured surfaces that help "trap" light within the cell. The necessary metal contacts on the Sun-facing surface block solar radiation, but the contact area can't be made too small without reducing the amount of electric current the cell can supply. Some electrons fail to reach the metal contacts, and their energy is lost. Varying the doping near the contacts helps steer electrons into the metal and thus reduce the loss. Such improvements enable the production of silicon PV cells with efficiencies as high as 25%, but with the downside of higher production costs.

Another approach to increasing PV efficiency is to use materials with different band gaps. A low band gap uses more of the solar spectrum, but then most of the photons have excess energy that is lost as heat. A high band gap minimizes this loss, but then most photons can't produce electron–hole pairs. So there's a compromise band gap that optimizes efficiency. For ground-based applications, the distribution of solar wavelengths puts this optimum gap at about 1.4 eV; in space it's closer to 1.6 eV. Both of these gaps are somewhat larger than that of silicon. Semiconductors with more appropriate band gaps are indium phosphate (InP; 1.35 eV), cadmium selenide (CdSe; 1.74 eV), and gallium arsenide (GaAs; 1.43 eV). However, the use of toxic elements such as arsenic and cadmium makes these less attractive because of possible environmental contamination.

FIGURE 9.21
A polycrystalline solar cell. The mottled appearance results from silicon crystals with different orientations. Horizontal structures and the wider vertical bands are the metallic contacts that channel current to an external circuit. This particular cell measures 12.5 cm by 12.5 cm and produces just over 2 W of electric power in direct sunlight, with a 14% conversion efficiency.

One way around the band-gap compromise is to use multilayer cells, so that light encounters several *PN* junctions with different band gaps. The topmost junction has the highest gap, allowing lower-energy photons to penetrate deeper before giving up their energy at a more appropriate gap. Obviously, multilayer PV cells are more expensive to produce, but their higher efficiency could make them economically competitive. The best multilayer cells produced to date are three-layer devices with efficiencies just over 41%. Achievement of this efficiency won the cell's German developers a 2010 Innovation Prize from EARTO, the European Union's association of research and technology institutions.

Although silicon is abundant, semiconductor-grade silicon is very expensive because semiconductor applications require exquisitely pure, crystalline silicon. In the semiconductor industry, large crystals are grown slowly from molten silicon; then they're sliced up into thin circular wafers that hold several hundred microcircuits such as computer processing units or memory chips. Growing crystals is expensive and energy intensive, and the slicing process wastes a lot of pure silicon; but given the tiny size and high profitability of the individual microcircuits, this is not an economic problem for semiconductor electronics. Solar cells require large areas of silicon, and this means high costs and significant environmental impacts for PV power systems produced from pure crystaline silicon. As a result, PV cells are often made from **polycrystalline silicon**, whose structure includes many individual crystals with different orientations. This characteristic reduces efficiency somewhat, but it permits PV cells to be manufactured with far less expensive processes that coat thin films of polycrystalline silicon onto a substrate. Even **amorphous silicon**—which, like ordinary glass, has no crystal structure—can be used to make inexpensive but low-efficiency PV cells. Most of today's commercial PV cells are polycrystalline, and a close look at the cell surface shows the multicrystal structure (Fig. 9.21).

FIGURE 9.22

SolarWorld's SW 240 solar panel measures approximately 1 m × 1.7 m, and produces 240 W of power in direct sunlight.

PHOTOVOLTAIC MODULES AND DISTRIBUTED APPLICATIONS

The properties of the *PN* junction limit the voltage of a single cell to around 0.6 volt. To achieve voltages capable of driving significant currents in most circuits, individual cells are connected in series and mounted together to make **photovoltaic modules**, also called **solar panels** (Fig. 9.22). Individual panels can be joined to make large arrays that are usually fixed in place but can have a Sun-tracking mechanism for optimal performance. A "light funnel," like the one shown in Figure 9.15, is sometimes used in conjunction with PV cells, allowing a smaller area of expensive solar cells to convert sunlight from a larger area at the funnel's entrance.

Photovoltaic panels have been used for decades in remote locations requiring reliable electric power. The most high tech of these applications is in space; nearly all spacecraft in Earth orbit or the inner Solar System are powered by solar PV panels. Lower-tech applications abound on Earth, wherever connections to the power grid aren't readily available. Solar-electric water pumps enhance life in the developing world, where women traditionally spend much of their time fetching water from distant sources. PV-powered lighting systems gather solar energy during the day and convert it back to light at night; they range from residential walkway lights to highway construction lighting. Field labs, remote weather stations, air-quality sensors, and a host of other scientific facilities use photovoltaics for remote power. For a home more than a few kilometers from the power grid, it's often more economical—not to mention environmentally friendly—to construct a PV system than to string wires. Environmentally conscious owners of transportation fleets use PV systems to charge electric vehicles; thus they can claim to have solar-powered cars and trucks. And PV cells find their way into novelty applications such as solar-powered calculators and watches. PV enthusiasts regularly race impractical but fully PV-powered cars (recall Fig. 9.2), and in 2010 the United Kingdom's Zephyr PV-powered aircraft completed a record 2 weeks aloft. Back on Earth, a PV-powered electric fence protects my garden from woodchucks.

GRID-CONNECTED PHOTOVOLTAICS

The largest and most rapidly growing use of photovoltaic panels is to supply electricity for power grids. This can be done either through systems on individual buildings, or through large-scale PV power plants. A typical home in most parts of the world can count on PV power to supply much of its electrical energy (Fig. 9.23). Some grid power may be needed on cloudy days, but at other times the PV system generates excess energy that's sold back to the grid. One issue with grid-connected photovoltaics is that PV cells produce direct current (DC), whereas power grids use alternating current (AC). However, the same semiconductor technology that gives us computer chips and PV cells now enables highly efficient **inverters** to convert DC to AC. This technology allows the house in Figure 9.23 to stay connected to the grid and sell excess energy to the power company. And it makes possible the use of photovoltaics for large-scale power generation.

The first grid-connected PV power plant was a 1-MW facility built in Japan in 1980. By 2000 about a dozen new plants were constructed annually throughout the world, and by 2010 that number had skyrocketed to more than 1,000.

FIGURE 9.23
Maine hardly qualifies as a "sunshine state," but the rooftop PV panels (arrow) on this Maine home generate more electric power than the home uses; the excess is sold to the local power company.

FIGURE 9.24
This 97-MW photovoltaic power plant in Sarnia, Ontario, Canada was the world's largest when it went on line in 2010.

The largest facilities now have peak power outputs approaching 100 MW (Fig. 9.24). Although concentrated in Europe, PV plants have been built throughout the world. Many are not even in the sunniest locations; cloudy Germany boasts more than a third of the world's PV power, and in 2010 Canada brought online what was then the world's largest single PV installation—an 80-MW facility comprising more than a million PV panels. Even larger plants were in the works, including the 550-MW Desert Sunlight project in southern California and a 16,000-acre, 2-GW Chinese plant in Mongolia.

In recent years world growth in PV installations has averaged a phenomenal 40% annually, driven by government incentives, falling prices, and concerns over climate impacts of fossil fuels (Fig. 9.25). By 2010, the world's peak

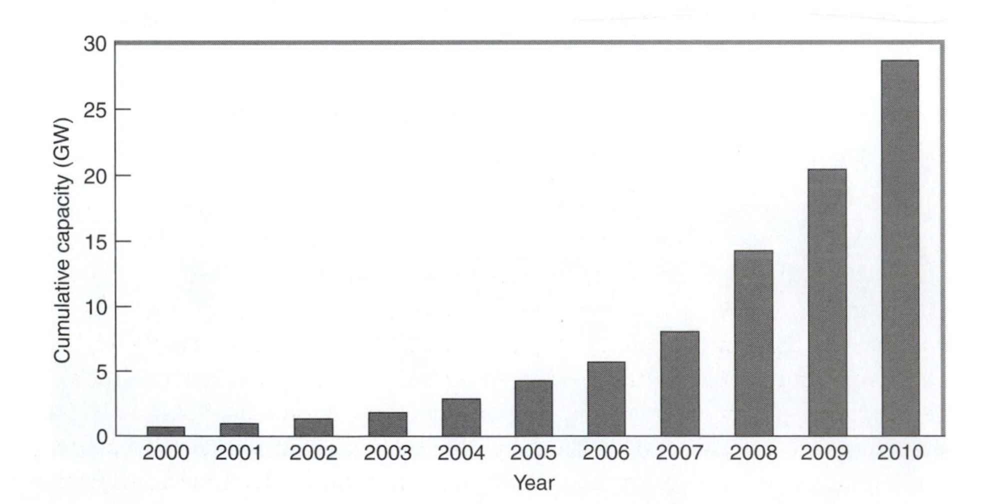

FIGURE 9.25
Global PV capacity grew at some 40% annually during the first decade of the twenty-first century. Grid-connected facilities comprised about two-thirds of the total in 2000, and by the end of the decade more than 99% of new installations were grid connected.

photovoltaic capacity approached 30 GW, the equivalent of 30 large coal or nuclear plants. Yet despite its rapid growth, PV still accounted for less than 0.1% of world electricity production.

EXAMPLE 9.4 | A Photovoltaic-Powered Country?

The average rate of U.S. electrical energy consumption is about 420 GW. Assuming commercially available PV cells with 15% efficiency, an average insolation of 254 W/m^2 (year-round, round-the-clock average for Arizona with collectors tilted at the local latitude), and a land-area requirement twice that of the actual PV modules, find the total land area needed to supply all U.S. electrical energy needs with Arizona-based photovoltaics. Compare with Arizona's total land area of 294,000 km^2.

SOLUTION

With 15% efficiency we average only (254 W/m^2)(0.15) = 38.1 W/m^2 from the PV modules. Accounting for the factor of 2 required for the extra land gives 19 W per square meter of land used for PV power plants. With U.S. consumption at 420 GW, we then need a total area given by $(420 \times 10^9\ \text{W})/(19\ \text{W/m}^2) = 2.2 \times 10^{10}\ \text{m}^2$. This sounds like a huge area, but with 1,000 m per kilometer, and therefore 10^6 m^2 per square kilometer, it's only 22,000 km^2. Given Arizona's 294,000-km^2 area, this is 22/294 or 7.5% of Arizona's land area. Of course I'm not advocating converting this much of Arizona's land to supply the United States with electricity. However, we've already flooded nearly 1,000 km^2 of Arizona to generate a paltry 2.7 GW from hydroelectric dams. In any event, the result of this entirely realistic example suggests it would take only a small fraction of the desert Southwest to power the entire United States with nonpolluting, climate-neutral solar PV energy. (However, distributing that power throughout the country is a challenge, to be discussed in Chapter 11.)

ENERGY STORAGE, GLOBAL PHOTOVOLTAICS, AND SOLAR POWER SATELLITES

Can you spot the flaw in Example 9.4? The average insolation in the example correctly accounts for nighttime and clouds, so the total energy produced from that 7.5% of Arizona is indeed enough for the entire country. The problem is that it's produced only during sunny daylight hours. So supplying a single country like the United States with photovoltaic electricity alone isn't going to work unless we develop better energy-storage technologies than we have today. We'll explore progress toward such technologies in Chapter 11. But even without energy storage, we can envision a global all-solar solution: Site PV plants around the globe with enough capacity to meet the world's electrical needs from the daytime side of the planet alone. This strategy requires an international grid capable of shipping power under the oceans—a major technical challenge. Furthermore, it demands a more stable and cooperative international political regime than the world knows today. But it may someday be the way to a PV-powered planet.

Another approach is to put PV arrays in orbit, where they would see nearly constant sunshine at levels undiminished by Earth's atmosphere or the night/day cycle that reduces average insolation on Earth. The electrical energy would be converted to microwaves and beamed to Earth for conversion back to electricity. The microwave intensity at Earth would actually be lower than the intensity of sunlight, but it could be converted to electricity using large, open-wire antennas. Unlike the other solar schemes I've described in this chapter, solar power satellites are a long way off and face stiff technological and economic challenges. Nevertheless, their advocates see solar satellites as a near-perfect long-term energy solution for humankind.

9.6 Other Solar Applications

Solar heating and electric power generation are the most important direct uses for solar energy, but there are plenty of other direct solar applications. Ironically, solar energy can be used for cooling. So-called absorption refrigeration systems use a heat source—in this case solar energy—instead of mechanical energy to drive a refrigeration cycle. Portable solar cooling systems of this sort have been developed to preserve medicines in developing countries that lack reliable electric power. So far, solar cooling has seen limited commercial use, although a handful of pilot projects in the 1-MW power range have been constructed.

On smaller scales, **solar ovens** have long been used in sunny climates to concentrate sunlight to temperatures appropriate for cooking. Such devices could stem the tide of deforestation in developing countries where wood for cooking is increasingly scarce. **Solar stills** use sunlight to evaporate seawater, leaving the salt behind and providing fresh drinking water. Small, inflatable versions are commercially available as emergency water supplies for boaters, and larger stills are supplying fresh water to communities on the Texas–Mexico border. **Solar ponds** are basically thermal collectors that use layers of salty water to inhibit convective heat loss; solar-heated water from such ponds can be used for conventional heating, industrial process heat, agriculture, or to run solar cooling units. A more high-tech application is **photolysis**, the use of solar radiation to break water into hydrogen and oxygen, producing clean-burning hydrogen fuel. A final use is so obvious it hardly bears mentioning: **solar lighting**. As with heating, passive solar design can eliminate much of a building's need for electric lighting during daytime hours. More sophisticated systems include optical fibers that "pipe" sunlight throughout a building. So-called hybrid lighting fixtures now being developed will merge electric light sources with fiber-optic sunlight, using only as much electricity as needed to maintain a chosen lighting level.

9.7 Environmental Impacts of Solar Energy

No energy source is completely benign, but solar energy comes close. Its most obvious impact follows from the roughly 200 W/m^2 of average insolation at Earth. That means large-scale solar energy installations necessarily occupy large land areas—although, as Exercises 14 and 15 show, the areas of open-pit coal mines or lakes backed up by hydroelectric dams can be considerably greater for the same power output. And large-scale solar installations aren't the only approach to solar power. Instead, we can use the tens of thousands of square kilometers of building roofs without requiring one additional acre of open land for solar energy.

While operating, solar energy systems produce no pollutants or greenhouse gases, although solar thermal electric power systems, like other thermal power plants, need to dispose of waste heat. Probably the greatest environmental impact from solar energy systems comes in their manufacture and eventual disposal. Passive and simple active heating systems use common building materials, and their environmental impacts aren't much different than with other construction projects. Photovoltaics, on the other hand, may require a host of harmful materials for their construction. Even benign silicon is doped with arsenic and other toxins, albeit at very low concentrations. During manufacture, however, these and other toxic substances are present in greater concentrations. The same, of course, is true throughout the semiconductor industry. Despite its high-tech reputation and its "clean rooms," this industry contends with serious pollution issues. Some alternative semiconductors, such as gallium arsenide or cadmium selenide, are downright toxic themselves. A fire in a house roofed with cadmium selenide PV cells would spread toxic cadmium into the environment, possibly resulting in significant contamination.

A subtler question—and one that's germane to every manufactured energy technology—involves the energy needed to produce a solar water heater or a PV electric system, or to build and launch a solar power satellite. We'll compare this production energy and its impacts for different energy sources in Chapter 16. If the production energy exceeds what a solar system will produce in its lifetime, then there's no point in going solar. Solar skeptics sometimes make this claim as a way of discounting solar energy, but they're wrong: Studies of the energy involved in the production of solar cells and other solar components are actually quite favorable to solar. The **energy payback time** quantifies the energy needed to produce a solar system and the time required for the system to produce as much energy as it took to manufacture. As long as the energy payback time is shorter than the system's anticipated lifetime, a solar energy system is energetically—although not necessarily economically—viable. Several studies in the United States and Europe point to an energy payback time for PV systems of typically 2 years, comparable to that of fossil energy facilities. This means the total energy budget for a PV system with a 30-year lifetime will be 28/30, or 93% pollution-free (Fig. 9.26)—unless, of course, the factory that manufactures the PV system is itself solar powered.

Finally, there's a question of possible climate effects in a world that makes wholesale use of solar energy. That's because solar collectors are necessarily

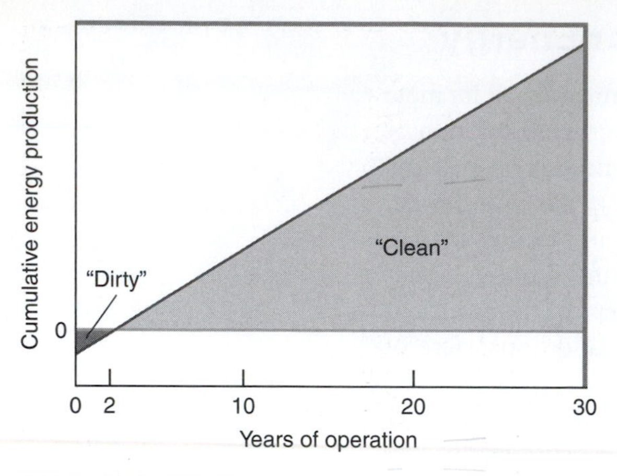

FIGURE 9.26

Energy payback time for a typical PV system is about 2 years. If the system is manufactured using fossil energy, then a system with a 30-year lifetime produces 2 years' worth of "dirty" energy followed by 28 years of "clean" energy. Note that the cumulative energy production after 2 years is zero, since at that point the system has produced as much energy as it took to manufacture.

black, and therefore they absorb more solar energy than most surfaces they cover. This alters Earth's net energy balance and could, therefore, affect climate. But with ten thousand times as much solar energy coming in as we humans use, it would take an enormous growth in energy consumption before this effect could be globally significant or even comparable to the climatic effects of fossil CO_2 emissions. There could be local warming effects from solar installations, but we're used to those already as we've covered vast land areas with black pavement and urban roofs. By the way, this same concern applies to *any* energy source—particularly fossil and nuclear—that brings to Earth's surface additional energy beyond what's carried in the natural solar flow.

9.8 Policy Issue: Solar Economics

Solar energy is abundant, widely available, and environmentally friendly. So why don't we live in a solar-powered world? The answer, in a nutshell, is economics. Most solar applications simply can't compete with traditional energy sources, in particular fossil fuels—at least under today's economic conditions. Where solar can compete, as in space and in remote power installations, it's been used for decades. One way to quantify the cost of energy systems is in terms of capital cost—dollars per watt, or the total cost needed to construct an energy plant divided by its power output in watts. Another is operating cost per unit of energy produced, typically in dollars per joule or kilowatt-hour. The cost of PV-produced electricity ranges from about 13¢/kWh to more than 30¢/kWh depending on the system size, climate, and PV technology. In contrast, wholesale electricity in the United States—mostly from fossil and nuclear sources—averages about 6¢/kWh. Solar costs continue to drop (Fig. 9.27), but

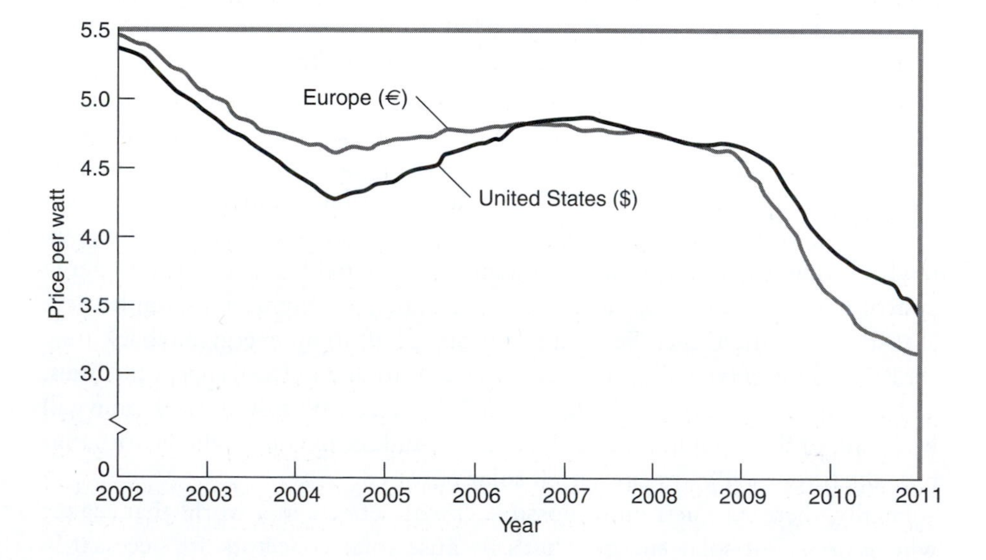

FIGURE 9.27

Retail prices for photovoltaic modules fell substantially in the first decade of the twenty-first century. Prices are in dollars (U.S.) or euros (Europe) per peak watt of power.

they still have a way to go before they're competitive in our current economic situation.

But economics is a human enterprise, not immutable like the laws of physics, and the costs of energy sources ultimately involve value judgments. Today's fossil energy technologies enjoy the benefit of more than a century of development, with few constraints on their environmental impacts for much of that time. They also enjoy—even today—massive government subsidies for research and development (Fig. 9.28) and for tax relief. And the cost of fossil energy doesn't include realistic measures of its environmental and health impacts or the cost of adapting to climate change resulting largely from fossil fuel combustion. Solar technologies, in contrast, have seen less government support, and they haven't had the time or the sales volumes to mature either technologically or economically. This situation can change as societies—governments, industries, and citizens—wake up to the realities of climate change and limited fossil fuel resources.

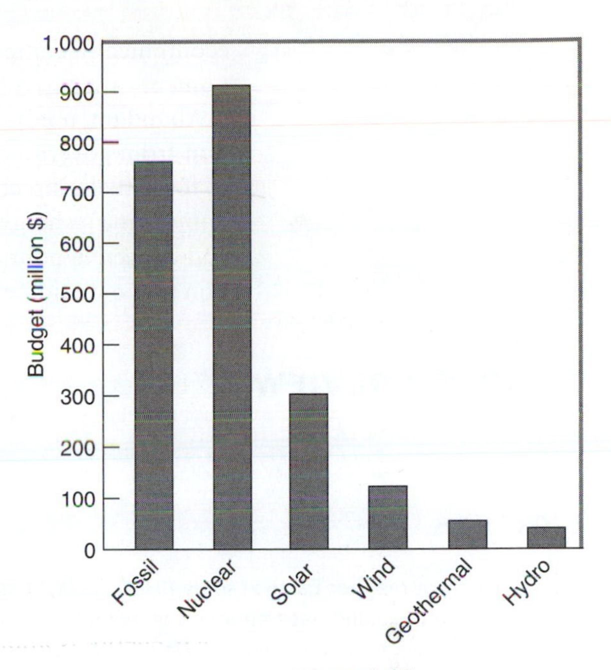

FIGURE 9.28

The president's budget request for U.S. Department of Energy research programs for fiscal year 2011. This budget represents substantial increases for renewable energy, but fossil and nuclear subsidies still dominate. Not shown is funding for research on energy-efficient buildings and vehicles.

If solar can't compete economically, why has the twenty-first century seen phenomenal growth in both solar-thermal and photovoltaic solar energy? The answer, in a nutshell, is government incentives. **Renewable portfolio standards (RPS)** require that a country or state get a specified portion of its electrical energy from renewable sources by a certain date. Most states in the United States have adopted RPSs; among the most ambitious are Maine (40% by 2017), California (33% by 2020), and New York (24% by 2013). Renewable sources generally include existing hydroelectric plants, so some states were already well underway when their RPSs were adopted. Others explicitly specify *new* renewable sources. Internationally, the European Union has adopted a standard of 33% renewable electricity by 2020, and had reached 22% renewable by 2010—by which year several individual EU countries were getting more than half their electricity from renewables.

Feed-in tariffs (FIT) require utilities to purchase renewable energy at above prevailing rates, ensuring a stable market for developers of renewable energy facilities. The downside, of course, is higher electricity rates for consumers. Much of Europe's recent renewable energy boom is a direct result of nationally adopted FITs. Spain was an early leader, adopting a renewable FIT in 1997. By 2008 Spanish utilities were paying the equivalent of more than 50¢/kWh of photovoltaic electricity, and PV development was booming. A reduction in the FIT then triggered a temporary collapse of Spain's solar industry. By 2010 other European countries were also reducing their FITs, although less precipitously. These reductions may slow renewable energy development, but in another sense they reveal the FITs' success: These subsidies are intended to encourage a renewable energy industry that can eventually stand on its own. As Figure 9.27 showed, decreasing costs have indeed accompanied a decade of dramatic growth in PV installations. PV is not yet economically competitive, but economic incentives and technological developments are narrowing the gap.

Abundant, nonpolluting, renewable solar energy is now getting serious attention from governments, industry, and citizens. Unlike some other alternatives to fossil fuels, direct solar energy really does have the potential to supply all of humankind's energy needs. But it's going to take a great deal more individual and collective political will before solar energy makes a substantial contribution to the world's energy mix.

CHAPTER REVIEW

BIG IDEAS

9.1 Solar energy reaches Earth at some 10,000 times the rate at which humankind uses energy. The available energy varies with location and season. We can capture solar energy directly as heat or to generate electricity.

9.2 **Passive solar heating** avoids active devices such as pumps or fans.

9.3 **Active solar heating** captures solar energy with **solar collectors**, and uses active devices to move the energy. Active systems can provide both space heating and hot water.

9.4 **Solar thermal power systems** use reflectors to concentrate sunlight, thereby heating a fluid that operates a heat engine to generate electricity.

9.5 Photovoltaic devices use **semiconductors** to convert solar energy directly into electricity.

9.6 Other solar applications include cooking, lighting, and **photolysis**—the use of solar energy to break water into hydrogen and oxygen.

9.7 No energy source is environmentally benign. Solar energy's impact includes land use, the energy used in manufacturing solar energy systems, and the use of toxic materials—particularly in some advanced PV devices.

9.8 Most solar applications aren't yet economically competitive with fossil fuels. Policy decisions to encourage solar energy can help establish an economically viable solar industry.

TERMS TO KNOW

absorber plate (p. 236)
active solar heating (p. 235)
amorphous silicon (p. 248)
compound parabolic concentrator (p. 242)
concentration ratio (p. 241)
diffuse insolation (p. 228)
direct insolation (p. 228)
energy payback time (p. 253)
evacuated-tube collector (p. 236)
feed-in tariff (FIT) (p. 255)
flat-plate collector (p. 236)
focus (p. 241)
heliostat (p. 243)
hole (p. 245)
indirect gain (p. 235)
insolation (p. 226)
inverter (p. 249)
N-type semiconductor (p. 246)
parabolic dish (p. 241)
parabolic reflector (p. 241)
parabolic trough concentrator (p. 242)
passive solar heating (p. 231)
photolysis (p. 252)
photovoltaic module (p. 248)
PN junction (p. 246)
polycrystalline silicon (p. 248)
power tower (p. 243)
P-type semiconductor (p. 246)
renewable portfolio standards (RPS) (p. 255)
selective surface (p. 236)
semiconductor (p. 245)
single-axis concentrator (p. 242)
solar chimney (p. 244)
solar collector (p. 235)
solar constant (p. 226)
solar lighting (p. 252)
solar oven (p. 252)
solar panel (p. 248)
solar pond (p. 252)
solar still (p. 252)
solar thermal power system (p. 241)
stagnation temperature (p. 236)
stratification (p. 238)
thermal mass (p. 233)
thermosiphon (p. 239)
Trombe wall (p. 235)
two-axis concentrator (p. 242)

GETTING QUANTITATIVE

Solar constant: ~1,364 W/m^2 above Earth's atmosphere

Solar power incident on Earth: 1.74×10^{17} W (174 PW), more than 10,000 times humankind's energy-consumption rate

Midday sunlight at Earth's surface: ~1 kW on each square meter oriented perpendicular to incident sunlight

Average insolation at Earth's surface: 239 W/m^2

Photovoltaic cell efficiency: typical ~15%; best ~25%; maximum in laboratory devices, 41%

QUESTIONS

1. Explain why the average rate per square meter at which solar energy reaches Earth is one-fourth of the solar constant.
2. Explain why we can't concentrate sunlight to temperatures hotter than the 5,800-K solar surface. *Hint*: Think about the second law of thermodynamics.
3. Why does stratification in a solar hot-water tank help increase the efficiency of solar energy collection?
4. At what time of year would you expect maximum midday insolation on (a) a solar collector tilted at latitude and (b) a horizontal collector?
5. How is a parabolic trough concentrator simpler than a parabolic dish?
6. In what way is the electrical output of a PV module not compatible with the electric power grid?
7. Some people claim that stone houses, because of their large thermal mass, don't need insulation. Discuss this claim.
8. What's the advantage of adding more layers of glass to a solar collector? What's the disadvantage?
9. Light with too long a wavelength can't excite electron–hole pairs in a PV cell. Why not? And why isn't light with too short a wavelength very efficient in producing PV electricity? Explain both answers in terms of the band gap.
10. Why isn't it possible to give a single number to describe the efficiency of a flat-plate solar collector?

EXERCISES

1. A precise value for Earth's average orbital distance is 149.6 Gm, and the best value for the Sun's power output is 3.842×10^{26} W, ±0.4%. Calculate the range of values for the solar constant that these figures imply.
2. Use the values of Earth's radius and the solar constant given in the text to verify the figure 174 PW for the total solar power incident on Earth's sunward side.
3. How long does it take the Sun to deliver to Earth the total amount of energy humankind uses in a year?
4. A passive solar house requires an average heating power of 8.4 kW in the winter. It stores energy in 10 tonnes (10,000 kg) of Glauber salt, whose heat of fusion is 241 kJ/kg. At the end of a sunny stretch, the Glauber salt is all liquid at its 32°C melting point. If the weather then turns cloudy for a prolonged period, how long will the Glauber salt stay at 32°C and thus keep the house comfortable?
5. A solar hot-water system stagnates at 85°C under insolation of 740 W/m^2 and an ambient temperature of 17°C. Find its loss rate in W/m^2·°C, assuming its glazing transmits 81% of the incident sunlight.
6. A household uses 100 gallons of hot water per day (1 gallon of water has a mass of 3.78 kg), and the water temperature must be raised from 10°C to 50°C. (a) Find the average power needed to heat the water. Then assume you have solar collectors with 55% average efficiency at converting

sunlight to thermal energy when tilted at your location's latitude. Use data from Table 9.1 to find the collector area needed to supply this power in Albuquerque in June.

7 Repeat part (b) of the preceding problem for Minneapolis in January, assuming a lower efficiency of 32% (because of higher losses at the lower temperature).

8 Find the average temperature in the passive solar house of Example 9.2 if it's located in (a) Fairbanks and (b) San Francisco.

9 A passive solar house has a thermal mass consisting of 20 tonnes of concrete. The house's energy loss rate is 120 W/°C. At the end of a sunny winter day, the entire house, including the thermal mass, is at 22°C. If the outdoor temperature is –10°C, how long after the Sun sets will it take the house's temperature to drop by 1°C?

10 My college's average electric power consumption is about 2 MW. What area of 15% efficient PV modules would be needed to supply this power during a summer month when insolation on the modules averages 225 W/m^2? Compare with the area of the college's football field (55 m by 110 m overall).

11 The Solar Two power tower used 42 MW of solar power to bring its molten salt heat transfer fluid to 566°C, and it produced 10 MW of electric power. Find (a) its actual efficiency, and (b) compare to its theoretical maximum thermodynamic efficiency, assuming T_c for its heat engine is 50°C.

12 The maximum temperature achievable with concentrated sunlight is the 5,800-K temperature of the solar surface. Find the maximum thermodynamic efficiency of a heat engine operating between this temperature and the 300-K ambient temperature typical of Earth.

13 A solar hot-water system for a U.S. home costs about $6,000 after government subsides. If the system reduces electrical energy consumption by 3,500 kWh of electrical energy each year, and if electricity costs 14¢/kWh, how long will the system take to pay for itself (neglect the economics of borrowed money and similar complications).

14 The Hoover Dam on the Colorado River generates hydroelectric power at the rate of 2,080 MW. Behind the dam, the river backs up to form Lake Mead, whose surface area is 63,900 hectares (1 hectare = 10^4 m^2). Using the assumptions of Example 9.4 (average insolation of 254 W/m^2, 15% PV efficiency, and an area twice that of the actual PV cells), find the area that would be occupied by a PV plant with the same output as Hoover Dam, and compare that number with the area of Lake Mead.

15 The controversial Cheviot open-pit coal mine in Alberta, Canada, occupies 7,455 hectares and produces 1.4 million tonnes (1 tonne = 1,000 kg) of coal per year. (a) Use the energy content of coal from Table 3.3 to find the power in watts corresponding to this rate of coal production, assuming it's burned in power plants with 35% efficiency. (b) Again using the assumptions of Example 9.4, find the area of a PV plant with this power output, and compare that number with the area of the Cheviot mine.

RESEARCH PROBLEMS

1 Download the latest version of the U.S. Energy Information Administration's "Existing Electric Generating Units in the United States" spreadsheet. Find out if your state has any PV or solar thermal power sources and, if so, determine the total capacity of each. If not, where are the closest solar thermal electric power plants?

2 Explore the National Renewable Energy Laboratory's PVWATTS solar calculator at www.nrel.gov/rredc/pvwatts/, and use it to find the performance of a PV system at your location. You can use the calculator's default values for the different parameters, or enter them yourself. Either way, prepare a graph showing monthly values of the AC energy produced, in kilowatt-hours. What is the value of the energy produced over the year?

3 If you're not in the United States, download the International Energy Agency's report *Trends in Photovoltaic Applications* (available at www.iea-pvps.org/trends/index.htm). If your country is listed, use the data (a) to find the total installed PV power in your country and (b) to prepare a plot of installed PV power over time in your country.

4 Describe any renewable portfolio standards and feed-in tariffs your state or country has adopted.

ARGUE YOUR CASE

1 A friend argues that subsidies for solar and other renewable energy sources make no sense because they force the public to pay extra for energy that just isn't competitive on the open market. What do you argue?

2 You're debating a representative of the coal industry, who claims that solar energy will never make a substantial contribution to our electrical energy supply because it's too diffuse and because the Sun doesn't shine 24 hours a day. Formulate a counterargument.

Chapter 10

INDIRECT FROM THE SUN: WATER, WIND, BIOMASS

Solar energy powers many of Earth's natural processes. As I showed in Figure 1.8, some 23% of the incident solar energy goes into evaporating water, which condenses to fall as rain, some on the continents where it returns via rivers to the oceans. Another 1% goes into the kinetic energy of atmosphere and oceans—the winds and ocean currents. A mere 0.08% of the incident solar energy is captured by photosynthetic plants, which store the energy chemically and provide the energy supply for nearly all life on Earth. These three energies—the energy of flowing water, the energy of moving air, and the **biomass** energy stored in living organisms—are thus indirect forms of solar energy. So is the thermal energy stored in soils and ocean water. Today, these indirect solar forms provide some 10% of humankind's energy (Fig. 10.1), a figure that could go much higher in the future.

10.1 Hydropower

Humankind's use of **hydropower**—the energy associated with flowing water—has a long history. More than 2,000 years ago the Greeks developed simple waterwheels to extract mechanical energy from moving water to grind grain.

Hydropower remained the unrivaled source of mechanical energy for industry until the development of steam engines in the eighteenth century. Waterwheels drove mechanisms for grinding grain, hammering iron, sawing wood, and weaving cloth. The factories of the industrial revolution often had elaborate belt-and-pulley systems to transfer mechanical energy from a waterwheel to machinery located throughout the factory. It's no coincidence that industrial towns sprang up along rivers with ample and reliable flows of water, especially where natural waterfalls concentrated the water's mechanical energy.

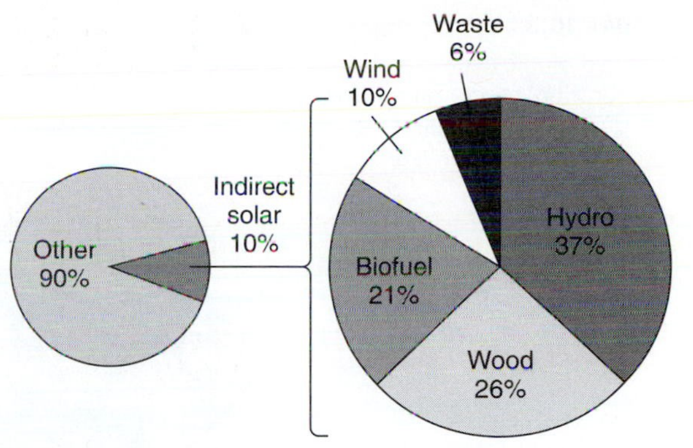

FIGURE 10.1
Small pie graph at left shows that indirect solar energy supplies 10% of U.S. net energy production. Larger pie shows the contributions from different forms of indirect solar energy.

The invention of the electric generator in the late nineteenth century provided a natural application for hydropower. Recall from Section 3.2 that a generator uses electromagnetic induction to convert mechanical energy into electrical energy; thus, connecting a waterwheel to a generator transforms the energy of moving water into electrical energy. Hydropower has competed with fossil-fueled generators since the inception of the electric power industry, and in the early twentieth century hydropower supplied 40% of U.S. electricity. By mid-century the increase in fossil-fueled generation had dropped that share to about one-third, and today only 7% of U.S. electricity comes from hydropower. Globally, however, hydroelectric power accounts for some 16% of total electric power generation, and for countries with large hydro resources, it's the main source of electrical energy (Fig. 10.2).

THE HYDROPOWER RESOURCE

Figure 1.8 showed that 23% of the solar energy incident on Earth goes into evaporating water. That's the source of the energy we harness with our hydropower facilities. As you can show in Exercise 1, the vast majority of that energy goes into the phase change from liquid water to vapor. Most of this energy heats the atmosphere when water vapor condenses to form clouds and precipitation,

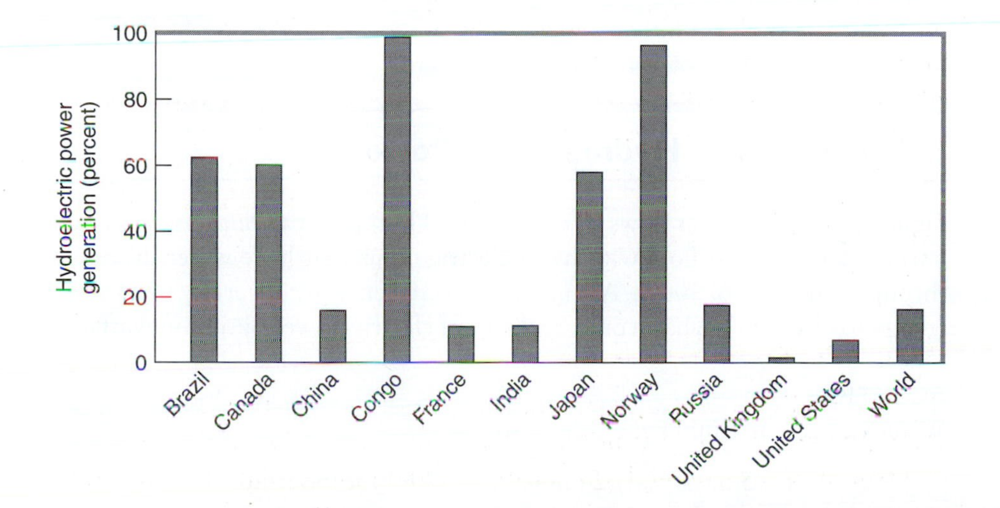

FIGURE 10.2
Percentage of electrical energy generated from hydropower for selected countries.

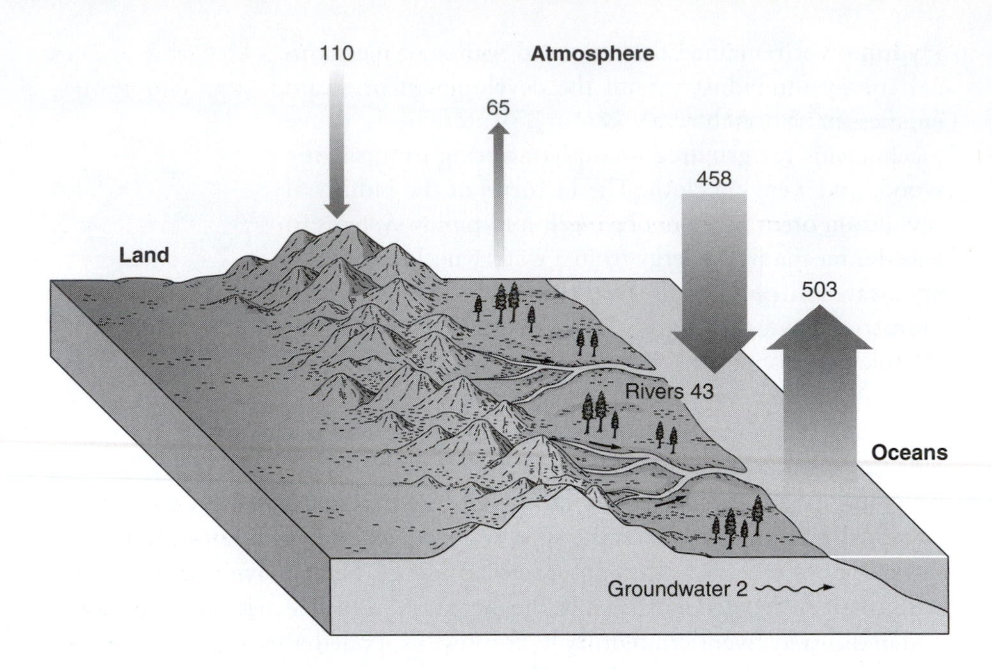

FIGURE 10.3
The hydrologic cycle, showing evaporation, precipitation, and return flows in thousands of cubic kilometers per year.

but a small fraction becomes gravitational potential energy of the airborne water (again, see Exercise 1). Most of this potential energy is dissipated as precipitation falls, but precipitation reaching higher ground still has considerable potential energy that becomes the kinetic energy of flowing waters.

The **hydrologic cycle** describes the ongoing process of evaporation, precipitation, and return flow. Most evaporation is from the oceans, and much of the evaporated water rains back into the oceans. But some wafts over the continents to fall as precipitation. About half the moisture that falls on the continents returns to the atmosphere via evaporation from rivers and lakes and via **transpiration**—the process whereby plants extract moisture from the soil and release it to the atmosphere. The remainder—some 45,000 cubic kilometers (km^3) per year, or about 8% of the total water cycle—returns to the oceans. Rivers carry most of that water, helped by underground flows. Figure 10.3 shows these quantitative details of the hydrologic cycle.

EXAMPLE 10.1 | Hydroelectric Power

Figure 10.3 shows river flows of some 43,000 km^3 per year. Suppose we could harness 10% of that flow with hydroelectric dams that have water dropping through a distance of 300 m. Assuming 90% efficiency in converting the water's energy to electricity, what would be the total electric power output in watts?

SOLUTION

Water's density is 1,000 kg/m^3; there are

$$(365 \text{ days/year})(24 \text{ hours/day})(3{,}600 \text{ seconds/hour})$$

or about 3.15×10^7 s in a year; and there are 10^9 m^3 in 1 km^3 (this follows because the number of cubic meters in a cube 1 km on a side is $1{,}000^3$, or 10^9). So that 43,000 km^3 per year is

$$(43{,}000 \text{ km}^3/\text{y})(10^9 \text{ m}^3/\text{km}^3)(1{,}000 \text{ kg/m}^3)/(3.15 \times 10^7 \text{ s/y}) = 1.37 \times 10^9 \text{ kg/s}$$

Now, the potential energy change of a mass m falling through a distance h is mgh, where $g = 9.8$ m/s^2 (Equation 3.4). Here we have 1.37×10^9 kg falling every second through 300 m; multiplying those two figures together with g then gives the *rate* of potential energy change. We're told we can extract 10% of that as electric power, with 90% efficiency, giving

$$\begin{aligned} P &= (1.37 \times 10^9 \text{ kg/s})(9.8 \text{ m/s}^2)(300 \text{ m})(0.1)(0.9) \\ &= 4.0 \times 10^{12} \text{ kg·m}^2/\text{s}^3 = 0.36 \text{ TW} \end{aligned}$$

This number is about 18% of the average global electrical energy-consumption rate of 2 TW. Since 16% of the world's electricity actually does come from hydropower, this example suggests that we're probably using somewhere on the order of 10% of all the world's river flows to generate electricity—possibly more because the dam height I specified applies only to the largest dams, and possibly less because many rivers go through more than one hydroelectric power plant. Those rivers are also cooling our fossil and nuclear power plants, so you can see that the energy industry really works the hydrologic cycle!

Example 10.1 confirms that global hydropower resources are certainly adequate for the 16% of the world's electricity that now comes from flowing water. The numbers suggest further that there's some room to increase hydroelectric power production, although not by orders of magnitude. In fact, the United Nations estimates that the world's technically exploitable hydroelectric resources are roughly equal to today's global electrical energy production—that is, a little over five times current hydroelectric power generation. This suggests that hydropower alone probably won't supply all the world's electricity, let alone meet the planet's entire energy demand. But it could come closer in some regions. As Figure 10.2 suggests, hydroelectric resources are far from evenly distributed. Regions with large land areas, mountainous terrain, and ample rainfall are blessed with substantial hydroelectric resources—enough that even some energy-intensive industrialized countries such as Canada and Norway can count on hydropower for most of their electrical energy (60% in Canada, 96% in Norway). But the hydroelectric resources of most industrialized countries are already substantially developed, and further increases are unlikely. Figure 10.4 shows estimates of hydroelectric potential for five continents, compared with what's

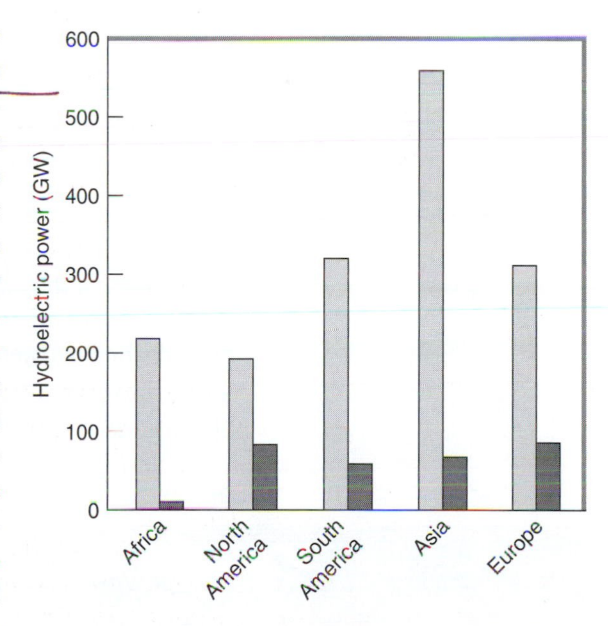

FIGURE 10.4
Estimates of average hydroelectric potential (tall bars) and actual average production (short bars) for five continents.

already been exploited in each. It's clear that hydropower has great potential in the developing world, but much less in developed countries.

ENERGY QUALITY

By now you're well aware that *amount* isn't the only criterion in evaluating energy sources; so is energy *quality*. Over and over I've stressed that the second law of thermodynamics limits our ability to convert lower-quality thermal energy into mechanical energy and electricity—whether that conversion takes place in an old, dirty coal-fired power plant or in a clean, renewable solar thermal plant. What about hydropower? Does the second law's efficiency limit apply to it, too?

The energy of moving water is pure mechanical energy, the highest-quality form, on a par with electricity. So in principle we can convert water's mechanical energy to electrical energy with 100% efficiency. We don't have to worry about thermodynamics because there's no thermal energy involved. In practice there are minor losses associated with mechanical friction, fluid viscosity, and electrical resistance, but these don't pose a fundamental limitation and clever engineers can reduce them toward zero. The result is that modern hydroelectric power plants achieve conversion efficiencies in excess of 90%—the highest of any large-scale energy-conversion technology.

Take a big-picture view, though, and the whole water cycle becomes a natural heat engine, driven by thermal energy in the form of solar radiation. Like every other heat engine, it's subject to the second law. Because it operates over a rather low temperature range and isn't specifically engineered to convert thermal to mechanical energy, the water cycle isn't a particularly efficient heat engine. But that doesn't matter when it comes to power generation, because we start from the mechanical energy of moving water that results after the natural heat engine has done its job.

HARNESSING WATER POWER

Prior to the twentieth century, water power was used to operate grain mills and industrial plants, generally using variations on simple waterwheels (Fig. 10.5). Today's hydroelectric power plants are descendents of these earlier technologies, and generally use vertical-axis turbines in which water spins a horizontal waterwheel (Fig. 10.6). Mounting any hydropower device at a natural waterfall or below a dam greatly enhances the available power, as gravitational potential energy turns into the kinetic energy of moving water. Most natural hydro sites in industrialized countries were already developed by the early twentieth century, so the majority of today's hydroelectric installations involve dams.

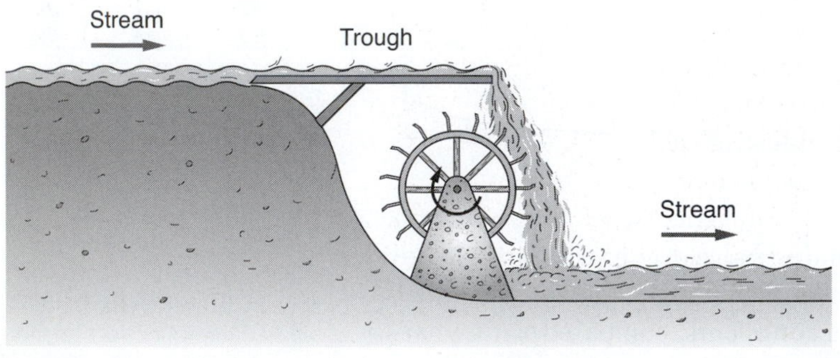

FIGURE 10.5
Overshot waterwheel, used historically to harness waterpower for grain mills and factories located at natural waterfalls.

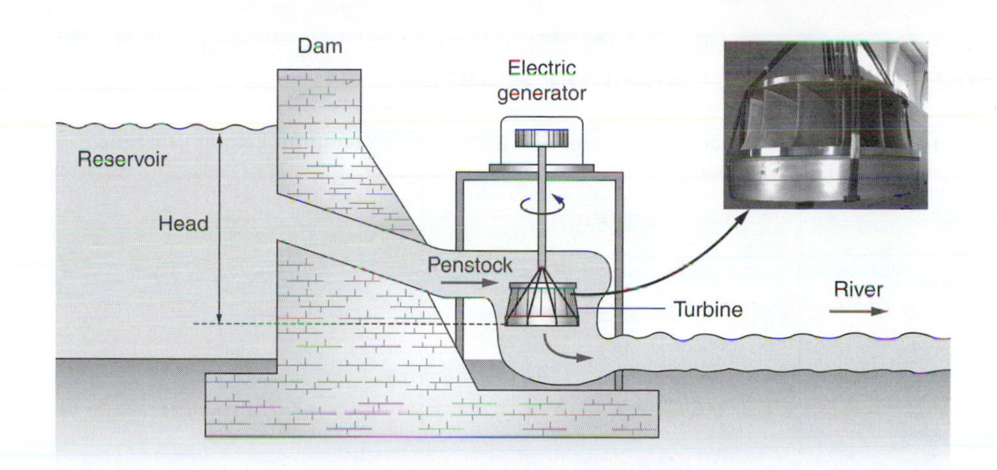

FIGURE 10.6
Cross section of a hydroelectric dam, showing one of the turbine-generators. Penstocks are huge pipes that deliver water to the turbines. The photo shows a turbine being installed at South Carolina's J. Strom Thurmond hydroelectric plant. (The figure is not to scale; the turbine-generator system would be considerably smaller.)

You saw in Example 10.1 that two factors determine the total power in a hydroelectric installation: the total flow and the vertical distance, called the **head**, through which the water drops. A given power output can come from either a large flow and low head or a high head and lower flow; Figure 10.7 and Exercises 4 and 5 explore this distinction.

Hydroelectric systems come in all sizes, starting with backyard "microhydro" units ranging from 100 W to a few hundred kilowatts. Small-scale hydroelectric plants up to a few tens of megawatts were built in abundance in the early and mid-twentieth century, and many are still in operation (see Research Problem 1). The largest hydroelectric installations contain multiple generators of typically several hundred megawatts each, and their total output exceeds that of the largest fossil and nuclear plants. Table 10.1 lists the world's largest hydroelectric generating facilities. Compare their power capacities with the typical 1 GW of a large fossil or nuclear plant, or even the 4 to 8 GW of some multiple-unit nuclear facilities, and it's clear that large hydroelectric installations are the world's largest electric power plants.

(a)

(b)

FIGURE 10.7
(a) The John Day Dam on the Columbia River in Oregon is a low-head dam with a high flow rate. (b) The Hoover Dam on the Colorado River at the Arizona/Nevada border is a high-head dam with a lower flow rate. Each produces just over 2 GW of hydroelectric power. Note the lock on the left side of the John Day Dam that permits navigation.

TABLE 10.1 | **TEN LARGEST HYDROELECTRIC POWER PLANTS**

Name	Country	Capacity when complete (GW)	Year of first power generation
Three Gorges	China	22.5	2003
Itaipu	Brazil/Paraguay	14.0	1983
Guri	Venezuela	10.2	1986
Tucuruí	Brazil	8.4	1984
Grand Coulee	United States	6.8	1942
Sayano-Sheshuensk	Russia	6.4	1989
Longtan	China	6.3	2007
Krasnoyarsk	Russia	6.0	1968
Churchill Falls	Canada	5.4	1971
La Grande 2	Canada	5.3	1979

ENVIRONMENTAL IMPACTS OF HYDROELECTRIC POWER

The actual generation of hydroelectric power is a clean, quiet, nonpolluting process that uses a completely renewable energy resource. Nevertheless, hydropower has substantial environmental impacts, nearly all of which are related to dams. Some of those impacts are obvious, while others are subtle and may surprise you.

The most obvious effect of dams is to block naturally flowing rivers and thus form lakes. In countries with heavily developed hydroelectric facilities, many rivers have become nothing more than a series of lakes separated by hydroelectric dams. Rivers best-suited for hydroelectric power are often among the most scenic waterways, prized for their wilderness qualities and for recreational uses such as fishing and whitewater rafting. Dams transform wild rivers into placid lakes that offer other recreational opportunities—most prominently fossil-fueled powerboats (see Research Problem 2 to explore the environmental impact of this aspect of hydroelectric power). The wild river issue alone has often pitted environmentalists against the development of clean, nonpolluting hydropower.

Large power dams inundate vast land areas and often necessitate the relocation of entire towns and villages. (Recall Exercise 14 in Chapter 9, which shows that the area inundated by a hydroelectric dam can substantially exceed that needed for a photovoltaic plant of the same capacity.) As recently as the late twentieth century, the Canadian government relocated Cree Indians and disrupted their hunting and fishing grounds in order to develop the vast hydroelectric resources of northern Quebec. China's massive Three Gorges hydroelectric project required the relocation of more than a million people. Worldwide, some 40 to 80 million people have been displaced by large dam projects.

Another obvious impact is the blockage of rivers to wildlife and navigation. These are less problematic with low-head dams, where fish ladders and locks are feasible. But fish used to the oxygen-rich waters of fast-flowing rivers may not find dam-impounded lakes all that hospitable. Indeed, hydroelectric development in the American Northwest and elsewhere is directly at odds with efforts to preserve or enhance the populations of salmon and other fish. Some dams in the Northwest have elaborate fish-transport systems, including ladders that enable adult salmon to travel upstream to spawn, and barges or trucks for transporting juveniles downstream around the dams. Other power dams are being removed to alleviate their impacts on fish.

Flowing rivers carry sediment that forms beaches and deltas. But when water stops behind a dam, it drops nearly all of its sediment and thus gradually reduces the reservoir's capacity. For this reason, large hydroelectric dams have limited lifetimes, on the order of 50 to 200 years, before their reservoirs become useless. Meanwhile, evaporation of impounded water lowers the overall flow and increases salinity, with impacts on river ecology and the water's suitability for agricultural irrigation. Damming of the Colorado River has reduced the number of sandbars in the Grand Canyon, which park visitors use as campsites. Cessation of natural floods also has resulted in silt buildup that adversely affects fish habitat in the canyon. In 1996 and 2004 the U.S. Bureau of Reclamation released flood-like flows from its Glen Canyon Dam, attempting to restore natural conditions in the Grand Canyon.

In Egypt's Nile valley, silt buildup behind the Aswan High Dam had another consequence: Cutting off the flow of rich silt starved agricultural land of needed nutrients and forced farmers to turn to chemical fertilizers. Ironically, much of the dam's electric power output eventually went to plants producing the fertilizer that was needed to replace the Nile's natural nutrient-delivery system. The loss of silt also reduced the replenishment of the Nile delta system, shrinking an area that is home to 30 million Egyptians and most of Egypt's agriculture. It reduced the nutrient flow into the eastern Mediterranean, harming local fisheries. And water losses from the dam and agricultural diversion lowered the freshwater flow into the Mediterranean, increasing salinity and affecting the marine ecosystem.

Dam building can also have an impact on human health. No longer subject to the Nile's regular flooding, Egypt's irrigation canals became breeding grounds for snails hosting parasitic worms that cause the debilitating disease schistosomiasis, whose prevalence in the Egyptian population increased dramatically following construction of the Aswan Dam. Another human impact comes from dam failures, which can cause catastrophic loss of life in downstream communities. In 1928, failure of the St. Francis Dam north of Los Angeles sent some 45 million m^3 (12 billion gallons) of water surging downstream, killing more than 400 people. A pair of dam failures in India in 1979 and 1980 killed 3,500 people. Large dam failures, like catastrophic nuclear accidents, are rare events, but like nuclear accidents they put large populations at risk. Indeed, the U.S. Federal Emergency Management Agency estimates that one-third of the 74,000 dams in the United States pose significant hazards to life and property should they fail. Causes of dam failure might include inadequate engineering or construction, earthquakes, excessive rainfall, and even terrorism.

A final impact of hydroelectric dams comes from a surprising source: greenhouse gas emissions. Indeed, advocates have long cited the lack of climate-changing emissions among hydropower's virtues. But research now suggests otherwise. The construction of a hydroelectric dam floods terrestrial vegetation, and when submerged vegetation decays under anaerobic conditions, it produces methane, which is a much more potent climate-change agent than CO_2. Rivers bring additional carbon into the reservoir, resulting in further greenhouse emissions. These emissions continue for decades. Canadian studies suggest modest greenhouse emissions from dams in cool climates, but tropical dams are another story. One study of a Brazilian hydroelectric dam indicates that its emissions may have more than three times the climate-warming potential of a fossil power plant of similar capacity. Emissions vary dramatically from dam to dam, depending on location, altitude, local ecology, and other factors. It will take further research, and probably site-specific climate impact studies, before we understand fully the climatic effects of existing and proposed hydroelectric installations.

10.2 Wind

Like waterpower, wind is an indirect form of solar energy. Winds range from local breezes to the prevailing air currents that encircle the globe, and all result from the differential solar heating of Earth's surface. This produces both horizontal and vertical air motions that, in the large scale, couple with Earth's rotation to produce regular patterns of airflow (Fig. 10.8). Figure 1.8 showed that a mere 1% of the solar input goes into wind and ocean currents, but that's still enough to give wind a serious place among humankind's energy sources.

Wind, like water, has a long history as an energy source. An Egyptian vase dating from 3500 B.C.E. depicts a sailing vessel, showing that wind-powered boats were in use more than 5,500 years ago. Millennia of technological advances culminated in the tall ships of the nineteenth century, which used wind energy at rates approaching 7.5 MW—more than today's largest wind turbines. Land-based windmills for pumping water and grinding grain were operating 4,000 years ago in China and 800 years ago in Europe; by 1750 there were 10,000 windmills in England alone. Picturesque four-blade Dutch windmills pumped water to help reclaim the Netherlands from the sea, and the familiar multivane windmills on American farms brought up water for agricultural use before rural electrification in the 1930s. Utility-scale electric power generation from wind began with a 1.25-MW unit on Grandpa's Knob in Vermont in the

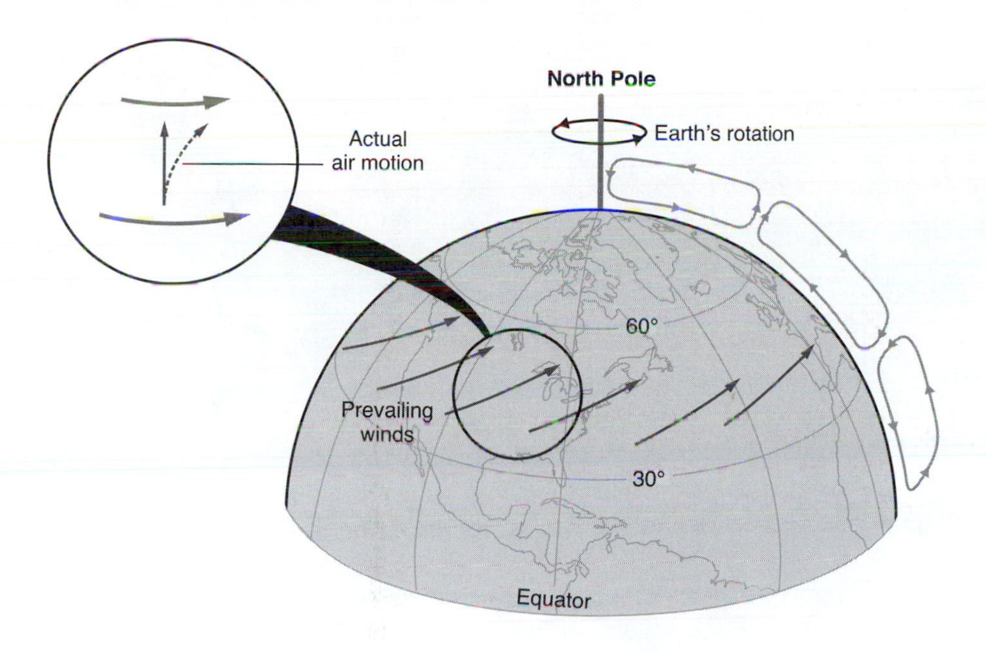

FIGURE 10.8
Origin of the prevailing westerly winds in the mid-latitudes of the northern hemisphere. Surface air is heated at the equator, rises, then cools as it flows northward, sinking back to the surface at about 30° north latitude. It then flows southward to the equator, forming a cell of circulating air. Additional cells form in the temperate and polar regions. Inset shows the effect of Earth's rotation, whose actual speed (curved gray arrows) is greatest at lower latitudes. As air moving northward passes over more slowly moving regions it deflects eastward because of its greater relative speed. This produces a west-to-east flow in the temperate zone.

1940s, and serious interest in wind-generated electricity followed the oil shortages of the 1970s. Today wind is the world's fastest-growing source of electrical energy, and it's already making significant contributions to the electrical energy supply in Europe and California.

THE WIND RESOURCE

The 1% of solar energy that becomes the kinetic energy of the wind amounts to about 2 PW—more than 100 times humankind's energy consumption. But its uneven distribution in both space and time reduces wind's potential, and technological considerations further limit our use of wind energy. Nevertheless, wind is capable of providing a substantial portion of the world's electrical energy, even in energy-intensive Europe and North America.

The energy available from wind increases as the cube of the wind speed—meaning that doubling the speed results in an eight-fold increase in power. To understand the reason for this dramatic increase in wind power with speed, consider a cubical volume of air moving with speed v (Fig. 10.9). The air has mass m and so its kinetic energy is $\frac{1}{2}mv^2$. The faster it moves, the more rapidly it transports this energy. Since the energy itself scales as the *square* of the speed, this means the rate of energy flow increases as the *cube* of the speed. Suppose our cubical volume has side s and that the air's density—mass per unit volume—is ρ kg/m^3. Then the mass of air is $m = \rho s^3$, and so its energy is $\frac{1}{2}\rho s^3 v^2$. Moving with speed v, it takes time s/v for the entire cube to pass a given point. Thus the rate at which this cubical volume transports energy is $\frac{1}{2}\rho s^3 v^2/(s/v) = \frac{1}{2}\rho s^2 v^3$. More significant is the wind power per unit area. Since the cube's cross-sectional area is s^2, this is

$$\text{Wind power per unit area} = \tfrac{1}{2}\rho v^3 \qquad (10.1)$$

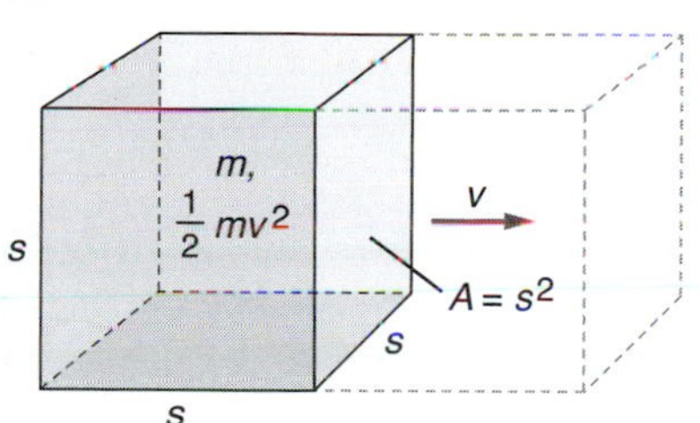

FIGURE 10.9
A cubical volume contains a mass m of moving air and therefore has kinetic energy $\frac{1}{2}mv^2$. The cube moves its length s in a time inversely proportional to its speed v; dashed lines show the new position. Therefore, the rate at which the wind transports energy across the area A is proportional to the cube of the wind speed v.

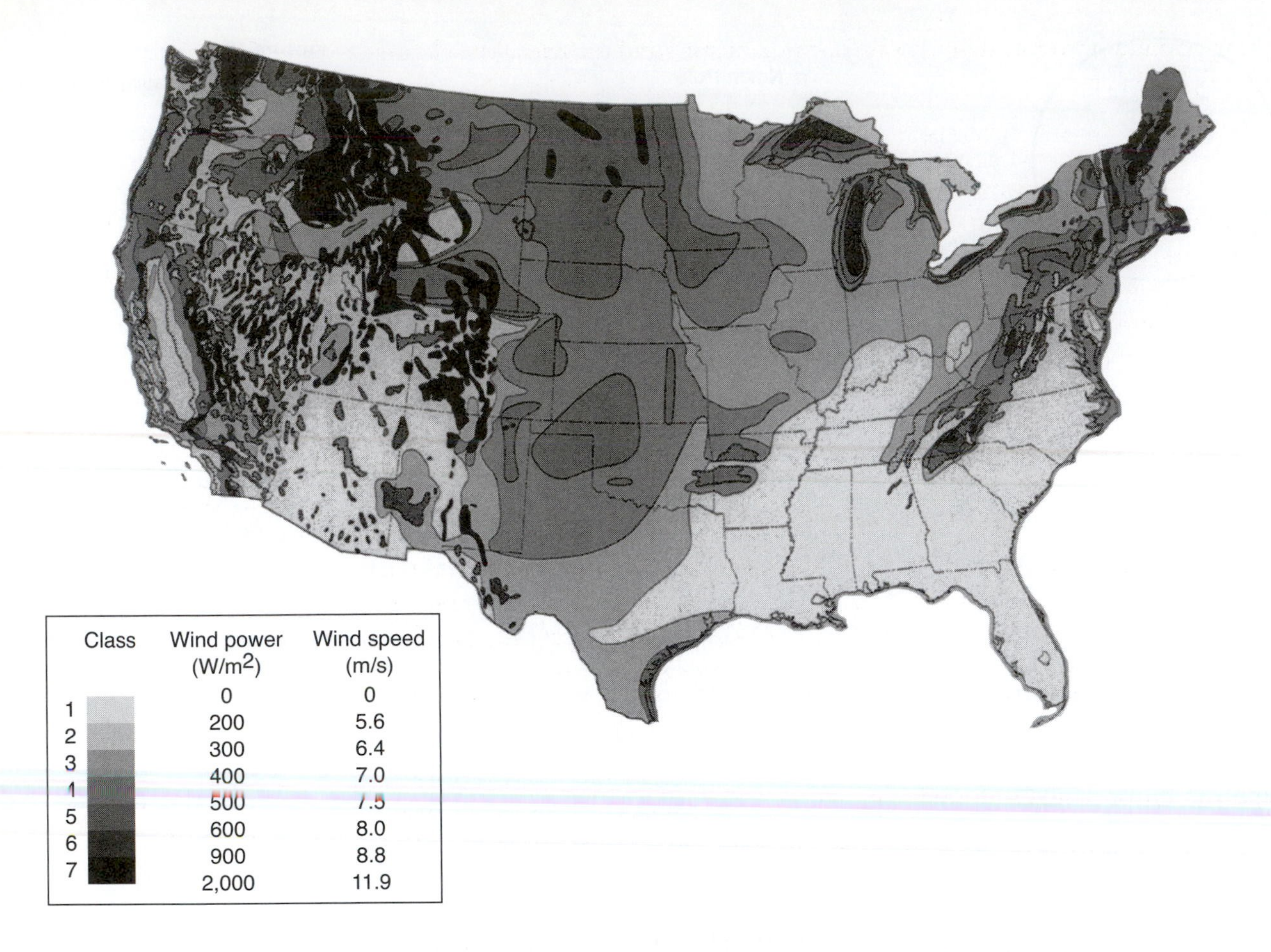

FIGURE 10.10
Wind energy resources in the United States, expressed as yearly average power per square meter and as wind speed at a height of 50 m. The wind industry recognizes seven wind energy categories. The average winds in class 3 and above are considered sufficient for large-scale power generation.

Equation 10.1 gives the actual wind power passing through each square meter of area; however, the amount of power we can extract with wind turbines is limited to a little over half of Equation 10.1's value.

Figure 10.10 shows the annual average wind power available in the United States, measured in watts per square meter. Friction reduces wind speeds close to the ground, so the values shown in Figure 10.10 are taken at the 50-m height typical of a medium-sized wind turbine. Average winds in categories 3 or 4 and above (greater than 6.4 m/s or 14.3 miles per hour for category 3) are generally considered sufficient for large-scale wind power generation with current technologies. A glance at Figure 10.10 shows that large areas of the American Midwest meet this criterion, with smaller regions of high average winds in mountainous regions, on the Great Lakes, in the Northeast, and offshore on both coasts.

EXAMPLE 10.2 | Wind Power Classes

Wind power class 3 has a minimum average speed of 6.4 m/s at 50 m elevation, whereas the minimum for class 6 is 8.0 m/s. (a) Verify that the corresponding wind powers in the legend for Figure 10.10 are in the correct ratio. (b) Assess the maximum power available to an 82-m-diameter wind turbine at the class 6 speed of 8.0 m/s.

SOLUTION

(a) Equation 10.1 shows that wind power scales as the cube of the wind speed. Therefore the ratio of wind powers for classes 6 and 3 should be the cube of their speed ratio—that is, $(8.0/6.4)^3 = 2.0$, to two significant figures. The actual powers listed in the legend are 600 W/m^2 and 300 W/m^2, respectively, in agreement with our factor-of-2 calculation. (b) An 82-m-diameter turbine sweeps an area $\pi r^2 = (\pi)(41\text{ m})^2 = 5{,}281\text{ m}^2$. At 600 W/m^2 for the class 6 winds, the total power carried by the wind is $(600\text{ W/m}^2)(5{,}281\text{ m}^2) = 3.2$ MW. An actual turbine with an 82-m blade diameter, the Danish Vestas V82, is rated at 1.65 MW in a 13-m/s wind.

Serious analysis of wind energy's potential requires knowing not only the wind's average speed but also its variability on timescales ranging from minutes to seasons. Because wind power varies as the cube of the wind speed, the average speed isn't a very good measure of the power one can expect from a wind turbine. A rough rule of thumb suggests that wind variability reduces the average output of a wind turbine to about 20% of its rated capacity in a region with category 3 wind resources, a figure that rises to more than 40% at category 7. Taking into account wind variability, the required spacing of wind turbines so they don't interfere with each other, and the wind resource as described in Figure 10.10 suggests that U.S. wind energy potential may be several times our total energy-consumption rate. A more practical estimate, with wind turbines on only 6% of the best wind areas in the United States, could still yield 1.5 times our electrical energy demand. However, wind's variability means that utilities today can't rely on wind to provide more than about 20% of their total capacity—a figure that could increase with the development of large-scale energy-storage technology. To supply 20% of U.S. electricity with wind would require wind farms on only 0.6% of the total U.S. land area, and most of that land would still be available for farming and ranching. Similar considerations apply worldwide. Europe, which in 2010 got some 5% of its electricity from wind, is on target for 14% to 17% wind-generated electricity by 2020, and 26% to 35% by 2030.

HARNESSING WIND ENERGY

Most modern wind turbines are propeller-like devices with three blades mounted on a horizontal shaft (Fig. 10.11). This design puts the blade system and electric generator atop a tower, and the entire turbine-generator and blade assembly can rotate as the wind direction changes. Some earlier systems used vertical-axis designs that respond to all wind directions, but these proved less efficient.

Equation 10.1 gives the actual power per unit area carried in the wind. But no wind machine can extract all that power, for a simple reason: If it did, then

FIGURE 10.11

This wind turbine at Carleton College in Minnesota sweeps out an area 82 m in diameter and is rated at 1.65 MW. The first commercial-scale wind installation at an American college, it supplies some 40% of Carleton's electrical energy.

the air that passed through the turbine would have no kinetic energy left. The air would be stopped, and there would be no place for incoming air to go. So incoming air would deflect around the turbine, which would then produce no power. You could avoid that deflection completely if you didn't extract any energy from the flow, but this of course defeats the purpose of your wind turbine. So there's a tradeoff: As you try to increase the energy extraction, you slow the air behind the turbine and thus increase the amount of air deflected around it.

A detailed analysis, first done by the German physicist Albert Betz in 1919, gives the maximum power extractable from the flow in terms of the ratio, a, of the air speed behind the turbine to the incoming wind speed. For the extreme cases discussed in the preceding paragraph, a takes the values 0 and 1, respectively. In general, Betz showed that the absolute maximum power of Equation 10.1 is tempered by a factor of $4a(1 - a)^2$, giving

$$\text{Maximum extractable wind power} = 2a(1 - a)^2 \rho v^3 \qquad (10.2)$$

The case $a = 0$ corresponds to air being stopped behind the turbine; at the other extreme, $a = 1$, the wind speed hasn't changed. In either case, as Equation 10.2 shows, the extracted power is zero. Figure 10.12 is a plot of Equation 10.2, and it shows that the maximum extractable power is about 59% of Equation 10.1's

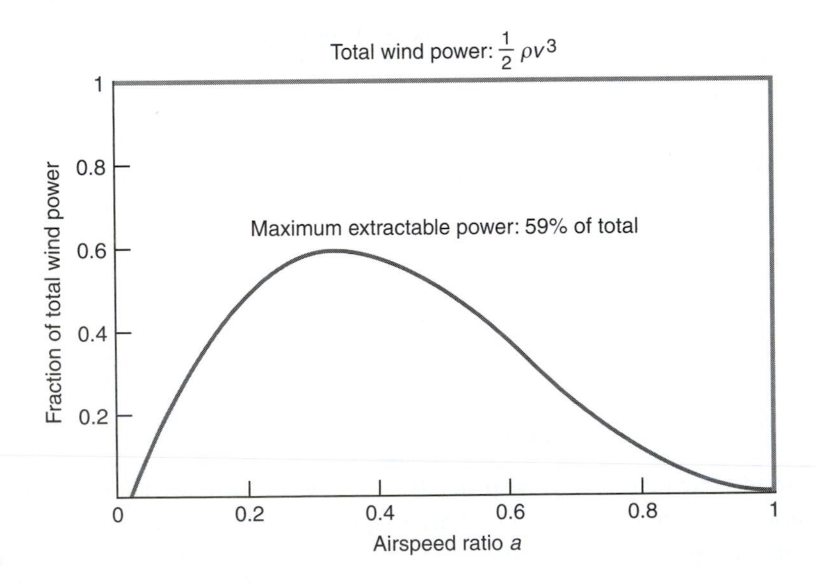

FIGURE 10.12

The fraction of extractable wind power as a function of the ratio a of the airspeed behind a turbine to the incident wind speed. The peak of the curve, at 59%, represents the maximum power any wind machine could extract.

total wind power. If you've had calculus, you can show in Exercise 11 that the exact value is 16/27 of the total power contained in the wind.

Real wind machines fall short of the 59% theoretical maximum for several reasons. Wind turbines have plenty of space between the blades where air can "slip through" without giving up its energy. Increasing the number of blades would help, in principle, but would make the device more expensive and more massive. Only a turbine with infinitely many blades would have the Betz **power coefficient** of 59%. Today's three-blade design is a compromise that can achieve power coefficients in excess of 45%.

The power coefficient also depends on the ratio of the blade tip speed to the wind speed. Practical turbine designs have an optimal wind speed at which their power coefficient is highest, and a designer will choose a turbine whose power coefficient peaks near the expected average wind speed. The longer the turbine blades, the higher the blade tip speed for a given rotation rate; therefore larger turbines can turn more slowly and still achieve their optimal power coefficients. This is the reason why you'll see small, residential wind turbines spinning furiously while their utility-scale counterparts turn with a slow, stately grace.

High winds can overpower a wind machine's electric generator or even damage the turbine blades, so it's necessary to shed wind—usually by changing the blade pitch—or even shut down the machine altogether in the highest winds. For that reason a practical wind turbine actually produces more power at speeds lower than the maximum it's likely to experience. There's also a minimum speed at which a turbine begins generating power. Putting these factors together gives a typical **power curve** like the one shown in Figure 10.13. Combining the power curve with a site's wind characteristics lets engineers assess the expected power output from proposed wind installations. There's a further complication with today's large turbines: Wind speed can vary substantially across the height of the turbine blades, and this *wind shear* results in reduced performance and increased mechanical stresses.

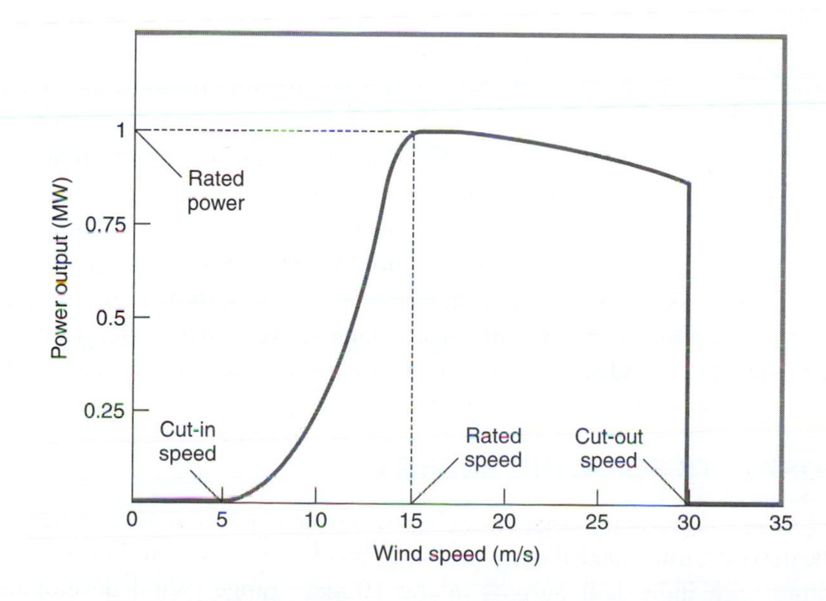

FIGURE 10.13
Power curve for a typical wind turbine. This unit has a rated power output of 1 MW, which it achieves at its rated wind speed of 15 m/s. The cut-in speed is 5 m/s, and power generation ceases at the cut-out speed of 30 m/s.

FIGURE 10.14
Some of the 4,000 wind turbines at California's San Gorgonio Pass.

ENVIRONMENTAL IMPACTS OF WIND ENERGY

Like hydropower, wind is a clean, nonpolluting energy source. Unlike hydropower, it doesn't require damming natural flows. Wind's energy payback period is even better than with photovoltaics; a Danish study suggests that the energy required to manufacture, install, maintain, and eventually decommission a wind turbine is produced in the first 2 or 3 months of turbine operation. There remain unanswered questions about the impact of large-scale wind installations on atmospheric circulation, with possible weather effects or impacts on downstream wind facilities. As wind's share of global electricity production grows, these issues will require further research.

Wind turbines generate some noise, although large, slowly rotating units can be surprisingly quiet. Standing directly below the blades of the 1.65-MW turbine in Figure 10.11, for example, one hears only a low, gentle "whoosh" as each blade swings through its lowest position. Turbines create a hazard for birds and bats, as studies of wildlife kills at wind farms attest. However, these losses are orders of magnitude below what we tolerate in bird kills from collisions with windows, vehicles, power transmission lines, cell-phone towers, and even domestic cats. Icing of turbine blades in winter can result in chunks of ice flying off the moving blades, but the danger is limited to a zone of roughly twice the blade diameter. Perhaps the greatest environmental impact of wind power is on wilderness values and aesthetics. In large installations, tens, hundreds, or even thousands of turbines are grouped in large wind farms like those shown in Figure 10.14. These obviously change the look and character of the landscape, although on open lands such as in the American Midwest they don't preclude continued farming or ranching operations. Offshore wind farms are more problematic, generating strong "not in my backyard" reactions from opponents who don't want their ocean views marred. In the eastern United States, the best wind sites are often on mountain ridges. Installing and maintaining turbines then means introducing roads and transmission lines into what was unspoiled wilderness, and ridge-top siting obviously maximizes the visual impact. Large turbines, furthermore, require flashing lights to ward off aircraft, and this increases the visual insult. These issues often split the environmental community between those who welcome clean, nonpolluting power and those who see industrial-scale desecration of wilderness. Both sides have their points. I wonder, though, if it might be better to see where our energy comes from than to have it out of sight in a distant coal-fired plant or hidden within the containment structure of a nuclear reactor.

PROSPECTS FOR WIND ENERGY

The first serious push for large-scale electric power generation using wind came in the 1970s. California led the world with the first wave of wind farms, using turbines with individual outputs in the 100-kW range. Wind development

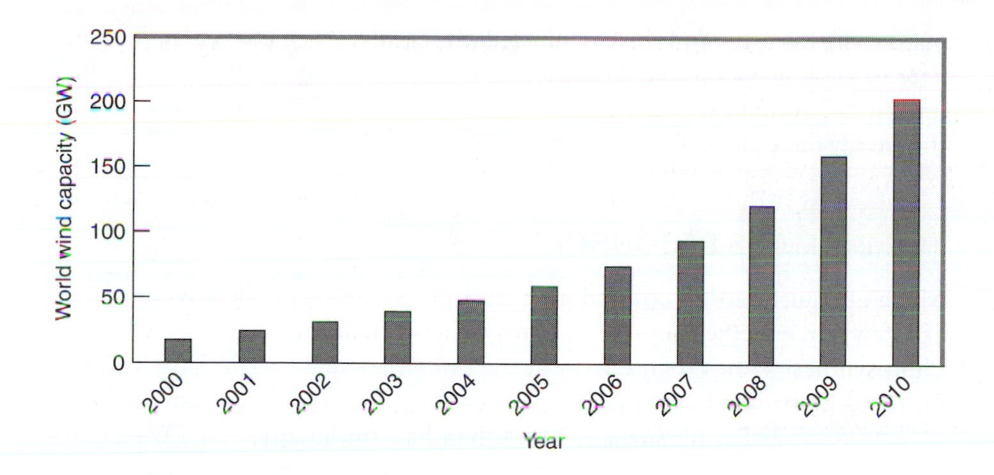

FIGURE 10.15
The world's wind capacity continues its rapid growth, with capacity in 2010 exceeding 200 GW peak, the equivalent of 200 large nuclear or coal-fired power plants. However, wind's variability reduces the global average wind power to around 60 GW.

declined in the 1980s, then picked up in the 1990s with Europe leading the way. Denmark became a major exporter of wind turbines, today typically rated at several megawatts each, and by 2010 Denmark was generating 20% of its electricity from wind. Worldwide, wind capacity has grown at more than 30% annually, to just over 200 GW in 2010 (Fig. 10.15). China, with its booming economy and growing energy demand, was doubling its wind capacity each year and by 2011 had become the world leader in wind energy.

Many of the same government incentives that have encouraged solar energy have also spurred the growth in wind. But, unlike solar, wind energy is already close to being cost-competitive with fossil fuels. Some estimates put wind's cost per kilowatt-hour in the same range as coal and natural gas; others show wind-generated electricity being up to 40% costlier. For wind, costs of electricity transmission and integration into the power grid tend to be higher than with less variable fossil and nuclear sources. Indeed, some utilities are considering charging suppliers a per-megawatt-hour fee to compensate for the difficulties of integrating the variable output of wind farms with conventional electric power generation—a practice that could lower wind's economic competitiveness. But with 2% of the world's electricity already coming from wind, this renewable energy source is here to stay, and will surely experience substantial growth in the coming decades.

10.3 Biomass

Waterpower is old, wind is older still, but biomass—fuel derived from recently living matter—is humankind's oldest energy source beyond our own bodies. Our ancestors domesticated fire somewhere between 1.6 million and 780,000 years ago, exploiting our first "energy servant" and starting us on the road to our modern high-energy society.

Biomass, predominantly in the form of wood, supplied nearly all of humankind's energy for most of our history. Other sources included dung, plant and animal oils, dried grasses, and agricultural waste. Wood remained the world's main energy source until the late nineteenth century, when coal came to dominate in most industrialized countries. Even today, as Figure 10.1 shows, biomass in the form of wood, waste, and biomass-derived fuels accounts for half the energy the United States gets from indirect solar sources.

THE BIOMASS RESOURCE

Much of Planet Earth is covered with natural solar energy collectors operating at a few percent efficiency—the photosynthetic plants and bacteria. Globally, photosynthesis stores solar energy at the rate of about 133 TW. That's almost 10 times humankind's average power consumption, and it amounts to some 0.26 W/m^2 over Earth's surface. More than half the energy—76 TW—comes from land plants, which store energy at about 0.51 W/m^2. At any time, the energy stored in Earth's biomass is some 1.5 ZJ (1.5×10^{22} J), or about 35 times the world's annual energy consumption.

Photosynthesis is a complex process that begins with chlorophyll absorbing solar energy. Details vary with different plant and bacterial species and their environments, but the net result is simple: Photosynthesis uses 4.66 aJ (4.66×10^{-18} J) of solar energy to combine 6 molecules of water and 6 of carbon dioxide, producing the sugar called glucose ($C_6H_{12}O_6$) and, as a byproduct, oxygen:

$$6H_2O + 6CO_2 + 4.66\ \text{aJ} \rightarrow C_6H_{12}O_6 + 6O_2 \qquad (10.3)$$

That oxygen byproduct is, of course, the atmospheric oxygen that began to accumulate after the appearance of the first photosynthetic organisms. It forever changed Earth's atmosphere and life's subsequent evolution.

Photosynthesis is not particularly efficient. Photon energy thresholds for chemical reactions limit the amount of the solar spectrum plants can use, just as band-gap energies limit the range of sunlight available for photovoltaic devices. Different plant and bacterial species respond to different parts of the spectrum, but for most green plants the energy absorption has two peaks, at blue and red wavelengths around 400 nm and 700 nm, respectively. There's less absorption in between, in the green region, so plants reflect these intermediate wavelengths and therefore appear green. In addition to spectral efficiency limits, chemical reactions themselves produce waste heat as well as the chemical energy stored in glucose. The net result is a theoretical maximum efficiency of around 14%. Most plants fall well short of this limit.

The actual rate at which plants store solar energy is called **primary productivity**, and it varies with plant type, insolation, temperature and other climatic factors, the availability of water and nutrients, and characteristics of the broader ecosystem. The total of all photosynthetic activity is Earth's **gross primary productivity**—the 133 TW figure just introduced. Although the planet's gross primary productivity exceeds humankind's energy demand, it's not possible to divert more than a fraction of this energy resource to power industrial society. For one thing, photosynthesis drives the entire biosphere—all of the living plants and animals except for rare communities that thrive on geochemical energy. In fact, plants themselves use about half the energy they capture, making the **net primary productivity** about half of the gross. Diverting too much of the planet's primary productivity to human uses endangers the rest of the biosphere. For another thing, we humans have to eat to fuel our own bodies. Unlike electric utilities, we can't eat uranium or coal or make direct use of sunlight, wind, or flowing water. So we need some of the planet's primary productivity to feed ourselves. We also use biomass for shelter (e.g., wood), clothing (e.g., cotton, leather), and a host of other products. In fact, it's estimated that humans already appropriate nearly 40% of the planet's net primary productivity. That's why, a few sentences ago, I didn't say that diverting too much of the planet's primary productivity "*would* endanger the rest of the biosphere" but "*endangers* the rest of the biosphere."

EXAMPLE 10.3 | Photosynthetic Efficiency

Insolation on a tropical rainforest averages around 210 W/m^2 year-round. For a temperate forest during the growing season of spring, summer, and fall, average insolation is around 190 W/m^2. Use these values, along with Figure 10.16, to estimate the photosynthetic efficiency of tropical rainforests and temperate forests.

SOLUTION

Figure 10.16 shows that the rainforest's productivity is about 1.2 W/m^2 and the temperate forest's is about 0.80 W/m^2. Their efficiencies are therefore 1.2 W/m^2 / 210 W/m^2 = 0.57% and 0.80 W/m^2 / 190 W/m^2 = 0.42%, respectively. These low numbers are typical of natural ecosystems, although they're underestimates for the actual photosynthetic process because they assume that all the incident sunlight manages to strike plant leaves.

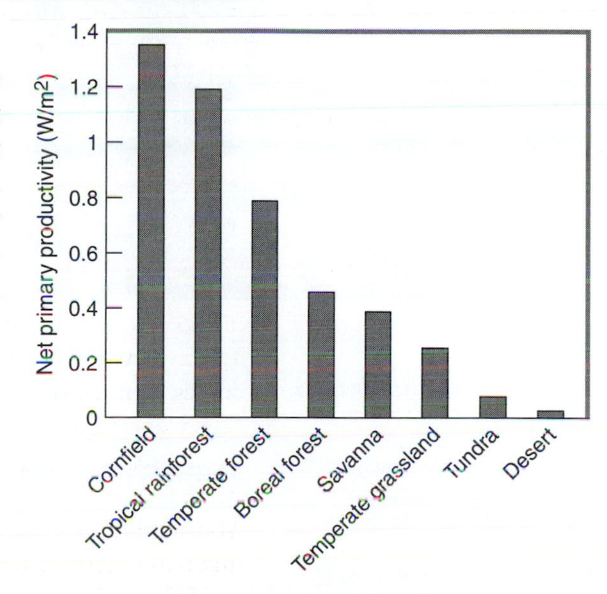

FIGURE 10.16 Net primary productivity for seven natural environments and a typical cornfield managed with modern agricultural techniques. The productivity of the cornfield is deceptive, because considerable energy goes into fertilizers, pesticides, and farm equipment.

The energy stored in dry biomass is nearly the same regardless of its origin. Pure glucose contains about 16 MJ/kg, roughly half the energy density of coal. Most wood contains very nearly 20 MJ/kg, while grasses and crop plants range from 16 to 19 MJ/kg. The adjective *dry* is important because the considerable energy it takes to drive out water reduces the net energy content of fresh biomass (see Exercise 16).

HARNESSING BIOMASS ENERGY: BURNING

The simplest and oldest way to harness biomass energy is to burn the biomass. Our ancestors used open fires for cooking and warmth, and individual fireplaces or woodstoves continued to be a major source of home heating until the widespread adoption of central heating systems in the late nineteenth and early twentieth centuries.

Burning wood in open fires is extremely inefficient. Heated wood exudes flammable gases that contain much of the wood's energy, and in an open fire these don't burn. Incomplete combustion of bulk wood results in a host of particulate and gaseous emissions, many of which are toxic or carcinogenic. The resulting indoor air pollution is a serious health threat in developing countries, where open fires are still used for cooking. The invention of the chimney around the twelfth century reduced indoor pollution, and the updraft created in the chimney helps to ensure a flow of fresh oxygen to the fire. The earliest stoves appeared around 1500, and in 1740 Benjamin Franklin invented his famous Franklin stove. Its controllable air intakes marked a significant advance in biomass burning efficiency, because a limited airflow keeps the stove interior hotter, allowing exuded gases to burn. Today's airtight woodstoves incorporate secondary combustion systems, catalytic converters, and other improvements engineered to ensure more complete combustion, thereby increasing efficiency and reducing pollution. Such technological improvements, along with regulations that specify the proper seasoning of firewood and the maintenance of steady combustion, can make wood burning a relatively clean, efficient process.

Wood burns more efficiently when it's in the form of uniform chips or pellets, especially in systems designed to gasify as much of the wood as possible prior to combustion. Wood chips and pellets can be burned much like coal to power woodstoves, industrial boilers, or turbine-generators. Many industries that use wood as a raw material obtain their heat, process steam, and even electricity from waste wood. Today there are nearly 200 wood-fired electric power plants in the United States, ranging from 500-kW to 72-MW units; their total capacity is just over 3 GW. Most have been built since 1980, many by companies that manufacture wood products or paper, and burn wood waste to produce power for manufacturing operations. Other wood-fired plants are owned by public and private utilities, supplying power to the electric grid (Fig. 10.17). In Europe, wood-fired electric power plants are often combined heat and power units that also supply hot water for community heating.

FIGURE 10.17
Burlington, Vermont's 54-MW McNeil Generating Station is the largest wood-fired power plant in the United States operated by a public utility. Here the plant is seen beyond a pile of its wood-chip fuel, much of which comes from logging waste.

Today, wood and wood by-products account for about 2% of U.S. primary energy consumption, including heating, industrial process heat, and electricity generation. Three-quarters of that wood is used in industry. About 1% of U.S. electricity comes from wood, most of which is generated by industries for their own use.

ENERGY FROM WASTE

Many materials we commonly discard as waste contain substantial energy, much of it derived from biomass. Animals don't extract all the energy from the food they eat, so manures and human sewage are sources of energy. Indeed, animal dung is an important cooking fuel in developing countries. And the solid waste that the industrialized world generates in vast quantities contains, when dry, about as much energy as an equivalent mass of wood, and nearly half as much as oil on a per kilogram basis. Most of that waste is paper, food scraps, and other materials derived from biomass; a smaller fraction (typically 20%) is plastic made from fossil fuels. The United States generates some 230 million tonnes of solid waste a year; as you can show in Exercise 12, this is equivalent to 750 million barrels of oil—enough to displace 2 months' oil imports. Burned in power plants at 20% efficiency, this waste could produce nearly 7% of the United States' electrical energy (see Exercise 13).

One change in the U.S. energy mix over the past few decades has been the emergence of municipal waste as an energy source; by 2010, garbage-to-energy plants contributed about 0.5% of the United States' electrical energy.

Even when solid waste is buried in landfills, it can still provide energy. Anaerobic decomposition of the buried waste produces methane, which can be captured and burned. Some landfills use their methane for heating and electric power, either for their own facilities or to sell to electric utilities. Others simply let methane vent to the atmosphere—an unfortunate approach, both because of the wasted energy and because methane is a potent greenhouse gas.

Municipal sewage is another source of methane, and many sewage treatment plants, like landfills, use sewage-produced methane to supplement their own energy needs. Cow manure in agricultural areas represents yet another methane source; my own home, in rural Vermont, gets half of its electricity from "cow power"—methane-generated electricity produced at local dairy farms. The 1,500 cows on one nearby farm yield electrical energy at the average rate of 200 kW. My electric company encourages this renewable energy production by paying farmers a premium for cow power and giving environmentally conscious consumers the option to purchase the power for a few cents more per kilowatt-hour than the standard rates.

BIOFUELS

With the exception of the fossil fuels, the energy sources I've described until now are used primarily to power stationary systems such as homes, industries, and electric power plants: Nuclear and tidal plants produce electricity, geothermal energy generates electricity and supplies heat, direct solar energy heats homes and buildings or is converted to electricity, and hydropower and wind are largely sources of electrical energy. With the minor exception of wind- and nuclear-powered ships, none of these energy sources is particularly useful for transportation. Yet transportation in the United States consumes well over one-fourth of our total energy.

Biomass is different. It's the one source today that can directly replace fossil fuels for transportation, because biomass readily converts to liquid **biofuels**. Gaseous biofuels such as methane are also possible transportation fuels, but their low density makes them less desirable than liquids.

A range of fuels, as well as other useful chemicals, derive from biomass through chemical or biochemical processes. This is hardly surprising; biomass is, after all, the precursor to the fossil fuels. The most important biofuels today are **ethanol**—ethyl alcohol (C_2H_5OH), the same chemical that is found in alcoholic beverages—and **biodiesel**. Ethanol results from the biological fermentation of plant sugars, or of other plant substances that can be converted to sugars. Yeasts and other microbes do the actual work of fermentation, and in addition to ethanol the process may yield coproducts such as sweeteners, animal feed, and bioplastics.

The United States is the world's leading ethanol producer, and gasoline mixed with 10% ethanol is common at U.S. gas stations; in 2010 the Environmental Protection Agency permitted an increase to 15%. These mixtures, designated E10 and E15, have less energy content than pure gasoline because ethanol's energy density is only two-thirds that of gasoline. Powering cars with ethanol is not a new idea; the original Ford Model T could run on any ethanol/gasoline mix, and gasoline with 12% ethanol was available in the 1930s. Ethanol in the United States comes from corn, with ethanol production now consuming one-third of the U.S. corn crop. Corn ethanol has been heavily subsidized, and without government subsidies, ethanol would probably not be an economically viable product in the United States. Ethanol subsidies in the United States have become increasingly unpopular, and by 2011 they faced an uncertain future, as the U.S. Senate voted to end tax credits for corn-based ethanol.

More important than ethanol's economic balance sheet is its energy balance. Producing ethanol in the United States means growing corn using fossil-powered mechanized agriculture, along with fossil-produced fertilizers and pesticides. Grinding corn and cooking it to convert the starch to sugar requires still more fossil fuel. So do fermentation, distillation, and removal of water. Quantitative studies suggest that the ratio of ethanol energy output to fossil fuel input in corn-to-ethanol conversion ranges from a low of about 0.75 to a high of around 1.7. Ethanol production with an energy ratio below 1 doesn't make much sense, since you have to put in more energy than you get out. A comprehensive study of the entire corn-to-ethanol production cycle puts the ratio at about 1.1, meaning that ethanol yields on average only 10% more energy than is used to make it. However, much of that energy comes from coal and gas, so ethanol ends up displacing a greater quantity of oil. Given these figures, you can see why ethanol policy in the United States today is controversial, with some decrying it as a "pork barrel" subsidy to agribusiness while others hail it as a step toward energy independence.

Ethanol, however, need not come from corn. A far larger bioresource is the cellulose that comprises much of the green parts of plants. A study led by the Natural Resources Defense Council argues that a vigorous program to produce ethanol from cellulosic plants—in particular, common native grasses—might by the year 2050 displace half the oil used in transportation. The energy balance of cellusose-derived ethanol could be substantially better than for the corn-derived ethanol produced today.

In Brazil, the world's second largest ethanol producer, the situation is quite different. Lacking in fossil fuels but with a warm climate and abundant sunshine, Brazil is committed to ethanol-based transportation fuels. The Brazilian government mandates at least a 25% ethanol content in gasoline, and many Brazilian "flex-fuel" vehicles can burn any ethanol–gasoline blend, including 100% ethanol.

Brazil's ethanol comes largely from sugarcane, which eliminates several fossil fuel–intensive steps needed to process starch or cellulose into sugar for fermentation. As a result, Brazilian ethanol has a much better energy balance, with the ratio of ethanol energy to fossil energy input in the range of 4 to as high as 10. Many Brazilian ethanol-conversion facilities use electrical energy generated on-site from burning sugarcane waste, thus wringing out still more energy and sending surplus electricity into the power grid. In Brazil, the cost of ethanol today is essentially competitive with fossil gasoline. Brazil's experience bodes well for other sunny, tropical countries. India, in particular, has significant potential to transport its billion-plus population with ethanol. World ethanol production has grown rapidly for several decades, with the United States and Brazil accounting for nearly 90% of the world's total production (Fig. 10.18).

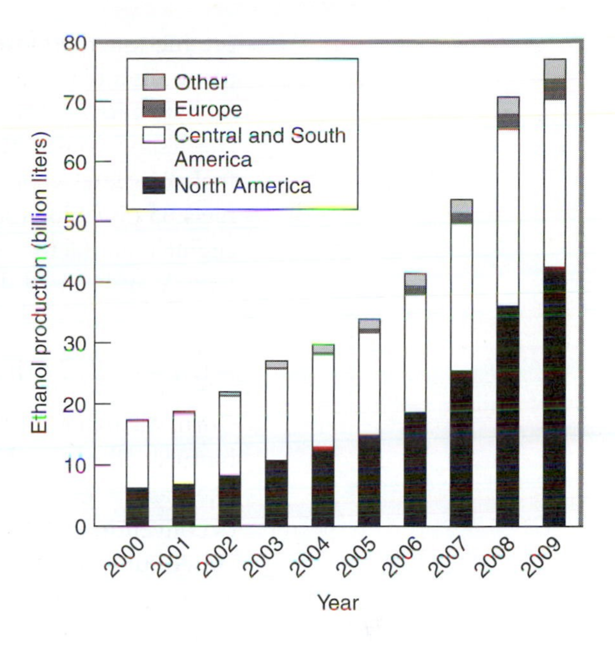

FIGURE 10.18
World ethanol production more than quadrupled in the early twenty-first century. The United States and Brazil are the dominant producers.

Biodiesel is the second biofuel in significant use today. In contrast to ethanol, it's most popular in Europe (Fig. 10.19). That's not surprising, because half the cars sold in Europe are diesels, and biodiesel can replace conventional fossil-based diesel fuel without engine modifications. Comparing Figures 10.18 and 10.19 shows that world biodiesel production is only a quarter that of ethanol—a situation that isn't likely to change, and for good reasons.

Biodiesel, like the oils used in cooking, derives from natural plant oils and fats. All plants contain these substances, which are concentrated in seeds and nuts. In most plant species this means the majority of the plant material is not available for biodiesel production, although nearly the entire plant can be made into ethanol. Therefore, any large-scale effort to capture solar energy for biofuels is likely to concentrate on ethanol, not biodiesel. Still, biodiesel remains a valuable and viable alternative fuel, originating not only in fresh plant material but also from waste oils and fats. In the United States, for example, the main sources of biodiesel are soybeans and so-called yellow grease, comprised of recycled cooking oils. Many U.S. biodiesel plants can switch among these two feedstocks with no equipment changes. Biodiesel's energy ratio is somewhat better than that of corn-based ethanol, with biodiesel returning about twice the fossil energy it takes to make it.

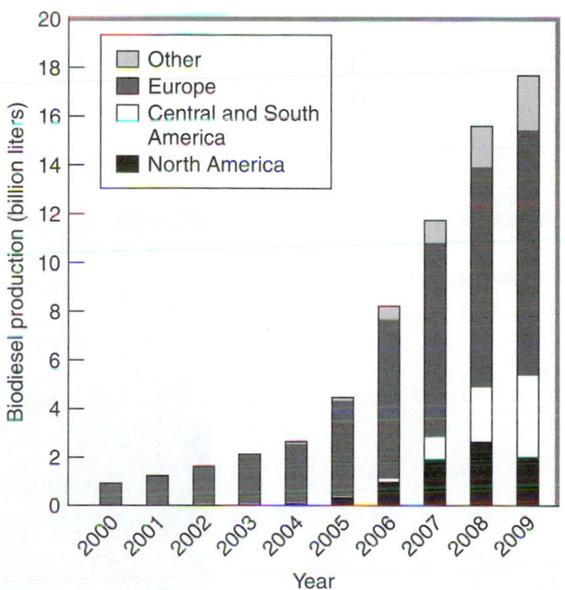

FIGURE 10.19
World biodiesel production is rising rapidly, with Europe the dominant producer.

It's possible to burn conventional cooking oils in modified diesel engines, and alternative-energy advocates are fond of vehicles powered by used oils from the deep-fat fryers of fast-food restaurants. However, cooking oils are more viscous than diesel fuel, due to the presence of glycerin, which gums up engines and reduces performance. Commercial biodiesel production therefore uses a process called *transesterification*, which reacts plant oils and fats with methanol or ethanol to remove the glycerin (itself a useful byproduct). Once broken down into smaller individual molecules, the result is a less viscous fuel with properties very similar to petroleum-derived diesel fuel. Biodiesel is often blended with conventional petrodiesel to produce the fuels B5 (5% biodiesel), B20 (20%), and B100 (pure or "neat" biodiesel). Some engine manufacturers currently approve blends up to B20, although many individuals and vehicle fleet owners use B100 with no harm to their engines.

ADVANCED BIOFUELS

Corn, sugarcane, and existing cellulosic plants aren't the only possible sources of biofuels. Photosynthetic algae produce oils that can be refined into petroleum-like fuels. Algae grow rapidly and could yield 4 to 10 times the biofuel per acre of conventional biofuel crops. They can be grown outdoors in open ponds or in closed tube-like structures that resemble solar collectors (Fig. 10.20). Algae are aquatic, but they can thrive on wastewater or other water unsuitable for crop or food production. Algae-derived biofuels remain experimental, with many technical and economic challenges to be solved. But the promise of algae is great enough that major oil producers are teaming with biotech companies and universities to explore algae-sourced biofuels.

A more futuristic approach is genetic engineering of algae and other species so that they produce biofuels directly. If successful, such an approach would bypass expensive and energy-intensive processing of plant material into biofuels. More

FIGURE 10.20
This experimental algae bioreactor resembles a solar collector—which, in a sense, it is.

modest genetic engineering could increase plants yields of biofuel feedstocks such as oils and cellulosic matter and/or modify plant-produced feedstocks so they require less processing into biofuels. In the same way that photovoltaic solar energy builds on the electronics industry, so would genetically engineered biofuel production build on the booming biotechnology industry.

ENVIRONMENTAL IMPACTS OF BIOMASS ENERGY

Burning biomass and its derivatives is like burning fossil fuels—converting hydrocarbons to carbon dioxide, water, and pollutants that include products of incomplete combustion and oxides of sulfur, nitrogen, and other substances. There are also substantial differences, however, nearly all of which favor biomass.

The most important distinction is in carbon emissions. Biomass burning necessarily produces CO_2, just like fossil fuel. But the carbon from biomass was only recently removed from the atmosphere by growing plants, and natural decay would have soon returned it to the atmosphere even if the biomass hadn't been burned. For biomass harvested on a sustainable basis and without fossil fuel required in harvesting and processing, therefore, the rate at which new plants remove carbon from the atmosphere would be equal, on average, to the rate of carbon emission from such biomass burning, which would make the biomass **carbon neutral**. *Sustainable* is the key word here; wholesale burning of biomass without replanting does result in net carbon emissions to the atmosphere—a continuing problem with "slash and burn" agriculture still practiced in tropical rainforests. Even sustainable practices initiated in forest ecosystems can take decades to achieve true carbon neutrality.

The combustion of wood in bulk can be a messy process unless burning is carefully controlled. Bulk wood is used mostly for residential heating, where high levels of particulate matter and organic chemicals can result from incomplete combustion, exceeding even the levels from coal. In the United States, newer stoves and fireplace inserts meeting EPA standards cut particulate emissions in half, to levels somewhat below coal but still far exceeding other fossil fuels. European standards go much further, mandating proper seasoning of firewood, thermal storage systems to minimize variations in combustion rate, separate heat-exchange and combustion chambers, and other design and operating techniques that make European wood burning a much cleaner process. Because stoves and fireplaces last a long time, it will be decades before all domestic wood burning is done in devices meeting modern standards.

Harvesting and transporting firewood consumes some fossil fuel, but on average wood used for heating returns some 25 times as much energy as its production requires in fossil fuels; thus fossil emissions associated with wood burning are minimal. However, cutting and burning of older trees can result in a rapid release of stored carbon that even vigorous new growth can't compensate for on timescales shorter than decades; as a result, firewood harvesting as it's now done in temperate forests isn't necessarily carbon neutral.

Wood burned as chips or pellets in industrial applications is another matter; more regulated burning conditions, wood gasification technologies, and the use of pollution-control devices reduces particulates and ensures more complete combustion. Wood burning—whether in domestic stoves or industrial boilers—has lower sulfur emissions than any of the fossil fuels, some 40% less than natural gas and over 90% less than coal. On the other hand, although chipping and pelletizing branches or other logging waste may sound like a good idea, taking entire trees robs the forest of nutrients that would be returned to the soil through natural decay, so the use of logging waste for energy may not be sustainable in the long term.

Liquid biofuels have clear environmental benefits. Unlike firewood, they're made from short-lived crops that quickly reach carbon neutrality through sustainable agriculture. Even modest ethenol blends such as E10 decrease many automotive emissions. The extra oxygen in ethanol (C_2H_5OH), compared with gasoline's octane (C_8H_{18}), helps to promote more complete combustion, thereby reducing carbon monoxide emissions by 25% or more. In fact, the original motivation for ethanol blends in the United States was not so much replacement of fossil fuels as controlling emissions. Ethanol also reduces particulate emissions, tailpipe emissions of volatile organic compounds (VOCs), and several carcinogenic compounds such as benzene. However, ethanol itself is more volatile (easily evaporated) than gasoline, so the use of ethanol increases somewhat the total emission of VOCs. That's because decreased VOC tailpipe emissions are more than offset by evaporation in fuel handling, dispensing, and storage.

Automotive emissions of NO_x don't change much with ethanol, but nitrogen-based fertilizer used to grow ethanol feedstock crops results in a net increase in NO_x emissions over the entire ethanol fuel cycle. Greenhouse-gas emissions don't fare much better. The considerable fossil fuel required to produce ethanol from corn results in substantial greenhouse emissions—so much that with current production methods in the United States, ethanol results, on average, in only a 19% reduction in greenhouse emissions compared with fossil gasoline. Sugarcane-based Brazilian ethanol does a lot better, reducing greenhouse emissions by nearly 80%.

The emissions benefits of biodiesel are more pronounced. Biodiesel contains virtually no sulfur, whereas even low-sulfur petrodiesel contains up to 350 ppm of sulfur. Particulates drop substantially, along with carbon monoxide and other products of incomplete combustion. Biodiesel is a better lubricant than petrodiesel, reducing engine wear. And biodiesel has a higher cetane number (the diesel equivalent of octane number), meaning it provides more reliable ignition in diesel engines. Producing biodiesel from waste grease disposes of a significant waste product; indeed, New York City could produce enough biofuel from its waste grease to power its municipal bus fleet five times over. One downside is slightly increased NO_x emissions with biodiesel, a level already high in diesels because of their higher operating temperature. And both biodiesel and ethanol can corrode the rubber seals used in automotive fuel systems. Like ethanol, biodiesel results in lower greenhouse emissions than petroleum-derived fuel, even accounting for the energy needed to make biodiesel. But for biodiesel the reduction is much greater than for ethanol, with greenhouse emissions as much as 80% below those produced by conventional petrodiesel.

Using existing cropland for biofuel crops runs up against the food needs of the world's still growing population, and developing new croplands brings all the attendant environmental problems of modern agriculture—increased use of fertilizers and pesticides, soil erosion, appropriation of limited water supplies, water pollution from runoff, and loss of biodiversity. If biomass were to make a serious contribution to the world's energy supply, then land-use issues would quickly become the primary environmental concern. You can show in Research Problem 4 that replacement of all U.S. gasoline with corn-based ethanol would require more than the country's entire cropland; Exercise 14 shows the same for European biodiesel. More efficient conversion processes, the use of grasses in place of corn, and greater use of plant waste from food-producing agriculture could all reduce the land area needed. Nevertheless, a realistic look at land requirements for biofuel production suggests that a plan to replace 10% of U.S. gasoline and diesel consumption by 2020 with ethanol and biodiesel might require more than 40% of U.S. cropland. Even the more optimistic projections run into land-use limitations unless they're coupled with substantial gains in vehicle energy efficiency. Table 10.2 summarizes some environmental impacts of biofuels as compared with fossil fuels.

TABLE 10.2 | ENVIRONMENTAL IMPACTS OF BIOFUELS COMPARED WITH GASOLINE AND PETRODIESEL

	Corn ethanol	Sugarcane ethanol	Biodiesel from oilseed rape
Particulate emissions	Lower	Lower	45% lower
Carbon monoxide emissions	25% lower with E10 blend	Lower	45% lower with B100
Volatile organics	Tailpipe: lower Fuel handling: higher Net: higher	Net higher	Lower
Sulfur emissions	~0	~0	~0 (much lower than the 350 ppm for low-sulfur petrodiesel)
Nitrogen oxide emissions	Higher	Higher	Higher
Toxicity	Lower	Lower	Lower
Energy gain $\left(\frac{\text{Energy from biofuel}}{\text{Fossil energy to produce biofuel}}\right)$	1.1	4	2
Greenhouse emissions (equivalent CO_2)	13% higher to 39% lower	20%–80% lower	55% lower

Note: All numbers are approximate, with greenhouse emissions strongly dependent on fuel production methods.

PROSPECTS FOR BIOMASS

Biomass enthusiasts see biomass energy, especially biofuels for transportation, as a major player in a renewable energy future. Maybe—but, at least with current technologies, the fossil energy needed to produce most liquid biofuels means that these fuels aren't truly renewable. And land-use conflicts cloud the future for biomass energy from plants grown exclusively as energy crops. It seems more likely that biomass will play a modestly significant role, more in some countries than in others. In the vital area of transportation fuels, we'll be well served by continuing to develop more efficient technologies for production of liquid biofuels. But in the end, the low efficiency of photosynthesis on an already-crowded planet means that biomass can't be our sole path away from fossil fuels.

10.4 Other Indirect Solar Energy

The same solar-induced temperature differences that power the wind also, in conjunction with differences in salt concentration and Earth's rotational energy, drive the great ocean currents. And wind itself produces ocean waves, providing a doubly indirect form of solar energy. I briefly described schemes for harnessing the kinetic energy of waves and currents in Chapter 8's section on tidal energy, because these sources of mechanical ocean energy have much in common with tidal energy, except that their ultimate origin is primarily in sunlight rather than the mechanical energy of the Earth–Moon system.

OCEAN THERMAL ENERGY CONVERSION

As the Sun warms the tropical ocean, it creates a significant temperature difference between the surface waters and deeper water to which sunlight doesn't penetrate. As shown in Chapter 4, any time there's a temperature difference, there's the potential to run a heat engine and extract mechanical energy. **Ocean thermal energy conversion (OTEC)** harnesses this energy, in most cases to generate electricity. The thermodynamic efficiency limit expressed in Equation 4.5, $e = 1 - T_c/T_h$, shows that we get the highest efficiency with the largest possible ratio of T_h to T_c. Practically speaking, this limits the OTEC energy resource to the tropics, where the temperature difference from surface to depth is greatest. Tropical surface temperatures can exceed 25°C, while a few hundred meters down the temperature is around 5°C to 6°C. You can show in Exercise 15 that the thermodynamic efficiency limit for a heat engine operating between these temperatures is only about 7%, but with no fuel to pay for, this number isn't the liability it would be in a fossil-fueled or nuclear power plant. And the OTEC energy resource is vast; after all, oceans absorb much of the 174 PW of energy that the Sun delivers to Earth. Most of that energy is out of our reach, but one serious estimate suggests that OTEC has the potential to produce as much as 10 TW, which is close to humankind's total energy-consumption rate. However, practical considerations suggest a far smaller OTEC contribution. The one significant advantage that OTEC has over solar-based schemes such as photovoltaic conversion and wind is its nearly constant availability. The large heat capacity of water smoothes out day/night and seasonal temperature variations, which would allow OTEC plants to operate continuously to supply baseload power.

OCEAN THERMAL ENERGY CONVERSION TECHNOLOGY

An OTEC plant, in principle, operates like any other thermal power plant: An energy source heats a working fluid, which boils to form gas that drives a turbine. Contact with a cool reservoir recondenses the working fluid, and the process repeats. With OTEC the heat source is warm surface water, and the cool reservoir is the colder water pumped from depths of up to 1 km. Because water boils at 100°C at normal pressures, it's necessary either to operate the heat engine at lower pressure or to use a working fluid with a lower boiling point.

Future OTEC plants have been envisioned as offshore platforms similar to oil-drilling platforms, or as free-floating structures (Fig. 10.21). Another possibility is OTEC-powered factory ships that would "graze" the tropical oceans, seeking out regions with the greatest temperature differences. The energy generated would be used for energy-intensive manufacturing done onboard the ship. Or OTEC-generated electricity could split water into hydrogen and oxygen, producing fuel for a hydrogen-powered economy. If used to desalinate seawater, OTEC technology could help the world satisfy its growing thirst for freshwater. On tropical coasts adjacent to deep ocean water, OTEC desalinization or electricity-generating plants might be constructed onshore, with warm and cold water piped in from the ocean.

OTEC faces numerous challenges that affect its practicality and economic viability. Designing structures to withstand the marine environment is difficult. The energy required to pump cold water from great depths reduces practical OTEC efficiencies to a mere 2% or so. Although fuel costs aren't an issue, such a low efficiency means a huge investment in physical structure returns a relatively paltry flow of energy. For these reasons, OTEC development to date has been limited to small (tens of kilowatts) pilot projects constructed, and mostly abandoned, in the late 1970s to early 1990s. Still, OTEC may someday provide a viable energy source, especially for tropical islands.

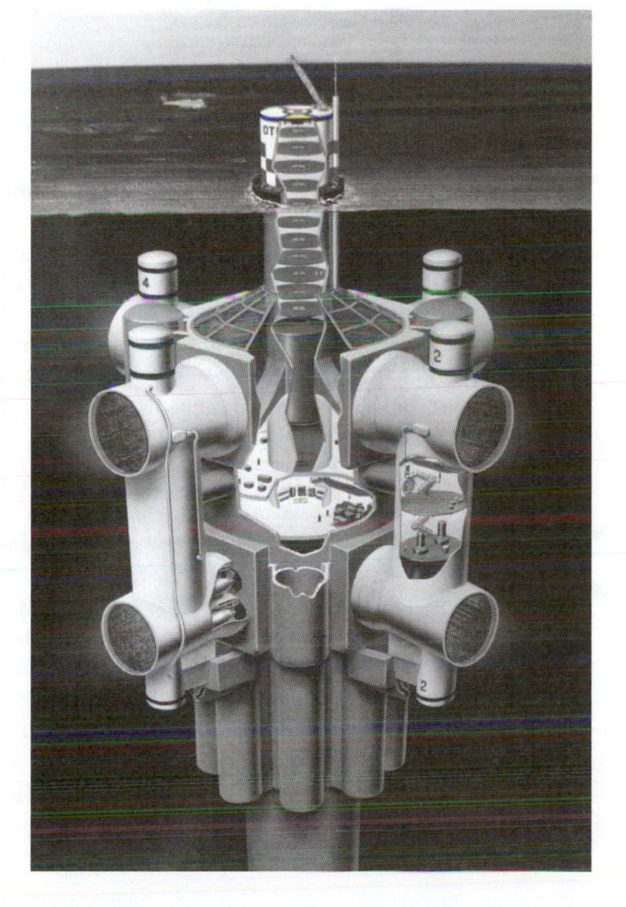

FIGURE 10.21
Artist's conception of a floating OTEC power station. The pipe at the bottom extends into deeper, cooler waters.

ENVIRONMENTAL IMPACTS OF OCEAN THERMAL ENERGY CONVERSION

Isolated OTEC facilities would have little environmental impact, although the mixing of warm and cool ocean waters changes the local temperature and could affect marine life. Nutrients brought up with the cooler deep water might enhance fish production, or lead to algae blooms. But large-scale OTEC power production could have serious effects by lowering the average ocean surface temperature and thus affecting weather, climate, and ocean circulation. A total of 60 GW of OTEC power worldwide—three times the power output of China's Three Gorges dam—would entail cold-water flows equal to all the world's rivers. We're unlikely to see anything like that level of OTEC development, which means we probably don't need to worry about global environmental impacts. But at the same time, it means that OTEC won't contribute substantially to the global energy supply anytime soon.

CHAPTER REVIEW

BIG IDEAS

Indirect solar energy includes flowing water, wind, and biomass.

10.1 **Hydropower** has a long history, and today it supplies some 16% of the world's electrical energy, mostly from dams built on rivers. Hydropower represents energy of the highest quality, and hydroelectric power generation produces essentially no pollution. But hydropower dams alter river ecosystems, and tropical installations may emit methane that contributes to climate change. The hydroelectric resource is almost fully utilized, especially in the developed world.

10.2 Roughly 1% of the incident solar energy goes into wind. The power available from wind increases as the cube of the wind speed, so high-wind areas are particularly promising sites for wind installations. Modern wind turbines have electric power outputs of several megawatts, and they're often grouped into wind farms with tens, hundreds, or thousands of turbines. Wind's predominant environmental impact is aesthetic, although wind turbines may also kill migrating birds and bats. Today, wind is the fastest-growing energy source, and provides some 2% of the world's electrical energy. Unlike direct solar alternatives, wind is close to economically competitive with conventional energy sources.

10.3 **Biomass** contains solar energy captured by photosynthetic plants. Biomass may be burned directly for heat or electric power generation. Decomposition of organic waste biomass releases combustible methane. Fermentation and chemical processing yield **biofuels**, of which **ethanol** and **biodiesel** are most important. Biomass and biofuel combustion produce air pollutants, although for some pollutants the amounts are lower than with fossil fuels. Depending on how it's grown and harvested, biomass may be **carbon neutral**, contributing no net carbon to the atmosphere. But unsustainable or energy-intensive harvesting and processing mean that some biomass operations entail significant carbon emissions.

10.4 Other indirect solar energy sources include waves and ocean currents, as described in Chapter 8. **Ocean thermal energy conversion** schemes operate heat engines between warm tropical surface waters and cooler deep water. None of these approaches has yet demonstrated large-scale practicality.

TERMS TO KNOW

biodiesel (p. 280)
biofuel (p. 279)
biomass (p. 260)
carbon neutral (p. 283)
ethanol (p. 280)
gross primary productivity (p. 276)
head (p. 265)
hydrologic cycle (p. 262)
hydropower (p. 260)
net primary productivity (p. 277)
ocean thermal energy conversion (OTEC) (p. 286)
power coefficient (p. 273)
power curve (p. 273)
primary productivity (p. 276)
transpiration (p. 262)

GETTING QUANTITATIVE

Solar energy used to evaporate water: 23%

Solar energy driving wind and currents: ~1%

Solar energy captured by plants: ~0.08%

Hydroelectricity: ~6% of world energy, 16% of world electricity

Power output of largest hydroelectric dams: 10–20 GW

Wind power per unit area: $\frac{1}{2}\rho v^3$ (Equation 10.1; p. 269)

Maximum power extractable with wind turbine: $2a(1-a)^2\rho v^3$ (Equation 10.2; p. 272)

Efficiency of wind turbine, theoretical maximum: 59%

Power output of single large wind turbine: ~5 MW

Photosynthesis, net result: $6H_2O + 6CO_2 + 4.66\ \text{aJ} \rightarrow C_6H_{12}O_6 + 6O_2$ (Equation 10.3; p. 276)

Gross primary productivity of the world's ecosystems: 133 TW

Net primary productivity: about half of gross

Energy content of dry biomass: 15–20 MJ/kg

QUESTIONS

1 As used today, are the indirect solar energy sources of water, wind, and biomass truly renewable? Answer separately for each.

2 Hydroelectric dams can, surprisingly, have significant greenhouse emissions. What's the source of these emissions?

3 Why can't a wind turbine extract all of the wind's kinetic energy?

4 Why is the curve in Figure 10.13 so steep in the region below the rated speed?

5 Tailpipe emissions of VOCs are lower for ethanol than for gasoline, but overall VOC emissions are higher with ethanol fuels. Why?

6 What accounts for the difference between gross and net primary productivity?

EXERCISES

1 It takes approximately 2.5 MJ to convert 1 kg of liquid water to vapor at typical atmospheric temperatures (this is the latent heat introduced in Section 4.5). Compare this quantity with the gravitational potential energy of 1 kg of water lifted to a typical cloud height of 3 km.

2 Given that 78 W/m^2 of solar energy goes into evaporating water, use the latent heat given in the preceding problem to estimate Earth's total average rate of evaporation in kilograms of water per second.

3 The average flow rate in the Niagara River is 6.0×10^6 kg/s, and the water drops 50 m over Niagara Falls. If all this energy could be harnessed to generate hydroelectric power at 90% efficiency, what would be the electric power output?

4 Estimate the water head at Hoover Dam shown in Figure 10.7b, given that it produces 2.1 GW of electric power from a flow of 1,800 m^3/s. Assume an energy conversion efficiency of 75%.

5 Repeat the previous problem for the John Day Dam (Fig. 10.7a), assuming comparable electric power, a flow rate of 9,100 m^3/s, and 82% conversion efficiency.

6 By what factor must the wind speed increase in order for the power carried in the wind to double?

7 What power coefficient would a wind turbine need in order to extract 350 W from every square meter of wind moving past its blades when the wind is blowing at the minimum speed for class 7?

8 The density of air under normal conditions is about 1.2 kg/m^3. For a wind speed of 10 m/s, find (a) the actual power carried in the wind; (b) the maximum possible power (Betz limit) extractable by a wind turbine with a blade area of 10 m^2; and (c) the actual power extracted by a wind turbine with a blade area of 10 m^2 and a power coefficient of 0.46.

9 What's the power coefficient of a 42-m-diameter wind turbine that produces 950 kW in a 14-m/s wind? The density of air under normal conditions is about 1.2 kg/m^3.

10 (a) Estimate the total energy produced by a wind turbine with the power curve shown in Figure 10.13 during a day when the wind blows at 2 m/s for 6 hours, at 10 m/s for 6 hours, at 15 m/s for 6 hours, and at 25 m/s for 6 hours. (b) What's the turbine's average power output over this day?

11 If you've had calculus, you know that you can find the maximum or minimum of a function by differentiating and setting the derivative to zero. Do this for Equation 10.2, and show that the resulting maximum occurs when $a = 1/3$ and that the maximum is equal to 16/27 of the total wind power $\frac{1}{2}\rho v^3$.

12 The United States generates some 230 million tonnes of solid waste each year, with an energy density (when dry) essentially the same as wood, around 20 MJ/kg. Find the equivalent in barrels of oil, and compare it with U.S. oil imports of around 4.5 billion barrels per year.

13 If all the solid waste of Exercise 12 were burned in waste-to-energy power plants that are 20% efficient, (a) how many kilowatt-hours of electrical energy could be produced in a year? Compare this quantity with the total U.S. electrical energy production of approximately 3.8 trillion kWh annually. (b) What would be the equivalent average electric power output from all those waste-burning power plants? How many 1-GW coal-fired plants could they displace?

14 Figure 10.19 suggests biodiesel production of around 10 billion liters (L) for Europe in 2009, which compares with a total diesel fuel consumption of about 150 billion L annually. European biodiesel production yields about 1.23 kL per hectare (ha) of cropland. (a) What fraction of Europe's 49 Mha of total cropland is now used for biodiesel production? (b) How much land would it take to replace all of Europe's petrodiesel with biodiesel, considering that the energy yield per liter of biodiesel is 10% lower than for petrodiesel?

15 Find the thermodynamic efficiency limit for an OTEC heat engine operating between surface waters at 25°C and deep water at 5°C.

16 Typical freshly cut "green" firewood contains about 50% water by weight; for seasoned wood the comparable figure is around 20%. Firewood is usually sold by volume, so compare the energy available from a given volume of green wood with that available from seasoned wood. Assume an energy content of 20 MJ/kg for perfectly dry wood, and 2.3 MJ/kg of energy needed to vaporize water. Also assume that the wood has essentially the same volume whether green or seasoned.

RESEARCH PROBLEMS

1 Download the latest version of the U.S. Energy Information Administration's "Existing Electric Generating Units in the United States" spreadsheet. Find (a) the largest hydroelectric installation in your state (you may need to add together multiple generating units at the same site); (b) the smallest hydroelectric installation in your state; and (c) the total hydroelectric generation capacity in your state.

2 Estimate the number of powerboats operating on Lake Mead, above the Hoover Dam, and assuming an average of 100 horsepower per boat, estimate the total power output of all these boats. Is it at all comparable to Hoover Dam's approximately 2 GW of electric power output? If it is, then the indirect environmental impact of this dam may be comparable to that of a fossil plant with the same output! You might start with the National Park Services Environmental Impact Statement for its Lake Management Plan for Lake Mead.

3 Has the growth of wind energy shown in Fig. 10.15 continued beyond 2010? Back up your answer with data from an authoritative source.

4 Find the total annual gasoline consumption in the United States, along with the country's total cropland. One study suggests that corn-to-ethanol production in the United States gives an annual yield of 71 GJ of ethanol energy per hectare (metric acre; 10^4 m^2) of cropland. Use your findings, along with Table 3.3, to estimate the percentage of U.S. cropland that would be required to replace all U.S. gasoline with corn-based ethanol.

ARGUE YOUR CASE

1 One environmentalist argues that wind turbines belong in the Great Plains but not on New England's scenic mountain ridgetops. Another claims that ridge-top wind turbines are preferable to the nuclear and gas power plants that supply much of New England's electricity. How do you argue?

2 Your university is planning to replace its oil-fired heating plant with a wood-fired boiler supplied by wood chips produced from logging waste. The university claims it will then be "100% carbon neutral." Formulate an argument supporting or opposing this claim.

3 Is it ethical to use cropland for biofuel production in a world where some of the population still does not have adequate food? Argue your view on this question.

Chapter 11

ENERGY CARRIERS: ELECTRICITY AND HYDROGEN

Chapters 5 through 10 explored fossil fuels, nuclear energy, geothermal and tidal energy, and solar energy in both direct and indirect forms. Those are the energy sources available from Earth's energy endowment, which I first introduced in Chapter 1. There are no others: All the energy sources available to humankind come ultimately from solar, geothermal, and tidal energy flows, or from fossil and nuclear fuels stored within our planet.

What about electricity? It's among the most familiar and versatile forms of energy, performing a wide range of tasks in our homes, in commercial establishments, in industry, and, to a lesser extent, in transportation. Chapter 3 described the fundamentals of generating electricity through the process of electromagnetic induction, and Chapter 4 emphasized thermodynamic limitations on the efficiency of electric power generation in thermal power plants. You saw fossil and nuclear electric power plants in Chapters 6 and 7, while Chapter 8 considered electricity from geothermal and tidal energy sources; Chapter 9 included photovoltaic and solar-thermal electricity, and Chapter 10 showed how we generate electricity from water and wind. So electricity has been very much with us as we've explored our energy-source options. But electricity is *not* an energy source, because it's not part of Earth's energy endowment. There's no ready-made natural flow of electricity on or inside Earth that we can tap into.

(There is a little atmospheric electricity, in a global circuit charged by thunderstorms, but it would be challenging to harness and is quantitatively far from sufficient to meet our needs.)

A clarification: When I'm talking here about electricity or electrical energy, I mean electricity in the form we commonly use it—namely, the flow of electric current through wires and other electrical conductors. As you saw in Chapter 3, the energy stored in electromagnetic fields is ubiquitous. Fossil and other chemical fuels store energy in the electric fields associated with their molecular configurations. Solar energy is ultimately in the electric and magnetic fields of the electromagnetic waves that comprise light. The energy associated with electric currents is also electromagnetic field energy, in this case in fields that surround conducting wires—and that's the form of electromagnetic energy I mean here when I say "electricity."

If electricity isn't an energy source, then what is it? It's an **energy carrier**—a form of energy that we generate from a true energy source and then deliver for our end uses. Electricity has many virtues. It's energy of the highest thermodynamic quality, on a par with mechanical energy. That means we can easily convert electricity into any other form of energy with, in principle, 100% efficiency. Electricity is also easy to transport—all it takes is a thin conducting wire—and it travels at nearly the speed of light. Ease of conversion and transport are what make electricity the most versatile form of energy, and one whose importance continues to grow (Fig. 11.1).

Another energy carrier is hydrogen, in the chemical form H_2. If you follow energy issues or are interested in alternative energies, then you've probably heard proposals for a *hydrogen economy*, or maybe you've seen advertising featuring futuristic cars filling up on clean, carbon-free, nonpolluting hydrogen. Hydrogen is an energy carrier, not an energy source, because H_2 doesn't occur naturally on or within Earth. We have to process hydrogen-containing compounds to make H_2, just like we have to generate electricity. And making hydrogen, like generating electricity, takes at least as much energy as we're going to get out when we use it.

So we have two energy carriers, electricity and hydrogen. They have in common that neither is naturally available, but each needs to be produced using other energy sources. Otherwise they're very different: Electricity is electromagnetic energy flowing at nearly the speed of light, while hydrogen is a storable, transportable, burnable chemical fuel. This chapter explores both.

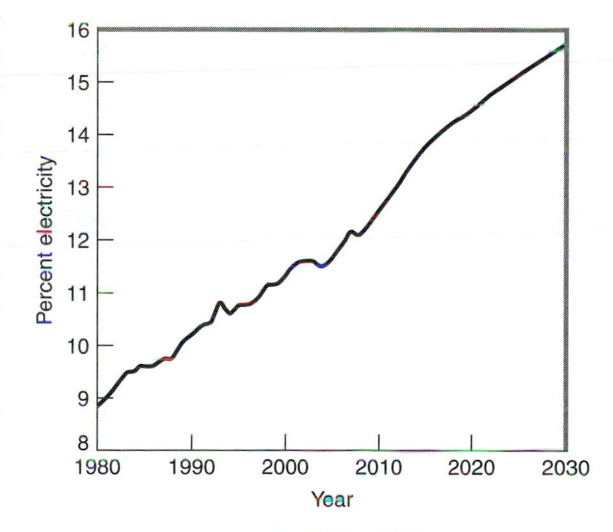

FIGURE 11.1
The percentage of the world's energy that's consumed in the form of electricity has grown steadily, and projections show that trend continuing.

11.1 Electricity

Some 36% of the world's primary energy is used to generate electricity, but inefficiencies and losses mean that we discard as waste some two-thirds of that 36%. As Figure 11.1 shows, the result is that only 12% of the energy we actually consume for end uses is in the form of electricity. On this measure the United States isn't much different from the world average: Generating electricity accounts for nearly 40% of U.S. primary energy, but electricity comprises only 13% of our end-use consumption.

You saw in Chapter 2 that the world runs on fossil fuels, and that's true as well for the electric power sector. But where oil is the most widely used fossil fuel overall (see Figure 2.6), coal dominates in electric power generation, followed by natural gas. Nuclear and hydropower play significant roles as well. But there's remarkable variation in the mix of sources used to generate electricity throughout the world, depending on a country's resources and on policy decisions—such as France's embrace of nuclear power. Figure 11.2 shows the mix of electrical energy sources in the world and in several countries and U.S. states. Those differing mixes mean that pollution, carbon emissions, and climate impact vary significantly among countries and states.

THE ELECTRIC POWER GRID: BEGINNINGS

You've seen that most of our electricity comes from generators—devices that convert mechanical energy into electrical energy. It doesn't matter, from an electrical standpoint, whether a generator is driven by steam from fossil fuel combustion, nuclear fission, or solar-thermal collectors, or by wind or water. Electricity can also come from other processes, like direct conversion of solar energy to electricity in photovoltaic panels. The source of electricity isn't all that important to a power engineer: What matters in today's interconnected electrical system is that electricity from these diverse sources can be joined into a common flow of energy.

"Interconnected" is a key word here. Today, entire continents function as unified electrical systems—complex interconnections of generators, transmis-

FIGURE 11.2

Sources of electrical energy vary dramatically from country to country, and from state to state in the United States. Pie graphs show electricity *generation* within a given country or state; *consumption* percentages can differ substantially because of electricity imports and exports. Not shown is the breakdown for fossil sources, which leads to greatly differing carbon intensities, and which you can explore in Research Problem 4.

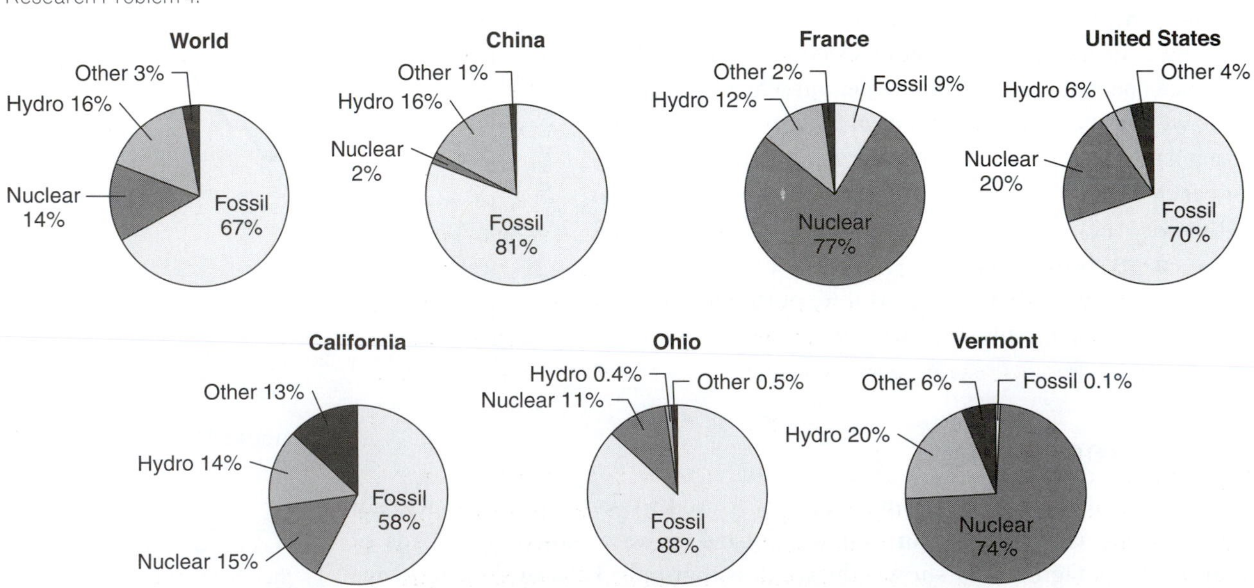

sion and distribution lines and associated equipment, and end users. This interconnected system is the **electric power grid**, or simply the grid. Understanding electricity and its efficient use means understanding how the grid works and how we might make it better and smarter.

The world of electricity wasn't always so interconnected. The earliest electric power systems consisted of individual generators directly supplying individual end users. If your generator failed, there was no backup. But inventor Thomas Edison had a better idea. In 1882 he completed the first real power grid—a system of six steam-powered generators, underground distribution lines, and end users. On September 4, 1882, Edison threw a switch that energized this first grid and brought electric light to lower Manhattan. With the advent of the grid, customers worried less about the source of their power. And the Edison Electric Illuminating Company—forerunner of New York City's present-day Consolidated Edison—had the flexibility and redundancy that came from multiple generators and wiring connections.

Edison's system used direct current (DC), a one-way flow of electrons in the wires. Although Edison's grid grew rapidly through the 1880s, his DC technology was soon superseded. Alternating current (AC), advocated by Nikola Tesla and George Westinghouse, had the advantage that it was easy to change voltage levels on the grid, stepping it up for long-distance transmission, then down to safer levels for end users. By the beginning of the twentieth century most electric power systems used alternating current.

VOLTS, AMPS, OHMS, WATTS

High-voltage electric transmission lines are a common sight in any industrialized country. Appreciating the reason for high-voltage transmission requires an understanding of a few simple electrical concepts and their quantitative relationships.

Voltage (symbol V) is analogous to the pressure in a water system; it provides the "push" that drives electrons through conducting wires. More precisely, voltage is a measure of energy gained or lost per unit of electric charge as charge moves between two points. The SI unit of voltage is the **volt** (symbol V), named in honor of the Italian physicist Alessandro Volta, who developed the electric battery in 1800. The actual flow of electrons or other charges is electric current, quantified as charge flow per time. In the SI system, current (symbol I) is measured in **amperes** ("amps," symbol A), named for the French physicist and mathematician André-Marie Ampère, a contemporary of Volta's who did pioneering research in electromagnetism. Multiply voltage by current, and you get (*energy/charge*) × (*charge/time*)—which reduces to *energy/time* or power:

$$P = IV \qquad \text{(electric power)} \qquad (11.1)$$

Equation 11.1 applies to *any* situation involving electric power. It could be power produced in a generator (see Example 11.1), lost to heat in a transmission line (Example 11.2), converted to mechanical power in a motor, or to light in a lightbulb. With voltage measured in volts and current in amperes, electric power comes out in the SI unit of power, namely watts (W). As I've emphasized before, the watt is a general unit expressing the rate of energy production or consumption; it's not restricted to electrical energy. But given Equation 11.1's convenient relation between voltage, current, and power, the watt is widely used to express energy rates in electrical systems.

EXAMPLE 11.1 | Electric Power

The Vermont Yankee nuclear power plant produces 650 MW of electric power from its single generator. The generator's output voltage is 22 kV. What's the generator current?

SOLUTION

Equation 11.1 gives the power in terms of voltage and current. In this case we want to solve for current:

$$I = \frac{P}{V} = \frac{650 \text{ MW}}{22 \text{ kV}} = \frac{650 \times 10^6 \text{ W}}{22 \times 10^3 \text{ V}} = 30{,}000 \text{ A}$$

That's 30 kiloamps (kA)—a huge current that requires enormously thick conductors to handle. By contrast, currents in typical household lamps and appliances range from a fraction of an amp to 10 amps or so. By the way, we could have just divided 650 by 22 to get 30, then multiplied by 1,000 from the division of mega (1 million) by kilo (1 thousand).

It takes voltage—energy per charge—to drive current because most electrical conductors exhibit **resistance** (symbol R) to the flow of current. Resistance arises from collisions between electrons and atoms in a conductor. These collisions degrade high-quality electrical energy into low-quality thermal energy. In most cases we want to avoid this energy loss; an exception is in electric heating devices like toasters, stove burners, and water heaters. For a given voltage across a conductor, a higher resistance results in lower current. This relation is quantified in **Ohm's law**:

$$I = \frac{V}{R} \quad \text{(Ohm's law)} \qquad (11.2)$$

The SI unit of resistance is the **ohm** (symbol Ω), after the nineteenth century German physicist Georg Simon Ohm.

We're now ready to see the advantage of high voltage for electric power transmission. Figure 11.3 is a simplified diagram showing an electric power plant supplying power to a city. Two wires comprise the transmission line connecting the power plant's generator to the city, forming a complete circuit. (Actually, most U.S. power distribution uses three wires, carrying alternating currents that are out of phase with each other. Figure 11.3 more closely resembles a DC transmission system.) The voltage across the generator is V, and the current in the wires is I. Their product, according to Equation 11.1, is the power $P = IV$ supplied by the power plant. We'd like all that power to end up delivered to the city, but in reality some will be lost in the resistance of the transmission line. How much? Applying Equation 11.2, Ohm's law, to the transmission line alone shows that the voltage between the generator end of each wire and the city end is IR, where R is the wire's resistance. Multiplying that voltage IR (which is *not* the same as the generator voltage V) by the current I in the wire gives the power dissipated in the wire: $P_{loss} = I^2R$. Since there are two wires, each with the same resistance and carrying the same current, the total power loss is twice this value. But here's the important point: For a given resistance, the power loss scales as the *square* of the current. Halve the current, and the power lost in the transmission line drops to one-fourth of its original value.

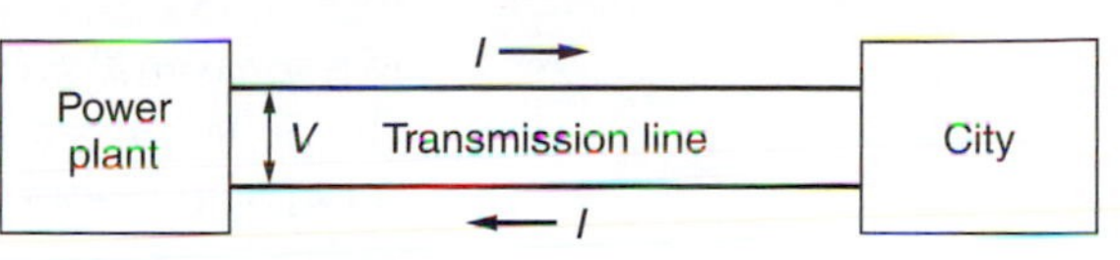

FIGURE 11.3
Simplified diagram showing a power plant, transmission line, and end user—in this case a city. Most AC systems carry "three-phase power" and are more complex, but DC transmission uses simple two-wire systems as shown here.

Equation 11.1 shows that we can achieve a given power output $P = IV$—set, in Figure 11.3, by the city's demand for electric power—with a low voltage V and high current I, or a high voltage V and low current I, or somewhere in between. But since the power loss in transmission is I^2R, we're clearly better off with a lower current I, and therefore a higher voltage V. That's the reason for high-voltage power transmission. You might wonder why we don't just lower the resistance R instead of the current I. We could—but then we only get a linear decrease in transmission loss rather than the quadratic (squared) decrease that results from lower current. More important, lower resistance means thicker wire or a better conductor—both of which are expensive. In practice, transmission-line voltages are determined by a mix of basic electrical considerations, practical limitations on handling high voltages, and economic factors.

EXAMPLE 11.2 | Transmission Losses

Suppose the power plant of Example 11.1 sends its power out over a two-wire transmission line like that in Figure 11.3, with each wire having resistance 2 Ω. What percentage of the plant's 650-MW output is lost in transmission if the voltage between the two wires of the transmission line is (a) 115 kV, the lowest level used for long-distance transmission in North America, and (b) 345 kV, typical of longer transmission lines?

SOLUTION

We've seen that the power loss in each wire is I^2R, where I is the current and R the wire's resistance; thus the total power loss from both wires is $2I^2R$. The power plant supplies power $P = 650$ MW at voltage V, for which we're asked to consider two values. By Equation 11.1, the current is $I = P/V$, so the power loss becomes

$$P_{loss} = 2I^2R = 2\left(\frac{P}{V}\right)^2 R$$

We're given $P = 650$ MW and $R = 2\ \Omega$; we then evaluate this expression to get the power loss at each of the two voltages: at 115 kV, $P_{loss} = 130$ MW, and at 345 kV, $P_{loss} = 14$ MW. As fractions of the total power, those losses are, respectively, an unacceptably high 20% and a more reasonable 2%. Actual long-distance transmission losses are typically 2% to 3% on the North American grid, rising to an average of 6.5% with lower-voltage distribution systems included. By the way, don't try to calculate the power loss for the generator's 22-kV output voltage—you'd get an absurd result because the voltage is too low to push the required 30 kA of current through the transmission line.

As Examples 11.1 and 11.2 make clear, the voltages produced by electric generators—and even more so the 120-V levels of household wiring—just aren't compatible with long-distance power transmission. So how do we change voltage levels? The answer lies in *electromagnetic induction*, a fundamental electromagnetic phenomenon introduced in Chapter 3 with the statement "induction entails the creation of electrical effects from *changing* magnetism." There I went on to show how electromagnetic induction is the basis of electric generators, which produce electric current by spinning coils of wire in the presence of magnets. In Chapter 3's introduction to electromagnetism, I also mentioned that electric currents give rise to magnetic fields—something you're familiar with if you've ever wound wire around a nail to make an electromagnet. So wind a coil of wire—call it the *primary coil*—and pass electric current through it; you've then got a magnetic field surrounding the coil. Make it *alternating* current, and you've got a *changing* magnetic field—a source of electrical effects. Put a *secondary coil* nearby, and the electrical effect consists of a voltage induced in the secondary coil, a voltage that's capable of driving current through a circuit connected across that coil. Your two-coil device is a **transformer**, used to change voltage levels on the power grid and in all sorts of electrical and electronic equipment. I'll spare you the mathematical details, but the voltage across the secondary coil is related to the voltage across the primary coil by the ratio of the number of turns in the secondary to that in the primary. Wind twice as many turns in the secondary as in the primary, and you've got a **step-up transformer**, whose output voltage is double the primary voltage. Wind half as many turns in the secondary, and your **step-down transformer** outputs half the primary voltage. In general, the secondary voltage V_2 is related to the primary voltage V_1 by

$$V_2 = \frac{N_2}{N_1} V_1 \tag{11.3}$$

where N_1 and N_2 are the numbers of turns in the primary and secondary, respectively.

Aren't we gaining (or losing) something here? Nope! You can get higher voltage out of your step-up transformer, but at the expense of lower current. In principle, the power $P = I_1 V_1$ supplied to the primary gets passed on to the secondary, where the same power emerges with higher (or lower) voltage and lower (or higher) current; the power $P = I_2V_2$ coming out of the secondary should be the same as what went into the primary. In practice, there's some loss in the resistance of the transformer windings, but modern power-grid transformers can hold that loss to less than 1%.

You're probably familiar with transformers, from the small plug-in "power bricks" that step down 120-V household electricity to operate your laptop computer, cell-phone charger, and other electronic devices; to the cylindrical transformer that hangs on a pole near your house and typically steps kilovolt-level power down to 120 V and 240 V for household use; to the large transformers at the utility substations that act as "off ramps" to deliver power from the grid to communities or industries. Figure 11.4 shows transformers and voltages on a typical grid.

You can see now why alternating current is essential to the operation of transformers and therefore of a power grid that steps voltages up and down: Transformers operate by electromagnetic induction, involving *changing* magnetism, and you don't get changing magnetism with steady, direct current.

THE MODERN GRID

Today's power grids are complex interconnections of generators, transformers, long-distance transmission lines and local distribution lines operating at a variety of voltages, supplying end users from homes to power-hungry industries. The interconnectedness of the grid brings redundancy and back-up capability,

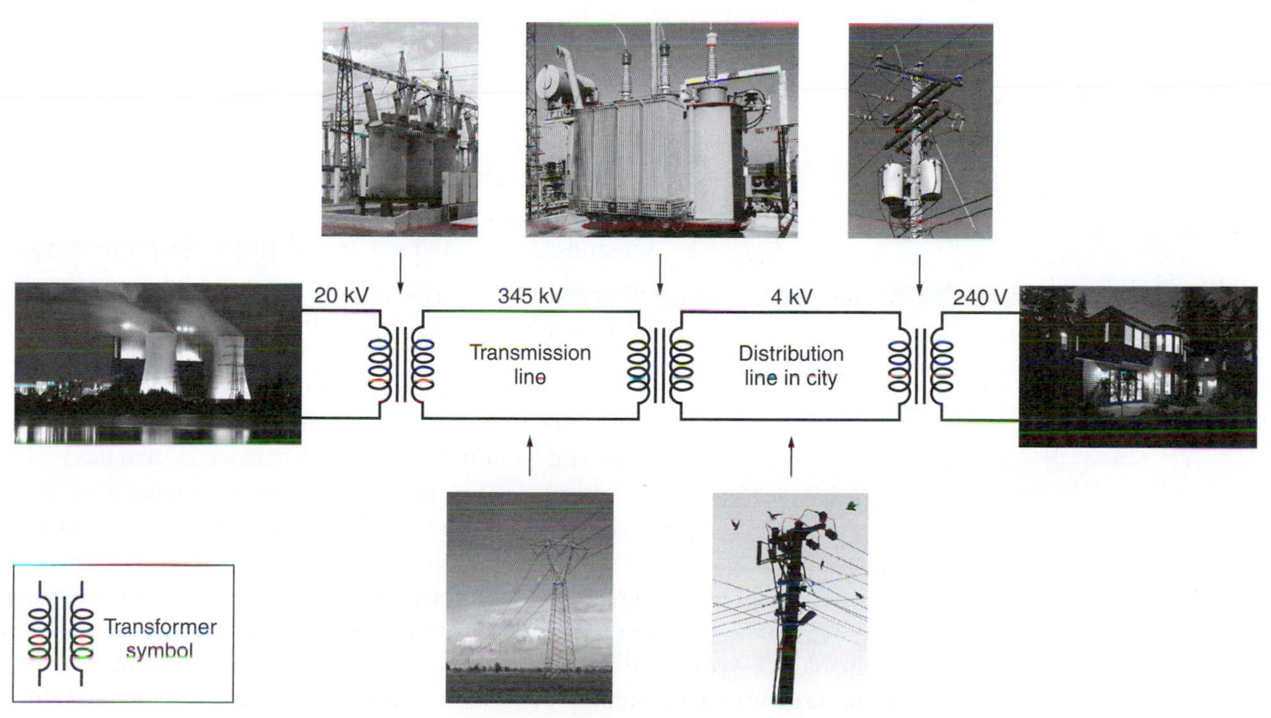

FIGURE 11.4
Transformers step voltage up and down along the power grid. The 240 V to the end user is actually a pair of 120-V circuits, with standard outlets getting 120 V in North America. As with Figure 11.3, this one is a simplification that neglects three-phase AC power.

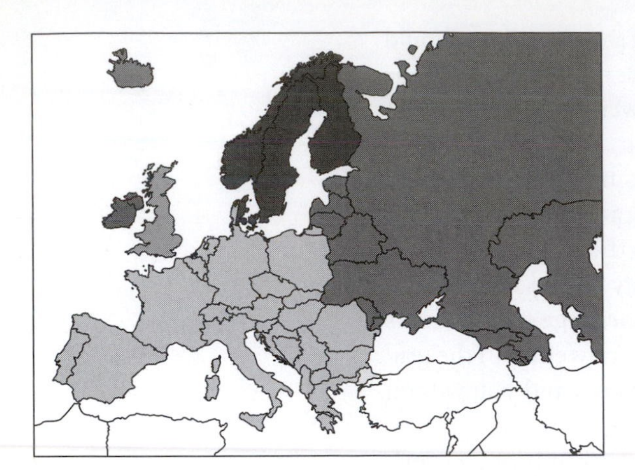

FIGURE 11.5

A total of six synchronous electric power grids serves all of Europe. On this map the grids are distinguished by different shadings.

allowing individual components to be disconnected from the grid for scheduled maintenance or upgrading. That redundancy usually minimizes the impact of unexpected failures, isolating the problem while providing alternative sources and transmission routes to keep power flowing to the affected region. But the grid's interconnectedness can also lead to massive power outages; should the system fail to contain an individual problem, a cascade of failures can ripple through the system and cut power to a vast region. In August 2003, hot weather and the resulting high demand for electric power caused an Ohio power line to overheat, sag, and short-circuit to a tree. A chain reaction of failures soon left 50 million people in the United States and Canada without power! Fortunately, power grids are engineered so that such massive failures are rare.

The advantage of ever-larger grids almost certainly outweighs their vulnerability—and that advantage will grow with increasing reliance on renewable energy. Although we speak of "the power grid," there are distinct grids serving different regions. Within each grid are smaller but interconnected power systems. Use of alternating current requires that all generators connected by AC transmission lines must be *synchronous*—meaning the alternating voltages they produce are in lockstep. So one way to define and visualize a power grid is by the area that encompasses synchronous generators. For example, Figure 11.5 shows that continental Europe has only three synchronous grids; three more separate grids power Great Britain, Ireland, and Iceland. Similarly, most of North America is served by three grids, shown in Figure 11.6.

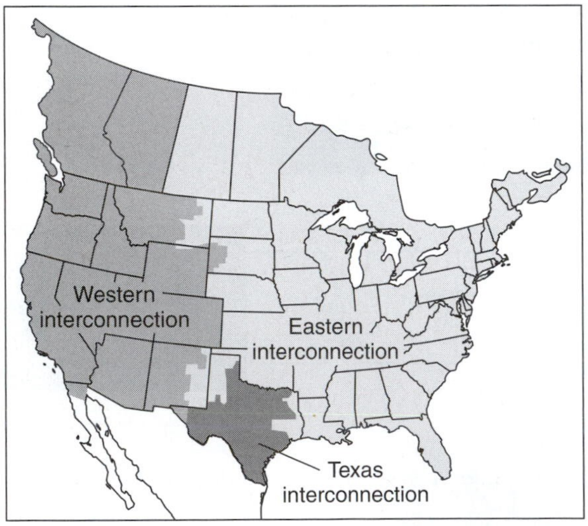

FIGURE 11.6

The three main grids in North America are called "interconnections."

Even the separate grids shown in Figures 11.5 and 11.6 are actually connected—but not by AC transmission lines. For very long distance power transmission, either within a grid or between distinct synchronous grids, high-voltage DC transmission has several advantages. These include lower losses, greater stability because voltage and current can't get out of phase—a problem that can lead to failures—and the ease of "piping" electricity directly from a distant source without involving the grid's complexity. The lower-loss advantage is partly offset by the cost, in money and power, of electronic equipment needed to convert AC to DC and then back to AC at the other end—but modern, high-efficiency electronics make DC the choice for the longest, highest-voltage lines. And for connections between different synchronous grids, DC avoids issues of synchronism; the so-called *inverter* converting DC to AC at the far end can synchronize its output with the receiving grid, regardless of what the sending grid is doing.

Use of high-voltage DC transmission is sure to expand in the coming decades, with the need to access remote renewable energy sources or for undersea cables from offshore wind farms or connections to island nations; or simply to move more power over greater distances. Indeed, the European Union has a grand vision of a so-called *supergrid* using DC to bring future photovoltaic power from the Sahara Desert under the Mediterranean to the population centers of Europe. On a more modest scale, countries bordering the North Sea plan a high-voltage DC supergrid that would connect Northern Europe, the United Kingdom, and Scandinavia with each other and with rapidly developing North Sea wind farms; a further benefit would be to use Norway's extensive hydroelectric resources as giant energy-storage systems—more on that shortly.

A fundamental problem in electric power transmission is electrical resistance, which results in power losses, limits the power a given line can carry before overheating, and dissipates power at every transformer and inverter. What if there were no resistance? Remarkably, there's a class of materials—called **superconductors**—that have truly zero electrical resistance. Unfortunately, all superconductors developed to date require temperatures below around 100 K—about –200°C. The energy and economic cost of refrigeration rule out long-distance power transmission with today's superconductors. But superconducting motors, electromagnets, and other devices are already in widespread use, and superconducting cables add capacity in crowded underground wiring channels in major cities. Development of an inexpensive room-temperature superconductor would spur a revolution in electric power transmission and throughout electromagnetic technology.

ELECTRICITY'S IMMEDIACY

Flip a switch, and your light comes on instantly. You don't have to order up additional electricity, or notify your power company of the extra load you're putting on the grid. Yet the extra energy has to come from somewhere, and it has to be available immediately. With electricity, it's not like there's spare electrical energy stored somewhere, awaiting your use. Instead, some generator on the grid has to produce more energy—right now, immediately, when you flip the switch. In that way electricity is different from, say, oil. Fill your car's gas tank, and no one needs to refine more gasoline or pump more oil from the ground right then to meet your demand. Conversely, an oil company can keep pumping oil, and a refinery can keep making gasoline, even if the demand isn't there. The excess gets stockpiled for future use, perhaps in times when production isn't keeping up with demand. In other words, production and consumption of oil aren't necessarily equal. But there's no way to store excess energy as electricity itself, so at any instant the power system must be producing electrical energy at exactly the rate it's being consumed.

This "immediacy" of electricity requires active balancing of supply and demand on the power grid. An important tool for this purpose is a hierarchy of electricity generators, starting with **baseload** power plants. These are typically large facilities, often coal-fired or nuclear, which take a long time to start up and that run at fixed power outputs. They supply the minimum electrical demand that's almost always present—the baseload—and they're the least expensive

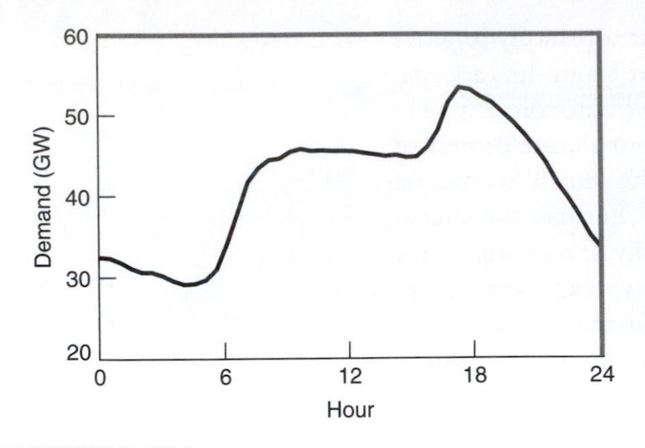

FIGURE 11.7

A typical demand curve, showing electric power demand as a function of time over a full day. Note the low demand at night, the rapid rise followed by steady demand into the workday, and the peak around dinnertime. This graph shows actual data for Great Britain (England, Scotland, Wales), for November 10, 2010.

plants to operate. Typically, demand fluctuates throughout the day, as shown in Figure 11.7, and baseload plants don't respond to those fluctuations. Other power plants, including gas-fired and hydroelectric plants, can vary their output somewhat in response to changing demand. Finally, at times of peak demand, utilities may turn on small gas turbines or other generating systems that are typically inexpensive to build and own, but expensive to operate. These **peaking power** facilities are run only as long as needed to meet peak demand.

But how does a utility know how much power it's going to need to supply, and how does it respond instantaneously to changing demand? Short-term, small-scale changes—like you turning on your light—get handled automatically. As with the hand-cranked generator shown in Figure 2.1, increasing load current makes generators harder to turn—and if nothing were done, they would slow down. So smaller units sense the increased load and burn a little more gas, or let a little more water through the turbines, to maintain speed. That's how your choice to turn on a light directly and immediately influences the consumption of energy resources. For the longer term, utilities schedule their generators using forecasts based on factors like the previous day's demand; weather; workdays versus weekends, season, and the like. They also factor in supply-side issues such as scheduled maintenance or nuclear-plant refueling. Forecasts are updated hourly, and various technical means are used for balancing between forecasts. If a utility finds itself with extra generating capacity, it can sell power over the grid to regions that need more; or if the utility comes up short, it may be cheaper to buy power from afar than to turn on expensive peaking generators. The interconnected grid enables these strategies, and makes scheduling and management a complex juggling act involving both automatic actions and human decisions, both physics and economics. In worst-case scenarios, where there simply isn't enough power to meet demand, a utility can choose to lower the voltage—a so-called **brownout**—or institute **rolling blackouts** that temporarily deprive individual neighborhoods of power. Such inconveniences are rare in industrialized countries, but they're common in the developing world, where electricity infrastructures are far from reliable.

As Figure 11.7 shows, demand is usually well below its daily peak. But utilities nevertheless need enough generating capability to meet not only the daily peak demand but also the highest demand they're likely to see. So utilities need to have money invested in generating capacity that sits idle much of the time, or they have to be prepared to pay top dollar for extra power from other utilities with surplus capacity. A flat demand curve would make their job much easier and improve their financial bottom line. In fact, most utilities have incentives designed to flatten the demand curve. If you look at your home electric bill, you'll find that you pay a fixed charge for your connection to the grid, along with a set amount for each kilowatt-hour of energy you use. But if you're an industry, a university, or other large institution, you also pay a **demand charge** based on the instantaneous peak electrical demand you hit at some instant during a given year or other time period. So it's in your advantage to keep that peak demand low. As you watch your electrical consumption rate increase toward a peak, you can temporarily shut down major consumers such as electric heating, air conditioning, and ventilating equipment. Many institutions have automated systems for such end-user-based **load management**.

Whatever strategies your utility and your institution use for load management today, they're probably far from optimum. Better load management is a significant step toward improving our overall energy efficiency—a step that's going to take much smarter electrical infrastructure, and a step that's complicated by renewably generated electricity.

RENEWABLES AND SMART GRIDS

Increased reliance on renewable energy—especially intermittent wind and solar—presents additional challenges for electric grids. Not only must they cope with varying and somewhat unpredictable demand, but now their sources become unpredictable as well. What happens when the wind stops, or a cloud passes between the Sun and a photovoltaic power plant? With today's modest contributions from wind and solar, grids are capable of handling these interruptions with strategies described in the preceding section. But when intermittent renewables rise to 10% or 20% of the electric power mix, handling their variable output becomes a real challenge.

Today, the challenge of renewable intermittency is typically met by having standby generating capacity—usually in hydroelectric or gas-turbine generators—that can be ramped up or down quickly as the supply of renewable energy varies. For example, more than 30% of Denmark's electric generation capacity is from wind—a situation made possible because Denmark's power grid can call on the vast hydroelectric resources of Scandinavia.

Large-scale use of wind and solar energy presents another problem: The best resources are located in remote areas far from population centers. In the United States, the desert Southwest is the best place for solar energy; for wind it's the Great Plains. So it's not enough to erect vast wind farms and acres of photovoltaic panels; we also need new transmission lines to deliver their energy to existing grids. That means a big investment in new long-distance, high-voltage transmission lines—most likely DC lines for reasons discussed earlier. However, tying distant renewable electricity sources together has a benefit: It helps mitigate the problem of intermittency, especially for wind. The wind may stop at one wind farm, but it can pick up at another, distant farm. Averaged over many installations, the total wind power output may be a lot steadier than that of any single wind farm. But that averaging helps only if distant wind farms are connected to a common grid—and, again, that means new long-distance transmission lines. Indeed, a University of Delaware study suggests that tying together wind farms along a 1,500-mile stretch of the United States' east coast would greatly reduce the overall variability, allowing wind to make a substantial contribution to the region's electricity supply.

While we're enhancing the grid to accommodate renewables, why not make it smarter as well? Today's grid is a one-way street, delivering power to consumers without taking anything back except a monthly reading of the electric meter for billing. A modest improvement, **net metering**, allows owners of small-scale photovoltaic and wind systems to sell power back to the grid, by effectively running their meters in reverse and thus reducing their electric bills. Still smarter is a grid that delivers not only energy, but also information. Peak power is expensive, and a **smart grid** coupled with a smart electric meter could charge you more for electricity during peak times. A smart meter might display the current price, and you could act accordingly to alter your electric demand. Or the smart grid could handle load management automatically, reaching into your home and shutting down devices like water heaters to shed load during times of high demand. Or the grid might decide it was smart to charge your electric car in the wee hours of the morning, when demand is low and utilities have surplus capacity. You'd save money and the power company would avoid building additional generators to supply a growing fleet of electric vehicles. These and many other smart-grid technologies are on the way or are being implemented now—in the United States thanks in part to $4.5 billion in economic stimulus funds aimed at modernizing the country's century-old electric grid. With a smarter grid we'll use energy more intelligently and we'll lower the economic, resource, and environmental impacts associated with constructing additional generating stations. We'll also make our power system more welcoming toward renewable energy.

ENERGY STORAGE

If only we could store and retrieve electrical energy, then we would break that challenging link between supply and demand—the link that requires electricity to be consumed as soon as it's produced. Gone would be the issues with intermittent renewable energy sources; gone would be many headaches of generator scheduling and load management; gone would be the need for excess generating capability; gone would be fears of brownouts and blackouts. But electricity is energy moving at nearly the speed of light, and there's no practical way to bottle it up and store it.

No way, that is, without changing electrical energy into a different form, then converting it back to electricity. And there are plenty of energy-storage systems that can do that. Batteries, capacitors, compressed air, flywheels, hydrogen—all these can store energy that was earlier electricity, and will later be electricity again. Some of these technologies have been with us for a long time: Your car battery, for example, stores as chemical energy the energy produced as electricity when the gasoline engine turns the car's generator. You use that stored energy to start the car and to run your headlights, windshield wipers, heated seats, and sound system. Hybrid cars take battery storage further, running their motors in reverse when braking, generating electricity that charges a large battery, then using that energy for motive power. Capacitors—devices that store energy in the electric fields of charged conductors—are similar to batteries but with shorter storage times. If you've ridden the San Francisco area's Bay Area Rapid Transit system, you've been the beneficiary of capacitor energy storage. As a BART train pulls into the station, its motors act as generators that charge onboard capacitors. The stored energy is used to accelerate the train as it pulls out of the station. Using electric motors to compress air provides another storage mechanism. Here, large quantities of air are compressed and pumped into underground caverns using surplus electricity; the air can then turn turbine-generators that convert their stored energy back into electricity. In another approach, electricity drives motors that set massive flywheels spinning on low-friction bearings, storing what was electrical energy as kinetic energy of rotational motion. Reverse the system, and the motor becomes a generator, converting kinetic energy back into electricity. Electrical energy can also be used to dissociate water into hydrogen and oxygen; recombining them through combustion can drive a steam generator to produce electricity again. And there are more sophisticated ways to get electricity from hydrogen; more on that in Section 11.2.

All these energy-storage technologies sound promising, and some find use today in applications like individual vehicles, not to mention cell phones, cordless tools and laptop computers—unthinkable without modern batteries. But, unfortunately, none has yet shown the technical and/or economic qualities necessary for large-scale storage of surplus electricity from the power grid. Part of the problem is **energy density**—the amount of energy stored per unit mass or volume of storage. Figure 11.8 shows energy densities, by both mass and volume, achievable with various storage schemes. Fossil-fuel energy densities are shown too, and the comparison goes a long way toward explaining why it's been so hard to produce pure electric cars with acceptable travel ranges, and why even the huge piles of coal outside a power plant would pale beside any system that could store a few days' worth of the power plant's electrical energy output.

There is one technology used today for large-scale storage of electrical energy. It's simple, relatively low-tech, and moderately efficient. This is **pumped storage**, a close cousin of conventional hydroelectric power. Pumped storage plants use electrical energy to run motors that pump water to an elevated reservoir, converting electrical energy into gravitational potential energy. Pumping occurs at times of low electrical demand, when there's surplus power to run the pumps. Then, when additional peaking power is needed, water is drawn from the reservoir to produce electricity. The same pumps and motors that sent the water uphill now act, respectively, as turbines and generators that make the reverse conversion back to electrical energy. Pumped storage systems share this reversibility with hybrid cars, whose motors act as generators when the car is braking, in this case storing energy in batteries. Figure 11.9 diagrams a typical pumped storage installation, and Example 11.3 explores an actual facility.

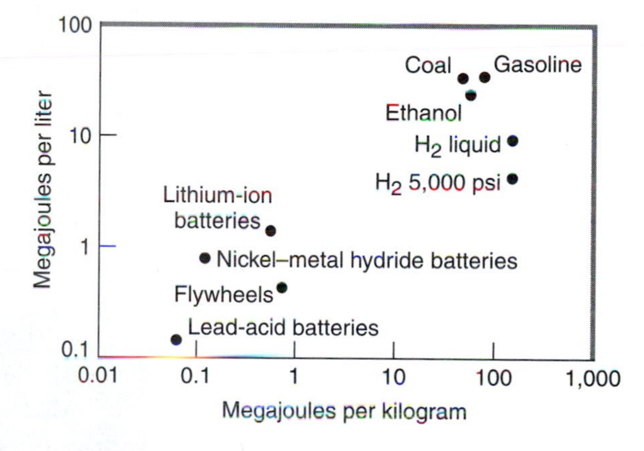

FIGURE 11.8
Orders of magnitude separate fossil and other chemical fuels from energy-storage technologies such as batteries. Graph shows both energy per unit mass (horizontal axis) and per unit volume (vertical axis). Note the logarithmic scales.

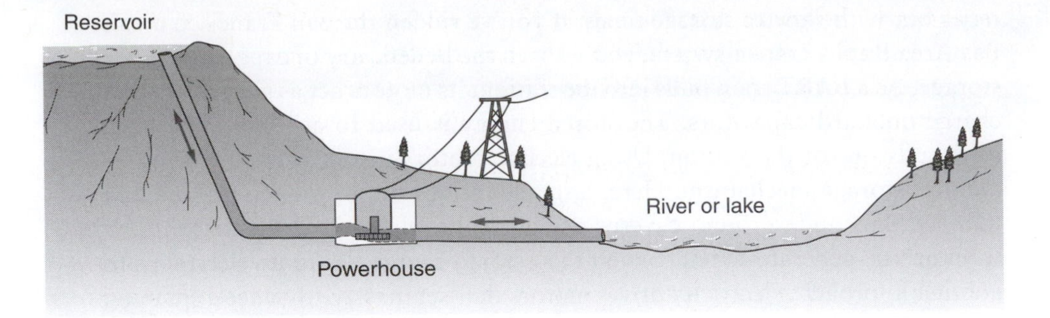

FIGURE 11.9
Cross section of a pumped storage facility showing the upper reservoir, piping, and underground powerhouse containing the turbine/generators that also serve as pumps/motors.

EXAMPLE 11.3 | Pumped Storage

The Northfield Mountain Pumped Storage Project in Massachusetts (Fig. 11.10) stores 21 million cubic meters of water 214 m above its turbine-generators. (a) Find the total gravitational energy potential energy stored in this water. (b) If all this energy could be converted back into electricity, how long could Northfield Mountain supply its rated power of 1.08 GW before emptying the reservoir?

SOLUTION

(a) Equation 3.4 shows that the gravitational energy stored in lifting a mass m through height h is mgh, where $g = 9.8\ \text{m/s}^2$ is the acceleration of gravity. Water's density is 1,000 kg/m^3, so Northfield Mountain's $21 \times 10^6\ \text{m}^3$ of water amounts to 21×10^9 kg. So the stored energy is

$$mgh = (21 \times 10^9\ \text{kg})(9.8\ \text{m/s}^2)(214\ \text{m}) = 4.4 \times 10^{13}\ \text{J} = 44\ \text{TJ}$$

(b) At 1.08 GW, or 1.08 GJ/s, this energy would last for a time given by

$$t = \frac{44 \times 10^{12}\ \text{J}}{1.08 \times 10^9\ \text{J/s}} = 40{,}740\ \text{s}$$

FIGURE 11.10
The Northfield Mountain Pumped Storage Project in Massachusetts can generate over 1 GW of electric power from the gravitational potential energy of water.

or about 11 hours. In fact, Northfield Mountain has four 270-MW turbine-generators that can be switched in or out as needed, so it can supply lower power levels for longer periods. About 75% of the energy used to pump water at Northfield Mountain is recovered as electricity.

Today, some 140 pumped-storage systems operate in the United States, with combined generating capacity of about 18 GW. Obviously, pumped storage requires both ample water and mountainous terrain. Absent continental-scale transmission lines, that makes pumped storage impractical for wind- and solar-generated electricity from the American Great Plains or southwestern deserts. A mountainous country like Norway, on the other hand, is already highly dependent on hydroelectric power and has the potential to store vast amounts of energy using pumped storage. That's one motivation for the proposed North Sea supergrid, which could send electrical energy from intermittent North Sea wind farms to Norway for pumped storage.

Like every other energy technology, pumped storage is not without environmental consequences. Natural mountaintop reservoirs are rare, so development of pumped storage involves significant disruption of mountain landscapes. The energy density of pumped storage is so low that I haven't included it in Figure 11.8 (but you can calculate it in Exercise 4). That low density suggests the need for large reservoirs. But where it's workable, pumped storage offers a way to store many gigawatt-hours of electrical energy—as much as a day's output of a large power plant—and its further development could help integrate renewable wind and solar energy into our power grids.

Energy storage need not involve large-scale systems like pumped-storage reservoirs or caverns of compressed air. Plug-in hybrid or all-electric cars could use their batteries as storage systems, taking energy from the grid at off-peak times and returning it during peak hours—if they're not being driven. So you might unplug your car in the morning, drive it to work, and leave it plugged in all day while a smart grid pumped energy into it or pulled energy off it to help manage fluctuating demand. Multiply by a hundred million plug-in vehicles, and you've got a very effective storage system for the electric grid. Will this vision come to pass? That depends on advances in battery technology, electric propulsion, and a smart grid.

11.2 Toward a Hydrogen Economy?

Molecular hydrogen, H_2, is a flammable gas that combines with oxygen to make water:

$$2H_2 + O_2 \rightarrow 2H_2O \tag{11.4}$$

Besides energy, water is the only product, which gives hydrogen its environmental reputation as a clean, nonpolluting, carbon-free energy source. The reaction in Equation 11.4 releases 0.475 aJ (0.475×10^{-18} J) for each water molecule formed, giving hydrogen an energy content of 142 MJ/kg—roughly three times that of petroleum-based fuels on a per mass basis. But hydrogen, the lightest element, is a low-density gas, so under normal atmospheric pressure this energy content represents only 12.8 MJ/m^3—about one-third that of natural gas on a per-volume basis.

Hydrogen energy has a long history. The nineteenth-century science fiction writer Jules Verne, in his lesser-known novel *The Mysterious Island*, envisioned a society powered by hydrogen rather than coal. Early twentieth-century tinkerers and engineers, some inspired by Verne's vision, proposed hydrogen-powered trains, cars, and aircraft. Hydrogen provided buoyancy for dirigibles that preceded transatlantic aircraft. In the 1930s, Germany and England tested hydrogen-powered buses, trucks, and trains, burning the hydrogen in modified gasoline or diesel engines. More recently, Russia converted a passenger jet engine to burn hydrogen. And the Space Shuttle ran on hydrogen, its notorious external fuel tank holding 100,000 kg of liquid H_2.

THE HYDROGEN RESOURCE

With every other energy alternative I've asked the questions: How much is there? How long will it last? How hard is it to extract? How's it distributed around the planet? . . . and so forth. For hydrogen this section is blissfully short. That's because *there is no hydrogen energy resource.*

On Earth, that is. Hydrogen is the most abundant element in the universe, and there's plenty of chemical hydrogen, H_2, floating around in interstellar clouds or in the atmosphere of Jupiter. Even Earth had atmospheric hydrogen in its early days. But smaller planets like Earth don't have strong enough gravity to hold light hydrogen molecules, which, at a given temperature, move faster than heavier oxygen and nitrogen. So Earth gradually lost its hydrogen, and today H_2 constitutes less than 1 part in 1 million of Earth's atmosphere. Exercise 7 shows that the associated energy resource is truly negligible. Of course, there's plenty of hydrogen in the water of Earth's oceans—but, in the form of H_2O, it's already "burned" and isn't a source of chemical energy. (Nuclear energy from hydrogen fusion is a different matter, as I showed in Chapter 7.)

So there's no chemical hydrogen energy resource on Earth. I want to make this very clear, because the public frequently hears "hydrogen energy" and imagines there's a vast new energy source out there, just waiting to be tapped. So I'll say it again: *There is no hydrogen energy resource on Earth.*

That's why hydrogen, the chemical fuel H_2, is not an energy source but an energy carrier. We can manufacture hydrogen using true energy sources, converting other forms of energy into chemical energy stored in H_2 molecules. We can transport the hydrogen, and burn or otherwise react it to produce energy for end uses. Hydrogen shares with electricity the role of energy carrier—but there the similarity ends. Hydrogen, unlike electricity, sees very little use in today's energy economy, and it's going to take a monumental effort to develop hydrogen infrastructure. On the other hand, hydrogen, unlike electricity, is an excellent energy-storage medium, and hydrogen can substitute for liquid and gaseous fuels in transportation.

PRODUCING HYDROGEN

Hydrogen in the form H_2 is produced by chemically processing hydrogen-containing compounds, commonly fossil fuels or water. Today, hydrogen is used mainly in petroleum refining and fertilizer manufacturing. Most industrial hydrogen comes from **steam reforming**, a reaction of natural gas with steam to produce a mix of hydrogen and carbon monoxide:

$$CH_4 + H_2O \rightarrow 3H_2 + CO \tag{11.5}$$

The mixture can be burned directly as a gas or processed further to convert the CO to CO_2, which is then separated from the hydrogen. Oil and coal, which contain less hydrogen than natural gas, also serve as feedstocks for hydrogen production. Some combined-cycle power plants (Section 5.4) include integrated coal gasifiers, which react coal with oxygen to produce a mix of hydrogen and carbon monoxide. In that sense hydrogen already plays some role in energy production.

A second approach is **electrolysis**, in which electric current passing through water splits H_2O molecules and results in hydrogen and oxygen gases collecting at two separate electrodes (Figure 11.11). Commercial electrolyzers operate at efficiencies over 75%, meaning that more than three-quarters of the electrical energy input ends up as chemical energy stored in hydrogen. But the normally high cost of electricity relative to fossil fuels makes electrolysis economical only where cheap, abundant hydroelectricity is available.

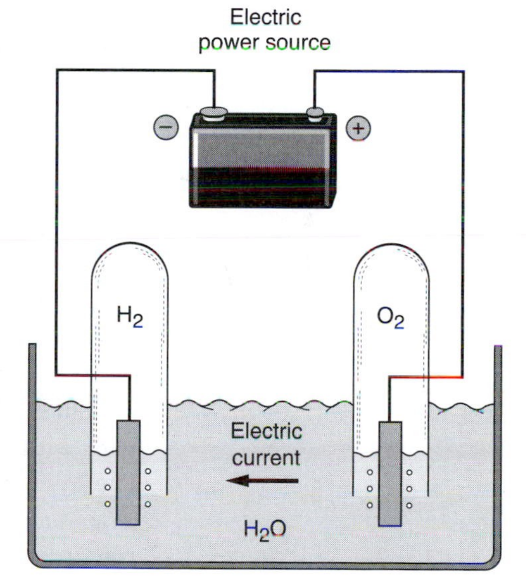

FIGURE 11.11
A simple electrolysis cell. Electric current passes through water, and bubbles of oxygen and hydrogen form at the positive and negative electrodes, respectively. The gas bubbles rise to fill collection chambers above the electrodes.

At sufficiently high temperatures water breaks apart into hydrogen and oxygen; high-temperature nuclear reactors or solar parabolic concentrators might someday produce hydrogen directly through such **thermal splitting**. Heating to less extreme temperatures can assist the process of electrolysis, giving a hybrid technique that reduces the electrical energy required at the expense of increased thermal energy.

Scientists are exploring several innovative hydrogen-production schemes, although none are near commercial viability. In photolysis, solar energy creates electron–hole pairs in a semiconductor immersed in a water solution. Water molecules lose electrons to the holes and split into hydrogen and oxygen. The process is akin to electrolysis, except that solar energy creates the hydrogen directly without the intermediary of electric current. Experimental photolysis cells have achieved energy conversion efficiencies up to 12%.

Biological processes also produce hydrogen; indeed, photosynthesis requires splitting water and rearranging the resulting hydrogen and oxygen to make carbohydrates. Under appropriate conditions, some algae and bacteria naturally exude hydrogen, and it should be possible to engineer organisms designed specifically to optimize hydrogen production. But making hydrogen biologically is likely to be inefficient, and in any event its practical realization is a long way off.

FUEL CELLS AND HYDROGEN VEHICLES

Although hydrogen could be burned in stationary facilities like power plants and industrial boilers, its real value is in transportation. Internal combustion engines in today's vehicles can be adapted to burn hydrogen, as can the gas turbines that power jet aircraft. There's good reason to burn hydrogen in such engines; not only are there fewer harmful emissions, but hydrogen-powered engines also tend to be more fuel efficient than their fossil counterparts.

However, combustion engines probably aren't the future of hydrogen-powered transport. It's more efficient to convert hydrogen energy to electricity using fuel cells—devices that combine hydrogen and oxygen chemically to produce water and electrical energy. An electric motor would then propel the vehicle. "Well to wheels" comparison with conventional gasoline-powered vehicles—including the energy required to produce both gasoline and hydrogen—suggests that fuel-cell vehicles could be as much as 2.5 times more energy efficient than conventional vehicles.

The first fuel cell was constructed 1839, predating the internal combustion engine. Fuel-cell development continued slowly through the nineteenth and twentieth centuries, getting a boost when NASA adopted fuel cells for the Apollo Moon missions. Today there's a modest market in commercial fuel cells; by 2010 world sales approached 200 MW of fuel-cell power and were growing at a rapid 40% annually. Industry, governments, and universities all engage in fuel-cell research, and applications range from stationary power generation through transportation to mini-cells that might someday power cell phones and laptop computers.

A fuel cell is like a battery that never runs out, because it's continually supplied with chemicals—usually hydrogen and oxygen—that it uses to make electrical energy. You can also think of a fuel cell as the opposite of an electrolysis cell. In electrolysis, you put in electrical energy and get out hydrogen and oxygen; with a fuel cell, you put in hydrogen and oxygen and get out electrical energy. There's a nice symmetry here, and it shows the significance of hydrogen as an energy carrier that can store electrical energy generated by other means, then turn it back into electricity for applications like transportation vehicles, for which the direct use of grid power isn't practical.

Figure 11.12 shows a **proton-exchange membrane fuel cell (PEMFC)**, a variety with promise for powering fuel-cell vehicles. Other fuel cells are conceptually similar. Two electrodes, the anode and cathode, sandwich a membrane that permits only protons to pass through. Anode and cathode are made of catalysts that promote chemical reactions. At the anode, these reactions dissociate the incoming hydrogen (H_2) into protons and electrons. Protons move through the membrane, while electrons flow through an external electric circuit. At the cathode, protons combine with incoming oxygen (O_2) and with electrons returning from the circuit to form water (H_2O), the cell's only material byproduct.

Most of the energy released in the fuel cell appears as electrical energy, a lesser amount as heat. PEM fuel cells are more than 60% efficient at converting chemical energy of hydrogen into electricity; coupled with more than 90% efficient electric motors, that gives a combined efficiency well over 50%—more than twice that of gasoline engines and about 20% better than diesels. The theoretical maximum efficiency for hydrogen fuel cells alone—set ultimately by the second law of thermodynamics—is 83%.

A complete fuel-cell system needs fuel storage, fuel-supply mechanisms, and heat removal. Catalysts and proton-exchange membranes require expensive materials. All these factors make present-day fuel cells expensive—about $4,000 per kilowatt of electric power capability. In contrast, modern gasoline engines cost about $30/kW. Given fuel-cell vehicles' greater efficiency, fuel-cell costs would have to drop to around $100/kW to compete with gasoline.

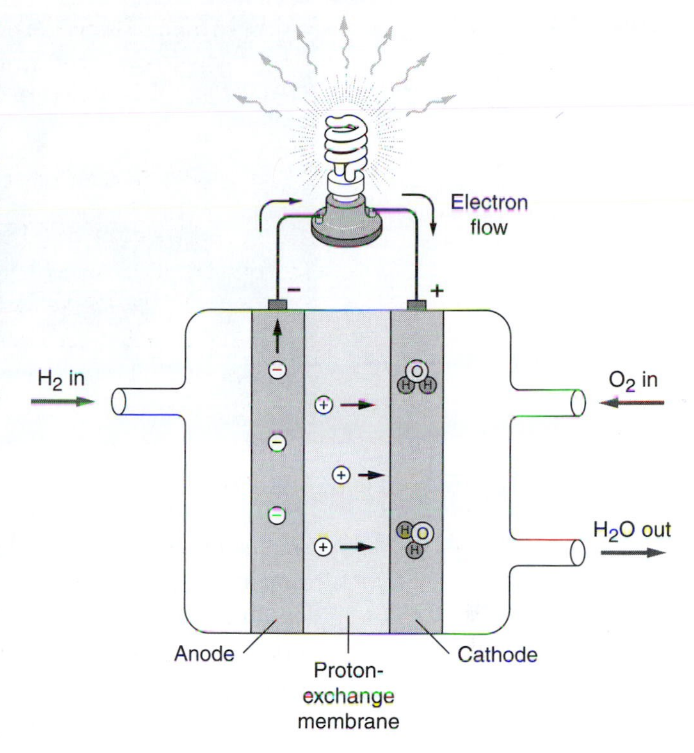

FIGURE 11.12

A proton-exchange membrane fuel cell. The anode separates hydrogen into protons and electrons. Protons pass through the membrane, while electrons go through an external electric circuit. At the cathode, protons and electrons combine with oxygen to form water.

Fuel-cell vehicles face other challenges, however. In its normal gaseous state, it would be impossible to store enough hydrogen to propel a vehicle even as far as 1 mile (see Exercise 9). To achieve the several-hundred-mile range typical of gasoline vehicles will require compressing hydrogen to hundreds of times atmospheric pressure—and that means heavy, expensive, and bulky storage tanks. Filling high-pressure hydrogen tanks presents technological, logistical, and safety challenges. An alternative is to liquefy the hydrogen, increasing its density but requiring heavily insulated tanks because hydrogen's boiling point is only 20 kelvins. Boil-off from even the best tanks results in a steady loss of hydrogen. There's also an energy penalty; the process of liquefying consumes some 40% of the hydrogen's energy.

A more promising approach may be to bind hydrogen chemically with other substances. At the low-tech extreme are fossil fuels themselves; for example, pressurized natural gas could be reformed onboard a vehicle, providing hydrogen to a fuel cell. Of course, this strategy doesn't get us away from fossil fuels, and it leaves carbon to be disposed of—which in a vehicle means releasing CO_2 to the atmosphere. A related approach involves methanol (wood alcohol, CH_3OH), which can be used directly in fuel cells, or converted to hydrogen. Again, there's CO_2 to dispose of. Methanol could be renewable, since it can be produced from biomass as well as from fossil fuels.

Hydrogen bonds with lightweight elements and their solid compounds, and the resulting **hydrides** represent another approach to hydrogen storage. Fueling involves flowing hydrogen through a porous structure with plenty of surface

FIGURE 11.13
Honda's FCX Clarity fuel-cell car is being leased to several hundred customers who are fortunate to live near rare hydrogen fueling stations. The FCX Clarity stores hydrogen gas at 5,000 psi pressure, as was indicated by a point on Figure 11.8.

area; some heat is released as the hydrogen bonds to the solid. The hydrogen remains bound until it's needed; heating the hydride then releases hydrogen. Finally, carbon or silicon *nanotubes*—a hot topic in materials research—might serve as "cages" or "miniature fuel tanks" capable of absorbing and releasing hydrogen. The greater efficiency of fuel cells means we won't need to duplicate gasoline's volumetric energy density to give fuel-cell vehicles the same range as today's cars. But we'll have to come within a factor of two or three—as Figure 11.8 (p. 305) showed is possible for some hydrogen-storage systems.

After years of experimentation, auto manufacturers have begun limited production of fuel-cell vehicles (Fig. 11.13). But these are far from competitive economically, and it's going to take much more technological development to bring down costs, particularly of fuel cells and onboard hydrogen storage. Even if we can do all that, hydrogen faces some broader problems.

INFRASTRUCTURE FOR THE HYDROGEN ECONOMY

Moving to hydrogen-based transportation presents a daunting challenge not shared by other energy alternatives we've considered. Develop effective photovoltaic power stations, and you have a new source of electrical energy to plug into the existing power grid. Make solar water heating competitive, and you change a minor aspect of home construction. Design efficient, economical wind turbines or advanced nuclear reactors and the grid awaits your electricity. In all these cases only the energy source changes. The distribution of electricity or hot water remains the same, as do the lightbulbs, computers, dishwashers, and showers using that electricity or heat. But with hydrogen we're talking about changing the entire energy picture—the fuel, the distribution infrastructure, the storage mechanisms, and the vehicles themselves. And the system we're moving from—oil wells, tankers, refineries, pipelines, gas stations, and cars powered by internal combustion engines—is highly developed, historically entrenched, and represents a huge capital investment. Even if it's no longer the best system, enormous inertia must be overcome to change it.

Hydrogen faces a kind of "chicken and egg" problem in trying to replace the fossil energy transportation system. As of 2010, there were only 60 hydrogen-fueling stations throughout the United States. A fuel-cell car wouldn't be much use if you weren't near one of those stations. But there were only a few hundred fuel-cell cars, so building more fueling stations doesn't make much economic sense. Somehow both the vehicles and the hydrogen infrastructure are going to have to develop together. But how?

The transition to a hydrogen economy might begin with modest growth of the existing hydrogen production and distribution system. The United States, for example, already produces some 9 million tonnes of hydrogen annually, much of it consumed on site but some distributed in liquid form by cryogenic tanker trucks. As demand grows, expensive truck-based distribution might be replaced by distributed hydrogen generation, with fossil fuel reformers or photovoltaic-powered electrolysers at individual filling stations. Another approach sees surplus power from baseload coal and nuclear plants being diverted to hydrogen production that could be delivered locally at reasonable cost. A maturing hydrogen economy might see a mix of centralized and distributed hydrogen production, with energy sources switching gradually from fossil to renewables. Expanded use of fuel-cell vehicles would accompany the growing hydrogen infrastructure. A report by the U.S. National Research Council and National Academy of Engineering projects optimistically that 100% of new U.S. vehicles could be hydrogen powered by the late 2030s, and that by 2050 virtually all vehicles on the road would use hydrogen (Fig. 11.14).

Realizing this vision of a renewable hydrogen economy will not be easy. It will take cooperation among government, industry, the public, and the research community. And it will take political courage, because it will certainly require subsidizing an energy system that, at first, won't be economically competitive with fossil fuels. At some point, rising fossil fuel prices and decreasing costs for renewables and hydrogen production will make nonfossil hydrogen competitive—but if we wait until then it may be far too late for a smooth transition.

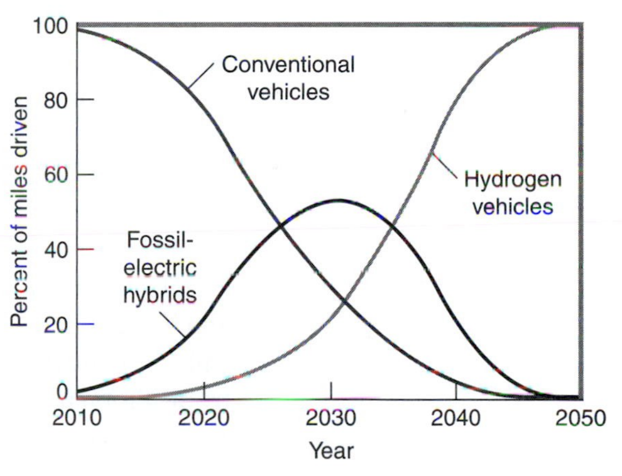

FIGURE 11.14
The U.S. National Research Council's projected transition to a hydrogen-based transportation system. The figure shows the percentage of miles driven by conventional vehicles, fossil–electric hybrids, and hydrogen vehicles. In this scenario, virtually all new vehicles would be hydrogen powered by 2030, but older non-hydrogen vehicles would remain on the road until 2050.

HYDROGEN SAFETY

For hydrogen to catch on, the public must perceive it as safe. Some still have visions of the 1937 *Hindenburg* disaster, when the world's largest airship burst into flames at the end of a transatlantic flight. More recently, the 1986 explosion of the Space Shuttle *Challenger*, although triggered by a solid-fuel rocket leak, involved the shuttle's huge tank of liquid hydrogen. Even the 2001 *Columbia* shuttle disaster was hydrogen-related, being caused by a loose chunk of the insulation needed to keep that same hydrogen tank cool. It's easy to imagine miniature *Hindenbergs* and *Challengers* with hydrogen-powered cars.

Hydrogen does present unique dangers, but in other ways it's safer than gasoline—a familiar but dangerously flammable product. As the smallest molecule, hydrogen is much more prone to leakage. But when it leaks into the open air, the buoyant gas quickly disperses. Not so with gasoline, which forms puddles of flammable liquid. On the other hand, hydrogen burns readily with a wider range of fuel–air mixtures than most other fuels, and its flame can spread with explosive speed. The high pressures needed for practical hydrogen energy densities exacerbate leakage and safety issues. Clearly, widespread hydrogen usage will require new safety procedures and technologies. The dangers those procedures will alleviate are different from, but not necessarily worse than, those we face with the fuels in widespread use today.

ENVIRONMENTAL IMPACTS OF HYDROGEN

Hydrogen's main appeal is that it's a clean, nonpolluting, carbon-free fuel. So what, if any, are the hydrogen's environmental impacts? When hydrogen combines with oxygen, whether in conventional combustion or in a fuel cell, the only product is harmless water. Drinking water on the Apollo missions actually came from the onboard fuel cells. Water vapor from hydrogen combustion is a greenhouse gas—but the water cycle removes it from the atmosphere in about a week, so that's not a significant issue. In hydrogen combustion, temperatures get high enough—as they do with fossil or biomass combustion—for some atmospheric nitrogen to combine with oxygen, forming nitrogen oxides. With fuel cells even that small impact vanishes. Hydrogen can be truly clean, nonpolluting, and carbon free.

But remember, *there is no hydrogen energy resource.* The environmental impacts of hydrogen therefore include whatever energy source was used to make it. If your hydrogen comes from electrolysis with electricity generated in a 35% efficient coal plant, then your 75% efficient electrolysis coupled with a 55% efficient fuel-cell vehicle would give an overall efficiency of only 14%—and you'd be burning more fossil fuel than with a gasoline-powered car. Your carbon emissions would be even higher, thanks to coal's high carbon content. If you make hydrogen by reforming natural gas, oil, or coal, then you're left with CO_2 to dispose of—and if you dump it into the atmosphere then you haven't gained much. You might gain on noncarbon emissions, because it's generally easier to control pollution from stationary sources like large-scale reforming plants. In a more technologically advanced future you might gain on the carbon front, too, because, as you'll see in Chapter 16, it's easier to capture and sequester carbon from reforming fossil fuels than from combustion in a power plant or internal combustion engine.

EXAMPLE 11.4 | Clean Hydrogen?

Hydrogen made by reforming natural gas (the reaction of Equation 11.5) is burned in a modified internal combustion engine operating at 38% efficiency. The carbon monoxide from that reaction is converted into CO_2 and released to the atmosphere. How do the carbon emissions in this scenario compare with those of a conventional gasoline engine operating at 22% efficiency? Gasoline produces about 70 g of CO_2 per megajoule of fuel energy.

SOLUTION

For comparison with the figure given for gasoline, we'll need to quantify hydrogen in terms of energy, and CO_2 in terms of mass. The reforming reaction produces 3 molecules of H_2 and 1 of CO: $CH_4 + H_2O \rightarrow 3H_2 + CO$; the CO molecule then gets converted to 1 CO_2. Burning H_2 according to Equation 11.4 produces 0.475 aJ per molecule of H_2O formed or, equivalently, of H_2 burned (see the discussion following Equation 11.4). So our 3 molecules of H_2 contain 3×0.475 aJ = 1.43 aJ of energy, and with that we get 1 CO_2 molecule. Given the atomic weights of 12 and 16 for C and O, respectively, that CO_2 has molecular weight $12 + 2 \times 16 = 44$ u, so its mass is $(44)(1.66 \times 10^{-27}$ kg/u$) = 7.30 \times 10^{-26}$ kg. So the ratio of CO_2 mass to hydrogen energy is

$$(7.30 \times 10^{-26}\ \text{kg})/(1.43 \times 10^{-18}\ \text{J}) = 5.1 \times 10^{-8}\ \text{kg/J, or } 51\ \text{g/MJ}$$

We're given gasoline's CO_2 yield as 70 g/MJ, so already we're better off with the hydrogen. Quantitatively, the lower CO_2 per unit energy coupled with the hydrogen engine's higher efficiency means that the hydrogen-associated emissions are lower by a factor of $(51/70)(22/38) = 0.42$, a reduction of 58%. The actual reduction would be less because we haven't accounted for the energy needed to make steam for the reaction in Equation 11.5. And the balance could well shift the other way if we used a more carbon-rich fossil fuel to produce the hydrogen (see Exercise 11).

If we want carbon-free hydrogen energy using proven technologies, we're left with electrolysis using electricity generated from nuclear or renewable energy. Hydrogen energy then carries whatever environmental cost goes with that electricity generation. Perhaps the most attractive hydrogen future, from an environmental if not economic standpoint, is to generate electricity from photovoltaics or wind, then produce hydrogen by electrolysis. That could be a win–win situation because renewable electricity's intermittency becomes a nonissue when it's used to produce hydrogen—itself an energy storage medium.

CHAPTER REVIEW

BIG IDEAS

Energy carriers are forms of energy that don't occur naturally in Earth's energy endowment, but can be made from other energy sources. Electricity is the most important energy carrier in use today, while molecular hydrogen (H_2) could play a significant role in the future. Electricity and hydrogen, although both energy carriers, have very different properties and uses.

11.1 Electrical energy in the form of electric current comprises electric and magnetic field energy moving at nearly the speed of light, guided by conducting wires. About 12% of humankind's end-use energy is in the form of electricity, but more than a third of our primary energy goes into generating electricity. Electric current, the flow of electric charge, can be either direct or alternating; most power systems use alternating current because it's easier to change **voltage** levels for low-loss long-distance power transmission. The network of generators, **transformers** for voltage conversion, transmission lines, and end users comprises the **electric power grid**. Using an interconnected grid helps with the balancing of supply and demand that must be rigorously satisfied in electric power systems. Better energy storage technologies would facilitate this balancing and would allow for integration of intermittent renewable energy sources like solar and wind.

11.2 Molecular hydrogen (H_2) is a gaseous fuel. The only byproduct of hydrogen combustion is water. But there is no hydrogen energy resource. We have to make H_2 from hydrogen-containing compounds, and that takes more energy than we would get out of it. Therefore chemical hydrogen isn't an energy source but an energy carrier. Hydrogen is extracted from fossil fuels or water. It can be burned in power plants and internal combustion engines, or reacted with oxygen in fuel cells to produce electrical energy. The environmental impact of hydrogen energy includes the impact of whatever energy source was used to produce the hydrogen.

TERMS TO KNOW

- **ampere** (p. 295)
- **baseload** (p. 301)
- **brownout** (p. 302)
- **demand charge** (p. 302)
- **electric power grid** (p. 295)
- **electrolysis** (p. 309)
- **energy carrier** (p. 293)
- **energy density** (p. 305)
- **hydride** (p. 311)
- **load management** (p. 303)
- **net metering** (p. 304)
- **ohm** (p. 296)
- **Ohm's law** (p. 296)
- **peaking power** (p. 302)
- **proton-exchange membrane fuel cell (PEMFC)** (p. 310)
- **pumped storage** (p. 305)
- **resistance** (p. 296)
- **rolling blackout** (p. 302)
- **smart grid** (p. 304)
- **steam reforming** (p. 309)
- **step-down transformer** (p. 298)
- **step-up transformer** (p. 298)
- **superconductor** (p. 301)
- **thermal splitting** (p. 309)
- **transformer** (p. 298)
- **volt** (p. 295)
- **voltage** (p. 295)

GETTING QUANTITATIVE

Electric power: $P = IV$ (Equation 11.1; p. 295)

Ohm's law: $I = \frac{V}{R}$ (Equation 11.2; p. 296)

Transformer equation: $V_2 = \frac{N_2}{N_1} V_1$ (Equation 11.3; p. 298)

Hydrogen combustion: $2H_2 + O_2 \rightarrow 2H_2O$ (Equation 11.4; p. 307)

Steam reforming: $CH_4 + H_2O \rightarrow 3H_2 + CO$ (Equation 11.5; p. 309)

Hydrogen energy resource on Earth: *Zero*

QUESTIONS

1 Explain why long-distance electric transmission lines operate at high voltages.

2 Why is the development of better energy storage technologies a more pressing problem for the electrical energy sector than for, say, fossil fuels?

3 Nearly 40% of humanity's primary energy goes into generating electricity, yet only 12% of the energy we consume is in the form of electricity. Explain the apparent discrepancy.

4 What's the advantage of using alternating current on the power grid?

5 Is molecular hydrogen a fuel? A source of chemical energy? Discuss.

6 A friend argues that there's plenty of hydrogen available for fuel-cell vehicles; just look at all the H_2O in the ocean. Refute this argument.

7 How is a fuel cell like a battery? How is it different?

EXERCISES

1 For a given length of wire, resistance is inversely proportional to the wire's cross-sectional area. A utility is replacing a 115-kV transmission line with one using wire whose diameter is half that of the line it's replacing. At what voltage must the new line operate if transmission losses are to stay the same?

2 Convert the 44-TJ of energy storage at Northfield Mountain (Example 11.3) into gigawatt-hours. Is your answer compatible with the answer to part (b) of that example?

3 A transformer that reduces 115 kV to 4.2 kV for distribution within a city has 1,000 turns in its secondary coil. How many turns are in its primary?

4 Where would a point representing pumped storage lie on Figure 11.8? Answer by calculating both the energy per kilogram and per liter for water stored at a height of 200 m.

5 In Example 11.2, find the voltage between the two ends of either wire—that is, the voltage between power plant and city across either wire, *not* the voltage between the two wires at the generator output. Answer for both choices of transmission voltage.

6 It takes one H_2 molecule to make one H_2O molecule. Use that fact to verify that the 0.475 aJ released in forming a water molecule corresponds to an energy content of 142 MJ/kg for H_2 when it's combusted with oxygen.

7 Earth's atmosphere contains 550 parts per billion of H_2 by volume. Estimate the total volume of the atmosphere, approximating it as a uniform layer 60 km thick. Then use Section 11.2's figure of 12.8 MJ/m^3 to estimate the total chemical energy content of atmospheric hydrogen. If we could do the impossible and extract all that hydrogen with minimal energy expenditure, how long would it power human civilization at our current energy consumption rate of 16TW?

8 What's the chemical energy content of the 100,000 kg of liquid hydrogen that was stored in the Space Shuttle's external fuel tank? How long could this energy power human civilization at our consumption rate of 16TW?

9 A typical car gas tank holds 15 gallons. (a) Calculate the energy content of this much gasoline. (b) Calculate the chemical energy content of 15 gallons of hydrogen gas under normal conditions (12.8 MJ/m^3 as described under Equation 11.4), and compare with the energy in the gasoline. Your answer shows why pressurizing, liquefying, or chemically combining hydrogen is necessary in hydrogen-powered vehicles.

10 Honda's FCX Clarity fuel-cell car stores 3.92 kg of hydrogen gas at 5,000 pounds per square inch pressure. Determine the energy content of this much hydrogen, and compare with that of a conventional car's 20 gallons of gasoline.

11 Burning gasoline releases about 10 kg of carbon dioxide per gallon. Compare the carbon dioxide emissions from a conventional car getting 20 miles per gallon, and a fuel-cell car getting 0.5 miles per megajoule of hydrogen energy if the hydrogen is produced through 75% efficient electrolysis using electricity from a 35% efficient coal-fired power plant, assuming that coal is all carbon.

RESEARCH PROBLEMS

1 Many years of electrical demand data for Great Britain are available at www.nationalgrid.com/uk/Electricity/Data/Demand+Data/; data points are every half hour. Explore these data and determine (a) whether there are significant differences between summer and winter demand and (b) how peak demand has grown over the 10 years of available data.

2 Read the article "Electric power from offshore wind via synoptic-scale interconnection" by W. Kempton et al.

(*Proceedings of the National Academy of Sciences*, vol. 107, pp. 7240–7245, 20 April 2010) and explain the article's Figure 2 in the context of using long-distance interconnections to achieve steadier wind power.

3 Find the energy mix used to generate electricity in your country, your state, or your local utility's service area, and make a pie graph for comparison with Figure 11.2.

4 Revise the state graphs in Figure 11.2 to include the breakdown of the fossil sectors among coal and natural gas.

5 Find the U.S. Department of Energy's 2015 goals for hydrogen energy storage densities. How close are different storage technologies to meeting these goals?

ARGUE YOUR CASE

1 The California Air Resources Board has certified Honda's FCX Clarity fuel-cell car as a Zero-Emissions Vehicle (ZEV). Do you agree with this certification? Argue your point of view.

2 Many commercial transactions feature discounts for higher sales volumes. Your electric utility is proposing the opposite: a lower price per kilowatt-hour for your first 200 kWh each month, and a higher price thereafter. Do you agree with this pricing policy? Why or why not?

Chapter 12

THE SCIENCE OF CLIMATE

In previous chapters we explored humankind's prodigious energy use, amounting to many times our own bodies' energy output. Because the vast majority of our energy comes from fossil fuels, I outlined many of the environmental impacts of fossil fuel extraction, transportation, and especially combustion. But no impact from fossil fuel consumption is as global in scope as the one I haven't yet discussed in detail—namely, global climate change. In Chapters 12 to 15 I'll describe the workings of Earth's climate and especially the role of the atmosphere; the changes resulting from fossil fuel combustion and other human activities; the evidence that Earth's climate is already changing; and the computer models and other techniques for projecting future climates. Then in the book's final chapter, I'll come full circle to explore the link between human energy use and its environmental impacts, especially climate change.

12.1 Keeping a House Warm

What keeps your house warm in the winter? A heat source of some kind—a gas or oil furnace, a wood stove, a solar heating system, a heat pump, or electric heat. You need that heat to replace the energy lost to the outdoors through the walls, windows, and roof. Your thermostat turns the heat source on and off as needed to keep energy gain and loss in balance, thus maintaining a constant temperature.

Suppose your thermostat failed, leaving the heat source running continually. Would the temperature keep increasing indefinitely? No, because heat loss depends on the temperature difference between inside and outside. The hotter it gets indoors, the greater the rate at which the house loses energy. Eventually it gets hot enough that the loss rate equals the rate at which the heat source supplies energy, even if the source is on continually. At this point the house is in **energy balance**, and its temperature no longer changes. You could find that temperature by equating the known input from the heating system to the temperature-dependent loss rate, as we did in Examples 4.2 and 9.2.

12.2 Keeping a Planet Warm

A planet stays warm the same way a house does: It receives energy from a source and loses energy to its environment. If the rate of energy input exceeds the loss rate, then the planet's temperature increases and so does the loss rate, until a balance is reached. If the loss rate exceeds the input, then the planet cools and the energy-loss rate goes down, until again a balance is reached. Once the planet reaches energy balance, its temperature remains constant unless the rate of energy input varies or the planet's "insulation"—that is, any factor that determines its energy loss—changes.

A house loses energy to the outdoors primarily through conduction, although convection and radiation are also important. But a planet's environment is the vacuum of space, and the only energy-loss mechanism that works in vacuum is radiation. Thus a planet's energy loss occurs by radiation alone. When I say "planet" here, I mean the entire planet, including its atmosphere. Other mechanisms move energy between the planetary surface and the atmosphere, as we'll see later, but when we take the big picture of a planet as an isolated object surrounded by the vacuum of space, then the only way the planet can lose or gain energy is through radiation.

The simplest estimate of a planet's temperature will therefore involve balancing incoming and outgoing radiation. Chapter 4 introduced the Stefan–Boltzmann radiation law, which gives the rate at which an object with absolute temperature T, area A, and emissivity e radiates energy:

$$P = e\sigma AT^4 \tag{12.1}$$

where $\sigma = 5.67 \times 10^{-8}$ W/m²·K⁴ is the Stefan–Boltzmann constant. P here is for *power*, since the radiation rate is in joules per second, or watts. For a planet losing energy by radiation alone, Equation 12.1 gives the total energy loss rate. To reach energy balance, the loss rate must equal the rate of energy input. That's an easy equation to set up, but first we need to account for the fact that planets are spheres.

For planets in our Solar System, the dominant energy source is the Sun. Figure 12.1 shows schematically the solar energy input and radiation loss for a planet, in this case Earth. As Figure 9.1 showed, sunlight reaches Earth's curved, Sun-facing hemisphere at the same rate it would reach a flat surface whose area is Earth's cross-sectional area, πR_E^2, where R_E is Earth's radius. The corresponding energy flow is the solar constant (S), about 1,364 W/m^2.

On the other hand, as Figure 12.1 shows, Earth loses energy from its entire surface area. Since the surface area of a sphere is $4\pi r^2$, or four times its cross-sectional area, the energy-balance equation for Earth looks like this:

$$\pi R_E^2 S = e\sigma(4\pi R_E^2)T^4$$

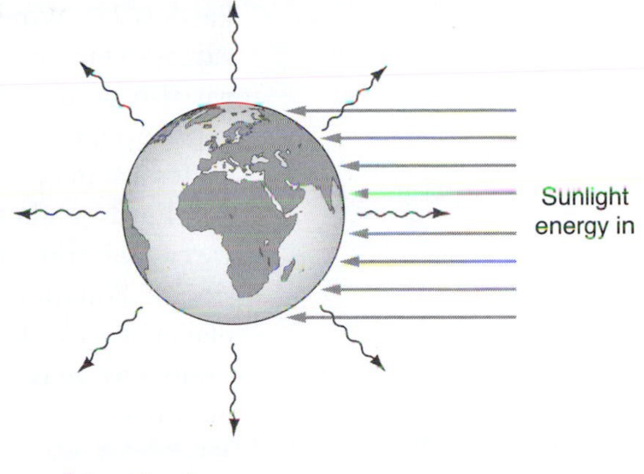

FIGURE 12.1
Earth's energy balance, showing the energy of incoming sunlight and the energy Earth radiates back to space.

Here the left-hand side of the equation is just the total energy reaching Earth—the product of the planet's cross-sectional area and the solar constant. The right-hand side is the radiation loss given by Equation 12.1, using $4\pi R_E^2$ for the area A. I've left this equation in symbols for now because we don't really need to know Earth's radius. If we divide both sides by $4\pi R_E^2$, the term πR_E^2 cancels and we're left with

$$\frac{S}{4} = e\sigma T^4 \quad (12.2)$$

This, too, is an energy-balance equation, but now it's on a per-square-meter basis. The left-hand side of Equation 12.2 gives the *average* rate at which solar energy reaches Earth. We found the same result in Chapter 9 when we recognized that the average insolation was one-fourth of the solar constant. That factor of 4 comes from the fact that Earth presents an effective area πR_E^2—its cross-sectional area—to the incident sunlight, but loses energy from its entire surface area, namely $4\pi R_E^2$. Physically, the factor of 4 accounts for Earth's spherical shape and for night and day.

Equation 12.2 thus describes any planet in energy balance. Given the solar constant for a particular planet—that is, the intensity of sunlight at that planet's distance from the Sun—we can use Equation 12.2 to find a measure of the planet's average temperature. We'll do that shortly for Earth. But first let me remind you that we're dealing here with two very different regions of the electromagnetic spectrum. The Sun, at a hot 5800 K, emits largely visible light and infrared at wavelengths adjacent to the visible. Earth, at its much lower temperature, emits infrared at much longer wavelengths. The way these two different wavelength ranges interact with the Earth–atmosphere system determines many of the subtleties of climate. In particular, reflection and absorption are very different for short- and long-wavelength radiation. As we delve deeper into climate science, it's important to keep in mind the wavelength range to which a given property—like reflection, emissivity, or absorption—applies.

We've just seen that, for Earth, the left-hand side of Equation 12.2 would be $S/4 = 341$ W/m^2 if our planet absorbed all the solar energy incident on it. But it doesn't: Some 30% is reflected directly back into space, mostly from clouds but also from the surface. As a result, the left-hand side of Equation 12.2 for Earth is 239 W/m^2. In energy balance, the right-hand side must be equal to this value—a fact that we're trying to use to solve for the temperature T. To do that, we also need Earth's emissivity e. Here's where that wavelength distinction is important. I've just said that some 30% of incident sunlight gets reflected. That's a statement about the short-wavelength solar radiation. But in the infrared, Earth (and other Solar System planets) aren't very reflective. That means we can take the emissivity e for infrared to be essentially 1. Since the right-hand side of Equation 12.2 describes the radiation emitted by the relatively cool planet, $e = 1$ is the value to use here.

Now we're ready to compute Earth's temperature. Setting the left-hand side of Equation 12.2 to 239 W/m^2, and using $e = 1$ we have

$$239 \text{ W/m}^2 = (1)(5.67 \times 10^{-8} \text{ W/m}^2\cdot\text{K}^4)T^4$$

This gives

$$T^4 = \frac{239 \text{ W/m}^2}{5.67 \times 10^{-8} \text{ W/m}^2\cdot\text{K}^2} = 4.22 \times 10^9 \text{ K}^4$$

so

$$T = (4.22 \times 10^9 \text{ K}^4)^{1/4} = 255 \text{ K}$$

That 255 K amounts to −18°C, or 0°F. Does this make sense for Earth's average temperature, lumping night and day, poles and equator, into a single number? It doesn't sound unreasonable—after all, it doesn't describe a planet whose oceans boil, or whose atmosphere freezes solid. But it's well below the freezing point of water, which sounds low for a global average. Earth's actual average temperature is about 288 K (15°C or 59°F)—some 33°C or 59°F warmer than our estimate. Why the difference? The answer lies in a profoundly important climate effect that I describe in Section 12.3.

The energy-balance equation we've set up here is, in fact, a very simple climate model. We solved that model to get a single quantity, Earth's average temperature. Because it treats Earth as a single point with no structure—no variation in latitudes, no oceans, no atmosphere—our model is a **zero-dimensional energy-balance model**. It's certainly not a perfect model, not only because it ignores the Earth's three-dimensional structure, but also because it doesn't even give the right value for Earth's average temperature.

EXAMPLE 12.1 | Martian Climate

Mars is about 50% farther from the Sun than is Earth. Since the intensity of sunlight drops off as the inverse square of the distance from the Sun, the solar constant at Mars is less than half that at Earth, about 592 W/m^2. Ignoring any reflection of solar energy from Mars, estimate the average Martian temperature.

SOLUTION

We need to solve Equation 12.2 using 592 W/m² for the solar constant. Rather than manipulating lots of numbers, let's first solve symbolically for the temperature T:

$$T^4 = \frac{S}{4e\sigma} \text{ or } T = \left(\frac{S}{4e\sigma}\right)^{1/4}$$

Now it's straightforward to put in the numbers:

$$T = \left(\frac{S}{4e\sigma}\right)^{1/4} = \left(\frac{592\ \mathrm{W/m^2}}{4(1)(5.67\times10^{-8}\ \mathrm{W/m^2\cdot K^4})}\right)^{1/4} = 226\ \mathrm{K}$$

This is cooler than our estimate for Earth, but not as cool as you might have expected given that the sunlight intensity at Mars is less than half that at Earth. That's because the fourth-power dependence of radiation loss on temperature means a relatively small change in temperature can balance a substantial change in sunlight intensity, making planetary temperatures only weakly dependent on a planet's distance from the Sun. Incidentally, this fact encourages astrobiologists looking for extraterrestrial life, because it means there's a rather broad habitable zone around Sun-like stars.

12.3 In the Greenhouse

Why does Equation 12.2 result in an estimate for Earth's average temperature that's too low? In my example of a home in energy balance, I treated the house as a single entity and ignored its detailed structure, just as I did when calculating Earth's temperature. But a house may be warmer in the living room than in an upstairs bedroom, for example, and air may flow from the warm living room to the cooler upstairs—an example of convective heat transfer. These and other processes act to maintain different temperatures in the different parts of the house. The house is, in fact, a complex system, whose overall "climate" involves not only the loss of energy to the outside but also exchanges of energy among its different rooms.

It's the same for Earth. Our planet has different "rooms," such as the tropical regions and the Arctic—or even finer gradations: a forest, a desert, a lake, an island. It has different "floors," including the ocean depths, the land surface, and the atmosphere. The oceans and atmosphere themselves have multiple levels with different properties. All in all, Earth is a pretty complex "house." The fullest understanding of climate requires that we account for all climatologically significant variations. That's the job of large-scale computer climate models, as we'll see in Chapter 15. But we can make a lot of progress by considering

Earth to be like a simple house with just two rooms, one downstairs and one upstairs—the surface and the atmosphere. In lumping our entire planet into just surface and atmosphere, we ignore a lot of structure—for example, latitudinal variations, oceans versus land, ice versus liquid water, desert versus forest, stratosphere versus troposphere, and the gradual decline in the temperature of the lower atmosphere with increasing altitude. What we gain is a simple, comprehensible model that does a surprisingly good job of explaining the global properties of Earth's climate, including changes resulting from human activities.

RADIATION AND THE ATMOSPHERE

I noted earlier that Earth and its atmosphere interact very differently with electromagnetic radiation of different wavelengths. For example, Earth's atmosphere is, obviously, largely transparent to visible light. I say "obviously" because we can see the Sun, other planets, and distant stars. This simple observation tells us that most visible light passes through Earth's atmosphere and reaches the surface. Clouds and particulate matter of natural or anthropogenic origin reduce atmospheric transparency somewhat, but transparency remains high for the short wavelengths of visible and near-visible infrared light.

The situation is very different for the long-wavelength infrared emitted by the cooler Earth. Certain gases with more complicated molecular structures than the dominant nitrogen (N_2) and oxygen (O_2) strongly absorb long-wavelength infrared. Most important of these are triatomic water vapor (H_2O) and carbon dioxide (CO_2). The many rotational and vibrational motions of these complex molecules happen to lie at frequencies that include those of Earth's infrared emissions, resulting in the molecules absorbing infrared energy.

We can summarize these observations about the short- and long-wavelength properties of the atmosphere with a statement that is of crucial importance in understanding climate and our influence on it: Earth's atmosphere is largely transparent to incoming solar radiation, but it's largely opaque to outgoing infrared.

THE GREENHOUSE EFFECT

Those infrared-absorbing gases in Earth's atmosphere are called greenhouse gases. The greenhouse gases inhibit outgoing infrared radiation, making it difficult for Earth to shed the energy it gains from the Sun. This is a simplistic description of the **greenhouse effect**. The same thing happens in a greenhouse, where glass plays the role of the atmosphere in admitting visible light but inhibiting outward heat loss. However, the term *greenhouse effect* is a misnomer, because in a greenhouse the dominant effect of glass is to block convective heat loss rather than radiation. It's true that glass is much less transparent to infrared than to visible light, but in a real greenhouse this is of secondary importance. However, the terms *greenhouse gas* and *greenhouse effect* are universally accepted, so I'll stick with them.

I'm now going to present three ways to understand the greenhouse effect, in increasing order of scientific sophistication. No one explanation does full justice to the complexity of the climate system and the role of greenhouse gases, but each provides insights into the process that keeps our planet habitably warm and that will likely warm it further in the near future.

An Insulating Blanket The most simplified description of the greenhouse effect might better be called the *blanket effect*. Atmospheric greenhouse gases have somewhat the same effect as an insulating blanket that you pull over yourself on a cold night, or extra insulation you might add to your house to cut your heating bills. The extra insulation initially reduces the outgoing energy flow, and as a result the temperature rises. As the temperatures rises, up goes the energy loss rate—until a new energy balance is established at a higher temperature. Your house and your body both have thermostatic mechanisms to lower their heating rate and maintain a comfortable temperature, resulting in less fuel consumption or lower metabolism with the more effective insulation. But Earth lacks such a thermostat, and as a result the presence of atmospheric greenhouse gases makes Earth's surface temperature higher than it would be in the absence of those gases.

The difference between our 255-K estimate of Earth's average temperature and its actual 288 K (a difference of 33 K, 33°C, or 59°F) is due to naturally occurring greenhouse gases, predominantly water vapor and, to a lesser extent, CO_2. (That's right: The dominant natural greenhouse gas isn't CO_2, but H_2O. Methane, ozone, and nitrous oxide also contribute.) This is the **natural greenhouse effect**, and without it our planet would be a chilly, inhospitable place. Life might still be here, but would civilization? During the last ice age, when there was a glacier 2 miles thick atop much of Europe and North America, the global average temperature was only about 6°C colder than today. Imagine a greenhouse-free planet averaging 33°C colder!

Although the "insulating blanket" description helps us understand the greenhouse effect in familiar terms, the physical mechanism of the greenhouse effect is very different from that of blankets and home insulation, which primarily reduce conductive and convective heat flow. A fuller understanding requires a more sophisticated look at the greenhouse effect.

Infrared, Up and Down Imagine an Earth without greenhouse gases and in energy balance at that cool 255 K. Sunlight delivers energy to the planet at the average rate of 239 W/m^2. This scenario assumes no change in the amount of reflected sunlight, which is unlikely given the probable ice cover under such cold conditions, but we'll make that assumption for the sake of illustration. Since it's in energy balance, our theoretical Earth is radiating infrared energy at the same rate, 239 W/m^2. Surface-emitted infrared escapes directly to space, since the greenhouse-free atmosphere is transparent to infrared as well as to visible light.

Now introduce greenhouse gases into the atmosphere. Water vapor and CO_2 begin absorbing infrared radiation coming up from the surface. The incoming

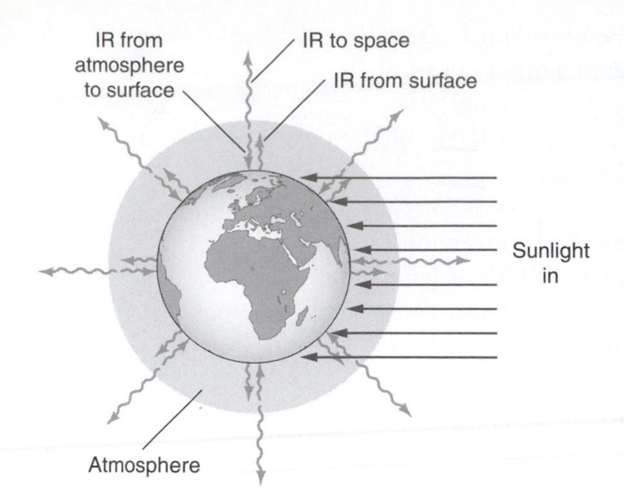

FIGURE 12.2

Greenhouse gases absorb infrared (IR) radiating up from Earth's surface, and the atmosphere radiates IR both upward and downward. The downward IR flow warms the surface, eventually establishing a new energy balance at a higher surface temperature. This drawing is highly simplified: The greenhouse process occurs throughout the atmosphere, which is much thinner than shown; also, not all of the incident sunlight reaches the surface.

solar energy hasn't changed, but now there's less infrared going out, so the planet is no longer in energy balance. As a result, the surface starts to warm up, and up goes its $e\sigma T^4$ rate of infrared emission, driving the planet back toward balance at a higher surface temperature. This is basically a more accurate description of my "blanket" explanation of the greenhouse effect, but the interactions are more complex. As the greenhouse gases absorb infrared, they warm up and share their energy with the surrounding atmosphere. So the atmosphere warms, and then it, too, radiates infrared—both upward to space and downward to the surface. Figure 12.2 modifies the simple energy-balance diagram of Figure 12.1 to include this infrared energy radiating both ways.

Because the greenhouse-warmed atmosphere is radiating infrared downward, the total rate at which energy reaches Earth's surface is considerably greater than it would be from sunlight alone. As a result, the surface warms. As the surface temperature increases, so does the rate, $e\sigma T^4$, at which the surface radiates infrared. Eventually that rate is high enough that the surface is losing energy at the same rate at which sunlight and the downward infrared flow from the atmosphere are delivering it. At that point the surface has re-established energy balance, but at a higher temperature than before. When all is settled, the atmosphere, too, is in a new state of energy balance.

Figure 12.2 and the discussion in the preceding paragraph might seem to suggest that we're getting something for nothing. After all, how can the total energy coming to Earth's surface exceed that of the incident sunlight, given that sunlight is, ultimately, essentially the only significant source of energy? The answer lies in the *net* energy flow. Sure, there's more energy coming to Earth's surface than there would be in the absence of greenhouse gases. But there's also more energy leaving the surface, as required for energy balance. Later we'll take a more quantitative look at Earth's overall energy budget, and you'll see how these various up/down flows all relate. If, after that, you're still bothered by that large downward flow from the atmosphere, Box 12.2 will give a helpful analogy with your checking account.

Warm Below, Cool Above Details of Earth's energy balance are complex, but concentrating for now on the loss of energy to space is sufficient to provide a still more sophisticated picture of the greenhouse effect—one that additionally shows why Earth's surface and atmosphere end up at different temperatures. In this simplified theoretical picture, I'll represent the surface and atmosphere as two distinct boxes that can exchange energy with each other. In addition, energy from the Sun arrives at this combined system, and the system loses energy by infrared radiation to space. Figure 12.3 diagrams the relevant energy flows. This isn't a complete energy-flow diagram, because I haven't shown reflection and absorption of the incident sunlight or all the energy exchanges between surface and atmosphere.

On the input side, Figure 12.3 shows a single arrow representing sunlight energy absorbed by the surface–atmosphere system. On the output side are two arrows, one representing infrared radiation from the atmosphere to space, the other infrared radiation from the surface. Each arrow's width reflects the size of the corresponding energy flow. At the top of the atmosphere, the widths of the two outgoing arrows combined give the same width as the incoming absorbed sunlight. This represents the planet in energy balance, with all the energy it absorbs from the Sun being radiated to space.

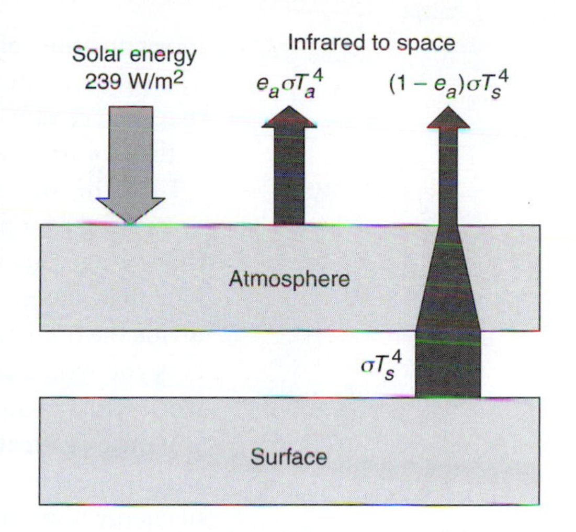

FIGURE 12.3
Simple two-box climate model showing only the incoming solar energy and the outgoing infrared radiation from the surface-atmosphere system. Greenhouse gases absorb most of the surface radiation, so very little gets through the atmosphere. Surface emissivity in the infrared is very nearly 1, so e_s isn't shown in the expression for surface radiation.

Why does the arrow representing the surface radiation taper as it passes through the atmosphere? Because greenhouse gases absorb some of the outgoing infrared. In fact, as the diagram suggests, very little of the surface radiation escapes directly; most is absorbed in the atmosphere. How much? That depends on the concentration of greenhouse gases. We can summarize the atmospheric absorption quantitatively with a number that indicates how effectively the atmosphere absorbs infrared radiation. Here's where fundamental physics comes in: As indicated in Chapter 4 when I introduced the Stefan–Boltzmann radiation law ($P = e\sigma AT^4$), an object's effectiveness at radiating energy—described by its emissivity, e—is the same as its effectiveness at absorbing energy. In general, emissivity depends on wavelength, and the statement about radiation and absorption effectiveness being equal holds at each wavelength. This means the emissivity may be quite different at visible and infrared wavelengths, but whatever its value is for, say, infrared, that value describes the object's effectiveness at both radiating and absorbing infrared. For that reason we can call e either *emissivity* or *absorptivity*, depending on which process we're discussing.

This is all reflected in the formulas I've put next to the arrows in Figure 12.3. The surface radiation is labeled σT_s^4, indicating that radiation from the surface depends on the *surface* temperature T_s. I didn't bother with e in this term because Earth's surface emissivity for infrared wavelengths is very close to 1. The radiation leaving the top of the atmosphere is determined by the atmospheric temperature T_a and emissivity/absorptivity e_a, whose value reflects the greenhouse gases' absorptive properties. Consequently I've labeled that arrow $e_a \sigma T_a^4$. Finally, there's the arrow for surface radiation escaping to space. It's smaller than the radiation leaving the surface, because much of the surface-emitted infrared has been absorbed in the atmosphere. How much smaller? If the atmospheric absorptivity were 1, then the atmosphere would be a perfect absorber and no radiation would escape to space. If the atmospheric absorptivity were 0.75, then three-fourths of the surface radiation would be absorbed and the remaining one-fourth would escape. In general, if the atmospheric absorptivity/emissivity is e_a, then the fraction of surface radiation absorbed in the atmosphere is e_a, and the remainder, namely $1 - e_a$, escapes to space (remember that e is between 0 and 1). So I've labeled the surface radiation escaping to space with the term $(1 - e_a)\sigma T_s^4$. This may seem odd, because it mixes an atmospheric property (e_a) with a surface property (T_s). But it's correct: The surface radiation depends on the *surface* temperature, but the amount of that surface radiation that escapes to space depends on how much is absorbed in the *atmosphere*.

I left the area A out of all of these expressions, which means I'm keeping everything on a per-square-meter basis; the terms and arrows shown represent the average radiation from a square meter of Earth's surface or atmosphere. Not that any given square meter radiates these amounts, but if you were to divide the total radiation from the planet by its surface area, these are the numbers you would get. In nearly all discussions of climate, it's more convenient to work with per-square-meter energy flows rather than total flows. And I remind you once again about that all-important difference between energy and power; strictly speaking, the arrows and terms in Figure 12.3 describe power—the rate of energy flow, in this case in watts per square meter—rather than energy itself.

Now let's add the two terms representing infrared radiation escaping to space:

$$\text{Power radiated to space} = e_a \sigma T_a^4 + (1 - e_a)\,\sigma T_s^4$$

Here the first term on the right-hand side is the infrared radiated from the atmosphere to space, and the second term is the surface radiation that made it through the atmosphere. Again, both are in watts per square meter. Now apply the distributive law to the second term:

$$\text{Power radiated to space} = e_a \sigma T_a^4 + \sigma T_s^4 - e_a \sigma T_s^4$$

Next, group the terms that involve the atmospheric emissivity, e_a, and pull out the common factor $e_a\sigma$:

$$\text{Power radiated to space} = \sigma T_s^4 - e_a \sigma (T_s^4 - T_a^4) \qquad (12.3)$$

The signs are correct; check them out! I could have used a plus sign between the two main terms and reversed the terms in parentheses, but the point I want to make is clearer the way I chose to express the math here.

Equation 12.3 gives the total power radiated to space from the Earth system—that is, from the combination of surface and atmosphere. (Well, not quite—it's the *average* power radiated *per square meter*. You could multiply it by Earth's surface area if you really wanted the total power.) For energy balance, that quantity must equal the rate at which the planet absorbs energy from the Sun. Absent any changes in absorbed solar energy—which could occur if the Sun's power output varied, or Earth's reflectivity changed—the quantity in Equation 12.3 remains constant if Earth is in energy balance.

To get comfortable with what Equation 12.3 says, consider first the special case when $e_a = 0$. This scenario would mean that no infrared absorption occurs in the atmosphere—in other words, there are no greenhouse gases. The energy radiated to space would all come from the surface. The rate of surface emission would be just σT_s^4, and if we equated that to the incoming solar energy, we would get back the cold 255-K average surface temperature from our earlier calculation that didn't take the greenhouse effect into account. So Equation 12.3 makes sense in the absence of greenhouse gases.

Now take a good look at the second term in Equation 12.3, the term that went away when we considered the case $e_a = 0$. Normally, Earth's surface is warmer than the atmosphere; averaged over the globe and the thickness of the atmosphere, this is always true. So $T_s > T_a$, and the second term in Equation 12.3 is *positive*. But this term is *subtracted* from the first term. So if nothing else changed, an increase in atmospheric infrared absorption—that is, an increase in greenhouse gases, expressed mathematically as an increase in e_a—would *decrease* the total outgoing infrared. This is hardly surprising, since greenhouse gases absorb infrared radiation. But in our model, the outgoing infrared has to balance the incoming solar energy, so it can't stay decreased once Earth reaches energy balance. The only way to keep the total outgoing infrared constant in the face of increasing e_a is to increase the first term on the right-hand side of Equation 12.3, namely the surface-radiation term σT_s^4. That corresponds to increasing the surface temperature T_s. Translation from math to English: As the concentration of atmospheric greenhouse gases rises, so does the surface temperature.

The second term in Equation 12.3 is called the **greenhouse term**, because it provides a mathematical description of the greenhouse effect as a reduction in outgoing infrared radiation, a reduction that requires a compensating increase in surface temperature in order to maintain energy balance. Physically, the greenhouse gases cause that reduction, and their effect is summarized mathematically in the value of the atmospheric emissivity/absorptivity e_a.

It isn't enough to have $e_a > 0$ for the greenhouse effect to occur. It's also essential that the atmosphere be, on average, cooler than the surface. As Equation 12.3 shows, the greenhouse term can be zero if *either* $e_a = 0$ *or* $T_a = T_s$. A greenhouse atmosphere at the same temperature as the surface would mean no *net* infrared flow between atmosphere and surface, hence no surface warming. It's the cooler upper atmosphere—a less effective radiator, given the σT^4 radiation-loss rate—that drives the greenhouse effect. In fact, for a given concentration of greenhouse gases (a fixed e_a), a *cooler* atmosphere *increases* the greenhouse effect. I emphasize this point to dispel one misconception about global warming resulting from increased greenhouse gases, namely the notion that we should expect warming at all levels of the atmosphere. In fact, as we'll see in Section 12.4, a strengthened greenhouse effect results simultaneously in *cooling* of the upper atmosphere and *warming* of the surface and lower atmosphere.

FIGURE 12.4
Infrared image taken from GOES-15, the 15th Geostationary Operational Environmental Satellite, just after its launch in 2010. Photo shows IR emission in the 10-μm wavelength band, characteristic of Earth's thermal radiation. The GOES satellites provide information for weather forecasting, and these images are especially useful for distinguishing cloud temperatures and hence cloud heights.

If you looked at Earth from outer space with infrared eyes (Fig. 12.4), you would see predominantly infrared radiation emitted from well up in the atmosphere (about 6 km altitude, on average). Very little of the radiation would come directly from the surface, because greenhouse gases absorb most surface

BOX 12.1 | A Greenhouse Complication

The real climate system is much more complex than my simple descriptions suggest. Here's just one example: As I've indicated, water vapor is the dominant natural greenhouse gas. In contrast to CO_2, whose concentration is more uniform, water vapor's concentration varies with geographical location, altitude, and time. Water vapor in the lower atmosphere, where the temperature isn't hugely different from that of the surface, makes for a higher effective T_a in Equation 12.3, which means a less substantial greenhouse effect. The same water vapor located in higher layers of the atmosphere absorbs and emits infrared radiation at a lower effective temperature, which strengthens the greenhouse effect and causes greater surface warming. Thus, subtle details such as the specific processes that carry evaporated ocean water into the atmosphere have a direct impact on the strength of the greenhouse effect. That's just one of many complex factors that serious climate modelers must take into account.

radiation. The radiation you would see averages 239 W/m^2, balancing the solar energy that's absorbed by the surface–atmosphere system at this same rate. You could infer an average planetary temperature, which would be close to our earlier 255-K estimate in the absence of greenhouse gases. But you'd be missing the greenhouse effect and other details that depend, ultimately, on the interaction between surface and atmosphere. The greenhouse effect plays the dominant role in that interaction, and it gives us a warmer surface.

Of course, my discussion of the greenhouse effect has been based on an overly simple model—that of Earth's surface and atmosphere as being just two distinct regions, each characterized by a single temperature. Even if we continue to ignore surface variations with latitude, longitude, water, land, or ice—that is, treat the surface as a single "box"—there are still obvious variations in the atmosphere related to altitude. So my analysis of the greenhouse effect is a bit naïve, and we must regard the symbol T_a as standing for some sort of effective average temperature of the upper atmosphere throughout the region where infrared emission has a good chance of escaping directly to space. We get the big picture right, but we've glossed over many details.

12.4 Earth's Energy Balance

Figure 12.1 gives a very simple picture of Earth's overall energy balance, showing just the incoming solar energy and the outgoing infrared radiated to space. It ignores the greenhouse effect and other interactions between Earth's surface and atmosphere. Figures 12.2 and 12.3 are improvements that account for the effect of greenhouse gases on infrared radiation from the surface. But none of these diagrams comes anywhere close to showing the full complexity of the interactions among incident sunlight, outgoing infrared, and Earth's surface and atmosphere. We can't hope to understand the climate system, and especially the climatic effects of human activity, unless we look more closely at these interactions.

Figure 12.5 represents our best understanding today of the global average energy flows that establish Earth's climate. Once again we ignore geographical variations, lumping Earth's entire surface into one "box" and the entire atmosphere into another. However, more detail is shown in the atmospheric box, with the presence of discrete clouds and processes involving them. Energy flows are labeled with the average rate of energy transport in, as usual, watts per square meter. Again, the width of a given flow reflects quantitatively the size of the flow. The numbers in Figure 12.5 come from a variety of observations and calculations, with satellite instruments providing the crucial flows at the top of the atmosphere. Although I've rounded the figures to integers, many are now known to within a few tenths of a W/m^2.

At the top center of Figure 12.5 is the incoming solar radiation at 341 W/m^2—a number that results, as I described earlier, from averaging the direct 1,364 W/m^2 above Earth's atmosphere over the full surface area of our spherical planet. I've alluded before to the 30%—some 102 W/m^2—of the incoming solar radiation that's reflected back into space without being absorbed by the Earth or its atmosphere. In Figure 12.5 you can see that most

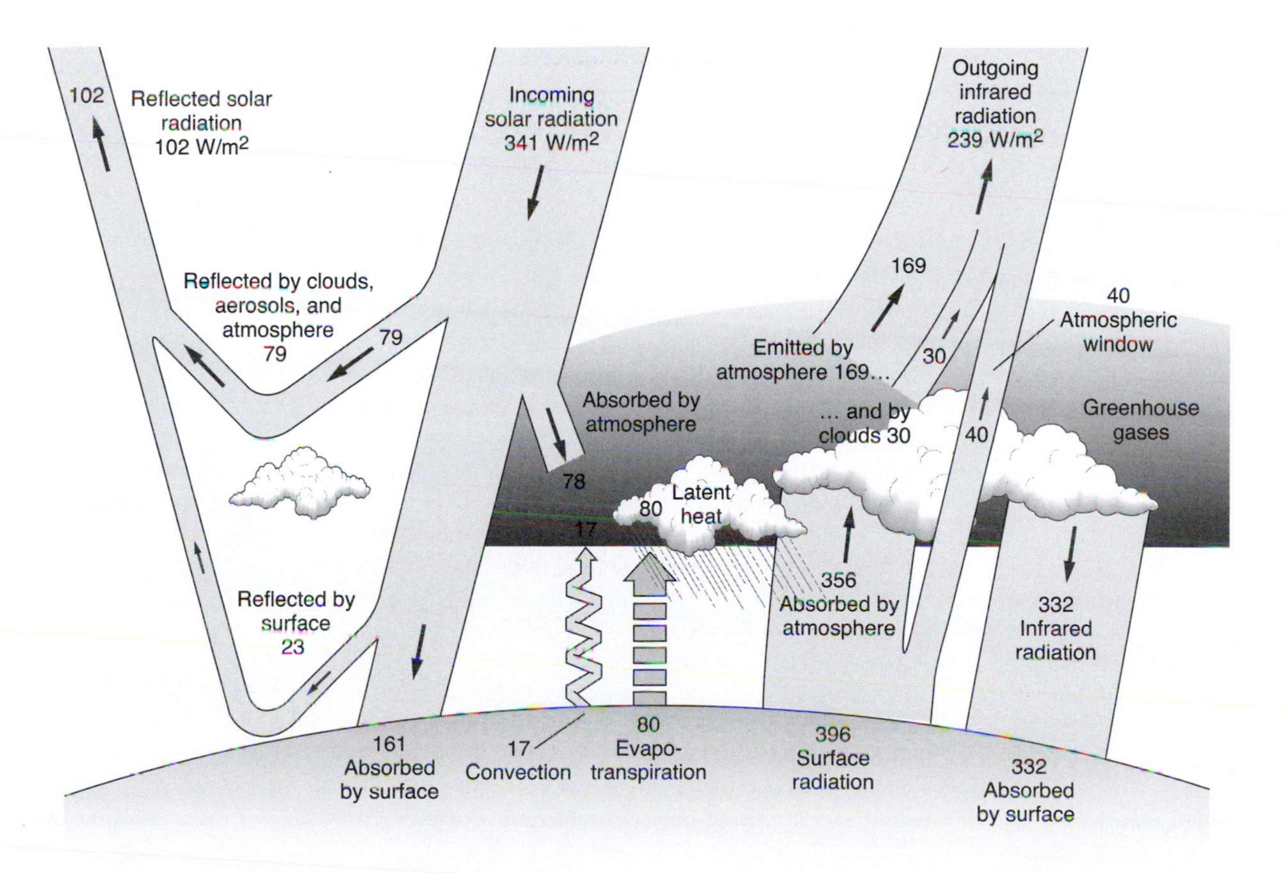

FIGURE 12.5

Interactions between Earth's surface and atmosphere are the determining factors in the planet's average energy balance. Numbers are rounded, and show a balance between ingoing and outgoing energy flows at both surface and the top of the atmosphere. Actually, Earth receives nearly 1 W/m^2 more than it emits—an indication of ongoing warming.

of that reflection—79 W/m^2—is from clouds and reflective particles (aerosols) in the atmosphere. Another 23 W/m^2 reflects from the surface, much of it from ice, snow, and desert sands.

The fraction of sunlight that's reflected is called **albedo**, and it's a major determinant of climate. That's because reflection reduces the energy absorbed by the Earth–atmosphere system and thus directly affects the overall energy balance. Figure 12.5 shows that Earth's average albedo is (102 W/m^2)/(341 W/m^2), or 0.30. Occurrences such as melting ice, advancing deserts, and vegetation changes all affect albedo and can therefore have significant effects on climate. Exercise 7 and Research Problem 2 explore Earth's albedo and its climatological significance in more detail.

Take away the reflected 102 W/m^2 from the 341 W/m^2 of incident sunlight, and you're left with 239 W/m^2 as the rate at which the Earth–atmosphere system actually absorbs solar energy. Although the clear atmosphere is largely transparent to sunlight, the presence of clouds and aerosols makes for a significant atmospheric absorption of 78 W/m^2. This leaves 161 W/m^2—just under half of the total incident solar energy—to be absorbed at the surface.

Absent any other effects, Earth's surface would have to rid itself of energy at that rate of 161 W/m^2. But there's more energy reaching the surface—namely, downward-flowing infrared from atmospheric greenhouse gases and clouds, which amounts to 332 W/m^2. So the surface has to rid itself of a total of 161 + 332, or 493 W/m^2. Most of this energy is in the form of infrared radiation, only 40 W/m^2 of which escapes directly to space.

BOX 12.2 | A Banking Analogy

Are you still bothered that the downward radiation reaching the surface is greater even than the solar input to the entire Earth system? Here's an analogy: Suppose you have $161 deposited into your checking account at the beginning of each month (analogous to the solar energy absorbed by Earth's surface), and you spend an average of $161 over the month (the energy the surface loses in order to maintain energy balance). Then, averaged over the month, your account balance would remain unchanged. But if that balance were large enough (analogous to the huge amount of energy stored in the Earth–atmosphere system), you could still lend your brother $332 each month—and as long as he paid you back before the month's end, you'd come out even. That $332 is like infrared going both ways between atmosphere and Earth. Your bank statement would show total deposits of $161 + $332 = $493 for the month, yet the only money coming in from outside the system would be the $161 that's analogous to the solar input to Earth's surface.

This analogy still doesn't capture all the complexity of Figure 12.5. For example, it doesn't include the three different ways you pay your brother, namely infrared radiation, convection, and evapotranspiration. And it doesn't include the fact that your brother pays some of your bills for you (energy from the surface that's "laundered" through the atmosphere before going out to space), so you actually give him even more than $332 each month. But even in its simpler version, the analogy shows clearly how the large flows of infrared between Earth and atmosphere are entirely self-consistent.

However, a significant portion of the surface energy loss isn't by radiation but through two related processes that amount to bulk motion of energy into the atmosphere. Convection, which I introduced back in Chapter 4, results when heated surface air rises and takes internal energy with it. That averages some 17 W/m^2 over Earth's surface. The second process, **evapotranspiration**, moves energy from the surface to the atmosphere in the form of the latent heat of water vapor. Recall from Chapter 4 that it takes energy to change a substance from liquid to gas, energy that's recovered if the gas recondenses to liquid. This process amounts to 80 W/m^2 of upward energy flow, associated with H_2O evaporated from surface waters as well as released into the air by plants in the process of transpiration, whereby soil moisture extracted by plant roots rises through trunks, stems, and leaves and evaporates into the atmosphere. When this atmospheric water vapor recondenses to form liquid droplets and clouds, the latent energy is released and heats the atmosphere.

Together, radiation, convection, and evapotranspiration from Earth's surface carry a total of 493 W/m^2 into the atmosphere. This is equal to the 493 W/m^2 flowing downward from both sunlight (161 W/m^2) and infrared from the atmosphere (332 W/m^2). So the surface is in energy balance.

What about the atmosphere? It absorbs 78 W/m^2 of sunlight energy, 17 W/m^2 coming up from the surface by convection, 80 W/m^2 carried upward by evapotranspiration, and 356 W/m^2 of infrared from the surface (the remaining 40 W/m^2 of the 396 W/m^2 total surface infrared is the small portion that escapes directly to space). That makes a total of 531 W/m^2 coming into the atmosphere. On the outgoing side, the atmosphere sends infrared energy down to the surface at the rate of 332 W/m^2 and up to space at 199 W/m^2 (169 W/m^2 from greenhouse gases and 30 W/m^2 from clouds). The downward radiation from atmosphere to surface is greater than the upward flux from atmosphere to space because the higher levels of the atmosphere, where the radiation to space originates, are cooler than the lower levels where the downward-directed radiation comes from (remember that radiated power depends on T^4). So the total rate of energy loss from the atmosphere is $332 + 199 = 531$ W/m^2. This number matches the atmosphere's incoming energy, so the atmosphere, too, is in energy balance.

Finally, the total upward radiation to space from the Earth–atmosphere system includes the 169 W/m^2 radiated upward from the atmosphere, another 30 W/m^2 radiated from clouds to space, and the 40 W/m^2 that manages to escape directly from the surface to space. Those figures add up to 239 W/m^2—precisely the rate at which the Earth–atmosphere system absorbs solar energy. So the entire system is in energy balance. Well, almost.

EARTH'S ENERGY *IMBALANCE*

Energy balance is never quite perfect: The distribution of incoming solar energy is always changing with seasons, or with longer-term fluctuations like ice-age cycles that alter Earth's albedo and hence the fraction of sunlight reflected, or with factors like volcanic eruptions that change atmospheric transparency.

Then there's the slight imbalance associated with the tiny fraction of incoming solar energy that's being stored as fossil fuels. These changing factors make for a planet that's generally very close to energy balance but never quite there. In this chapter I've presented a theoretical understanding of climate science that assumes energy balance; given that natural deviations from the balanced state are very small, that's an excellent approximation and one that provides a clear picture of how climate is established.

However, satellite measurements such as those that go into Figure 12.5 suggest an unusually large imbalance today, with some 0.9 W/m^2 more energy coming in than going out. This excess energy is warming the planet, although the surface warming is moderated because much of the excess energy goes into the oceans. Today's anomalous energy imbalance results largely from increased greenhouse gas concentrations, due primarily to fossil fuel combustion. That 0.9 W/m^2 is uncertain within a few tenths of a watt per square meter, and in any event it's tiny compared with the hundreds of watts per square meter associated with most flows in Figure 12.5, so energy balance remains an excellent approximation for understanding Earth's overall climate. But the imbalance is real, and it's arguably the most fundamental indication of a human influence on climate.

Recall that I rounded the numbers in Figure 12.5 to integers; that's so I could discuss Earth's energy flows in the context of this chapter's emphasis on energy balance. Had we looked to the right of the decimal point, though, we would have found that things don't quite add up—and the difference, at the top of the atmosphere, would be 0.9 W/m^2 of excess energy coming in compared with what's going out. At this point a global warming skeptic might claim that 0.9 W/m^2 is so small compared with the actual energy flows that we can't be very sure of it. That's true, and although climate scientists remain confident in the reality of the measured energy imbalance, I hasten to add that this imbalance is not our main line of evidence for anthropogenic global warming. That evidence comes from increases in greenhouse gases that we'll discuss in the next chapter; from warming trends, ice melt, sea-level rise, and ecosystem changes that we'll consider in Chapter 14; and from computer models to be described in Chapter 15. Confirmation of an actual energy imbalance is a recent addition to the growing evidence for anthropogenic climate change, and is more like nonessential "icing on the cake" than something we must have in order to be convinced of the reality of climate change.

EXAMPLE 12.2 | Energy Balance and the Greenhouse Effect

(a) Use Figure 12.5 to determine the value of the atmospheric emissivity e_a. (b) Assuming energy balance and the actual surface temperature of 288 K, use Equation 12.3 to determine the effective atmospheric temperature.

SOLUTION

(a) Figure 12.5 shows the surface radiating 396 W/m^2, of which only 40 W/m^2 get through the "atmospheric window" directly to space. So 356 W/m^2 must be absorbed in the atmosphere. This gives an absorptivity, and thus an emissivity, of 356/396, or $e_a = 0.899$. (b) In energy balance, the radiated power given in Equation 12.3 must equal the 239 W/m^2 of unreflected solar input, so Equation 12.3 becomes $239\ \text{W/m}^2 = \sigma T_s^4 - e_a\sigma(T_s^4 - T_a^4)$. We then solve for T_a to get

$$T_a = \left[\frac{1}{e_a\sigma}(239\ \text{W/m}^2 - \sigma T_s^4 + e_a\sigma T_s^4)\right]^{1/4}$$

Using our values $e_a = 0.899$ and $T_s = 288$ K gives $T_a = 250$ K. As expected, this number is significantly lower than the surface temperature. It's also close to the 255 K we got from our first simple energy-balance calculation that ignored the whole surface–atmosphere interaction. That's not surprising either: Seen from above the atmosphere, Earth is in energy balance with the 239 W/m^2 of incoming solar energy. Since the atmosphere radiates most of the outgoing energy, we should expect an atmospheric temperature close to that of our earlier calculation. The difference between our 250-K result here and the earlier 255 K comes from the 40 W/m^2 that sneaks through from the surface, implying an effective surface emissivity of about 0.9 rather than the 1.0 we assumed in our earlier and simpler calculation.

12.5 A Tale of Three Planets

JUST A THEORY?

Is the greenhouse effect the only viable explanation for Earth's surface temperature? Or is it just a theory that might or might not account for Earth's climate? First, a word about scientific theories: There's a widespread misconception, fueled in part by those who don't accept Darwin's theory of evolution, that a theory is not much more than a scientific guess, hunch, or plausible but unproven hypothesis. In fact, scientific theories are solidly established and coherent bodies of knowledge, based typically on a few key ideas, and reinforced by diverse observations that provide credence and self-consistency. In physics, the theories of relativity and quantum electrodynamics provide our most firmly established understandings of physical reality. In biology, the theory of evolution gives the most solidly consistent picture of how Earth's diverse species arose. In geology, the theory of plate tectonics—a relative newcomer to the great theories of science—ties together widely diverse geological evidence to give a clear and solidly established picture of Earth's history. And in climate science, the theory of the greenhouse effect, which is grounded in long-established, basic principles of physics and chemistry and corroborated with observations, explains why our planet's surface temperature is what it is. Are these theories guaranteed to be correct? Of course not—nothing is 100% certain. But they're the best we have, and in the case of long-established theories, "best" means very certain indeed.

The theory of the greenhouse effect dates to the early nineteenth century, when the French mathematician Jean-Baptiste Fourier was the first to recognize the role of atmospheric greenhouse gases in warming Earth's surface. (Fourier is better known for his technique of analyzing waves into their constituent frequencies.) Later in the century, the Irish-born John Tyndall—a versatile scientist who studied everything from the germ theory of disease to glaciers to the dispersion of light in the atmosphere—made actual measurements of infrared absorption in CO_2 and water vapor and explored the possibility that a decline in the greenhouse effect might be the cause of ice ages. Late in the nineteenth century the Swedish chemist Svante Arrhenius, better known as the 1903 Nobel laureate for his work on electrolytic chemical decomposition, quantitatively explored the effect of increased CO_2 concentration on Earth's temperature. Arrhenius carried out tens of thousands of hand calculations and reached the conclusion that Earth's temperature might increase by 5°C to 6°C with a doubling of atmospheric CO_2—a result generally consistent with modern predictions. However, Arrhenius erred in his timescale, believing it would take humankind 3,000 years to double atmospheric CO_2. His error was in predicting human behavior, though, not in the science underlying the greenhouse effect. The fact is that, by the beginning of the twentieth century, the greenhouse effect was well-established science, although hardly common knowledge.

NATURE'S EXPERIMENT

Still, it would be nice if we could do controlled experiments to verify the greenhouse theory. That might involve dumping some greenhouse gases into the atmosphere, observing the climatic response, removing the greenhouse gases and then starting a new experiment, which is of course impossible. We've got only one Earth, and it's a complex system that's also our home, so we can't use it for controlled experiments. We are, however, doing an uncontrolled experiment with the greenhouse gas emissions from fossil fuel combustion. The problem is that we may not know the outcome until it's too late.

Nature, however, has provided us with a ready-made greenhouse climate experiment in the form of the three planets Venus, Earth, and Mars (Fig. 12.6). Being different distances from the Sun, each has a different value for the rate of incoming solar energy. But as Example 12.1 indicates, that may not make as big a difference as one might expect. Considering Earth in the absence of greenhouse gases, but accounting for the reflection of 30% of the incident sunlight, we found a predicted temperature of 255 K. For Mars, Example 12.1 gives 226 K in the absence of reflection. Including reflection with Mars' planetary albedo at 0.25 gives 210 K for the predicted Martian temperature (see Exercise 2). For Venus, as you can show in Exercise 3, a calculation analogous to Example 12.1 yields 328 K for Venus' temperature without reflection and only 232 K when Venus' substantial albedo of 0.75 is taken into account.

However, the three planets differ in a more climatically significant way than their different distances from the Sun and their different albedos: They have very different atmospheres. Earth, as we know, has an atmosphere that's largely nitrogen and oxygen, but with enough greenhouse gases to raise the average

FIGURE 12.6
Venus, Earth, and Mars are different distances from the Sun and have different atmospheric compositions. Together they provide a natural "experiment" that verifies the theory of the greenhouse effect.

temperature by 33°C. Mars, in contrast, has an atmosphere with only 1% the density of Earth's. Mars' atmosphere is mostly CO_2, but it's so diffuse that the Martian greenhouse effect is negligible. Venus, on the other hand, has an atmosphere 100 times denser than Earth's, and it's 96% CO_2. Venus' surface temperature is a sizzling 735 K—way beyond the boiling point of water and hot enough to melt lead! Table 12.1 summarizes the very different conditions on these three neighboring planets. I've listed predicted temperatures that include the effect of reflection as calculated from the tabulated planetary albedo, since that's how we did the calculation for Earth.

The huge greenhouse effect on Venus demands a closer look. How did a planet that in many ways is much like Earth end up with such a huge surface temperature? The answer lies largely in two factors: Venus' proximity to the Sun and a positive feedback that enhanced its greenhouse effect. I've shown that Venus'

TABLE 12.1 | **THREE PLANETS**

Planet	Albedo	Calculated temperature	Actual temperature	Greenhouse effect*
Venus	0.75	232 K −44°C	735 K 462°C	503°C
Earth	0.31	255 K −18°C	288 K 15°C	33°C
Mars	0.25	210 K −63°C	210 K −63°C	0°C

* The greenhouse effect is listed in °C only because it's a temperature *difference*, and so has the same numerical value in both K and °C.

smaller distance from the Sun wouldn't make it all that much hotter than Earth in the absence of an atmosphere. But the difference is enough that in Venus' case all surface water evaporated early in the planet's history. Water vapor is a strong greenhouse gas, so as Venus' atmospheric water vapor increased, so did the surface temperature, which drove greater evaporation. As a result, Venus suffered a runaway greenhouse effect that led rapidly to its inhospitable temperature. (See Research Problem 1 for more on Venus' climate.) Today, Venus' CO_2 atmosphere continues the strong greenhouse effect that originally resulted from water vapor. On the cooler Earth, in contrast, surface water maintains equilibrium with atmospheric water vapor, limiting further evaporation and making a runaway greenhouse effect unlikely on our planet.

There are other climatically significant differences between Earth and its neighbors. Earth has a large Moon, which helps stabilize its rotation axis and hence its climate. Earth is more active geologically, and over the long term the geologic cycling of carbon between Earth's crust and atmosphere provides a thermostatic effect that moderates temperature swings. Finally, Earth's stronger magnetic field protects the planet's atmosphere from the solar wind, which would strip away the lighter molecules. Note that two of these three factors directly influence Earth's atmosphere.

So our neighboring planets confirm what basic science already tells us: The composition of a planet's atmosphere has a major influence on climate. In particular, the presence of greenhouse gases in the atmosphere necessarily leads to higher surface temperatures. Anything that changes the concentration of greenhouse gases therefore changes climate. On present-day Earth, the greenhouse gas concentration is indeed changing, increasing largely as a result of CO_2 emissions from our fossil fuel combustion. So basic science suggests that climate should be changing as well. Just how much and how fast are more subtle questions that we'll explore in the next three chapters.

CHAPTER REVIEW

BIG IDEAS

12.1 Keeping a house warm involves a balance between energy input from the heating system and energy lost through the walls.

12.2 Keeping a planet warm involves a balance between incoming solar energy and infrared radiation to space. Since the rate of energy loss increases with increasing temperature, a planet naturally achieves this state of energy balance.

12.3 Earth is warmed further because of infrared-absorbing greenhouse gases in its atmosphere. These make the atmosphere partially opaque to outgoing radiation, and therefore increase the surface temperature needed to maintain energy balance. This is the **greenhouse effect**.

12.4 Earth's complete energy balance involves additional processes, including reflection of sunlight; convection, evaporation, and **evapotranspiration**; and the role of clouds.

12.5 The three planets Venus, Earth, and Mars corroborate the theory of the greenhouse effect. Mars has a tenuous atmosphere and very little greenhouse effect; Earth's atmosphere contains enough greenhouse gases to increase its temperature by 33°C; and Venus has a runaway greenhouse effect of some 500°C.

TERMS TO KNOW

GETTING QUANTITATIVE

Stefan–Boltzmann radiation law: $P = e\sigma AT^4$ (Equation 12.1; p. 320, introduced in Chapter 4)

Stefan–Boltzmann constant: $\sigma = 5.67 \times 10^{-8}$ W/m²·K⁴

Solar constant: $S = 1{,}364$ W/m² (introduced in Chapter 9)

Earth's energy balance, no atmosphere: $\frac{S}{4} = e\sigma T^4$ (Equation 12.2; p. 321)

Natural greenhouse effect, Earth: 33°C = 33 K = 59°F

Power radiated to space: $\sigma T_s^4 - e_a\sigma(T_s^4 - T_a^4)$ (Equation 12.3; p. 328)

Earth's albedo: ~0.30

QUESTIONS

1. Ignoring the complications of the greenhouse effect, explain why changing a planet's distance from its star has only a modest effect on temperature.
2. In what sense is radiation the only heat-transfer process affecting a planet's energy balance? In what sense do other processes play a role?
3. Explain why the greenhouse effect requires that the atmosphere be cooler than the surface.
4. Which arrow in Figure 12.5 is most directly related to the presence of greenhouse gases?
5. How does the study of Mars and Venus help establish the validity of the greenhouse effect?

EXERCISES

1. Use a simple zero-dimensional energy-balance model without atmosphere to estimate Earth's average temperature, but also assume there is no reflection of solar energy.
2. Rework Example 12.1, now assuming a Martian albedo of 0.25.
3. Use a simple zero-dimensional energy-balance model without atmosphere to estimate Venus' average temperature, (a) ignoring reflection and (b) assuming an albedo of 0.75. Venus is 108 million km from the Sun, versus Earth's 150 million km. *Hint*: Since sunlight intensity falls as the inverse square of the distance, the solar constant at Venus is a factor of $(150/108)^2$ larger than at Earth.
4. What's the solar constant at Jupiter, which is 5.2 times as far from the Sun as is Earth? See the hint in Exercise 3.
5. What should be the temperature on a spherical asteroid located between Mars and Jupiter, twice as far from the

Sun as Earth? The asteroid has no atmosphere, and its albedo is 0.15. See the hint in Exercise 3.

6 What albedo would give a body at Earth's distance from the Sun a surface temperature of 220 K, assuming it had no atmosphere?

7 If Earth's albedo dropped from 0.30 to 0.28, what would happen to the surface temperature? Ignore the greenhouse effect (this makes your absolute temperature incorrect but still gives a reasonable estimate of the change).

8 A hypothetical planet receives energy from its star at the average rate of 580 W/m^2. Its surface temperature is 340 K and its surface emissivity in the infrared is 1. Its atmospheric temperature is 290 K. Find (a) the rate at which the surface radiates energy and (b) the atmospheric emissivity.

9 A planet with a surface temperature of 380 K and a surface emissivity of 1 has an atmospheric temperature of 320 K and an atmospheric emissivity of 0.82. If it's in energy balance, what's the average rate of energy input from its star?

10 A planet's atmospheric emissivity is 0.25, and its atmospheric temperature is 240 K. It receives energy from its star at the rate of 47 W/m^2. (a) Does it have a greenhouse effect? (b) What's its surface temperature?

RESEARCH PROBLEMS

1 Read the article "Global Climate Change on Venus" by Mark A. Bullock and David H. Grinspoon in the March 1999 issue of *Scientific American* (also available in *Scientific American*'s September 2003 special edition "New Light on the Solar System"). Write a brief paragraph contrasting Venus' climate history with Earth's. Include an interpretation of Bullock and Grinspoon's comment that Venus' "atmospheric processes are one-way" (in the caption on page 57 of the original article).

2 Earth's albedo plays a major role in climate because it determines the proportions of solar energy reflected and absorbed by the Earth-atmosphere system. Table 12.1's albedo value of 0.30 is a global average; albedo varies with latitude, surface conditions, season, and other factors, and it's varied significantly with time over the past few decades. Find more detailed data on Earth's albedo, and make either (a) a graph of albedo as a function of latitude (its "zonal average," meaning an average over longitude), (b) a table of different surface conditions (desert, open ocean, ice, snow, cropland, forest, etc.) and their respective albedos, or (c) a graph of Earth's average albedo versus time over several decades.

ARGUE YOUR CASE

1 A member of Congress gives a speech claiming that the greenhouse effect is "just a theory" and that there are plenty of other explanations for global warming. Give a counterargument.

2 A more precise look at the numbers that go into Figure 12.5 shows that Earth isn't quite in energy balance, with the planet receiving almost 1 W/m^2 more than it returns to space. Formulate an argument that will convince a friend who isn't very science literate that this fact implies that global warming is occuring. You might consider developing a more familiar everyday analogy.

Chapter 13

FORCING THE CLIMATE

Chapter 12 showed how Earth's energy balance results from a complex set of energy flows from Sun to Earth, from Earth to space, and between Earth's surface and atmosphere. Those flows ultimately establish surface and atmospheric temperatures. Any deviation from energy balance eventually leads to a new balance, with different values for the energy flows and therefore different temperatures. This chapter explores the causes and implications of such changes in Earth's energy balance.

13.1 Climate Forcing

Suppose some abrupt change occurs to upset Earth's energy balance. It could be an increase in the Sun's power output, an increase in greenhouse gases, a change in surface albedo, a volcanic explosion that dumps particulate matter into the atmosphere, or any of a host of other changes to Earth's surface or atmosphere that affect energy flows. How are we to characterize this change and understand its impact on climate?

In principle, the correct way to account for a change in Earth's energy balance and to determine its climatic implications is to explore quantitatively the physical, chemical, geological, and biological processes that are affected, and to work out how those altered processes affect the entire Earth system. This is a job for large-scale computer models, which even today have trouble handling every conceivable process of climatic importance.

But a simpler, approximate approach is to characterize changes by the extent to which they upset Earth's energy balance. Suppose, for example, that the average rate of solar energy input were to increase. This change would clearly result in planetary warming. Or suppose that the concentration of greenhouse gases increases; this, too, should result in warming. Although these two processes are physically very different—one involves a change in the Sun, the other a change in Earth's atmosphere—they both have the same general effect of warming the planet.

An upset in Earth's energy balance is a **climate forcing**. The word *forcing* comes from the scientific idea of applying a force—a push or a pull—to a mechanical system and then seeing how it responds. The force appears as a mathematical term in the equations describing the system. Even when scientists aren't talking about mechanical systems and push–pull forces, they still use the word *forcing* for analogous terms in their equations. Thus an increase in greenhouse gases or a change in solar radiation adds a forcing term to equations describing the climate system. Numerically, the forcing term is the value of the resulting energy imbalance in watts per square meter. The forcing is considered positive if the imbalance results in a net energy inflow to the Earth system and negative if it results in a net outflow. Thus an increase of 5 W/m^2 in solar input is a forcing of +5 W/m^2, and an increase in greenhouse gases that blocks 5 W/m^2 of outgoing radiation is also a forcing of +5 W/m^2. An increase in aerosol particles that reflect to space an average of 2 W/m^2 of solar energy is a forcing of –2 W/m^2. A positive forcing results in warming and a negative forcing in cooling.

In the crudest approximation to the climate system, namely the zero-dimensional model I introduced in Chapter 12, a 5-W/m^2 forcing would have the same effect whatever its cause. But that's a gross oversimplification. For example, an increase in solar radiation warms all levels of the atmosphere; however, as you saw with Chapter 12's two-box climate model, an increase in greenhouse gases warms the troposphere and cools the stratosphere. So we look at the positive or negative sign of a forcing to determine whether that forcing results in a warming or a cooling, but we consider the magnitude of the forcing as giving only a rough quantitative estimate of its climatic effect.

It's simplest to think of energy balance as occurring only between the Earth–atmosphere system and space. However, climate scientists use forcing to mean specifically an upset in energy balance at the boundary between the troposphere (the lower atmosphere) and stratosphere (upper atmosphere). This location gives the best correlation between forcing and effects on climate because, when the climate system is in balance, there's as much energy flowing upward across the troposphere–stratosphere boundary as there is flowing downward. Any new forcing upsets that balance, at least temporarily. You may also see what I call *climate forcing* referred to as **radiative forcing**, a term that reflects the fact that forcing amounts to an imbalance between incoming and outgoing radiation.

Suppose an instance of forcing occurs—say, an abrupt increase in the incident solar radiation to a new, constant value. Then, if nothing else changes, there's a net incoming energy flow to the Earth system, and the planet warms up. The outgoing infrared increases accordingly, and eventually we reach a new energy balance at a higher temperature. The forcing is still there, in that the solar radiation has increased over its original level. The value of the forcing still describes the upset in the original energy balance, but in the new equilibrium state there's no longer an imbalance. So when I'm talking about a forcing of, say, 5 W/m², I'm talking about a change relative to some earlier state of energy balance. Depending on how much time has elapsed since the forcing "turned on," the energy flows may still be out of balance or the climate may have reached the new state in which they're again in balance.

EXAMPLE 13.1 | Climate Forcing in a Simple Model

Use a simple zero-dimensional model to estimate the effect on Earth's average temperature of a 5-W/m² forcing associated with an increase in solar radiation.

SOLUTION

In Chapter 12, we used a solar input of 239 W/m² to calculate an effective temperature for Earth of 255 K. Our solar input accounted for reflection but we ignored the greenhouse effect. A forcing of 5 W/m² adds an additional 5 W/m² to the solar input. So we can set up the same equation, now including the forcing term:

$$239\ \text{W/m}^2 + 5\ \text{W/m}^2 = e\sigma T^4 = (1)(5.67 \times 10^{-8}\ \text{W/m}^2\cdot\text{K}^4)T^4$$

which gives

$$T^4 = \frac{244\ \text{W/m}^2}{5.67 \times 10^{-8}\ \text{W/m}^2\cdot\text{K}^4} = 4.30 \times 10^9\ \text{K}^4$$

or $T = 256$ K. Thus the temperature goes up about 1 K, or 1°C. I say "about" because I didn't keep any significant figures beyond the decimal point in calculating the 255-K result in Chapter 12. Exercise 1 leads to a more accurate value for the temperature increase for this example, namely 1.3°C.

The results of Example 13.1 would be the same if the 5-W/m² forcing resulted from a greenhouse-induced decrease in outgoing infrared. This equivalence occurs in the simple zero-dimensional model because it doesn't matter whether you add 5 W/m² to the left-hand side of the energy-balance equation or subtract

5 W/m^2 from the right-hand side. But I emphasize that this equivalence is only an approximation; with a more sophisticated model, the effects of the two numerically equal but physically distinct forcings aren't quite the same.

13.2 Climate Sensitivity

Climate scientists, policymakers, environmentalists, and others concerned with climate change often ask "what if" questions such as, "What if this or that action or policy were to result in a climate forcing of such-and-such a value? What would be the resulting temperature change?" To answer these questions we could repeat a calculation like the one in Example 13.1 or, better, run a computer climate model. But it's simpler to characterize the climate system's response to a forcing in terms of **climate sensitivity**. Because increased atmospheric CO_2 is the dominant anthropogenic forcing, most scientists define climate sensitivity as the temperature change expected for a doubling of atmospheric CO_2 from its preindustrial level of about 280 parts per million. (Here "preindustrial" means before 1750.)

Scientists estimate climate sensitivity from experiments with computer climate models and from data on past climates. There are some subtleties involved, especially with sensitivities derived from climate models. For example, the climate response to a sudden doubling of CO_2 is different from the response to a gradual increase to the same level. After a sudden change, it takes a while for the climate to reach equilibrium at the doubled concentration. Sometimes climate sensitivity is qualified as to whether it describes the immediate response—sometimes called the *transient response*—or the long-term *equilibrium response*. This distinction is important: For example, if we humans were to stop all our climate-forcing activity immediately, we would still be committed to several decades of warming at a rate of around 0.1°C per decade as Earth's climate comes to equilibrium with present-day concentrations of greenhouse gases and other forcing agents.

Today's best estimates of Earth's climate sensitivity are in the range of 2°C to 4.5°C, with values below 1.5°C considered very unlikely. These are equilibrium estimates, meaning Earth's surface temperature might rise 2°C to 4.5°C after the climate came to equilibrium with a doubling of CO_2 from its preindustrial 280 ppm. However, the statistical distribution of possible climate sensitivities has a long "tail" at the high end, so there's a significant chance that the actual sensitivity is considerably higher than 4.5°C. Furthermore, these values don't account for long-term equilibration associated with melting of ice sheets—meaning that over timescales of thousands of years, the climate sensitivity could be even higher.

You may also see climate sensitivity described as the change in temperature (K or °C) per unit change in forcing (W/m^2)—a more general definition that attempts to quantify the climatic effects of all forcings, not just those of CO_2. Expressed this way, climate sensitivity is believed to lie in the approximate range of 0.25 K to 0.75 K per watt per square meter. To be quantitatively useful, however, this more general definition of climate sensitivity must be adjusted for different forcings, especially those that are geographically localized. Exercises 2, 3, 4, and 6 explore this second definition of climate sensitivity.

The concepts of forcing and climate sensitivity are useful on the assumption that the climatic response to a given forcing is proportional to the magnitude of the forcing. Double the forcing, and the temperature goes up twice as much; triple the forcing, and the temperature rise triples. This is known as a **linear response**. The assumption of linearity is usually a good one for relatively small changes, but isn't guaranteed.

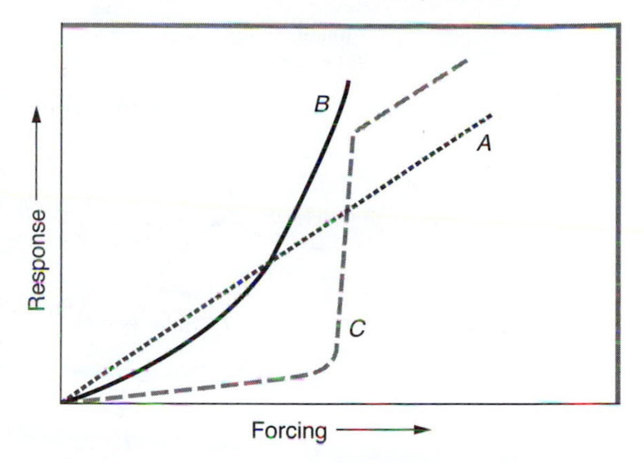

FIGURE 13.1
Linear (*A*), nonlinear (*B*), and highly nonlinear (*C*) responses to increasing forcing. At the vertical portion of curve *C*, a tiny change in forcing produces a huge change in the response. With climate, the response could be temperature change.

LINEAR VERSUS NONLINEAR

Under what conditions would a response (e.g., a temperature rise) not be linearly proportional to its cause (e.g., a forcing)? For a simple example from everyday life, think of a light switch. The motion of the switch lever is the forcing, and the brightness of the light is the response. If you push slowly on the lever, for a while nothing happens; there's no response whatsoever. But suddenly you reach a point where the switch flips almost instantly from off to on and the light comes on immediately and at full brightness. That's a **nonlinear response**, and the switch is a nonlinear device. Figure 13.1 shows examples of linear and nonlinear responses.

Could the climate behave like the light switch, exhibiting a sudden, dramatic change with just a small change in forcing? Climate scientists believe it could, but it's difficult to predict extreme nonlinear effects. One example of a nonlinear climate response would be a sudden shutdown of the ocean current system that carries warm water from the tropical western Atlantic toward Europe. This event could be caused by a decrease in ocean salinity driven by the infusion of fresh water from melting arctic ice. Ironically, the now-benign climate of Britain might rapidly cool in response to warming-induced ice melt.

Many processes in nature and in technological systems are nonlinear, albeit less dramatically so than a light switch or ocean-circulation "switch." But it's almost always the case that the response to small disturbances is essentially linear. For this reason, the linear approximation is widely used throughout science and engineering, and it's used wisely as long as one keeps in mind that deviations from linearity may become significant and even dramatic with larger disturbances.

13.3 Feedback Effects

The energy balance that determines Earth's climate involves interactions among the surface, atmosphere, and incident sunlight; biological, chemical, and geological systems; human society; and a host of other components of the complex

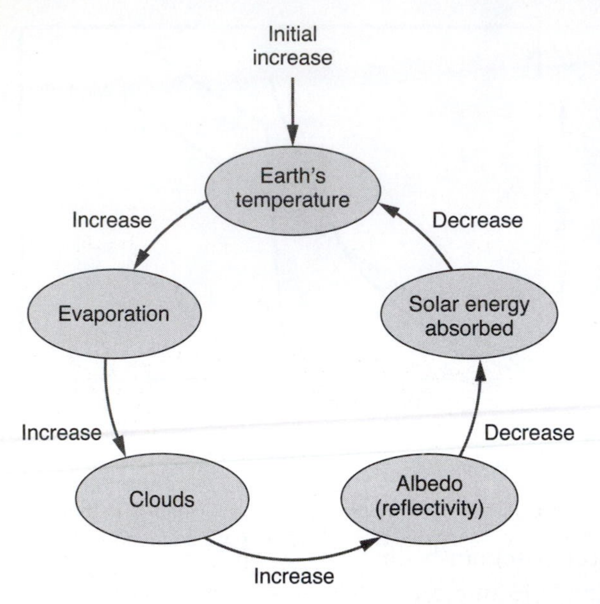

FIGURE 13.2
Negative feedback associated with clouds. An initial increase in Earth's temperature results in increased evaporation and therefore more clouds. Because clouds are highly reflective, they increase the planet's albedo. More reflected sunlight means less solar energy is absorbed by the Earth system. The final effect, therefore, is to moderate the initial temperature increase. This is only one of many cloud feedbacks. The feedback works in both directions, so the same diagram applies if the increases and decreases are interchanged.

system that is Planet Earth. When one aspect of those interactions changes—for example, an increase in the surface temperature—other changes may follow. Those, in turn, may further alter the climate. These additional effects on climate are called **feedback effects**.

NEGATIVE FEEDBACK

Feedback effects fall into two broad categories. Negative feedback effects tend to reduce the magnitude of climatic change. A household thermostat is an example of negative feedback: If the temperature in the house goes up, the thermostat turns off the heat and the temperature then drops. If the temperature falls, the thermostat turns on the heat and warms the house back up. The increase in Earth's radiation with increasing temperature also provides negative feedback: If the energy balance is upset by a small increase in the surface temperature, then the infrared energy loss $e\sigma T^4$ goes up, thus restoring balance, although at a higher temperature.

Many other negative feedbacks operate in the climate system. For example, increased atmospheric CO_2 can stimulate plant growth, and growing plants remove CO_2 from the atmosphere. With this feedback alone, an increase in CO_2 emissions has a lesser effect on climate than it otherwise would. Clouds can provide another example of negative feedback: When Earth's surface warms, the result is more evaporation of surface waters and therefore more cloud formation. Clouds reflect incident sunlight, lowering the energy input to the climate system. Again, the effect of the initial warming becomes less than it would have been without the feedback effect (Fig. 13.2). Here's another negative feedback: An increase in surface temperature may result in decreased soil moisture in continental interiors; these dry conditions, in turn, may reduce vegetation and increase surface albedo, which means less sunlight is absorbed and therefore the initial warming is mitigated. By the way, negative feedback works for either an initial rise or a decrease. A drop in temperature, for example, means less evaporation, less cloud formation, less sunlight reflected, and therefore greater energy absorption in the climate system, which thus tends to mitigate the initial temperature drop.

Could a negative feedback effect be strong enough to reverse an initial warming or cooling? You might think so, but in fact this is not possible—at least in the small-change realm where the climate's response is linear.

POSITIVE FEEDBACK

Positive feedback effects exacerbate, rather than mitigate, initial changes in a system. The most important positive feedback in the climate system is **water vapor feedback**. Water vapor is itself a greenhouse gas—the dominant factor in the natural greenhouse effect. Suppose increased atmospheric CO_2 enhances the greenhouse effect, resulting in surface warming. Again, there's more evaporation of surface waters, resulting in a still higher concentration of atmospheric greenhouse gas, in this case from the evaporated H_2O. This further enhances the greenhouse effect, and the temperature rises more than it would have without the feedback. Water vapor feedback is difficult to measure directly, but computer climate models suggest that it's very significant, increasing the response to radiative forcing changes by about 50% over what would occur without this feedback.

Ice and snow provide a second important positive feedback, the **ice–albedo feedback**. Suppose Earth's surface warms a bit, causing some Arctic sea ice to melt. The darker liquid water has a much lower albedo than the shiny white ice, so more incident solar energy is absorbed and less is reflected. The result is additional warming. As with any feedback, the opposite occurs, too: If the planet cools a little, more ice forms. Then more sunlight is reflected and the energy input to the climate system drops, exacerbating the initial cooling (Fig. 13.3). A runaway ice–albedo feedback is an essential factor in the snowball Earth episodes described in Chapter 1.

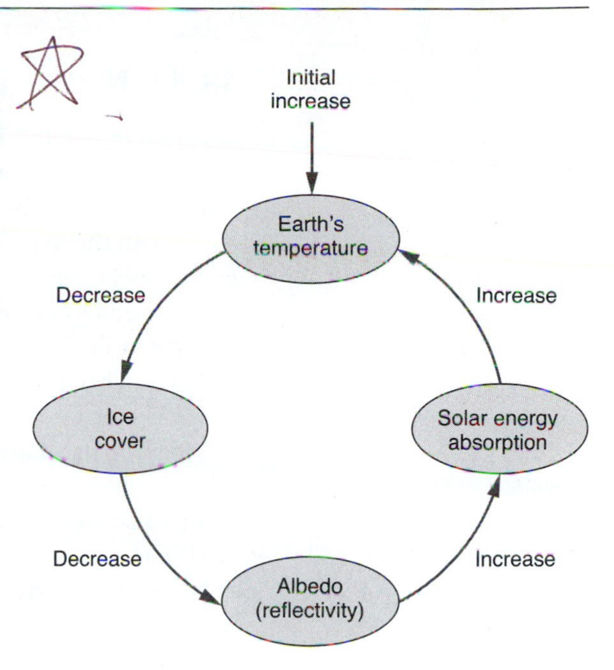

FIGURE 13.3
Ice–albedo feedback is a positive feedback effect. A temperature increase results in melting sea ice, exposing dark seawater. Earth's albedo is reduced, resulting in more solar energy absorption. This, in turn, exacerbates the initial temperature increase. Ice–albedo feedback would also exacerbate an initial cooling.

Clouds exert positive as well as negative feedbacks. Clouds, like greenhouse gases, absorb outgoing infrared radiation. So an increase in surface temperature, leading to more evaporation and more clouds, results in a stronger greenhouse effect and still more warming. So what's the overall effect of clouds? The answer is still a toss-up. The formation of clouds and their interaction with incoming solar radiation and outgoing infrared remains an active area of climate research. The details depend on the type of cloud, its three-dimensional shape, whether it's made of water droplets or ice crystals, the size of the droplets or crystals, the cloud's altitude, and a host of other factors. Cloud effects are further complicated by the geographical and temporal distribution of clouds. For example, computer models and observations from satellites both suggest that clouds result in a positive forcing in the Northern Hemisphere during the winter and a negative forcing in the Southern Hemisphere at the same time. Therefore cloud feedback can be either positive or negative, depending on location and time of year. The overall global average effect of clouds on climate is a balance between positive and negative feedbacks that's so close we don't yet know for certain whether the net effect is positive or negative.

Here's yet another positive feedback, one that illustrates the role of human society in the climate system: When the weather gets a little warmer, people buy more air conditioners and more coal is burned to generate the electricity to run all those air conditioners, thus increasing CO_2 emissions, enhancing the greenhouse effect, and causing still more warming.

13.4 Natural and Anthropogenic Forcings

What do this chapter's new concepts of climate forcing, climate sensitivity, and feedbacks have to do with Earth's real climate, especially with climate change that's occurring now or will occur in the future? Above all, what do they have to do with the main theme of this book—the environmental impact of human energy use, especially as it affects climate?

Let's begin by considering climatic conditions just before the start of the industrial era, around 250 years ago. We can use this preindustrial state as a baseline and ask what changes have occurred in Earth's energy balance since then—changes that we might expect would result in an altered climate. I'm not suggesting here that Earth's climate was unchanging before the industrial era began. Climate change has both natural and anthropogenic causes, and natural climate change has been an ongoing feature of Earth's history. But choosing the preindustrial state as our baseline lets us focus particularly on anthropogenic forcings.

Figure 13.4 shows some changes in forcings that climate scientists believe have occurred since about the year 1750. All but one of these forcings is anthropogenic; the lone natural forcing is a somewhat uncertain increase in the Sun's energy output since preindustrial times. Volcanic activity is another natural forcing, this one negative, but it's so highly variable that it's difficult to assign an average value for the industrial era and so volcanic forcing isn't shown in Figure 13.4. The values in Figure 13.4, like much of the climate data in this book, are from reports of the Intergovernmental Panel on Climate Change, whose work is described in Box 13.1.

GREENHOUSE GASES

The dominant forcing shown in Figure 13.4 is from **well-mixed greenhouse gases.** These gases remain in the atmosphere long enough to become thoroughly mixed and thus have nearly equal concentrations around the globe. The most

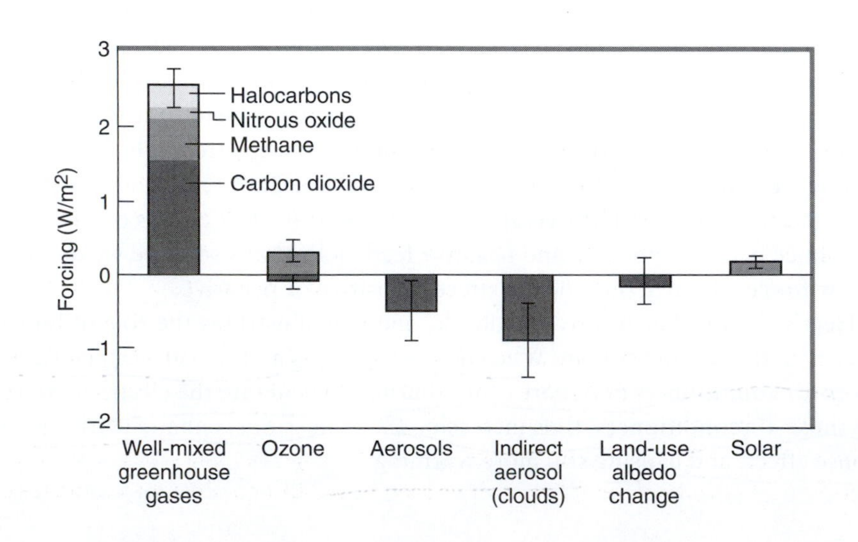

FIGURE 13.4

Changes in climate forcings since 1750. Positive forcings result in warming; negative forcings result in cooling. Error bars show uncertainties, some of which are relatively large. The well-mixed greenhouse gases are stacked in a single bar. The only natural forcing shown here is solar forcing, resulting from changes in the Sun's power output. Not shown are anthropogenic forcings due to aviation-induced clouds (contrails) and the sporadic natural forcing associated with individual volcanic eruptions. This graph, based on 2007 IPCC data, may underestimate the positive forcing due to black carbon aerosol.

BOX 13.1 | The Intergovernmental Panel on Climate Change

The **Intergovernmental Panel on Climate Change (IPCC)** was established in 1988 as an outgrowth of the World Meteorological Organization and the United Nations Environment Program. Its purpose is to assess our scientific knowledge of climate change, the impacts of that change on human and natural systems, and the steps we might take to adapt to climate change or to mitigate its effects. The IPCC comprises hundreds of scientists from a wide range of disciplines, as well as policymakers from the world's governments. The IPCC compiles and evaluates the results of published scientific research, and expresses its findings in comprehensive assessment reports published about every 6 years.

IPCC reports include volumes on climate science, climate impacts, and adaptation to or mitigation of changing climate. They also provide Technical Summaries, themselves comprehensive introductions to climate science, climate impacts, and adaptation and mitigation. Finally, the reports include Summaries for Policymakers, which distill the IPCC's findings succinctly in a way that's designed to inform intelligent policy decisions.

The IPCC itself does not conduct scientific research, and it has no laboratories, climate-observing satellites, or computer climate models. The scientists who contribute to IPCC reports are independently active in climate research. They're experts in such diverse areas as atmospheric chemistry, oceanography, botany, cloud physics, computer science, paleoclimatology, geology, glaciology, solar physics, fluid dynamics, remote sensing, and myriad others. But as participants in the IPCC process, their role is not to do research but to synthesize research that's already been done.

important of the well-mixed greenhouse gases—and indeed the single greatest source of anthropogenic forcing—is carbon dioxide. CO_2 results largely from fossil fuel combustion, although cement production and land-use changes such as deforestation are also significant (see Fig. 13.5).

Following CO_2 in its contribution to greenhouse-gas forcing is methane. You've met methane before; it's the principal component in natural gas, one of the three fossil fuels. Atmospheric methane emissions result from a wide range of natural and anthropogenic sources. Among the anthropogenic contributions to the roughly 0.5 W/m^2 of methane forcing shown in Figure 13.4 are natural gas releases from coal mining, oil and gas drilling, and pipeline leaks. There's considerable uncertainty in our estimates of anthropogenic methane emissions, and fossil fuel–related methane might account for more than half the total, or for a lot less. Other sources include sewage treatment plants and landfills, where methane forms when organic waste decays under anaerobic conditions. Cows and other ruminants produce methane in their digestive tracts, and the decay of animal waste contributes additional methane. Rice paddies are another significant source, again because of anaerobic conditions in the flooded paddies where rice is grown. And, as we found in Chapter 10, even hydroelectric dams can result in significant methane emissions.

Lesser contributions to well-mixed greenhouse gas forcing come from nitrous oxide (N_2O) and halocarbons. You've seen how N_2O results when atmospheric

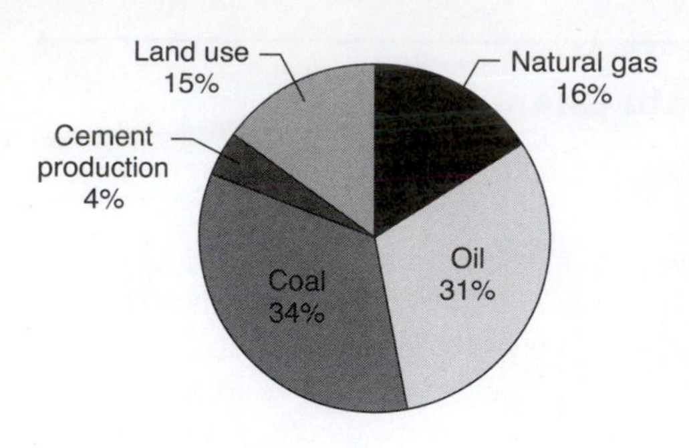

FIGURE 13.5
Major sources of global anthropogenic CO_2 emissions in the early twenty-first century. Four-fifths of the CO_2 comes from fossil fuel combustion. The CO_2 emissions from land-use changes are largely from deforestation in the tropics. All of these emissions total about 10 Gt of carbon annually.

nitrogen and oxygen combine during fossil fuel combustion. Industrial processes also emit N_2O, but the dominant source is the use of synthetic nitrogen-based fertilizers in agriculture. The halocarbons are synthetic chemicals; one group of these, the **chlorofluorocarbons (CFCs)**, was widely used in spray cans and refrigeration systems until they were phased out beginning in 1987 after it was discovered that they destroy the stratospheric ozone that protects us from solar ultraviolet radiation. Halocarbons also include the hydrochlorofluorocarbons (HCFCs), which were introduced as ozone-safe replacements for CFCs. Unfortunately, the HCFCs also contribute to greenhouse warming.

GLOBAL WARMING POTENTIAL

Greenhouse gases differ in their ability to absorb infrared radiation. For example, a methane molecule is 26 times more effective at infrared absorption than is a molecule of CO_2. This means a given amount of methane released to the atmosphere has a greater forcing effect and therefore causes greater warming than the same amount of CO_2. Determining how much more is a bit subtle, for reasons having to do with **atmospheric lifetime**. Methane remains in the atmosphere for about a decade before chemical reactions destroy it, while CO_2 lasts much longer. So on short timescales, a given amount of methane emission has a much greater effect than the same amount of CO_2. But wait a hundred years and the cumulative effect of the lingering CO_2 is greater. The effectiveness of a greenhouse gas relative to CO_2 is its **global warming potential (GWP)**. For the reason I've just outlined, the GWP isn't a single number but depends on one's time frame. Table 13.1 lists the GWPs of some important greenhouse gases over three different time frames. The gases include a common CFC, an HCFC, and a hydrofluorocarbon (HFC). Note that these synthetic substances have far higher GWPs than the more widespread but simpler CO_2, methane, and N_2O. The GWPs in Table 13.1 are per unit of mass, rather than per molecule. Since a kilogram of methane has many more molecules than a kilogram of CO_2, the GWP of methane on a per-mass basis is higher than on a per-molecule basis (see Exercise 8).

CONCENTRATION AND FORCING

Because of their differing GWPs, a low atmospheric concentration of methane or a CFC can have a much greater impact on climate than a larger concentration of CO_2. On the other hand, there's much more CO_2 in the atmosphere—and in anthropogenic emissions—than there is of the other greenhouse gases. That's why CO_2, despite its lower GWP, contributes the greatest greenhouse gas forcing.

You might expect the forcing of a given gas to depend directly on its concentration, but that isn't generally the case. The reason for this discrepancy is the **saturation effect**, which occurs when a given molecule absorbs 100% of the

TABLE 13.1 | GLOBAL WARMING POTENTIALS

Gas	Atmospheric lifetime (years)	Global warming potential relative to CO_2 — Time frame: 20 years	100 years	500 years
Carbon dioxide (CO_2)	~1,000*	1	1	1
Methane (CH_4)	11	67	23	6.9
Nitrous oxide (N_2O)	114	291	298	153
CFC-11 (CCl_3F)	45	6,700	4,750	1,620
HCFC-22 ($CHClF_2$)	12	5,200	1,800	550
HFC-23 (CHF_3)	270	12,000	14,800	12,200

*The lifetime of CO_2 is ambiguous; see Section 13.5.

infrared at certain wavelengths. An increase in gas concentration can't increase absorption at those wavelengths, since there's nothing left to absorb. Other wavelengths aren't saturated and so, overall, an increase in a given greenhouse gas does increase infrared absorption and thus climate forcing. But because of the saturation effect, the increase isn't linear. Exercise 5 explores the relationship between concentration and forcing for CO_2.

The forcings shown in Figure 13.4 reflect the cumulative effect of changes in greenhouse gases and other forcing agents since preindustrial times. In the case of the greenhouse gases, much of that increase has occurred very recently. Figure 13.6 shows the industrial-era increases in atmospheric concentrations of methane and HCFC-22, the former from a preindustrial level of around 700 parts per billion and the latter from zero because HCFCs are synthetic chemicals first created only a few decades ago. Atmospheric gas concentrations are generally expressed in parts per million, billion, or trillion by volume (ppmv, ppbv, or pptv, often without the "v"). To picture methane's preindustrial concentration of 700 ppb, for example, imagine assembling a billion 1-gallon milk jugs full of air. If you separated out the preindustrial methane, it would occupy 700 of those jugs.

Figure 13.7 shows atmospheric concentrations of the most important anthropogenic greenhouse gas, CO_2, since the beginning of the industrial era. Values prior to about 1950 come from gas bubbles trapped in ice cores drilled from the Antarctic ice cap. Since 1958, a monitoring station at an altitude of nearly

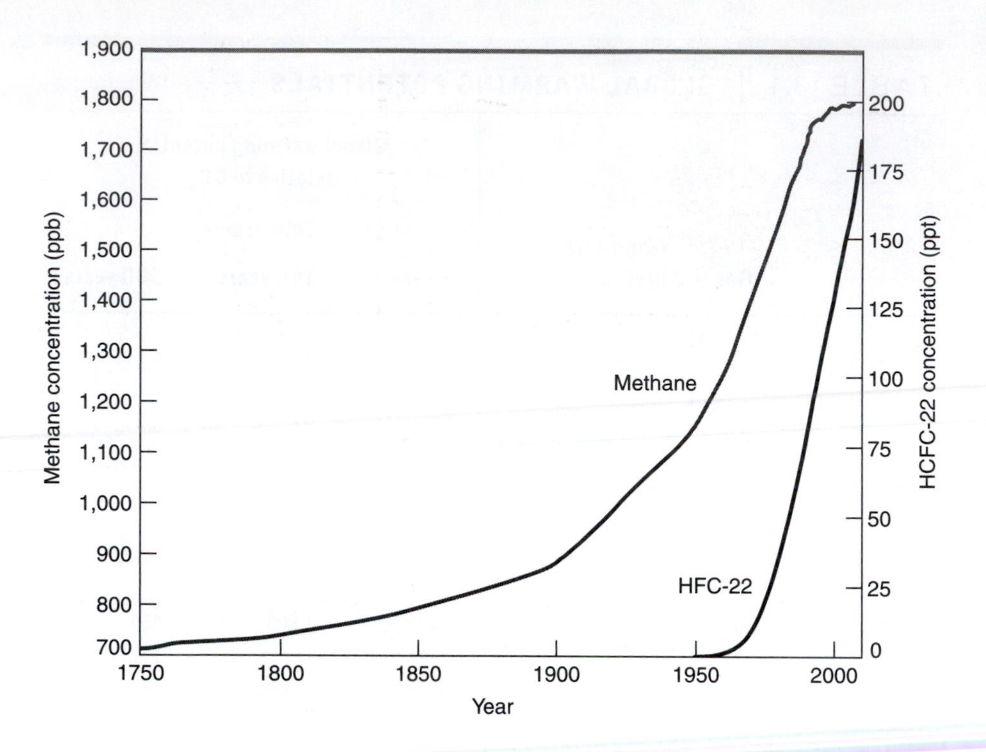

FIGURE 13.6
Concentrations of methane and HCFC-22 since preindustrial times. The methane curve starts at its preindustrial value of around 700 parts per billion (ppb). HCFC-22 starts from zero since this synthetic chemical became widespread only in the second half of the twentieth century. Although HCFC-22's concentration is measured only in parts per trillion, it's still a significant greenhouse gas because, as Table 13.1 shows, its global warming potential is more than 5,000 times that of CO_2.

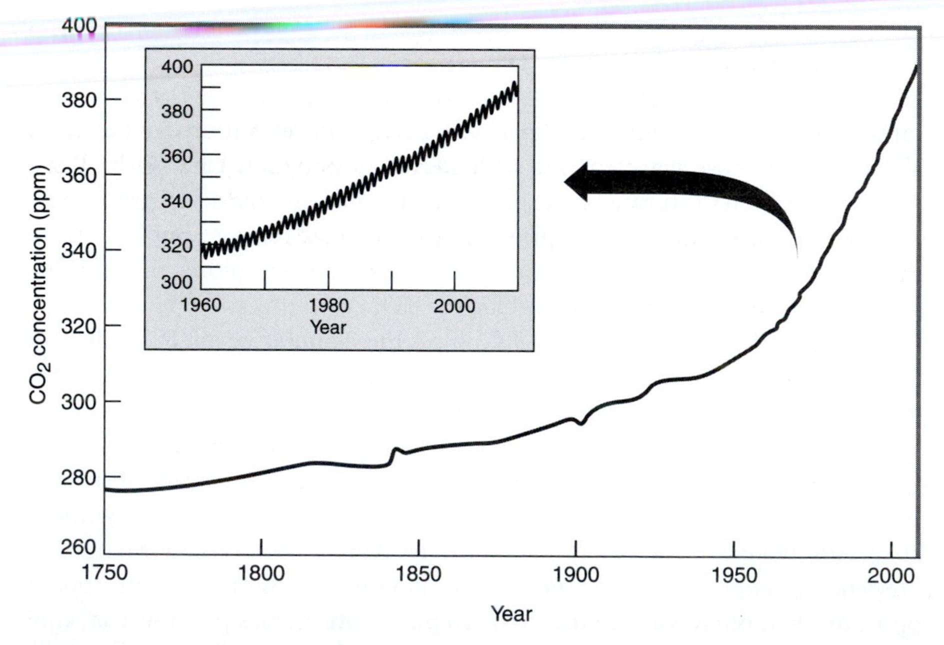

FIGURE 13.7
Increase in atmospheric CO_2 since preindustrial times. The inset shows monthly averages recorded at Mauna Loa in Hawaii since 1960, with measurements sensitive enough to record the seasonal variations as plants take up CO_2 during the Northern Hemisphere's growing season and then reduce their uptake during the winter.

2 miles on Hawaii's Mauna Loa volcano has provided detailed measurements of atmospheric CO_2 in the clean, well-mixed air over the central Pacific Ocean. The Mauna Loa data are so precise that, as the inset in Figure 13.7 shows, they track the seasonal variations in CO_2 each year, reflecting the increased uptake of CO_2 by trees and plants during the Northern Hemisphere's growing season, and the corresponding reduction of this uptake during the winter. Research Problem 1 explores this effect further.

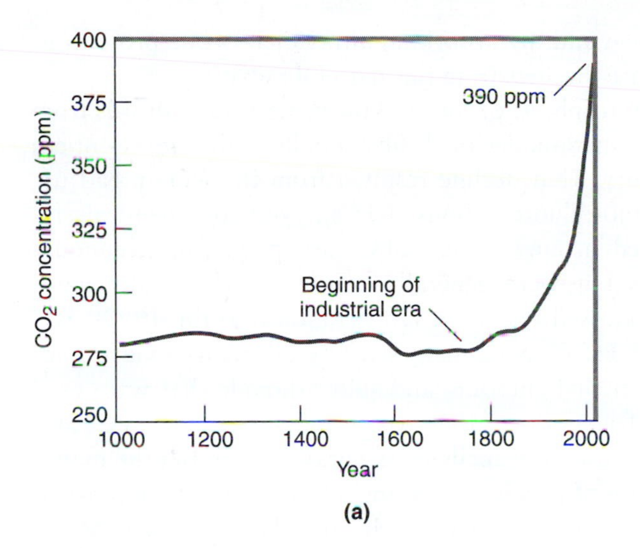

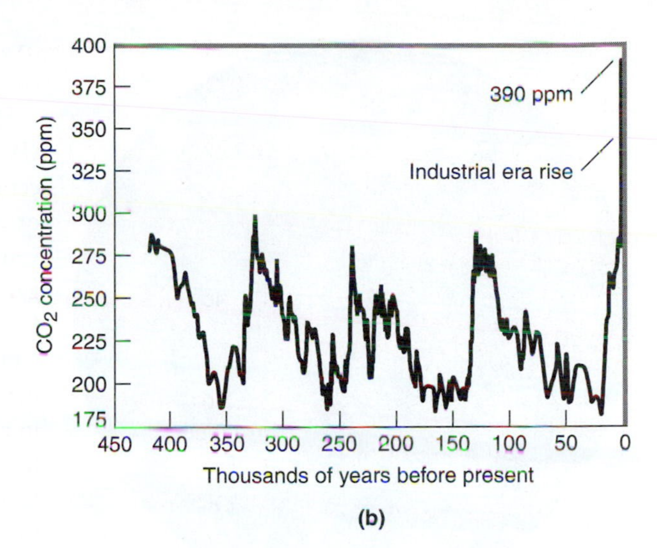

FIGURE 13.8

(a) Carbon dioxide concentration over the past thousand years shows a sharp rise that coincides with the industrial era: by 2010 the level had reached 390 ppm. (b) The rise is even more dramatic on a longer time scale. Clearly, Earth has not seen anything close to today's CO_2 levels in at least half a million years. The regular patterns in (b) correspond to ice ages and interglacial warm periods, to be discussed in Chapter 14.

How do we know that the CO_2 rise shown in Figure 13.7 is the result of anthropogenic CO_2 emissions? Several independent pieces of evidence lead to this universally accepted conclusion. (1) Because fossil fuels are commercial commodities, we have a good quantitative handle on the rate at which we burn them. The buildup of atmospheric CO_2 is consistent with the known emissions. (2) Carbon dioxide is a well-mixed greenhouse gas, but its concentration is marginally higher in the Northern Hemisphere, where most fossil fuel combustion takes place. (3) The ratio of the cosmic ray–formed radioactive isotope carbon-14 to ordinary carbon-12 has been decreasing, as would be expected if the atmosphere were being flooded with ancient carbon in which C-14 had long since decayed. That ancient carbon is the carbon trapped underground in fossil fuels. Additional evidence comes from a decline in the ratio of stable C-13 to the more abundant C-12. This ratio is lower in plants than in the atmosphere, so the percentage of atmospheric C-13 is dropping as plant-derived carbon enters the atmosphere due to fossil fuel combustion. Figure 13.8 shows that the rise in atmospheric CO_2 during the industrial era has been dramatic and unprecedented on timescales of thousands to hundreds of thousands of years. Today's CO_2 concentration is 40% higher than the preindustrial value and is thought to be significantly higher than it's been for millions of years.

OZONE

Ozone (O_3) is another greenhouse gas, but it's included separately in Figure 13.4 because it isn't mixed evenly throughout the atmosphere. Furthermore, it's shown with both positive and negative forcings, because the effects of low-level (tropospheric) ozone and high-level (stratospheric) ozone are very different. Tropospheric ozone results from photochemical reactions involving air pollution. Tropospheric ozone is both a noxious, toxic pollutant and a greenhouse gas. Anthropogenic ozone in the troposphere is responsible for a positive forcing of about 0.35 W/m^2. Stratospheric ozone is a different story altogether. It forms naturally from the action of solar ultraviolet radiation on atmospheric

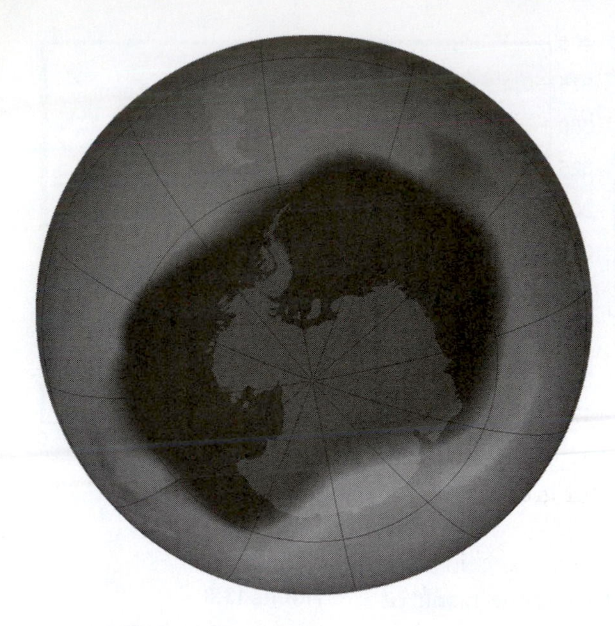

FIGURE 13.9
The ozone hole appears as the dark region surrounding Antarctica in this NASA image, taken in 2006 when the hole had its greatest extent.

oxygen and, by absorbing ultraviolet, ozone protects us surface dwellers from this harmful radiation.

Stratospheric ozone, and the protection it affords from UV radiation, declined substantially in the late twentieth century. That decline resulted from the widespread use of chlorofluorocarbons (CFCs), synthetic chemicals invented around 1930. CFCs are powerful greenhouse gases but were generally considered chemically inert. Freon, used widely in refrigeration into the 1990s, is a CFC. Freon saved many lives by replacing toxic ammonia, methyl chloride, and sulfur dioxide that were used in refrigerators of the early twentieth century. CFCs also found use as propellants in spray cans and in the manufacture of plastic foams, including energy-saving insulation. In 1974, however, chemists Paul Crutzen, Mario Molina, and F. Sherwood Rowland showed that CFCs rising into the stratosphere would decompose under the influence of solar ultraviolet, and that the chlorine freed in this decomposition would destroy ozone without itself being used up. Their work explained the depletion of stratospheric ozone over the Antarctic, first observed in the 1970s and popularly referred to as the *ozone hole* (Fig. 13.9). Crutzen, Molina, and Rowland shared the 1995 Nobel Prize in Chemistry—the first time a Nobel was awarded for environmental research. Alarm over ozone depletion led to a remarkable international agreement, the 1987 Montreal Protocol, which requires a gradual phase-out in the production and use of CFCs. As a result, stratospheric ozone should be back to its natural levels by about 2070.

Still, ozone is a greenhouse gas, so depletion of stratospheric ozone reduces the greenhouse effect and thus contributes a negative forcing. That's why the bar for stratospheric ozone in Figure 13.4 extends downward. Overall, however, the net effect of anthropogenic processes that produce or destroy ozone is a positive forcing due to the greater influence of tropospheric O_3 as shown in Figure 13.4.

The public often equates the problems of ozone depletion and global warming. Actually, they're very different, and the remarkable international cooperation of the Montreal Protocol is resolving the ozone problem. We're far from anything remotely similar with global warming. However, ozone depletion and global warming are not unrelated, as I've just discussed. But the relationship is a subtle one, involving atmospheric chemistry and two distinct atmospheric layers.

AEROSOLS

Fine particulate matter constitutes a serious form of air pollution. But particulates and liquid droplets—collectively, aerosols—also affect Earth's energy balance. Our chart of industrial-era forcings shows that, overall, aerosols produce a negative forcing—that is, a cooling effect—with magnitude about 0.5 W/m^2.

Aerosols are less well understood than greenhouse gases, so there's considerable uncertainty in this value. Aerosol forcing is further complicated by the presence of different types of aerosols that provide positive or negative contributions to the overall aerosol forcing equation.

Most significant are the sulfate aerosols, which are produced largely from coal combustion. These substances reflect sunlight back to space, decreasing the solar energy input to the climate system. Some have argued that this negative forcing might produce enough cooling to compensate for CO_2-induced warming, but this view is incorrect for two reasons. First, the overall aerosol forcing simply isn't great enough to counteract greenhouse warming. Second, sulfate aerosols have relatively short atmospheric lifetimes, so they don't get well mixed and thus their climatic effect tends to be localized. Direct forcing by sulfate aerosols may indeed cause cooling in regions downwind of heavily industrialized areas, but this can't compensate globally for greenhouse gas warming.

Other aerosols result from incomplete combustion of fossil fuels and biomass, which produces black carbon (soot) and dozens of organic compounds. Black carbon absorbs sunlight and contributes to warming—an effect that's especially significant when it falls on snow. Recent research suggests that the warming effect of black carbon may be some three times greater than IPCC estimates, perhaps contributing 60% as much forcing as carbon dioxide. Organic carbon, on the other hand, exerts a small cooling effect. Aerosols associated with nitrate fertilizers, mineral dust from industrial activities, and land development also contribute to aerosol forcing, although their effects are small and not well quantified. Increasingly, aerosol pollution results in so-called brown clouds that cover vast regions of the globe for months at a time, impacting both climate and human health. The most prominent of these, the Asian brown cloud, forms during the winter months over much of the Indian Ocean and the surrounding landmass—an area occupied by nearly half the human population.

Sulfate and other aerosols act as nuclei on which atmospheric water vapor condenses, forming clouds. Clouds, as you've seen, have significant forcing effects, both positive and negative. Cloud formation is an indirect effect of atmospheric aerosols, and with it comes an indirect aerosol-caused forcing. As Figure 13.4 suggests, this indirect effect is negative so it, too, results in cooling. Although its magnitude remains somewhat uncertain, indirect aerosol forcing is likely greater than the direct effect of aerosols.

OTHER ANTHROPOGENIC FORCINGS

Land-use changes affect climate in several ways; for example, CO_2 forcing includes CO_2 from deforestation. More directly, land-use changes alter Earth's albedo. Crop plants are generally more reflective than trees, so replacing forests with agriculture increases albedo and thus decreases the absorbed solar energy. This is a negative forcing. High-latitude winters exacerbate this effect, since snow cover is more reflective on cropland than when snow falls in trees. A variety of other factors related to land use also affect albedo, both positively and negatively. Overall, land-use albedo forcing is not well quantified, although it's most likely negative.

Aviation creates a small but potentially important climatic effect, which is not shown in Figure 13.4 because at present it's only a few hundredths of a watt per square meter. Jet contrails are clouds that wouldn't be there if airplanes weren't flying, and we know that clouds affect climate. When commercial aviation in the United States was grounded for 3 days in the wake of the September 11, 2001 terrorist attacks, scientists found subtle changes in the diurnal temperature range that could be attributable to the absence of forcing by jet contrails.

THE SUN AND CLIMATE

Finally, there's the Sun. Our star undergoes a complex cycle of magnetic activity that results in the magnetic field of its north and south poles reversing every 11 years. The best known of the many manifestations of this solar cycle is the changing number of sunspots, those cooler, darker areas on the Sun's surface that are associated with strong magnetism. The presence of sunspots, and areas of higher than normal temperatures surrounding them, result in a variation of the Sun's overall energy output. This, in turn, causes changes in the solar constant—the intensity of sunlight just outside Earth's atmosphere—of about 1 W/m^2, or about 0.1%. The direct effect of such a small change on Earth's energy balance should produce a temperature change measured in hundredths of a degree Celsius. Solar-cycle changes in global and regional climate have been detected, but it's very difficult to isolate them from the larger fluctuations associated with natural climate variability.

More interesting for climate is the question of long-term variations in the Sun's energy output. Astrophysicists understand very well just how stars evolve, so we know that on billion-year timescales the Sun's energy output has been increasing and will continue to do so. That long-term trend has significant climatic implications over the lifetime of our planet, but it's irrelevant on scales of decades to centuries. However, there's some evidence for solar variations on these shorter timescales. This evidence includes records of sunspot numbers and their assumed correlation with solar activity and hence energy output; measurements of radioactive isotopes formed by cosmic rays whose intensity is affected by solar activity; and observations of Sun-like stars. Such evidence suggests a modest increase in the solar output since preindustrial times, giving the forcing of +0.12 W/m^2 shown in Figure 13.4. However, this quantity remains uncertain by a factor of about 2 in either direction.

Although the Sun's overall energy output changes very slightly on timescales of current climatic interest, there's much greater variability in the shortest wavelengths of the solar spectrum—that is, in the ultraviolet. Because stratospheric ozone absorbs solar ultraviolet radiation, the impact of solar variability on the stratosphere is significant. Climatologists are exploring subtle mechanisms whereby changes in the stratosphere could influence climate in the lower atmosphere and even at Earth's surface. Such effects might amplify the very modest direct influence of solar variability on climate. But given the diffuse nature of the stratosphere, with its very low mass and low energy content, any change in the stratosphere is unlikely to increase significantly the surface climate response to solar variability. The bottom line is that solar forcing, while a significant factor in climate change even through the early twentieth century, is inadequate to explain the rapid warming observed in recent decades.

13.5 Carbon: A Closer Look

Figure 13.4 shows a variety of climate forcings since preindustrial times, but one factor stands out as the largest single anthropogenic contribution to net climate forcing: carbon dioxide. One can argue that CO_2 emissions are a rough measure of human energy consumption in our fossil-fueled world, and therefore of industry, economic activity, and population. We might then expect other anthropogenic forcings to scale with CO_2. For example, much of the anthropogenic CO_2 comes from burning coal, which also produces sulfate aerosol. Absent stricter air-pollution regulations, sulfate aerosol forcing should scale roughly with that of CO_2. So might other forcing agents, such as N_2O and tropospheric ozone, since these result from human activities that increase with population, industrial growth, and the spread of automobiles. So in a crude sense, CO_2 becomes a proxy for nearly all anthropogenic forcings.

However, CO_2 also has some unique properties that distinguish it from other forcing agents. In the natural world, many substances—including carbon, water, nitrogen, sulfur, phosphorus, and so forth—cycle through the system of atmosphere, land, ocean, and living organisms. Many of these natural cycles influence climate, so any human-caused disturbance can lead to anthropogenic climate change. Most important in this respect is the **carbon cycle**.

THE NATURAL CARBON CYCLE

Figure 13.10 is a simplified diagram of the carbon cycle. The figure shows several **reservoirs**—parts of the Earth system where carbon is stored. The quantity of carbon in each reservoir is given in gigatonnes (Gt) of *carbon*, not *carbon dioxide*; Box 5.1 showed that the difference between these two is a factor of 44/12 = 3.67, so there's 3.67 times as much mass of CO_2 as of carbon. The reason we measure carbon rather than CO_2 is that carbon occurs in different chemical forms as it cycles through the system. Figure 13.10 shows that the preindustrial atmosphere held about 560 Gt of carbon (the number 560 in the atmosphere box, followed by the year 1750 that marks the rough beginning of the industrial era). By 2010 that figure was about 800 Gt, as the atmosphere box also shows. This atmospheric carbon is largely in the form of carbon dioxide. Living organisms on land hold another 500 Gt or so of carbon, nearly all of it in plants, and soils hold some 2,000 Gt. We know the atmospheric figure quite accurately, but the others are less certain. A far larger amount of carbon, some 39,000 Gt, resides in the ocean. This carbon takes various forms, including dissolved CO_2 (a small amount) and other inorganic carbon compounds

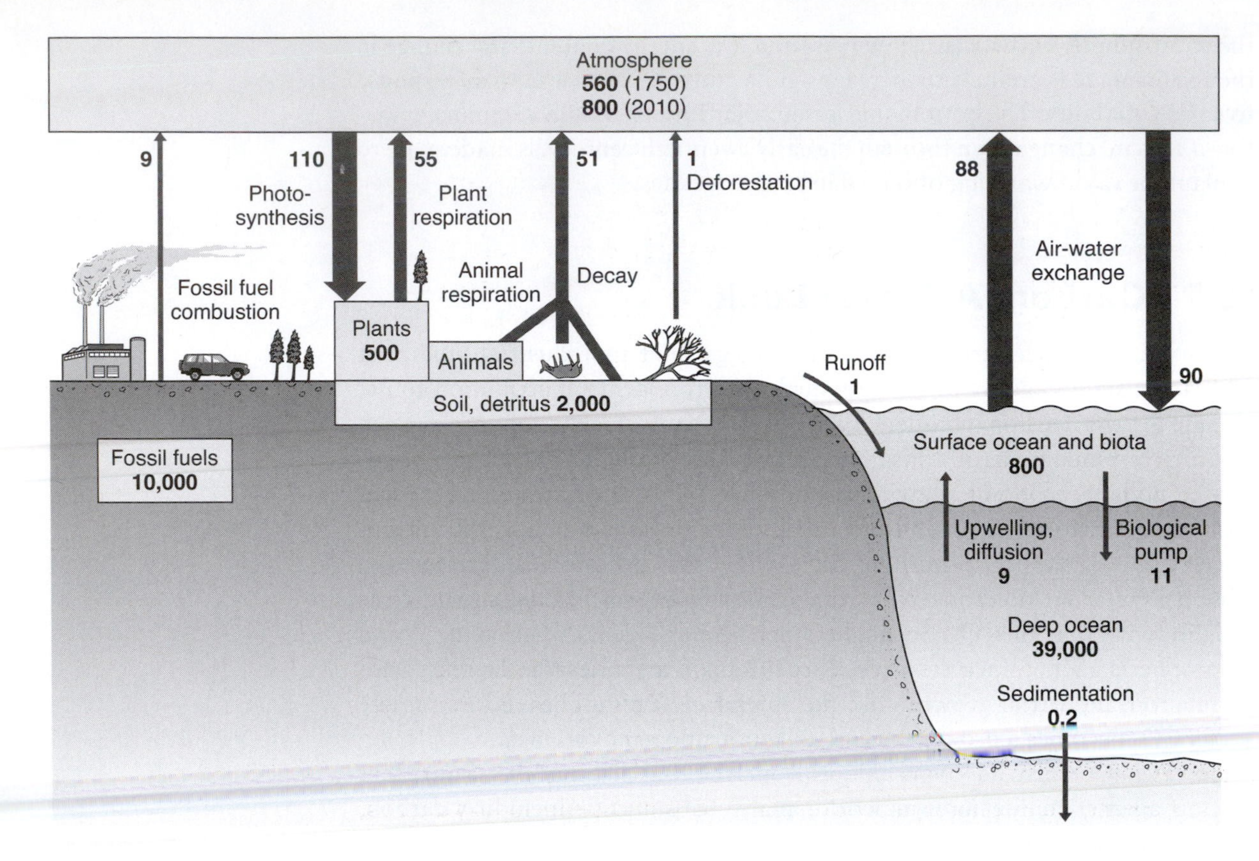

FIGURE 13.10

The carbon cycle. Carbon reservoirs are in gigatonnes (Gt) of carbon; flows (arrows) are in gigatonnes per year. All numbers are approximate and have been rounded to integers. Fossil fuel combustion and deforestation are the dominant anthropogenic factors upsetting the carbon cycle. Today, carbon is accumulating in the atmosphere at about 4 Gt per year.

(comprising most of the dissolved carbon), calcium carbonate ($CaCO_3$) in the shells of marine organisms, and organic carbon in plants and animals. However, most of that 39,000 Gt of oceanic carbon lies deep in the ocean and plays little role in the short-term carbon cycle. Only the carbon in the surface layers of the ocean cycles rapidly, and the amount of this carbon is comparable to what's in the atmosphere.

Carbon doesn't just sit in its reservoirs, but cycles around between them. Figure 13.10 also shows the associated carbon flows, quantified in gigatonnes per year. For example, terrestrial plants remove 110 Gt of carbon from the atmosphere each year; that's indicated by the downward arrow from atmosphere to plants. Through the process of respiration, both plants and animals "burn" their carbon-containing food to produce energy, water vapor, and CO_2. This process returns some 55 Gt of carbon to the atmosphere each year. The difference between these two flows is the net primary productivity I introduced in connection with biomass energy in Chapter 10. The rest of the biotic carbon eventually ends up as waste or as dead matter, although some of it temporarily becomes animal biomass. Nearly all of it eventually decays and returns to the atmosphere as CO_2. Rivers carry a small fraction (less than 1 Gt per year) to the oceans, and an almost infinitesimal amount is buried and begins the long process of turning into fossil fuels. That amount is so small that I'm not even showing it in Figure 13.10.

Meanwhile the oceans exchange carbon directly with the atmosphere, as CO_2 from the atmosphere dissolves in ocean water or as dissolved CO_2 comes out of solution and is released to the atmosphere. Flows in this process amount to about 90 Gt per year in both directions. Additional cycling occurs within the oceans, as marine organisms take up carbon to build their bodies and shells, and emit CO_2 through respiration. Again, much of this cycling occurs in the surface layers, but there's a modest exchange with the deeper ocean. Physical processes such as diffusion and upwelling transport carbon upward. Meanwhile a continuous "rain" of dead microorganisms and other formerly living things falls from the surface waters to the depths, and this so-called **biological pump** carries carbon downward. The upward and downward flows aren't quite balanced, giving a small net flow downward. This flow removes carbon from the ocean surface more or less permanently, contributing to that huge store of carbon in the deep ocean. Some of the deep carbon is incorporated into ocean floor sediments and is eventually returned to the atmosphere by volcanoes, but this process involves geological timescales of hundreds of millions of years. On shorter timescales, we can consider that carbon reaching the deep ocean is lost from the carbon cycle.

So here's the natural carbon cycle in a nutshell: Carbon is stored in the atmosphere, surface biota and soils, and ocean surface. This carbon cycles rapidly back and forth between the atmosphere and other reservoirs. How rapidly? So rapidly that a typical CO_2 molecule spends only about 5 years in the atmosphere before it's removed by photosynthesis or dissolved in the ocean. So why did I list CO_2's atmospheric lifetime as approximately 1,000 years in Table 13.1? To resolve this question we need to look at what happens when we alter the carbon cycle.

ANTHROPOGENIC PERTURBATION OF THE CARBON CYCLE

We humans have upset the balance of the natural carbon cycle, largely by burning fossil fuels, with the net effect that carbon is accumulating in the atmosphere. Figure 13.10 indicates the fossil carbon flow with an arrow from the buried fossil fuel reserves to the atmosphere. Today, fossil fuel combustion (with a little help from cement production, which releases carbon stored in rocks) accounts for almost 9 Gt of carbon released to the atmosphere each year—a number that continues to rise (Fig. 13.11). Another gigatonne comes from land-use changes, mostly deforestation. Now, you might argue that 10 Gt isn't much compared with the natural flows of 110 Gt per year back and forth between the surface biota and the atmosphere, and so the human impact should be negligible. After all, isn't the anthropogenic carbon also removed on that 5-year timescale? Yes, but here's the problem: The cycling of carbon between the atmosphere, land surface, and ocean takes place on roughly

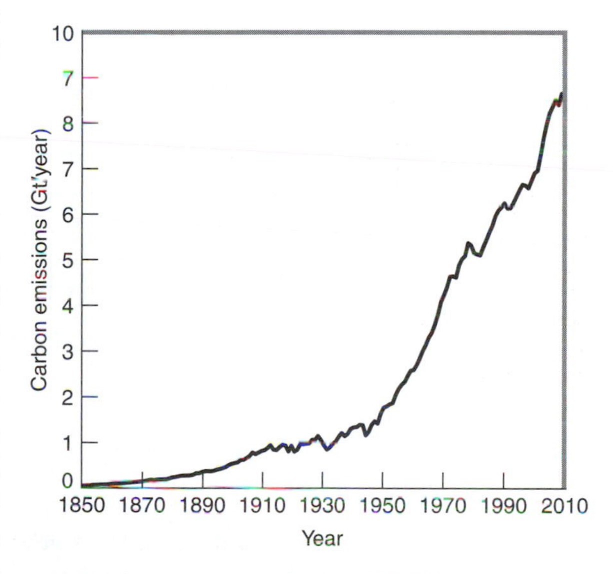

FIGURE 13.11
Annual global carbon emissions from fossil fuel combustion and cement production since 1850. The slight downturn in 2009 resulted from the global economic recession, but 2010 emissions were at a record high.

that same timescale. Thus, much of the carbon we add to the atmosphere continues to cycle through the system and ends up back in the atmosphere, resulting in a net increase in atmospheric CO_2.

CARBON SINKS

A close look at Figure 13.10 shows that the 9 Gt per year from fossil fuel combustion and the gigatonne from deforestation aren't the only imbalances in the carbon cycle. On the terrestrial side, photosynthesis removes about 110 Gt of carbon per year from the atmosphere, but the processes of plant and animal respiration and decay return only about 106 Gt per year. Given that anthropogenic deforestation sends approximately another gigatonne per year to the atmosphere, this means there are mechanisms that remove carbon from the atmosphere to the terrestrial biosphere and soils. The size of this terrestrial **carbon sink** averages about 3 Gt per year, although it varies substantially from year to year. This terrestrial sink is poorly understood, but is thought to include the incorporation of carbon into increasing forest biomass as well as directly into soils. There's a similar imbalance on the ocean side, with the ocean carbon sink removing 2 to 3 Gt more carbon each year than is returned to the atmosphere. Overall, the effect of these two sinks is to remove about half of the carbon that we humans put into the atmosphere.

The operation of both terrestrial and ocean carbon sinks depends on climate, and there's no guarantee that the sinks will continue in a warming world. Here are two examples: A warming-caused increase in forest fires could return to the atmosphere much of the carbon that the terrestrial sink has removed through increased forest growth. And acidification of the oceans as they absorb more CO_2 will have a deleterious impact on marine organisms, many of which play important roles in the ocean carbon cycle.

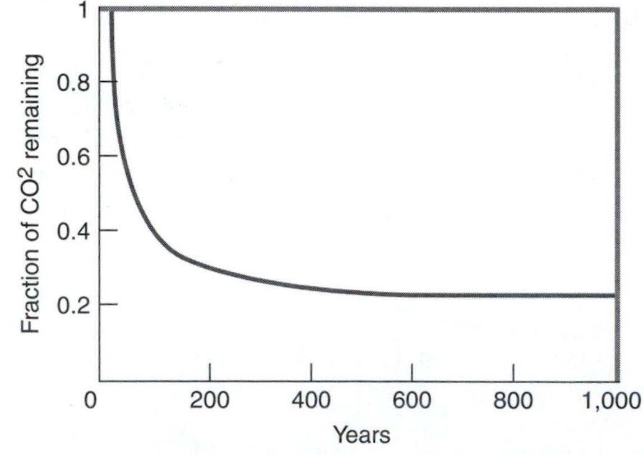

FIGURE 13.12
Removal of CO_2 injected into the atmosphere takes place on a variety of timescales. The concentration drops rapidly at first, but significant CO_2 remains in the atmosphere even after 1,000 years.

CARBON'S LONG-TERM FATE

How long would it take to return to the natural CO_2 concentration if we humans stopped dumping CO_2 into the atmosphere? We can answer this question by considering what would happen to a quantity of CO_2 injected suddenly into the atmosphere, as shown in Figure 13.12. We've seen that terrestrial and ocean sinks would remove about half of this anthropogenic CO_2 on a relatively short timescale, accounting for the steep drop seen in Figure 13.12. But the remaining carbon would continue to cycle through the atmosphere–land surface–ocean system, sustaining a higher atmospheric CO_2 concentration. Only on the much longer timescale of centuries to millennia does the rest of the carbon leave the system more or less permanently, mostly through processes that take it into the deep ocean. That's why the decline shown in Figure 13.12 slows, leaving a significant chunk of anthropogenic carbon in the atmosphere even after 1,000 years. So here's what's unique about carbon: Even though it's rapidly removed from the atmosphere, much of it is returned just as rapidly. It takes far longer for carbon to leave the entire system, so the relevant lifetime for anthropogenic carbon in the atmosphere is far longer than the 5 years that a typical carbon atom spends in the atmosphere. Because carbon-removal processes operate on different timescales, it's impossible to pin down an exact lifetime for atmospheric carbon. A rough figure of 300 to 1,000 years is often used, but this estimate is a compromise between the rapid removal by terrestrial and ocean sinks and the much longer timescales associated with processes such as the flow of carbon to the deep ocean and into ocean floor sediments.

In this way carbon is different from other forcing agents. Methane, for example, is removed from the atmosphere on a roughly 10-year timescale, largely by chemical reactions that convert it to CO_2. Once the methane is destroyed, it can't cycle back into the system. The same is true of aerosols, N_2O, and most other forcing agents. But CO_2 is different, in that we're stuck with it for a very long time. This means the climatic effects of anthropogenic CO_2 will be difficult to reverse, and in fact they may not become fully apparent until it's too late to prevent additional and very likely deleterious climate change.

In the next two chapters we'll explore the effects of anthropogenic CO_2 and other forcing agents on present and future climates.

CHAPTER REVIEW

BIG IDEAS

13.1 A **climate forcing** is any factor that upsets Earth's energy balance.

13.2 **Climate sensitivity** describes the climatic change that results from forcings. A common measure of climate sensitivity is the global average temperature change for a doubling of atmospheric CO_2.

13.3 **Feedback effects** can either exacerbate (**positive feedback**) or reduce (negative feedback) the response to a given forcing. Water vapor and ice–albedo are important positive feedbacks; clouds may provide negative or positive feedback.

13.4 Anthropogenic climate forcings have increased throughout the industrial era, dominated by greenhouse gases. Some forcings, including greenhouse gases, lead to warming; others, such as sulfate aerosols, result in cooling.

13.5 Carbon dioxide is a particularly significant forcing agent. It alters the **carbon cycle** by adding carbon to the atmosphere, biosphere, and ocean; the additional carbon remains in the system for hundreds to thousands of years.

TERMS TO KNOW

atmospheric lifetime (p. 350)
biological pump (p. 359)
carbon cycle (p. 357)
carbon sink (p. 360)
chlorofluorocarbon (CFC) (p. 350)
climate forcing (p. 342)
climate sensitivity (p. 344)
feedback effect (p. 346)
global warming potential (GWP) (p. 350)
ice–albedo feedback (p. 347)
Intergovernmental Panel on Climate Change (IPCC) (p. 349)
linear response (p. 345)
nonlinear response (p. 345)
positive feedback (p. 346)
radiative forcing (p. 342)
reservoir (p. 357)
saturation effect (p. 350)
water vapor feedback (p. 346)
well-mixed greenhouse gas (p. 348)

GETTING QUANTITATIVE

Preindustrial CO_2 concentration: ~280 ppm

Climate sensitivity estimate: ~2°C to 4.5°C for CO_2 doubling

Current CO_2 concentration (2010): ~390 ppm

Water vapor feedback: ~50% enhancement of warming

Total anthropogenic carbon emissions: ~10 Gt per year

Anthropogenic carbon emissions from fossil fuels: ~9 Gt per year

Anthropogenic forcing due to well-mixed greenhouse gases: 2.6 ± 0.3 W/m^2

Anthropogenic forcing due to aerosols, direct and indirect: −1.2 ± 0.5 W/m^2

Net anthropogenic forcing, all sources: 1.6 W/m^2 (uncertainty +0.8 W/m^2, −1 W/m^2)

QUESTIONS

1 Give two examples of positive feedback effects and two examples of negative feedback effects that operate in the climate system.

2 What is meant by a climate forcing? What are its units, and why?

3 Sulfate aerosols exert a cooling effect on climate, and they're produced in the same coal-burning facilities that account for a significant portion of anthropogenic CO_2 emissions. Why, then, doesn't the cooling effect of aerosols cancel CO_2-induced warming?

4 When you hold a microphone near a loudspeaker, you get feedback—a shrill screeching sound that may grow in intensity. How is this phenomenon related to the feedback mechanisms discussed in this chapter? Is this acoustic feedback positive or negative?

5 A CO_2 molecule remains in the atmosphere for only about 5 years. Why, then, are we stuck for centuries with anthropogenic increases in atmospheric CO_2?

6 A change of 1 W/m^2 in the solar constant results in a forcing of 0.25 W/m^2, or less when albedo effects are included. Why is the forcing only one-fourth of the change in the solar constant?

7 Most of the greenhouse gases listed in Table 13.1 show declining GWPs as the time frame increases, but N_2O and HCFC-23 actually increase at the 100-year time frame before decreasing at 500 years. Why?

EXERCISES

1 Repeat the calculations used in Example 13.1, carrying enough significant figures that you can be sure of one figure to the right of the decimal point in your final answer. You should find a temperature rise of 1.3 K.

2 This exercise is for those of you who have had calculus. (a) The zero-dimensional model of Chapter 12 expresses energy balance through the equation $F = e\sigma T^4$, where we now take F to be the solar input plus any additional forcing. Use calculus to take the derivative dF/dT. Invert to get dT/dF, which corresponds to my second definition of climate sensitivity as the change in temperature per change in forcing. Evaluate your expression for climate sensitivity, using the value T = 255 K that we found in Chapter 12. (Your answer is an underestimate because

it doesn't include feedback effects.) (b) The climate forcing due to anthropogenic CO_2 now in the atmosphere is about 1.65 W/m^2. Use your value for climate sensitivity to estimate the temperature rise that would have occurred since preindustrial times if CO_2 were the only forcing at work in the climate system.

3 Repeat Exercise 1, now using the climate sensitivity in kelvins per watt per square meter found in Exercise 2.

4 Using a climate sensitivity of 0.5 K per watt per square meter (in the middle of the range given in the text), estimate how much of the 0.9°C global temperature increase since 1900 could be due to solar forcing, whose value in Figure 13.4 is in the range of 0.06 to 0.24 W/m^2.

5 Because of the saturation effects described in Section 13.4, the forcing due to CO_2 doesn't increase linearly with CO_2 concentration C. Rather, the change in forcing relative to some reference concentration C_0 is given approximately by $\Delta F = \alpha \ln (C/C_0)$, where α is a constant equal to 5.35 W/m^2 and ln is the natural logarithm. (a) Use this equation to find the change in CO_2 forcing associated with the increase from the preindustrial concentration of 280 ppm to today's approximately 390 ppm, and compare it with the appropriate quantity in Figure 13.4 (your answer will be higher because the data in the figure are for the year 2004). (b) Find the forcing we can expect if CO_2 concentration reaches twice its preindustrial level.

6 Exercise 5 shows that a doubling of preindustrial CO_2 amounts to a forcing of about 3.7 W/m^2. Given a best-guess climate sensitivity of 3°C for a doubling of CO_2, find the corresponding climate sensitivity when expressed in kelvins per watt per square meter.

7 The density of gasoline is about 6 pounds per gallon, and carbon accounts for nearly all the weight of gasoline. Show that combustion of 1 gallon of gasoline produces about 20 pounds of CO_2 (the exact value is closer to 22 pounds).

8 On a per-molecule basis, methane is 26 times more effective as an infrared absorber than is CO_2. Calculate the corresponding ratio on a per-unit-mass basis, and compare it with the 20-year GWP for methane from Table 13.1. What's the reason for any discrepancy you find?

9 Figure 13.10 shows global fossil carbon emissions of about 9 Gt per year. Given that the United States accounts for about one-fourth of global emissions, estimate the U.S. annual per capita emissions of CO_2 (not carbon; see Box 5.1).

10 Figure 13.10 shows that we humans have added about 240 Gt of carbon to the atmosphere during the industrial era. If the dominant removal mechanism is the transfer of carbon to the deep ocean, use the flows shown in Figure 13.10 to obtain a crude estimate of the time it would take to remove all this anthropogenic CO_2, and compare your answer with the 300- to 1,000-year CO_2 lifetime discussed in Section 13.5.

RESEARCH PROBLEMS

1 Locate monthly CO_2 concentration data from the Mauna Loa observatory at www.esrl.noaa.gov/gmd/ccgg/trends/, and plot the data for any five successive years *on the same graph*, using the same horizontal axis running from January through December. Explain the common seasonal pattern shown in each year's data, and explain why the individual years' plots don't end up right on top of each other.

2 Go to the data source for Figure 13.5 (listed in the Credits and Data Sources) and add up the total carbon emissions from fossil fuel combustion and cement production since 1750. Assuming that roughly half of this carbon stays in the atmosphere, add your result to the 560 Gt of preindustrial carbon shown in Figure 13.10 to get an estimate of today's total atmospheric carbon, and compare it with Figure 13.10. (Your answer will be lower since the data in Figure 13.5 don't include carbon from land-use changes.)

3 Exercise 9 shows that annual CO_2 emissions in the United States average about 20 tonnes of CO_2 per capita. Find the corresponding figure for your state or country.

4 The National Oceanic and Atmospheric Administration's Earth System Research Laboratory makes a host of climate-relevant data available at its FTP Data Finder: www.esrl.noaa.gov/gmd/dv/data/. Use data from this site to make a graph similar to Figure 13.6 for any two other greenhouse gases, not including CO_2.

ARGUE YOUR CASE

1. A skeptic argues that the 9-Gt-per-year flow of carbon from fossil fuels to the atmosphere is insignificant compared with natural flows shown in Fig. 13.10, and that we humans therefore can't be having a significant affect on climate. Present a counterargument.

2. The ice-age cycles depicted in Figure 13.8b show that large changes in atmospheric carbon dioxide occur naturally, and a skeptic seizes on this data to argue that today's 390-ppm CO_2 concentration could be of natural origin. Formulate an argument showing that this CO_2 level is the result of anthropogenic emissions.

3. Proposals have been made to counter global warming by injecting sulfate aerosols or their precursor chemicals into the upper atmosphere. What's your opinion of such so-called geo-engineering schemes? Formulate an argument in favor of or in opposition to sulfate aerosol injection.

Chapter 14

IS EARTH WARMING?

Fossil fuel consumption and other human activities have increased atmospheric CO_2 by some 40% since preindustrial times. We've changed a host of other climate forcing agents, too. Some changes are positive, some negative—but the net effect is a positive forcing that means more energy is coming into the Earth system than is leaving. So Earth should be warming. Is it?

14.1 Taking Earth's Temperature

It's not easy to take a planet's temperature, even when we live on it. Temperature varies with geographical position, with altitude and ocean depth, and with time. Air, water, and land temperatures are generally different. So there are many measures of Earth's temperature. Most immediately relevant to us is the temperature at Earth's surface, although you've seen enough of how the climate works to know that it's also important to understand what's happening to the temperature in the atmosphere and oceans.

THE THERMOMETRIC RECORD

Glance out the window at your thermometer and you've got a temperature measurement at a single time and place. In principle, enough such measurements, spread over the planet, could be combined to provide an average global temperature. Tracking that temperature over time would then answer our question about whether Earth is warming.

We do, in fact, have enough thermometer-based temperature measurements to calculate global average temperatures going back to about the mid-nineteenth century. Although the mercury thermometer was invented in the early eighteenth century, records before about 1850 are simply too sparse to compute a meaningful global average from direct temperature measurements. And once we do have enough data, combining those temperature measurements requires care.

The thermometric temperature record consists of observations from weather-station thermometers that measure the surface air temperature, and measurements of sea-surface temperature and marine air temperature from ships and automated buoys. Several groups around the world maintain independent analyses of global temperature records, each using different combinations of data and different techniques for calculating the global average. Each group applies corrections to the raw data to compensate for factors that could lead to artificial trends in the calculated temperature. For example, today's weather stations mount thermometers at a standard height somewhat over 1 m above the surface to avoid steep temperature gradients just above the ground. But early stations weren't standardized, which introduces uncertainties. Sea-surface temperature measurements suffer analogous inconsistencies. Early sea-surface temperature measurements were made by hauling buckets of water on board ships and then measuring the water temperature. Obviously, some time elapsed between collecting the water sample and getting it onto the ship, during which time its temperature could change. Both the distance to the ship deck and the material from which the bucket was made could affect that change. Later, ships began measuring sea temperature more directly by taking samples at the engine cooling water inlets. An accurate record of sea-surface temperature requires corrections for these changes.

Urban energy use and albedo changes (e.g., more black pavement, fewer plants) make cities warmer than the surrounding countryside. As cities grow, weather stations located on their outskirts become increasingly urbanized, which means they'll begin to record higher temperatures. Absent any real global temperature change, this **urban heat island effect** would produce a warming trend in temperature records from urban weather stations. Fortunately, we can separate urban stations from rural and marine records; when that's done, it appears that the urban heat island effect accounted for no more than about 0.05°C of warming in the twentieth century—less than 10% of the total observed warming.

Climatologists consider these and other effects as they merge individual temperature measurements into time series of global average temperatures and associated uncertainties. It's harder to establish unambiguously the actual temperature than it is to calculate **temperature anomalies**, or deviations from an average temperature. That's partly because temperature changes are well correlated over large distances, but absolute temperatures aren't. It's also because two thermometers that might not agree perfectly on the actual temperature will nevertheless give the same response to a given temperature change. So the various time series calculated by different groups all give temperature anomalies relative to some average; for most of the data I use in this chapter, it's the average temperature for the period from 1961 to 1990. A negative anomaly indicates a global temperature below this average, and a positive anomaly indicates a higher temperature.

Figure 14.1 shows the global temperature record from the University of East Anglia's Climatic Research Unit (CRU) in the United Kingdom, one of the several groups around the world that maintains such records. You can explore the CRU's data further in Research Problems 1 through 3, in which you'll find that the Northern and Southern Hemispheres show very similar trends, and you'll see the increase in global coverage of the temperature reporting stations. In Research Problem 4 you can compare the CRU data with those of the U.S. National Climatic Data Center (NCDC), and you'll find close agreement.

The temperatures in Figure 14.1 fluctuate a lot from year to year, but nevertheless the trend is obvious: There's been a significant rise over the period from 1850 to the present. A more detailed look shows distinct trends in different eras: (1) a fairly steady average temperature during the second half of the nineteenth century, (2) an obvious rise beginning around 1910 and ending around 1945, (3) a steady or slightly declining temperature from 1945 to about 1975, and (4) a steep rise beginning in the mid-1970s and continuing to the present. The overall temperature increase for the past hundred years averages about

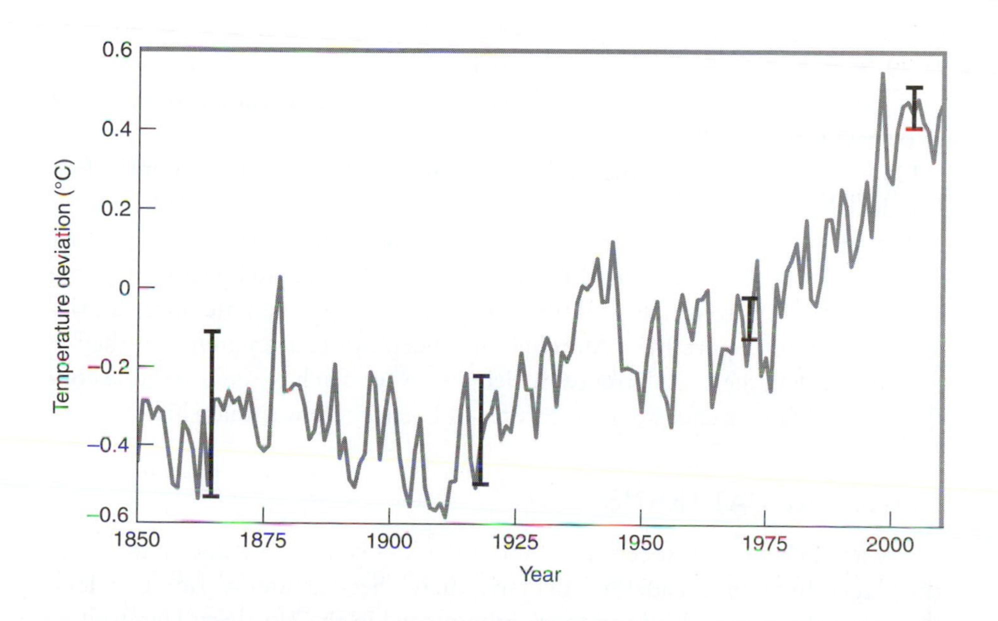

FIGURE 14.1
Global temperature variation from 1850 through 2010, using the 1960–1991 average as a baseline. Error bars show approximate uncertainty at four different times. Data are from the University of East Anglia's Climatic Research Unit; Box 14.1 and Research Problem 4 compare other temperature records.

BOX 14.1 | The Warmest Year?

What's the warmest year on record? Different temperature records disagree on this, because of differences in their statistical methods (Figure 14.2). The University of East Anglia's Climatic Research Unit shows 1998 as the warmest—likely the result of a very strong El Niño that year coupled with the CRU's underrepresentation of the rapidly warming Arctic. For the U.S. National Climatic Data Center record, 2005 and 2010 are essentially tied for warmest, while NASA's Goddard Institute for Space Studies (GISS) shows 2010 as the warmest year so far. Within the margins of error, it's impossible to say with certainty which is the warmest. But it's clear that 2010, 2005, and 1998 are the strongest candidates. And all three datasets agree that 9 of the top 10 warmest years occurred in the first decade of the twenty-first century, the exception being the El Niño year 1998.

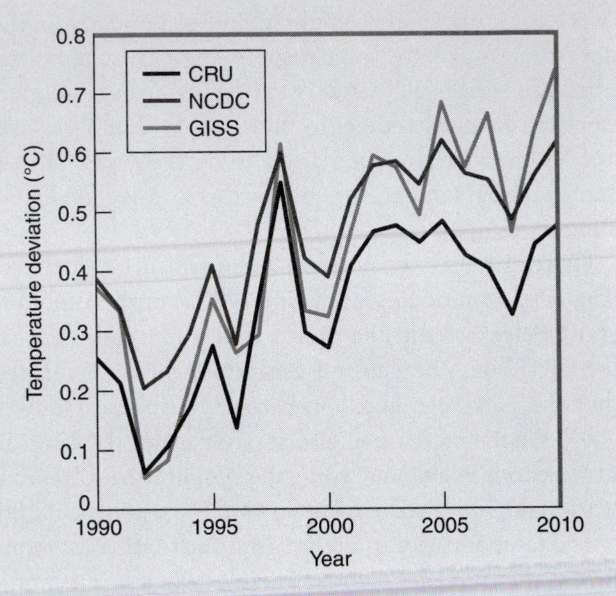

FIGURE 14.2
Comparison of the CRU, NCDC, and GISS datasets for the period 1990–2010. Deviations are relative to the 1961–1990 average.

0.076°C per decade, increasing to nearly 0.2°C per decade for the period since 1990. Although such global numbers mask a lot of details, they're useful figures to keep in mind when we talk about longer-term temperature trends or try to project future climate.

How confident are we in the temperatures shown in Figure 14.1? For the first few decades, the estimated uncertainty is on the order of ±0.2°C—a little larger than typical year-to-year fluctuations. By 1950, with an increase in the number of reporting stations and their global distribution (see Research Problem 2), the uncertainty drops to roughly ±0.05°C. I've marked approximate uncertainties at four points in Figure 14.1. Although the uncertainties are significant, they're far smaller than the overall trends, which means we can have considerable confidence that those trends are real features of Earth's recent climate history.

REGIONAL PATTERNS

The warming of the past century—about 0.8°C—is unevenly distributed over the planet. In recent decades, particularly, there's been greater warming on land than over the oceans. It's been more pronounced in the Northern Hemisphere

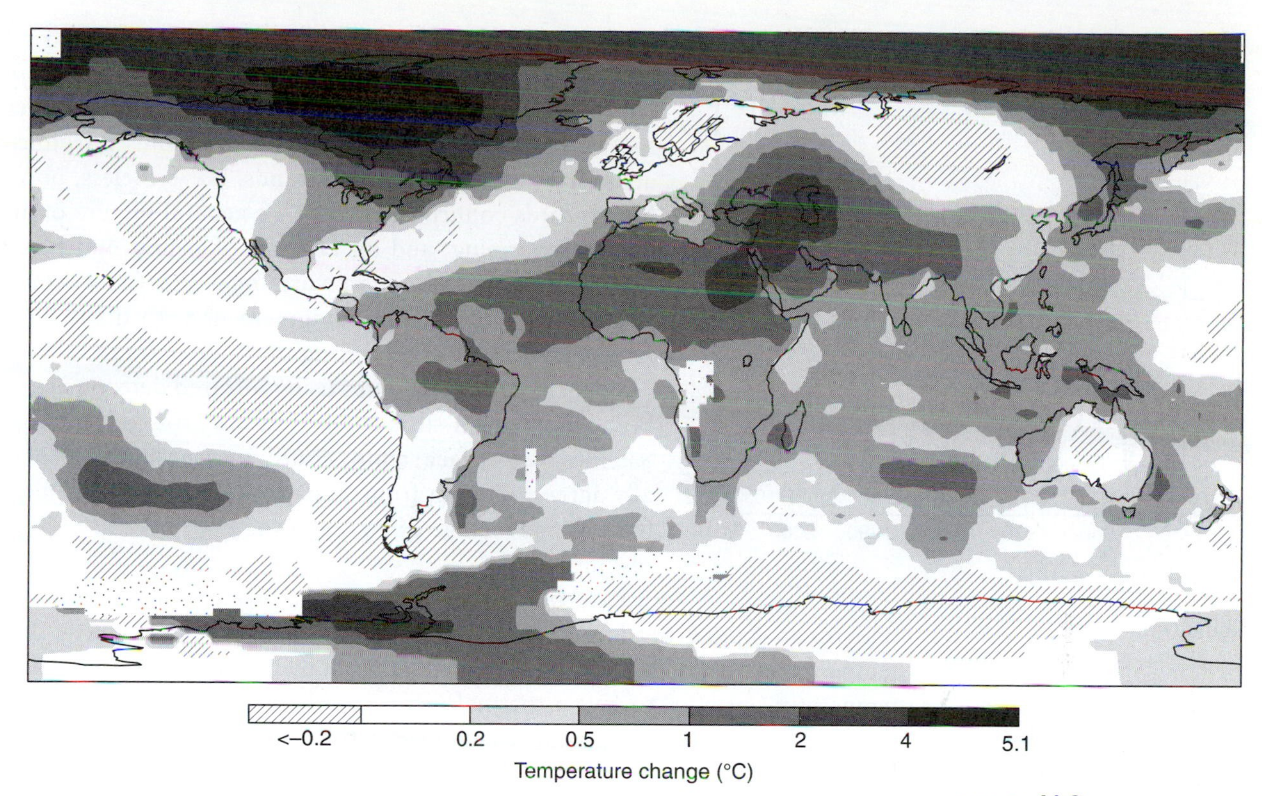

FIGURE 14.3

Average temperature change in 2010 relative to the 1961–1990 average. Note the greater warming over land, in the northern hemisphere, and especially at high latitudes. White areas show no significant change, hashed areas show cooling, and stippled areas have insufficient data.

than in the Southern Hemisphere. And the Arctic has warmed at about twice the global rate, which is to be expected as the ice–albedo feedback introduced in Chapter 13 exacerbates Arctic warming. In addition, there's less evaporation in the ice-covered Arctic, so more of the incident solar energy goes directly into surface heating. Evaporation changes water's state rather than its temperature, so if less of the incident solar energy is being used for evaporation, then more is available for warming. Several other processes also contribute to enhanced Arctic warming, all of which make the Arctic an especially sensitive indicator of climate change. Figure 14.3 shows the distribution of warming as of 2010, relative to the standard 1961–1990 averages.

OTHER MODERN TEMPERATURE RECORDS

The temperature record of Figure 14.1 derives from air and water temperatures measured with thermometers at Earth's surface; it's therefore a good indication of surface temperature. We can also get a record of surface temperature over time by boring into the Earth and measuring the temperature as a function of depth. That's because a year's average temperature is reflected in the near-surface soil temperature, and that "signal" of yearly temperature propagates downward as new temperature "information" flows into the Earth from above. Temperature records from such **boreholes** have the advantage of smoothing over the rapid fluctuations seen with surface temperatures, but they have the

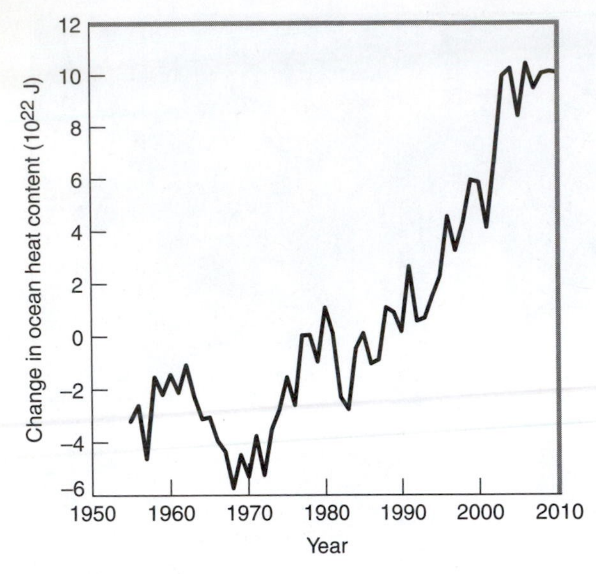

FIGURE 14.4

Increase in ocean heat content since the mid-twentieth century, in units of 10^{22} joules, relative to an arbitrary zero. The increase results from the imbalance between Earth's incoming and outgoing energy flows, caused largely by anthropogenic greenhouse gases and other forcing agents.

disadvantage of offering far less global coverage. And it's important to ensure that vegetation over the borehole site hasn't changed; otherwise, changing albedo can affect the rate of solar energy absorption and thus introduce artificial warming or cooling trends. Nevertheless, borehole records confirm Figure 14.1's general picture of an ongoing warming, and an analysis of nearly 400 boreholes worldwide indicates a warming of 0.5°C during the twentieth century—in good agreement with the surface record's 0.6°C warming during the same period.

Since the 1980s, satellites using infrared detectors have been mapping sea-surface temperature around the globe. Satellite sea-surface temperature measurements generally agree with data from ships and buoys, and satellites have the advantage of covering the entire ocean at high resolution.

Like borehole temperatures, profiles of temperature with ocean depth give additional information about what's happening to global temperature. In the early 2000s researchers combined millions of individual temperature profiles to give a thermal picture of the top ocean layers. These data yield a measure of the total heat content in the upper ocean. This quantity, like temperature itself, has been increasing (Fig. 14.4), an indication that the oceans are absorbing some of the greenhouse-trapped energy. This trend provides supporting evidence for the 0.9 W/m^2 imbalance in Earth's energy flows that I described in Chapter 12.

Air temperature is harder to measure, except right near Earth's surface. For decades, balloon-borne **radiosondes** have probed the atmosphere, reporting back temperature as a function of altitude. But these devices are designed to help with immediate weather forecasting, not long-range climate studies. Variations among instrument designs casts doubt on the reliability of radiosonde-derived climate data, and improved shielding of radiosonde thermometers from sunlight has resulted in a phantom cooling trend in the radiosonde record. This is important to note, because some radiosonde analyses claim a recent cooling in the lower troposphere, especially over the tropics, which contrasts with climate model predictions that the lower troposphere should warm somewhat more than the surface.

A more consistent approach to air temperatures comes from satellite instruments that measure atmospheric microwave emissions. These **microwave sounding units (MSUs)** look down on vast volumes of atmosphere, and the data from different microwave wavelengths yield temperatures at different levels in the atmosphere. The satellite temperature record is short, extending back only to 1979. Extraction of temperatures from MSU data is complicated, requiring corrections for variations in satellite orbits and for the measurement time in relation to the daily cycle of atmospheric temperatures. Furthermore, the instruments that measure tropospheric temperature also sample the lower stratosphere, "contaminating" the tropospheric record with stratospheric temperatures that

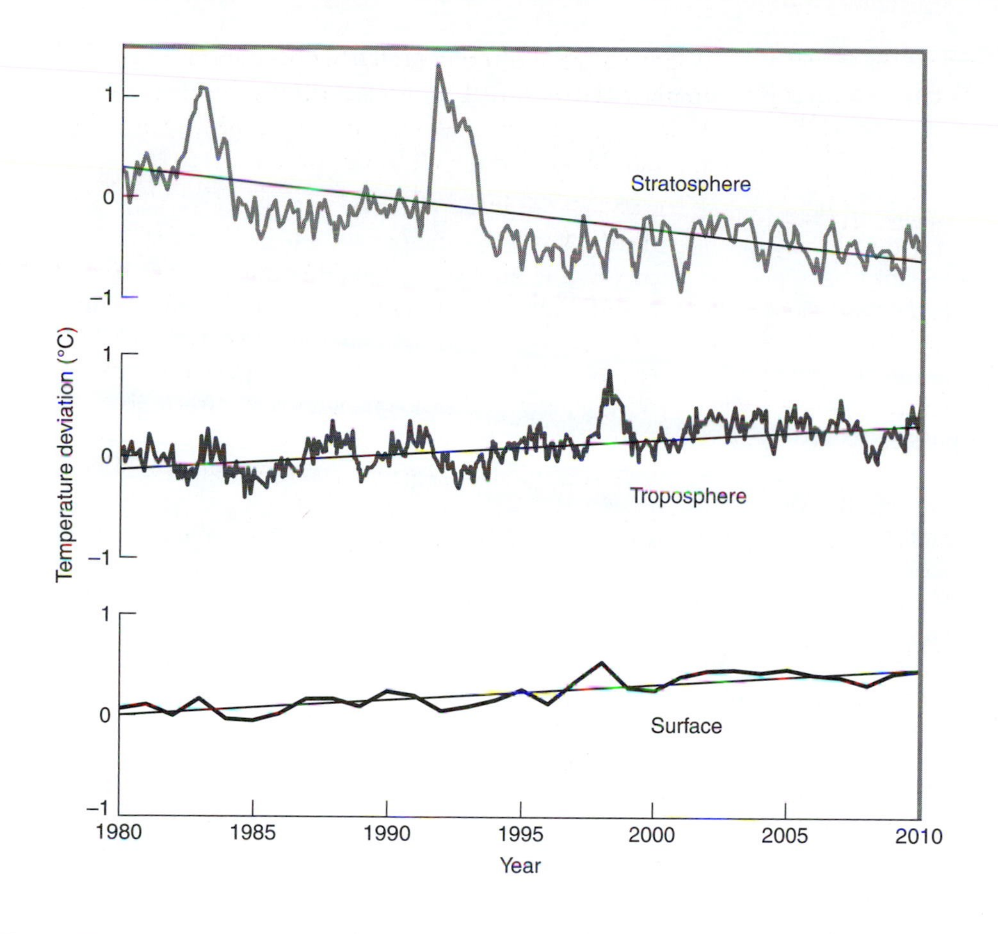

FIGURE 14.5
Temperature records for the lower stratosphere (top), lower troposphere (middle), and surface (bottom), relative to the 1979–1999 average. Lines show general trends. Atmospheric temperatures are from satellite MSU instruments; surface is adapted from Figure 14.1. Note that the stratosphere cools, consistent with Chapter 12's discussion of the greenhouse effect. Spike in stratospheric temperature in the 1990s corresponds to the Pinatubo volcanic eruption, which cooled the lower atmosphere but warmed the stratosphere.

behave oppositely. Nevertheless, careful analyses of MSU data make clear that observations are consistent with model predictions of an enhanced warming in the lower troposphere relative to the surface. Figure 14.5 compares temperature records from the surface, the troposphere, and the stratosphere.

GOING FURTHER BACK

The global temperature of Figure 14.1 shows a general rise since the record began in the mid-nineteenth century, and an especially steep rise in the past several decades. Borehole records confirm this trend, and more recent satellite-based measures of sea-surface and lower atmosphere temperatures agree that recent decades have seen substantial warming. The rise in ocean heat content is a further indication of ongoing global warming. But is this unusual? Is it related to the increasing anthropogenic climate forcings—especially the rise in greenhouse gas concentrations? Or is it a natural climatic fluctuation? We can answer that question in several ways, the most obvious being to push the temperature record further back in time.

Unfortunately, we don't have enough thermometric measurements to determine an accurate global average temperature before about 1850. Instead, we use **proxies**—quantities that "stand in" for temperature. To be useful, a proxy

BOX 14.2 | Isotopic Temperature Reconstruction

Isotopes (Chapter 7) are versions of an element that differ by number of neutrons and hence by mass. Although chemically similar, lighter isotopes are more mobile and therefore participate more readily in physical and chemical processes. That's why plants take up carbon-12 more readily than carbon-13—which helps confirm (Chapter 12) that growing atmospheric CO_2 comes from fossil fuels. Oxygen, like carbon, has several stable isotopes: Some 99.8% of it is O-16; nearly all the rest is O-18. Water with O-16 ($H_2{}^{16}O$) is lighter and evaporates more readily, making atmospheric water vapor higher in O-16 compared with the oceans. Conversely, heavier $H_2{}^{18}O$ condenses more readily. As water evaporated in the tropics rides the atmospheric circulation toward the poles, $H_2{}^{18}O$ precipitates out at lower latitudes, leaving polar precipitation depleted in O-18. How depleted depends on the prevailing temperature; the cooler the climate, the sooner the heavier water precipitates out, and the less O-18 ends up in polar ice.

The Greenland and Antarctic ice sheets contain hundreds of thousands of years' accumulation of snow, compacted into ice by the weight of the overlying layers. In Greenland, where the deepest ice is over 125,000 years old, distinct bands within the ice mark each season's snowfall and thus allow scientists to date ice cores by year. Antarctic ice goes back as far as 800,000 years, but it generally lacks the yearly banding, so its dating isn't as finely resolved. In either case, though, measurement of the O-18 to O-16 ratio provides an estimate of temperature versus time. The relative amount of O-18 is specified by a quantity called $\delta^{18}O$, the fractional deviation of the $^{18}O/^{16}O$ ratio from a standard value, expressed in parts per thousand (‰.) The deep ocean has $\delta^{18}O$ about zero, whereas in warm tropical waters it's around 1‰. Antarctic ice, in contrast, can have $\delta^{18}O$ as low as –55‰, as water vapor is enriched in O-16 by preferential evaporation in the tropics and depleted of O-18 by precipitation during its journey to the Antarctic.

Oxygen isotope ratios also yield the total amount of water locked up as ice. This is evident, again, because the water that evaporates and eventually becomes ice is enriched in O-16, leaving behind seawater with higher $\delta^{18}O$ values. $\delta^{18}O$ in seawater therefore provides a measure of the total ice volume. Measuring $\delta^{18}O$ values in sediments containing calcium carbonate ($CaCO_3$) from shells of marine organisms provides a record of ice volume over time and therefore, indirectly, of sea-level variations.

Hydrogen isotopes provide a similar approach to climate reconstruction: Water containing the heavy isotope deuterium evaporates less readily and condenses more readily than normal water. Therefore, measurement of the ratio of deuterium to hydrogen in ice cores provides a measure of temperature at the time the water fell as precipitation.

has to be widely available and we have to understand quantitatively how its value relates to temperature. Among the frequently used proxies are tree rings, whose width and density reflect climatic conditions on a year-to-year basis; lake sediments, whose thickness and composition reflect the thermal energy available to produce snowmelt streams that carry sediments into lakes; coral reef bleaching, which occurs when coral organisms are stressed by rising temperatures; and isotope ratios (see Box 14.2). Proxies sensitive to precipitation or other quantities may also convey information about regional temperature variations even though they don't translate directly into local temperatures.

Figure 14.6 compares 10 different proxy-based studies of temperature change over the past millennium. Also shown is the instrument record of Fig-

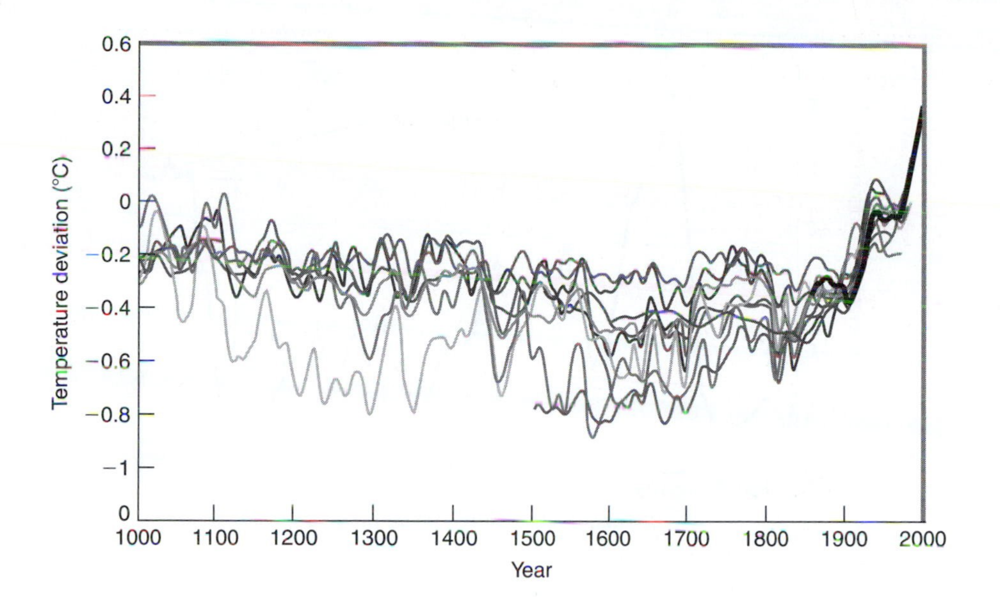

FIGURE 14.6
Results of 10 different reconstructions of temperature for the past millennium (gray curves). The instrumental record, dating to the mid-nineteenth century, is shown in black. Deviations are relative to the 1961–1990 average temperature.

ure 14.1 for the twentieth century and beyond. It takes only a quick glance to spot the general trend in all the studies—a gradual decline for the first 900 years, followed by a sharp upturn from 1900 to the present. The different studies use a variety of statistical techniques to reconstruct historical temperatures from different sets of proxies. As a result, there's considerable variation among these temperature reconstructions, especially prior to the twentieth century. Nevertheless, all concur in suggesting that the warming of the last 100 years is unprecedented in both its magnitude and rapidity, at least on a millennial timescale.

GOING EVEN FURTHER BACK

The warming of the twentieth and twenty-first centuries appears unusual in the context of the millennial reconstructions shown in Figure 14.6. Does it still look unusual if we go further back? Ice-core data provide an answer. Figure 14.7 shows a temperature reconstruction of the past 160,000 years made from hydrogen isotope ratios in Antarctic ice cores. This particular data set goes back some 420,000 years; a similar core completed in 2003 extends nearly 800,000 years.

There's a lot of variation in Figure 14.7, but there's also a hint of a pattern. At the modern end of the ice-core record, we're at a much higher temperature than the average. Roughly 130,000 years ago, Earth also saw a short (10,000 years or so) spell of warmer than normal temperature. Between these brief, warm **interglacials** is a much longer **ice age**. During the ice age, Earth's climate was profoundly different, with 1.5 to 3 km of ice covering a good part of North America and northern Europe (Fig. 14.8).

What causes this pattern of long, cold ice ages punctuated by briefer warm spells? The answer lies in variations in the distribution of solar radiation

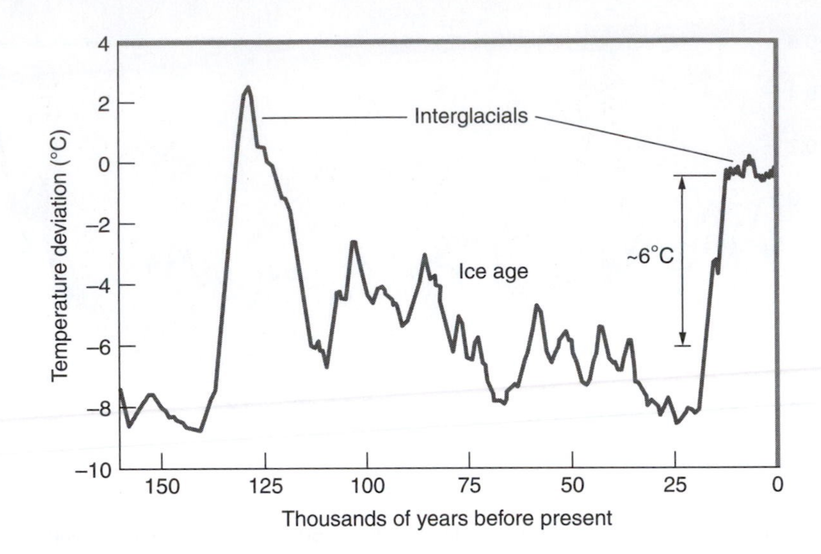

FIGURE 14.7

The ice-core temperature record from Vostok, Antarctica, shows a pattern of warm interglacial periods separated by ice ages lasting roughly 100,000 years. This pattern persists for nearly a million years into the past. Note that only about 6°C separates the modern climate from the average ice-age temperature. The plot is a running average over 25 data points, with deviations relative to the preindustrial global temperature. The industrial-era temperature rise is not shown.

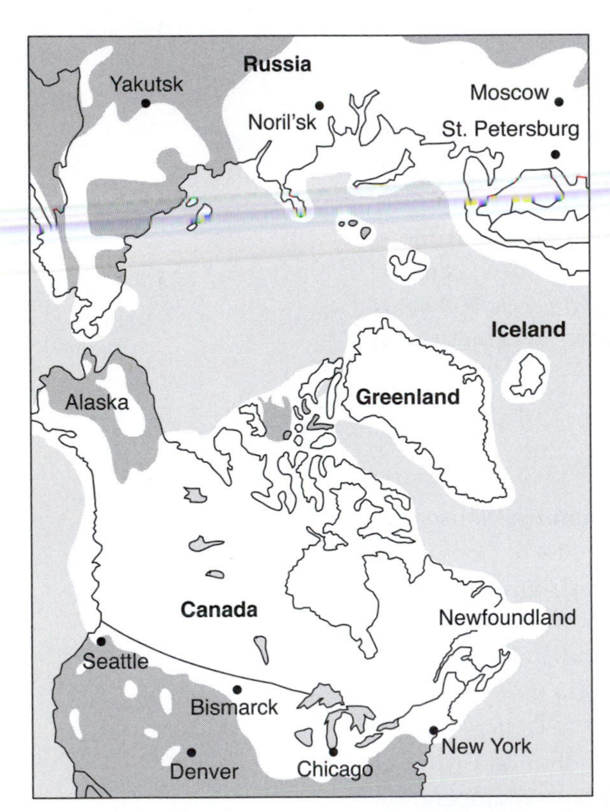

FIGURE 14.8

Maximum extent of northern hemisphere glaciation during the most recent ice age. Ice sheets were typically 1.5 to 3 km (about 2 miles) thick over much of the region shown in white.

that result from periodic changes in Earth's orbit and the tilt of its axis. However, these effects alone are too small to account for the climate variations shown in Figure 14.7. Rather, it's believed that small temperature changes resulting from orbital variations trigger feedback effects that enhance the orbitally induced heating and cooling trends. Combining the temperature record of Figure 14.7 and the CO_2 concentration from Figure 13.8b illustrates this point. Figure 14.9 is the result: a remarkable correlation between temperature and CO_2 concentration. But be aware that this isn't a simple cause-and-effect relationship in which, for example, variations in CO_2 concentration cause a similar pattern of temperature changes. Rather, imagine a small temperature increase associated with orbital changes. This, in turn, increases sea-surface temperatures; since warmer water can hold less dissolved CO_2, the result is a flow of CO_2 into the atmosphere. The additional atmospheric CO_2 creates a stronger greenhouse effect and exacerbates the warming. On land, warming temperatures increase the respiration rate of soil microbes, returning more CO_2 to the atmosphere and again enhancing the warming. Although the details aren't yet fully understood, these and other feedbacks are ultimately responsible for both the overall temperature and CO_2 patterns shown in Figure 14.9, and for the correlation that's so dramatically evident.

It's possible to push climate reconstructions back millions and even billions of years by examining oxygen isotope data, the shape of fossil leaves, and other proxies. But more interesting here is the question of natural climate variability in recent prehistoric times. As Figure 14.7 shows, there's been considerable

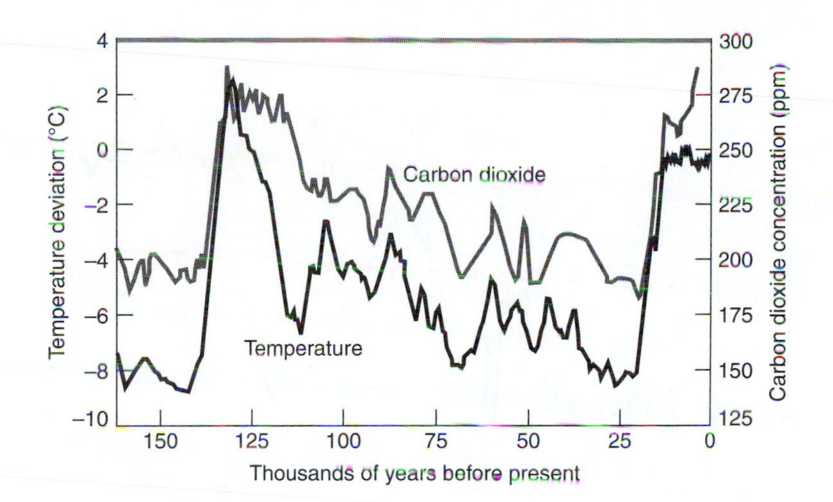

FIGURE 14.9
Temperature and atmospheric CO_2 show a tight correlation through the ice-age cycles, as evident in this record from the Vostok ice core in Antarctica. This correlation involves complex feedbacks between temperature and CO_2 concentration. Both plots stop before the industrial era.

variability on finer timescales within the overall ice-age cycle. A closer look, using data from Greenland ice cores, reveals some especially rapid variations in the roughly 20,000 years since the depths of the last ice age (Fig. 14.10). A particularly dramatic example is the so-called Younger Dryas event, which swung the Arctic back into ice-age conditions for some 1,500 years starting around 12,000 years ago. This cool period ended abruptly with a 7°C warming of Greenland in less than a century. The Younger Dryas and similar rapid climate fluctuations are nonlinear effects believed to result from sudden changes in ocean circulation that occur when an influx of fresh water from melting ice alters the salinity and hence the density of ocean water. Although the Younger Dryas fluctuation may seem as dramatic as the late-twentieth-century warming in Figure 14.6, I caution you against inferring that the latter must also be a natural fluctuation. The large, rapid temperature rise at the end of the Younger Dryas is clearly evident only in Greenland—an Arctic region that tends to exaggerate global trends. Although the Younger Dryas event appears in other ice cores, pollen records, and ocean sediments across the globe, its climate signature is much less obvious in these other areas, suggesting an event that was concentrated largely at higher latitudes in the Northern Hemisphere.

Figure 14.10 shows that the climate of roughly the past 10,000 years has been substantially more stable than that of the last ice age and the beginning of the current interglacial warm period. This is not to say that there haven't been significant variations, but such variations have been gradual and modest. This stable period coincides with the development of agriculture, and many would argue that a stable climate was essential for this major advance in human civilization. Although most climatologists accept predominantly natural causes for climate behavior until the past half-century or so, some argue that humans, as early as 8,000 years ago, began modifying the climate substantially in ways that enhanced stability and forestalled the slide into the next ice age. This anthropogenic modification purportedly results from the greenhouse gases (CO_2 and

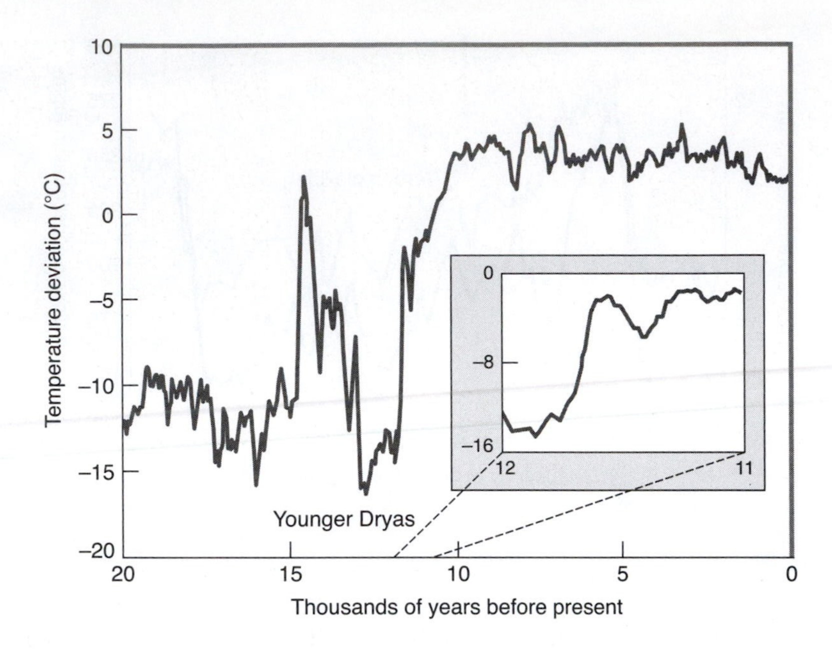

FIGURE 14.10 Detailed record of Greenland ice-core temperatures over the past 20,000 years, relative to the average for the period. The Younger Dryas event marked a sudden, brief return to ice-age conditions. The inset depicts the period from 12,000 to 11,000 years ago, showing the extremely rapid temperature rise at the end of the Younger Dryas period. The graph indicates that climate has been much more stable over the past 10,000 years.

methane) generated by land clearing and agricultural practices. Whether natural or anthropogenic, the climate of recent millennia has been benignly constant compared with that of earlier times.

This picture of earlier rapid climate fluctuations followed by the more stable climate of recent times suggests two points: First, the rapid increase in global temperature during the late twentieth and early twenty-first centuries is abnormal in the present climatic context. Second, slight changes in the climate system can result in highly unstable conditions, with swings of several degrees in average temperature occurring within centuries or even decades over large regions of the globe. Such changes would not be beneficial to a highly developed civilization on an overcrowded, underfed planet.

14.2 Other Climatic Changes

Temperature isn't the only measure of global climate change. A variety of other indicators suggest a general global warming trend accompanied by more subtle changes in temperature and weather patterns. All reinforce the picture of a climate that's undergoing abnormally rapid change.

ICE AND SNOW COVER

When Montana's Glacier National Park was established in 1910, it had well over 100 glaciers. By 2010, only about 25 of them remained. At the current rate of glacier retreat, the park's name will be meaningless in several decades, as permanent glaciers disappear altogether.

FIGURE 14.11
The South Cascade Glacier in Washington State has shrunk dramatically as seen in these photos taken in 1928 (left) and 2000 (right).

Mountain glaciers are shrinking across the globe, with many exhibiting dramatic reductions (Figure 14.11). A few glaciers are growing, but that's because glacial behavior is governed by a balance between snowfall and melting. Snowfall depends on evaporation that occurs elsewhere, and evaporation, as we've seen, can increase with global warming. But on average, glaciers worldwide are in retreat, losing both area and thickness. As they melt, mountain glaciers leave behind an altered landscape with lower albedo and thus increased solar energy absorption. They also increase freshwater runoff that ends up in the oceans, where it contributes to rising sea level.

Polar ice is also melting, both the floating sea ice and the large continental ice sheets of Greenland and Antarctica. On average, Arctic temperatures have increased in recent decades at about twice the global average (recall Figure 14.2). The picture for Antarctica is less clear, because the Antarctic climate has behaved differently in different regions and seasons. Interior Antarctica, for example, has cooled somewhat in recent decades, while the Antarctic coast has generally warmed. The Antarctic summer has seen little change, while autumn has cooled. Winter and spring exhibit complex regional patterns of warming and cooling.

Among the most dramatic ice changes is the shrinking of summertime Arctic sea ice, which has declined some 40% since 1979, with more modest but still significant declines in other seasons (Fig. 14.12). It now appears that the Arctic Ocean may be ice-free in summer by the end of the twenty-first century. Sea-ice area is obvious when viewed from above, but thickness isn't. However, data from nuclear submarines operating under the polar ice contain information on ice thickness, and these data, along with later oceanographic studies, suggest that Arctic ice thickness has declined by as much as 40% in recent decades. In contrast to these large Arctic ice declines, the climatologically complex Antarctic has seen a modest 3.6% increase in sea ice extent since 1979.

Loss of sea ice affects the local ecology and in particular endangers wildlife such as polar bears and seals; this in turn affects indigenous peoples of the Arctic. Melting sea ice floods the ocean with freshwater, whose lower density can alter large-scale ocean circulation—an example of a nonlinear climate effect.

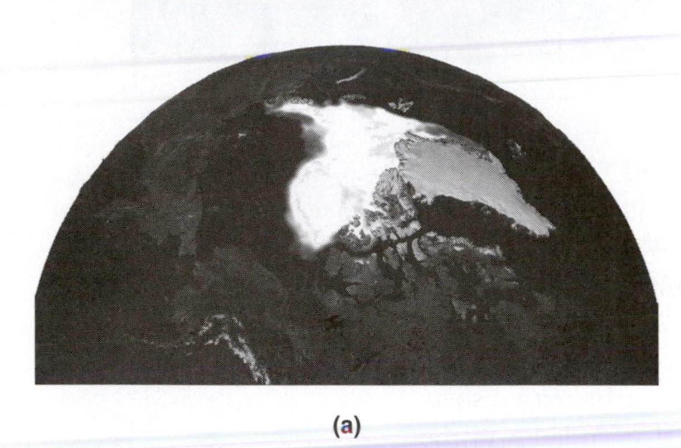

(a)

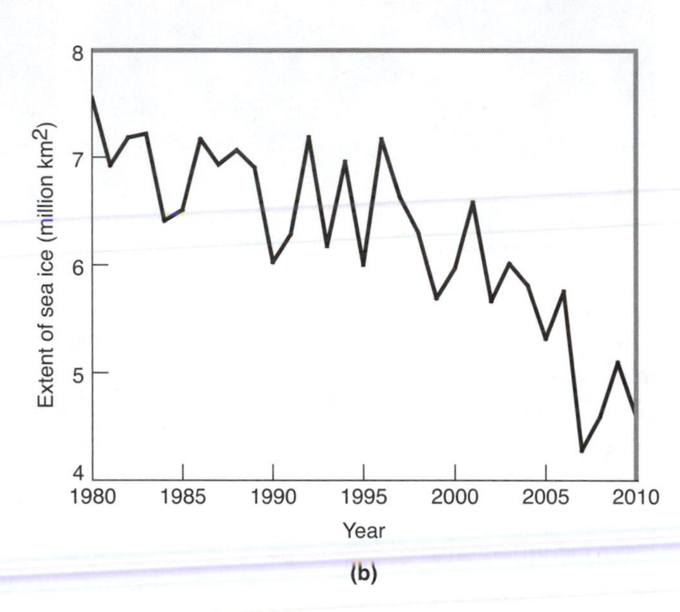

(b)

FIGURE 14.12

Decline in the area covered by Arctic sea ice, based on satellite measurements taken annually in September. (a) Composite satellite images of the minimum ice cover in 1980 and in the record low year of 2007. (b) A plot of the decline over time. Sea-ice extent has declined by some 40% since 1980, with the decline accelerating in recent years.

When the area of sea ice is reduced, its high albedo diminishes, leaving dark seawater to absorb more solar radiation. The resulting ice–albedo feedback is one of the reasons for increased warming in the Arctic. Unlike melting land ice, however, one thing melting sea ice doesn't do is raise sea level significantly. That's because floating ice displaces the same mass of ocean water whether it's in a solid or liquid state—although there is a tiny effect because of the density difference between the freshwater that comprises ice and the salt water that remains in the oceans.

SEASONAL TEMPERATURE CHANGES

Figure 14.2 shows that global temperature increases aren't spread evenly around the globe; for example, they're greater on land and at high latitudes. They're also not evenly distributed in time. Although the average Arctic warming over the past hundred years is about twice the 0.8°C global average for that period, Arctic winter temperatures have increased as much as 3°C to 4°C. A warming climate also affects the frequency of extreme temperature excursions, such as heat waves or extreme cold. Because extremes are by nature rare, it's more difficult to obtain firm statistical evidence. However, several studies point to a decrease in the number of extremely cold days over some regions, thereby lengthening the growing season (the frost-free period) by as much as several weeks. There's also evidence for increases in the number of extremely hot days.

WEATHER CHANGES

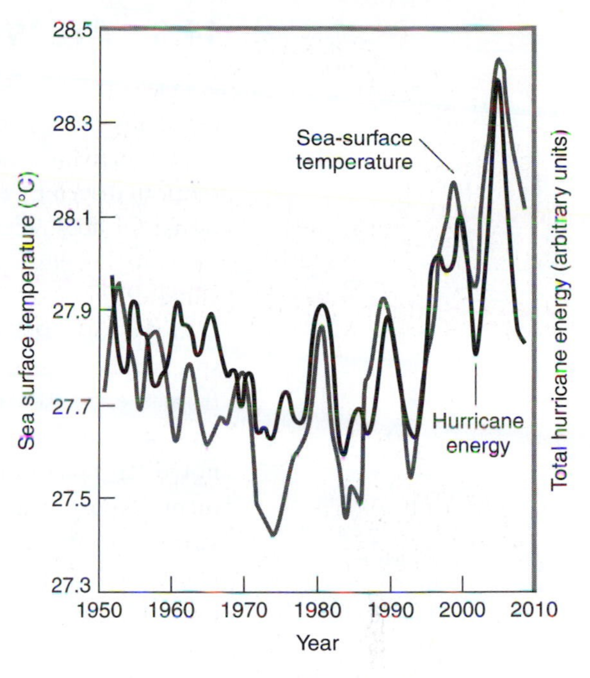

FIGURE 14.13
Total energy dissipated per year in all North Atlantic hurricanes (solid curve) has increased substantially in recent decades, in step with rising sea-surface temperatures (dashed curve).

Rising temperatures should affect Earth's weather in several ways. Increasing sea-surface temperatures mean more evaporation, cloud formation, and precipitation. There's also more energy available to power tropical storms. Large-scale patterns such as El Niño—a warming in the eastern Pacific that affects global weather—could change in frequency and intensity. Weather's intrinsic variability in both time and space makes it difficult to confirm such changes with high levels of statistical confidence. Nevertheless, there's strong evidence that precipitation at midlatitudes increased some 5% to 10% during the twentieth century, with lesser increases in the tropics and subtropics. A related change is that more precipitation falls in shorter, more intense events. Atmospheric water vapor and clouds have increased slightly, although we're less confident of those measures. Hurricane studies show increases in the number of category 4 and 5 hurricanes and in measures of overall hurricane energy, especially in the North Atlantic (Figure 14.13). The latter scales as the cube of the wind speed, for reasons I discussed in Chapter 10 in connection with wind energy, and is a rough measure of the total destructive potential of a season's worth of hurricanes. Figure 14.13 shows that the rise in hurricane energy correlates closely with rising sea-surface temperatures. All these weather changes are broadly consistent with increasing global temperature.

SPECIES RANGES

As Earth warms, you might expect temperature-sensitive species of plants and animals to migrate so as to remain in their optimum climatic zone. However, climate is one of several factors that affect species' geographical ranges, so studies of individual species don't give conclusive evidence for global warming. But a method known as *meta-analysis*—a statistical survey of many individual studies—can reveal significant trends that indicate a biological response to global warming.

One such meta-analysis examined hundreds of different plant and animal studies, and the conclusion was that species were moving northward at an average of 6.1 km per decade, or upward in altitude at 6.1 m per decade. Significant springtime events in the lives of species—breeding, nesting, flowering, budburst, ending migratory journeys—were happening earlier in some 62% of species studied; in one subset of 172 species, the spring events were advancing at the rate of 2.3 days per decade. Overall, some 87% of species studied showed changes in timing or range consistent with global warming; only 13% showed contradictory trends. So it appears that plants and animals know what our thermometers are telling us: We live on a warming planet.

14.3 Are We to Blame?

It's clear that Earth's climate is changing, with a temperature rise that's been especially pronounced in recent decades. But is the observed change a result of human activities, especially our greenhouse emissions? Or is it a natural fluctuation, driven by external factors such as solar variability and volcanic eruptions? Or does recent climate change result from natural internal variations of the complex physical, chemical, and biological system that determines Earth's climate?

In the early 1990s, when concern about climate change first became widespread, the "signal" of anthropogenic effects hadn't unambiguously emerged from the "noise" of natural climate variability. But by 1995 the scientists and policymakers writing the IPCC's Second Assessment Report cautiously noted "a discernible human influence on climate." The IPCC's Third Assessment Report, published in 2001, stated more boldly, "There is newer and stronger evidence that most of the warming observed over the past 50 years is attributable to human activities." When asked for an independent assessment of the IPCC's findings, the U.S. National Academy of Sciences concluded that "greenhouse gases are accumulating in Earth's atmosphere as a result of human activities, causing surface air temperature and subsurface ocean temperature to rise." The U.S. Global Change Research Program, in a 2009 report, put it more bluntly: "Global warming is unequivocal and primarily human-induced."

So what's the evidence that today's climate change results largely from human activities? First, there's the basic physics of the greenhouse effect, confirmed by the natural "experiments" of Mars and Venus: We know that greenhouse gas concentrations have an impact on climate. In particular, increasing CO_2 in itself should result in increasing temperature. In that context, take another look at Figure 13.8b, which shows that today's atmospheric CO_2 concentration is nearly 40% higher than at any time during the past half-million years. And we know, from the evidence described in Chapter 13, that most of the excess carbon comes from fossil fuel combustion. So we humans have taken Earth's atmosphere into a regime that our planet probably hasn't seen for millions of years. Now look back at Figure 14.9. Although the interplay between CO_2 and temperature here is complex and not necessarily predictive, their obvious correlation suggests that we might expect a significant climatic response to the industrial-era spike in fossil-derived atmospheric CO_2.

Comparison of Figures 13.7 and 14.1 shows somewhat similar rises in temperature and CO_2 over the past 150 years. Again, this isn't enough to establish cause and effect. However, the coincidence is certainly noteworthy, especially given that these are quantities basic physics says should be related and because both increases are probably unprecedented, at least in recent millennia. To explore this relationship further, we can ask how the recent global temperature rise correlates with several known forcing agents that we expect to influence climate. This, in fact, was done for the most recent four centuries of one millennial temperature reconstruction shown in Figure 14.6. The authors of that study checked their results for correlations with variations in solar radiation, with the so-called dust veil index that measures volcanic particulates in the atmosphere, and with atmospheric CO_2 concentration taken as a proxy for all anthropogenic greenhouse gases. The volcanic correlation is negative, because dust exerts a cooling effect, and it was strongest in the early 1800s—a time of exceptionally active volcanism. For the majority of the four-century interval, the global temperature correlates most strongly with variations in solar radiation. Especially significant is the so-called Maunder minimum, a time when solar activity diminished substantially and northern Europe experienced a period of unusually cool conditions sometimes called the "little ice age." The solar influence continued into the twentieth century, and it probably explains much of the rise that Figure 14.1 shows occurred in the early part of the century. But by the second half of the twentieth century, anthropogenic greenhouse gases emerge as the factor most strongly correlated with global temperature. The correlations associated with millennial temperature reconstructions are noteworthy because they involve only observational data; there are no hidden model assumptions and no need to understand complex feedback mechanisms or details of cloud physics. If anything, the correlation approach is a bit conservative because it compares temperature and forcing agents contemporaneously, without allowing for "inertia" associated with, for example, the slow warming of the oceans that should delay the temperature response relative to the forcing. So the late-twentieth-century correlation between temperature and CO_2 may be even stronger than these studies suggest.

Additional evidence for a human influence on climate comes from computer models designed to project future climate. I'll have more to say about these models in Chapter 15, but for now I'll describe one way of testing their validity, which is to start at some time in the past, run the model forward, and see if it reproduces today's climate. A variant on this test lets modelers assess the factors influencing climate. By running their models without volcanism, or without solar variations, or without greenhouse gases and other anthropogenic forcings, modelers can explore the role of each factor in determining present-day climate. The results of several distinct studies all point to the same conclusions: that solar variability and volcanism account for much of the climate change seen into the early twentieth century, but that only with anthropogenic forcings can models reproduce the climate of recent decades. Figure 14.14 shows the results of one such study.

I noted at the beginning of this section that climate change might result not from external forcings, either anthropogenic or natural, but from natural internal variability of the complex climate system. Once again, though, computer models help rule out the possibility that internal variability could explain recent climate change. By "turning off" all changes in external forcings, climate modelers can focus on internal variability alone. Thousand-year runs with different models exhibit internal variability that, although significant, never shows trends as large as the observed warming of recent decades.

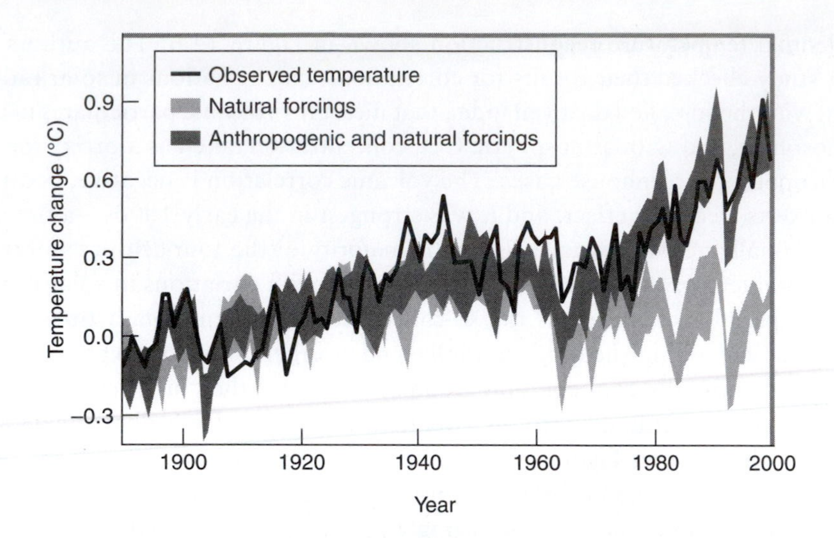

FIGURE 14.14
Results of climate model experiments to determine the roles of natural and anthropogenic forcings on present-day climate. Only the models that include anthropogenic effects can account for the temperature rise observed in recent decades. The two shaded bands show the results from an ensemble of four model runs conducted with and without anthropogenic forcings. Natural forcings are volcanism and solar variability; anthropogenic forcings include greenhouse gases, sulfate aerosols, and ozone.

A third piece of evidence is more subtle. In Chapter 13 I noted that different forcing mechanisms can have different effects on climate, even if they result in the same numerical upset in Earth's energy balance. In particular, some forcings have different geographical and temporal distributions, and their influence varies with region and season. So there are unique patterns in the climate response to different forcings, and these provide "fingerprints" that climatologists can use to identify the causes of climate change. Among the fingerprint evidence for anthropogenic climate change are regional patterns of global temperature change, including increased warming over land and regional variations in precipitation. The cooling of the stratosphere shown in Figure 14.5 is another fingerprint of anthropogenic greenhouse warming. Recall from the discussion surrounding Equation 12.3 that the greenhouse effect requires a temperature difference between the surface and upper atmosphere. Figure 14.5 shows that this difference is increasing and thus is evidence that recent warming is due to a strengthening greenhouse effect. Coupled with the anthropogenic greenhouse gas increases documented in Chapter 13, this is solid evidence for anthropogenic warming.

Thus climatologists have a host of evidence—statistical inferences, past climate reconstructions, computer model experiments, and pattern-based fingerprint studies—that all point to a substantial human influence on climate since the mid- to late twentieth century. That influence, as we saw as early as Chapter 2 and again in Chapter 13, results primarily from our species' prodigious consumption of fossil fuels and the resulting atmospheric greenhouse emissions. In the next chapter we'll see where the human impact on the atmosphere is likely to lead, and in the final chapter we'll address the steps we might take to moderate future climate change.

CHAPTER REVIEW

BIG IDEAS

14.1 Thermometers provide a global temperature record back to the mid-1800s, and their accumulated data show rapid warming in recent decades. **Proxy**-based climate reconstructions suggest that this warming is unprecedented over the past 1,000 years. Ice-core data extending back nearly 1 million years show a pattern of long, cool periods—the **ice ages**—alternating with briefer, warmer **interglacials**.

14.2 Other indicators of global warming include melting ice and decreased snow cover, changes in weather patterns and storm intensities, and poleward movement of plant and animal species.

14.3 Solar variability and volcanic activity account for much of the climate variation observed until the mid-twentieth century. But only by including anthropogenic effects, especially greenhouse emissions, can climate scientists account for the unprecedented warming of the late twentieth and early twenty-first centuries.

TERMS TO KNOW

boreholes (p. 369)
ice age (p. 373)
interglacials (p. 373)
microwave sounding unit (MSU) (p. 370)
proxies (p. 371)
radiosondes (p. 370)
temperature anomalies (p. 367)
urban heat island effect (p. 366)

GETTING QUANTITATIVE

Global warming over the past one hundred years: ~0.8°C

Temperature difference between present day and ice age: ~6°C

Poleward migration of species ranges, average: ~6 km per decade

QUESTIONS

1. By roughly how much did global temperature increase during the twentieth century?
2. How do oxygen isotopes help scientists extract temperature information from ice cores?
3. What is the ultimate origin of the periodic climatic changes—ice ages and warmer interglacial periods—that are so obvious in Figure 14.7?
4. Why is climate change more rapid in the Arctic than elsewhere?
5. How does Figure 14.5 provide evidence that recent global warming results, specifically, from increasing greenhouse gas concentrations?

RESEARCH PROBLEMS

1 Go to the University of East Anglia Climatic Research Unit's web page, located at www.cru.uea.ac.uk/cru/data/temperature, and locate the Northern Hemisphere (NH) and Southern Hemisphere (SH) temperature data from the HadCRUT3 dataset. Copy these into a spreadsheet and put each entry in its own column (in Excel use the Text to Columns function). Plot the annual average temperature versus time for both of these datasets, and compare.

2 Look at the CRU's global temperature dataset (GL, from the HadCRUT3 dataset; see Research Problem 1 for instructions) and note that the second line for each year gives the percentage of global coverage provided by the reporting stations. How has that increased from the early years of the dataset to the present?

3 Use the CRU's global temperature dataset (see Research Problem 2) to plot an updated version of Figure 14.1 using the latest available data. Does the warming trend continue, or does the global temperature decline? Or isn't there an obvious trend?

4 Using the data sources for Figure 14.2 (see Credits and Data Sources), plot the temperature data from the CRU, the U.S. National Climatic Data Center (NCDC), and the Goddard Institute for Space Studies (GISS) on the same graph, and compare.

5 Has the accelerated loss of Arctic ice cover evident in Figure 14.12b continued? To find out, go to the National Snow and Ice Data Center's web site (http://nsidc.org/) and find the data to update Figure 14.12b to the present.

6 Go to the Carbon Dioxide Information Analysis Center's web site (http://cdiac.esd.ornl.gov) and find the Vostok temperature data used in Figure 14.7. Plot the entire dataset. How many ice-age cycles does it show? Are the temperatures and durations of the warmer interglacials roughly the same, or do they vary significantly?

7 A newer ice core from the European Project for Ice Coring in Antarctica (EPICA) goes back 740,000 years. You can find the EPICA data on the Carbon Dioxide Information Analysis Center's web site (see Research Problem 7), listed under "Deuterium Record in an Ice Core from Dome C, Antarctica." Does the earlier data show the same cyclic pattern of ice ages and interglacials as the Vostok data of Figure 14.7 and of the previous problem?

8 The U.S. Historical Climatology Network (HCN) provides monthly temperature and precipitation data for more than 1,000 stations across the United States, with many station records going back over 100 years. If you're a U.S. resident or a student in the United States, go to the HCN portal at http://cdiac.ornl.gov/epubs/ndp/ushcn/ushcn.html and use the web interface link to find the station nearest you. You can follow the "Get Monthly Data" link and download monthly data for the period 1910–2010. Plot the data, fit a straight line, and use the line's slope to estimate your station's 100-year temperature rise. Compare with this chapter's estimate of 0.8°C for the global temperature rise for this period.

ARGUE YOUR CASE

1 A friend argues that climate naturally varies, so the temperature increase of recent decades might well be a natural fluctuation. How might you counter this argument?

2 A climate skeptic claims that warming trends like that shown in Figure 14.1 actually result from the urban heat island effect, and that the planet as a whole is not warming. How do you reply?

3 Some have argued that temperature records derived from weather stations, ships, and other thermometer-based measures of surface temperature aren't available in sufficient numbers to yield meaningful changes in the global average temperature. Formulate a counterargument, including mention of global warming indicators that are independent of these temperature records.

Chapter 15

FUTURE CLIMATES

Chapter 13 described how humans have altered the global atmosphere, particularly through a substantial increase in atmospheric CO_2 caused largely by our fossil fuel consumption. Chapter 14 documented a century-long rise in global temperature, and recounted climatologists' arguments as to why humans are behind at least the unusual warming of the most recent half-century. So can we predict accurately the global climate response to our ongoing greenhouse gas emissions and other climate forcings?

Predict *accurately*? No, and for two main reasons. First, we don't know enough about every detail of the climate system—the myriad feedbacks, the surprising nonlinear effects, the behavior of clouds, and the complex interactions of land, sea, air, ice, and biosphere—to make precise predictions of what will happen when and where. Second, we can't predict future human behavior, yet humans are now major players in the global climate system. But despite our imperfect knowledge, we do know a great deal about the physical, chemical, and biological processes that govern climate. And we can make assumptions about human behavior that, together with our knowledge of climate science, let us project scenarios of how Earth's climate might evolve in the future.

Although the simple zero-dimensional energy-balance model I've been using in previous chapters can give rough estimates of the global average temperature change associated with a given forcing, it can't do justice to the complexity of the climate system or the geographical diversity of our planet. Nor can it account for the variations in precipitation, storm intensity, melting ice and snow, altered vegetation, and other changes resulting from shifts in global climate. To project global changes with higher confidence, or to project any regional changes at all, we need sophisticated, multidimensional models that account for the myriad interactions affecting climate. Then we need powerful computers to run those models.

15.1 Modeling Climate

Figures 12.3 and 12.5 hint at a possible next step in modeling sophistication beyond our zero-dimensional model: Treat the climate system as if it consisted of two "boxes"—surface and atmosphere—and describe the flows of energy between them. Figure 12.3 is, in fact, part of such a model, complete with equations describing some of the energy flows. Figure 15.1 is a full **two-box model** that includes all the processes shown in Figure 12.5's diagram of the surface–atmosphere energy flows. Describing each arrow with a numerical value or an equation involving surface and/or air temperature would then let us write separate equations for energy balance in both the surface and atmosphere. Solving these two equations simultaneously would give the model's values for the temperatures of surface and atmosphere. You can explore this two-box model in Research Problems 1 and 2.

We know that temperature varies with altitude, so an obvious next step is to represent the atmosphere with many boxes. This gives a **one-dimensional model**, where quantities vary with position in a single dimension. A one-dimensional model could, for example, explore the simultaneous warming of the lower atmosphere and cooling of the upper atmosphere in response to increasing quantities of greenhouse gases.

Climate also varies with geographical location, and **two-dimensional models** account typically for different latitudes as well as altitudes (Fig. 15.2). These models can simulate such things as seasonal changes and the atmospheric circulations that transport energy from the tropics toward the poles. For the latter, a model needs more than just energy balance at each box; modelers also have to account for flows of matter and momentum. Matter includes the air itself, along with the all-important moisture that carries latent energy and is responsible for cloud formation and precipitation; a more advanced model might also track separately the movement of greenhouse gases and other forcing agents. Models also follow momentum—the "quantity of motion" first introduced by Isaac Newton—whose flows describe the forces one box of air exerts on its neighbors, as well as the frictional forces the ground exerts on the air.

Three-dimensional models have it all. They divide the Earth–atmosphere system into boxes whose positions vary with latitude, longitude, and altitude

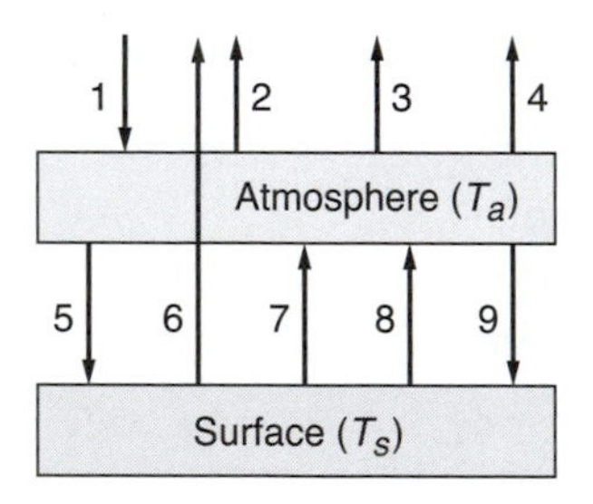

1 = Incident solar energy
2 = Sunlight reflected from atmosphere
3 = Infrared transmitted through atmosphere to space
4 = Infrared emitted by atmosphere to space
5 = Sunlight absorbed by surface
6 = Sunlight reflected from surface
7 = Convection and evapotranspiration
8 = Infrared emitted upward from surface
9 = Infrared emitted downward from atmosphere

FIGURE 15.1
A two-box climate model featuring energy exchanges between Earth's surface, atmosphere, Sun, and space. The model is a simplified version of Figure 12.5.

(Fig. 15.3). Three-dimensional models can therefore account for land and water surfaces; for the actual configurations of the continents; for snow, ice, and vegetation cover and their associated albedos; for atmospheric circulation in both latitudinal and longitudinal directions; for the transport of moisture from the oceans to the continents; and for a host of other realistic processes that affect climate. Large-scale models that include atmospheric circulation are called **GCMs**—originally standing for **general circulation model** but now taken also to mean **global climate model**.

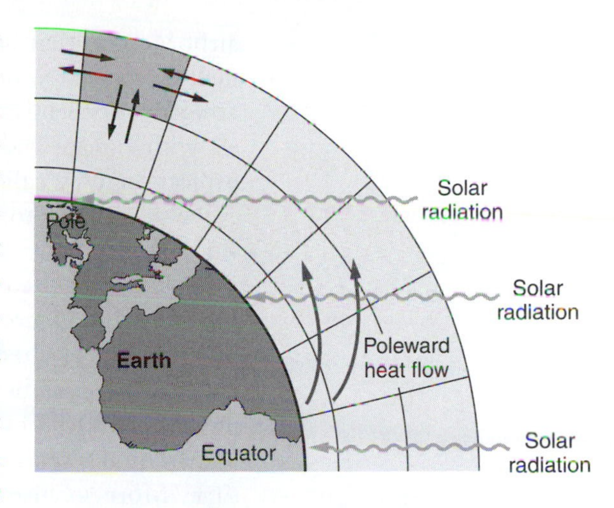

FIGURE 15.2

A two-dimensional climate model that accounts for the varying intensity of solar radiation from equator to poles. One of the two-dimensional grid boxes is shaded in darker gray, with arrows representing the interchanges of matter and energy between the box and its neighbors.

Physical processes in the oceans are sufficiently different from those in the atmosphere that they're usually handled with separate models. Ocean and atmosphere models are then coupled, with the output of each model feeding the other to describe processes at the air–water interface—processes such as evaporation; CO_2 dissolving or coming out of solution; wind and waves producing

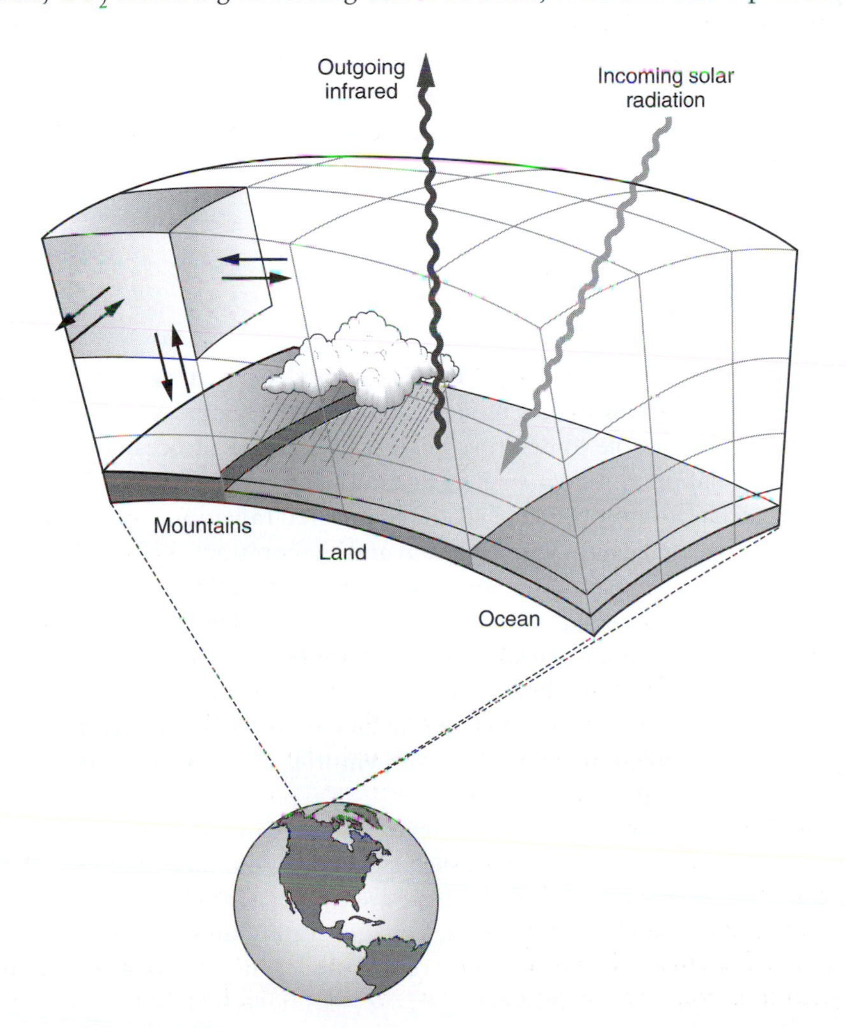

FIGURE 15.3

A three-dimensional climate model includes land and ocean surfaces. One of the three-dimensional grid boxes is shaded in darker gray, with arrows representing the interchanges of matter and energy between the box and its neighbors.

airborne salts that become cloud condensation nuclei; and ice formation and melting. Similarly, separate specialized models often handle the carbon cycle, atmospheric chemistry, and biological systems. Coupling such specialized models with climate and ocean models gives today's most sophisticated climate-projection tools, called **Earth system models**, or **ESMs**.

We're now up to three-dimensional models. But there's another "dimension" to the physical universe, namely time. The simplest models, like our zero-dimensional energy-balance model of Chapters 12 and 13, or the two-box model of Figure 15.1 and Research Problem 1, are **equilibrium models**. Their results give temperatures that would prevail only after the climate system reaches equilibrium under conditions chosen for the model. Equilibrium models can't tell how long it would take to reach equilibrium or how the climate would respond to, say, varying concentrations of greenhouse gases over time. More sophisticated models, in contrast, are **time-dependent models**. Not only do they consider flows of energy and matter among a great many individual boxes, but they also advance the model system forward in time, solving the entire many-box model over and over as time progresses and model conditions—temperatures, greenhouse gas concentrations, ice and vegetation cover, and other quantities—change. Thus, time-dependent models not only project future climate, but they also show the path from present to future. As such, they let climate modelers experiment with different scenarios for future greenhouse emissions, solar variability, land-use changes, and other factors that might alter Earth's climate.

MODEL RESOLUTION

The more boxes—also called **grid cells**—a model has, the finer its spatial **resolution**. A high-resolution model can simulate regional variations in climate and is better able to describe small-scale features of oceanic and atmospheric circulation. But high resolution comes at a price. Dozens of processes go on in each cell—absorption of solar and infrared radiation, inflows and outflows of matter and energy, condensation and evaporation, chemical reactions, photosynthesis and respiration, and many more. It takes at least one equation to describe each process, and equations for a given cell contain terms linking adjacent cells. So a model consists, mathematically, of an enormous number of equations to be solved simultaneously, over and over again with each successive time step. As a result, the most elaborate climate models tax even the fastest computers. A simulation covering several centuries of climate can take months of computer time to run.

Even today's highest-resolution climate models can't capture every important climate process. Many clouds, such as the puffy cumulus that form in fair weather, are smaller than any grid cell. So are cities, farm fields, and many lakes and forests. Yet clouds, urbanization, local geography, and land-use variations have definite effects on climate. As a result, models use techniques known as

BOX 15.1 | An Earth System Model

The Hadley Centre of the U.K.'s Met (for "meteorological") Office is home to some of the world's most sophisticated climate modeling hardware and software. HadGEM2-ES—for Hadley Centre Global Environment Model 2 Earth System—is a state-of-the-art climate model performing calculations to be featured in the IPCC Fifth Assessment Report. HadGEM2-ES's atmospheric component features a grid of 192 points in longitude, 144 points in latitude, and 38 vertical levels. That's over a million separate boxes to describe the atmosphere alone! The ocean component includes a horizontal resolution ranging from 0.33° at the equator to 1° at higher latitudes, and 40 levels of depth. In a major advance over earlier models, HadGEM2-ES treats 9 different types of vegetation, each of which responds to changes in the model climate—including seasons. A coupled carbon-cycle model accounts for carbon flows on timescales up to centuries. A sulfur-cycle model handles sulfate aerosols, whose negative forcing exerts a cooling effect, while non-sulfate aerosols are handled by a separate model component. The atmospheric chemistry component follows 46 different chemical species. With all that calculating, HadGEM2-ES takes a full hour of supercomputer time to simulate a month's climate variation. If you'd like to contribute some computer time to climate modeling efforts, check out Research Problem 3; it describes a global project that harnesses spare computing power in tens of thousands of personal computers to run large-scale climate models, including some from the Hadley Centre.

subgrid parameterization to account for these and other small-scale effects. A subgrid parameterization of clouds, for example, might ascribe average cloud-related albedo and absorption to an entire grid cell, based on a calculation that suggests the formation of small-scale clouds providing only partial sky coverage.

Can't modelers handle every aspect of climate by simply shrinking their grid cells further? In principle, perhaps, but today's largest climate models challenge even the fastest supercomputers. Halving each dimension of every grid cell in a three-dimensional model increases the number of cells eightfold (since $2^3 = 8$), and in a time-dependent model this requires halving the time step as well, which makes for a sixteenfold increase in computer time! Modelers do use smaller grid cells to study regional effects—the climate of, say, the United Kingdom, the northeastern United States, or the Sahel region of Africa. Embedding such a regional model in the coarser grid of a global model provides values of climate variables at the boundaries of the region under study. With its fewer but finer cells, the regional model then runs in reasonable computer time to give a detailed picture of regional climate. However, computer models are advancing rapidly in resolution and in the number and detail of the physical, chemical, and biological effects they include. Models improve with our increasing understanding of climate processes, with advances in computational techniques, and especially with increases in raw computer power.

A HIERARCHY OF MODELS

With supercomputers grinding out results from comprehensive coupled climate models at the world's major climate research centers, you might think there's no place for simpler models such as Figure 15.1's two-box model, one- and two-dimensional models, or equilibrium models that don't include time dependence. But that's not the case. Large Earth system models take so much computer time that it's simply impossible to experiment with as broad a range of climate situations as modelers might want. And since these models are nearly as complex as the real climate system itself, interpreting the results and determining the causes of modeled phenomena are sometimes difficult.

Simpler models, in contrast, run quickly and so allow researchers to carry out numerous experiments with different initial conditions, assumptions about greenhouse emissions, and other parameters. Since simpler models involve fewer details, it's often easier to interpret the results or focus on a particular physical process of interest. Sometimes simpler models are calibrated by comparing their results with those of complex ESMs and adjusting the model as needed for agreement. Then modelers can be confident that, for small changes in model conditions, the simpler model will give results similar to those of the full ESM (Fig. 15.4). So there's room in climate studies for a hierarchy of models, from the simplest energy-balance models to the full, computing-intensive Earth system models.

VALIDATING CLIMATE MODELS

Why should we trust climate models? There are several reasons. First, climate models are based on long-established scientific principles and measured properties of basic materials—science that governs the behavior of matter and energy in realms far beyond climate. Examples include the $e\sigma T^4$ radiation law

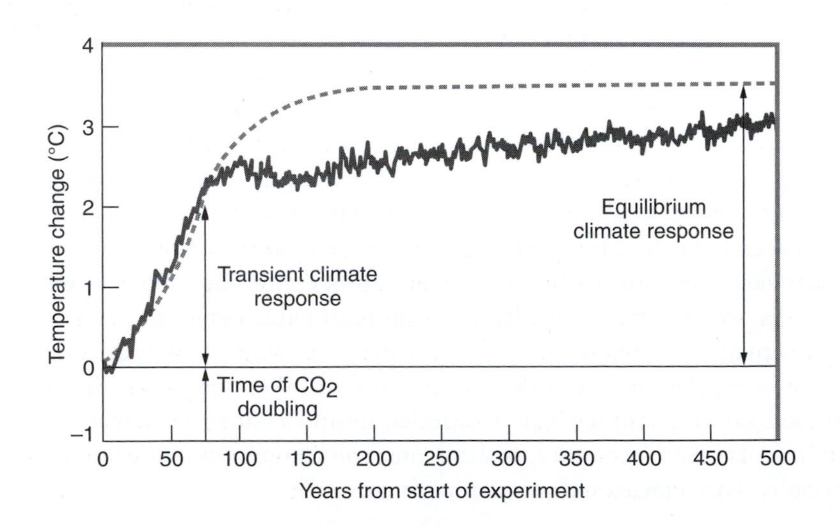

FIGURE 15.4
Model runs of a simplified model (dashed curve) and a coupled complex atmosphere–ocean climate model showing transient climate response to a 1% per year increase in atmospheric CO_2, resulting in a CO_2 doubling at about 70 years. After that point, CO_2 is held constant. The more complex model takes longer to reach equilibrium because it includes heat exchange between the ocean surface layer and the deep ocean, a process that delays equilibrium. Like the real climate, the complex model also exhibits intrinsic variability.

introduced in Chapter 4 and used throughout the book; the infrared absorption spectrum of CO_2 that ultimately determines CO_2's greenhouse characteristics; the thermodynamic properties of water as it changes between solid, liquid, and gaseous phases; the laws of electromagnetism as they describe how electromagnetic radiation (light) scatters off small particles such as atmospheric aerosols; and even more fundamental principles such as conservation of matter and energy. Where modelers don't know all the details—for example, just how plants respond to changes in CO_2 concentration; the physics of cloud formation and cloud behavior; the strengths of some feedback effects—they incorporate the results of laboratory and field experiments as empirically based equations in their models. Climate models are based on solid science despite uncertainties in some of the details.

Second, climate models successfully reproduce present and past climates. Figure 14.14 is one example: Starting with nineteenth-century conditions and given the subsequent changes in natural and anthropogenic forcings, it reproduces the observed temperature record of the twentieth century. Confidence that models work outside the present-day climate regime comes from model experiments using very different conditions from ancient times. For example, changes in Earth's orbit led to different distributions of solar radiation thousands of years ago; when these changes are incorporated into climate models, the resulting model climates show features similar to those found in paleoclimate data.

Such verification isn't limited to global temperature. Climate models reproduce the distribution of temperature changes with geographical location, and they do an excellent job of describing the observed temperature structure throughout the atmosphere (Fig. 15.5). They give reasonable agreement with observed variations in precipitation, cloudiness, and ice cover, and the ocean components of coupled models correctly follow the spread of anthropogenic trace chemicals throughout the oceans. The models also reproduce changes in patterns of natural climate variability.

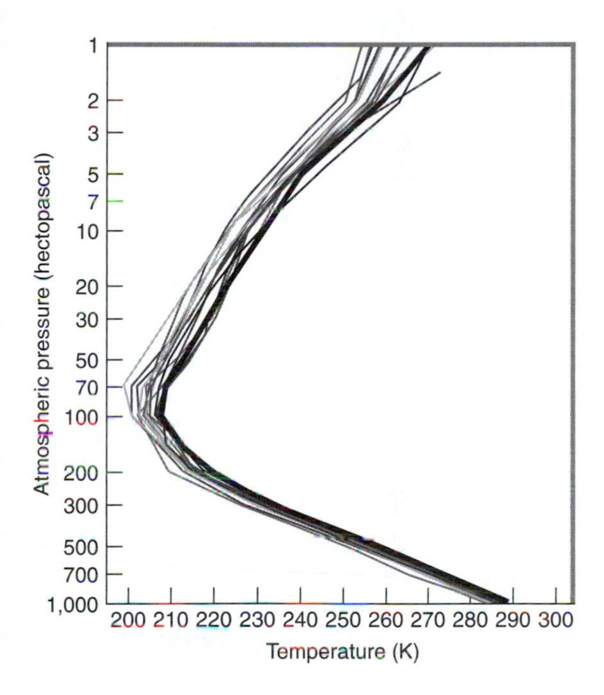

FIGURE 15.5
Temperature structure of the atmosphere, from actual observations (thick black curve) and 13 different climate models (gray curves). The vertical axis is atmospheric pressure, measured in hectopascals, which climatologists use as a proxy for altitude (1,000 HPa is approximately atmospheric pressure at Earth's surface). Decreasing pressure corresponds to greater altitude.

A helpful test would be to do controlled experiments with Earth's climate, making changes to the actual climate and seeing if models reflect those changes. We can't do that, of course, although our greenhouse emissions amount to an uncontrolled climate experiment. However, nature provides us with real-world experiments in the form of volcanic eruptions. The 1991 eruption of Mount Pinatubo in the Philippines provided one such experiment, putting enough dust into the atmosphere to cause a brief global cooling. Average global temperature quickly dropped about 0.5°C following this eruption, then recovered over several years. Figure 15.6 shows that climate models successfully reproduced the pattern of Pinatubo-related global temperature change.

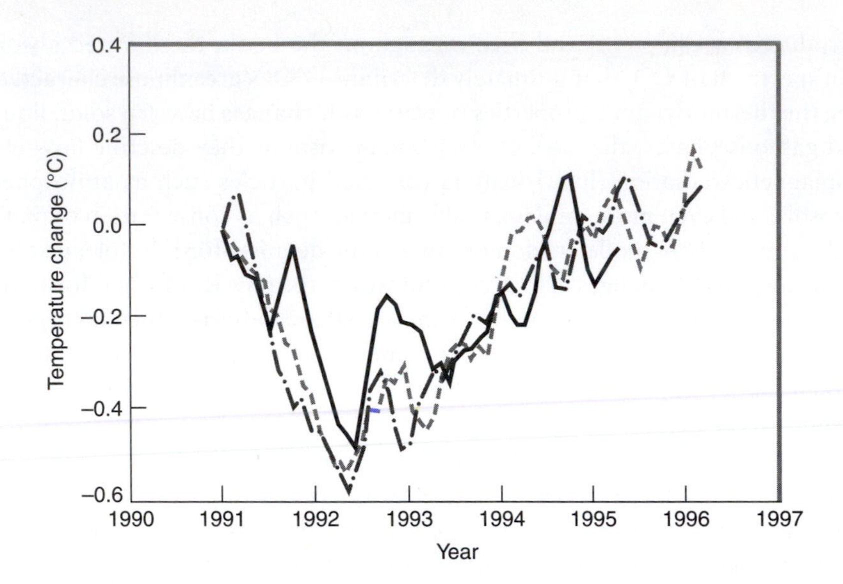

FIGURE 15.6
Global temperature change following the 1991 eruption of Mount Pinatubo in the Philippines. The solid curve is the observed temperature, and the dashed curves show two model results. Temperature changes are relative to the average temperature from April through June, 1991.

Finally, there are a great many climate models in use today. Climate groups around the world develop their own independent models, using different computational techniques, coupling schemes, submodels, and, where details are less certain, different assumptions about such things as feedback strengths or subtle climate processes. Yet, as the multiple plots in Figure 15.5 show, the diverse climate models exhibit good agreement, even on such details as atmospheric structure. Our most robust conclusions about future climate come from ensembles of many individual model runs. At the same time, the variation among models helps to quantify the range of uncertainty about future climate.

15.2 Climate Projections

So we have a variety of climate models, and we have some confidence that they can give us reasonably realistic projections of future climate change in response to both anthropogenic and natural factors. Note that I'm using the word *projection* rather than *prediction*. This is partly because climate models advancing decades and centuries into the future can't claim the precision of tomorrow's weather prediction. It's also because future climate depends on factors that are unpredictable, such as volcanic eruptions and human behavior. So model-based studies are more like "what if" stories—for example, What's the climate likely to do if our CO_2 emissions change in such and such a way?—than they are forecasts of what *will* happen.

DOUBLING OF ATMOSPHERIC CARBON DIOXIDE

One benchmark for assessing future climates is a doubling of atmospheric CO_2 from its preindustrial level of 280 ppm to 560 ppm, a condition often abbreviated as $2\times CO_2$. Given the inertia in the fossil-fueled global economy and the fact that we're already at nearly 400 ppm, we're likely to reach 560 ppm of

CO_2 sometime in the mid- to late twenty-first century. Only a massive, determined, and almost immediate effort to replace fossil fuels or sequester carbon could stabilize atmospheric CO_2 at a lower level. So $2\times CO_2$ provides a model input that we can realistically expect to reflect conditions later in the current century.

Figure 15.4's equilibrium temperature increase of 3.5°C in response to CO_2 doubling is typical of model projections for a world with 560 ppm of atmospheric CO_2. Since Earth warmed by about 0.8°C over the past hundred years, the particular models of Figure 15.4 suggest an additional warming of somewhat over 2°C if we manage to stabilize CO_2 at twice its preindustrial level.

Figure 15.4 is based on just two models, one of them highly simplified. But our confidence in climate projections comes from ensembles of different, independent models. So how do other climate models do? Figure 15.7 shows the results of 19 different runs calculating the same 1% per year CO_2 increase used in the first 70 years of Figure 15.4. At 70 years—the time of doubled CO_2—the models show a range of temperature increases from about 1°C to 3°C, with most clustering at just below 2°C. Remember, however, that these are just the transient responses; if the CO_2 increase were to stop at 70 years, then, as Figure 15.4 suggests, we might expect another 1.5°C or so of warming as the climate equilibrates.

What if we don't manage to halt our CO_2 emissions at 560 ppm? Then, as Figure 15.8 shows, we can expect correspondingly higher temperature increases in the current century, with still higher equilibrium values being reached several centuries hence.

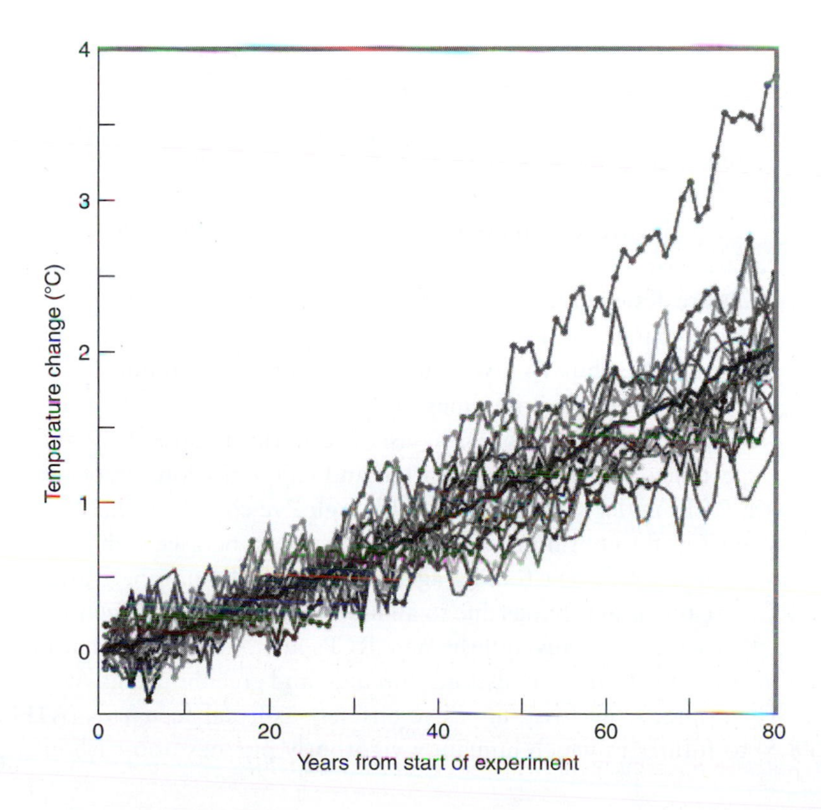

FIGURE 15.7
Results of 19 different climate model runs showing response to a 1% per year increase in atmospheric CO_2, resulting in a CO_2 doubling at about 70 years. These runs compare with the first 70 years of Figure 15.4. Individual model runs are shown in gray, and the thick black curve is the mean of all runs.

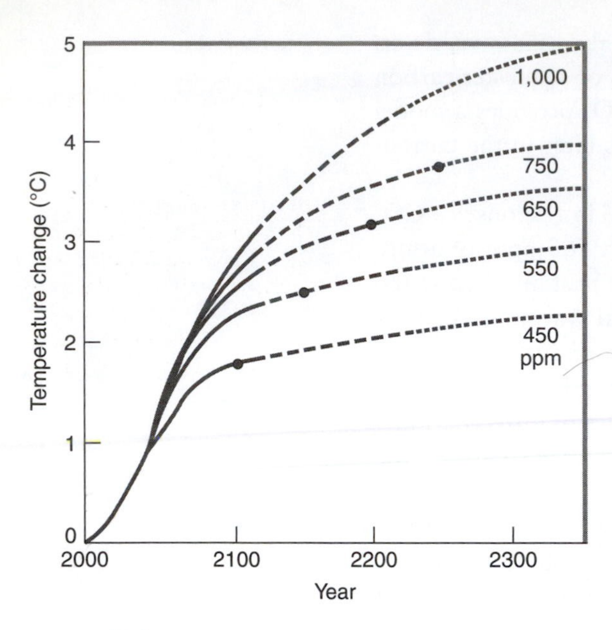

FIGURE 15.8

Projected global temperature increases for CO_2 stabilization at five different concentrations. The black dots indicate when stable CO_2 concentrations are reached. Results after 2100 (dashed curves) are increasingly uncertain.

THE HUMAN FACTOR

Changes in greenhouse gases and other climate forcings depend on human behavior; so, therefore, does climate change. Will we make a concerted effort to end our dependence on fossil fuels? Will we develop new carbon-free energy technologies, or learn to sequester fossil carbon emissions? Or will it be business as usual? Will developing countries follow traditional energy paths, or will they embrace low-carbon sources as their energy demand grows? Will population growth continue indefinitely?

For its Third and Fourth Assessment Reports, the IPCC developed a set of scenarios for future greenhouse emissions, which served as inputs to climate models. Scenarios designated A describe a world emphasizing economic growth, while B emphasizes environmental quality. The number 1 describes an increasingly homogenized world whose population peaks in mid-century. Number 2 scenarios emphasize regional differences and have population growing past the year 2100. Combining the two designations gives scenarios such as B1—a globalized world that emphasizes environmental quality over economic growth, or A2, which maintains regional distinctions but emphasizes economic growth. Within the globalized, economic-oriented A1 category are the fossil-fuel-intensive, business-as-usual A1FI; the more balanced A1B that includes gradual introduction of alternatives to fossil fuels; and the A1T scenario that rapidly deploys non-carbon energy technologies. You'll find these scenarios still widely discussed in the context of climate projections.

For the IPCC's Fifth Assessment Report, the climate research community is developing **representative concentration pathways (RCPs)** as alternatives to the emissions scenarios used in earlier IPCC reports. The RCPs specify concentrations of forcing agents over time, or, equivalently, the associated radiative forcing. The pathways are "representative" in the sense that alternative emissions scenarios could lead to the same overall climate forcing. Among other goals, RCPs are designed to explore the impacts of different strategies for mitigating climate change. RCPs are designated by their approximate industrial-era forcing in 2100; thus RCP8.5 ends the century with about 8.5 W/m^2 of forcing relative to preindustrial times (today it's about 2 W/m^2), and is somewhat like the earlier A1FI business-as-usual scenario. Figures 15.9 and 15.10 show, respectively, the CO_2 concentration and CO_2 emissions associated with four main RCPs under consideration. Although I've chosen in these figures to emphasize CO_2, for continuity with earlier chapters, be aware that the RCPs account for many different forcing agents, and that the numbers attached to each RCP are the total forcings due to all industrial-era forcing agents.

Both the earlier scenarios and the new RCPs are "storylines" that describe plausible futures in terms of emissions, forcings, and greenhouse-gas concentrations. They range from fossil-intensive business-as-usual scenarios (A1FI and RCP8.5) to futures in which humanity vigorously pursues non-carbon energy

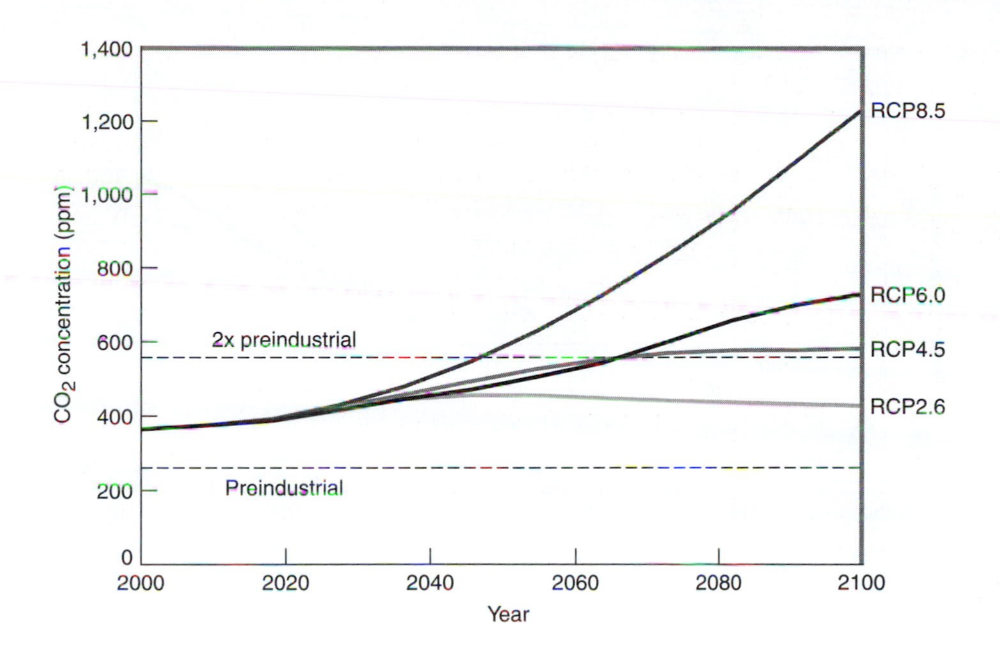

FIGURE 15.9 Carbon dioxide concentrations through the year 2100 for four representative concentration pathways (RCPs). Also shown are preindustrial and twice preindustrial CO_2 concentrations; note that only RCP2.6 avoids a CO_2 doubling. Numbers on the curve labels are the total industrial-era forcings in 2100, in W/m^2, for each pathway.

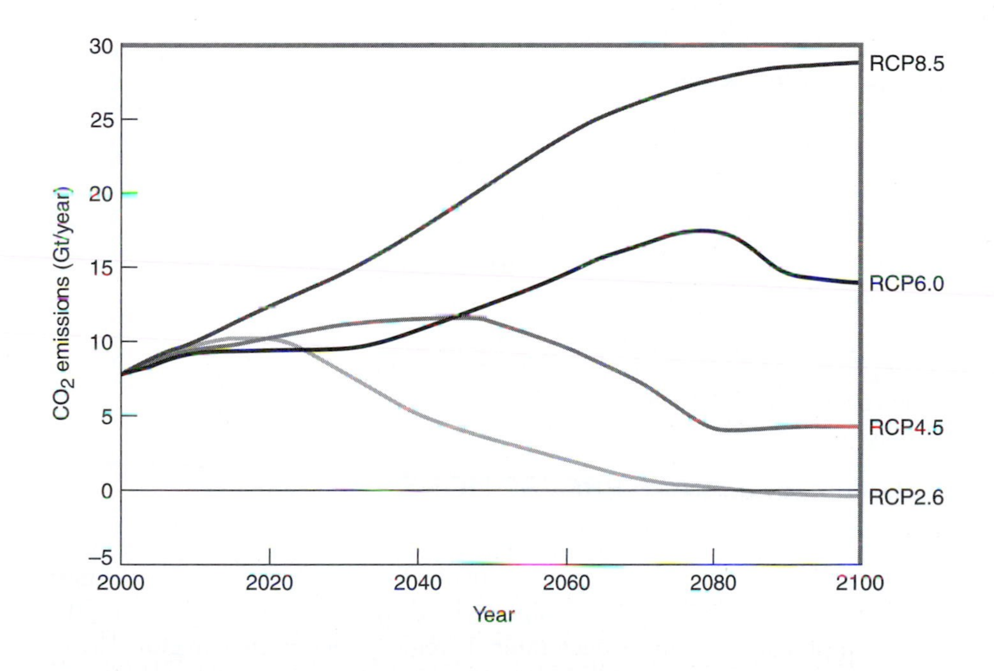

FIGURE 15.10 Carbon dioxide emissions for each of the RCPs. RCP8.5, a business-as-usual scenario, ends the century with some three times today's CO_2 emissions. Emissions in RCP2.6 go negative near the end of the century—requiring actual removal of CO_2 from the atmosphere.

sources and other strategies to mitigate global warming and limit the buildup of greenhouse gases (A1T, B1, and RCP2.6). RCP2.6 is so ambitious as to assume an actual reduction in atmospheric CO_2, shown in Figure 15.10 where emissions go negative near the century's end—a result of as-yet-unproven technologies that would remove CO_2 from the atmosphere.

The range of emission and concentration scenarios generates a range of projections for future warming. In the 2007 IPCC Fourth Assessment Report, best

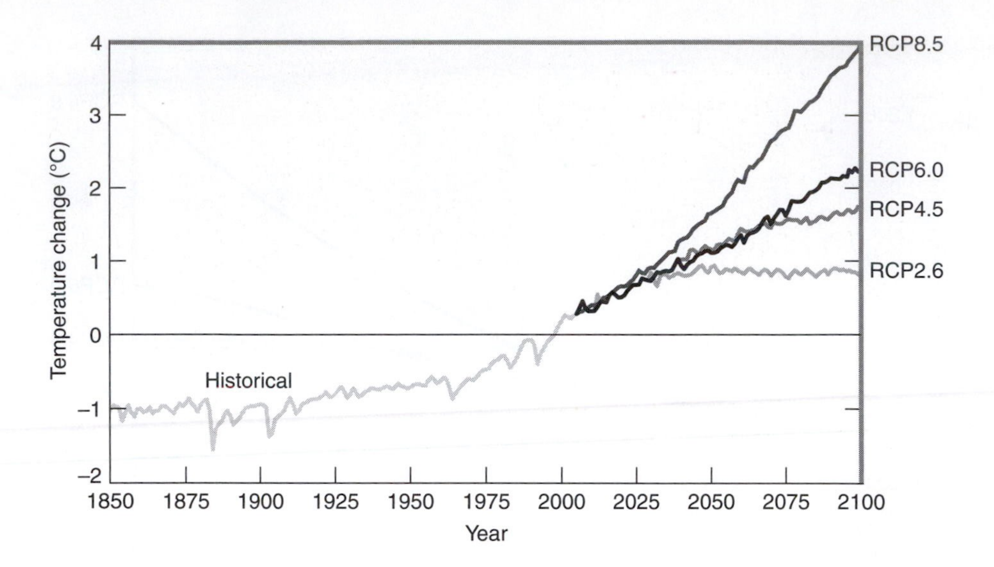

FIGURE 15.11 Projections for global surface air temperature through 2100, from multiple simulations with the Community Climate System Model, version 4, of the National Center for Atmospheric Research. Temperatures are relative to the 1990–2009 average. Results are shown for the four representative concentration pathways also used in Figures 15.9 and 15.10. Only RPC2.6 holds the global temperature increase by 2100 to the widely accepted goal of less than 2°C above its preindustrial level (which is at about −1°C on this graph).

estimates for twenty-first-century warming went from a high of about 4°C for the fossil-intensive A1FI scenario to a low of just under 2°C for the globalized but environmentally friendly B1. Scenario A1B, perhaps the most realistic, gave a rise of about 3°C. Taken with their uncertainties, the results of multiple runs with many different climate models suggested a likely twenty-first-century temperature rise in the range of 1.1°C to 6.4°C, with 3°C a good "best guess." As of 2011, work was still underway using ensembles of models to project climate under the RCP scenarios, but it's already clear that these newer scenarios yield future climates that are broadly consistent with the earlier IPCC results. Figure 15.11 shows the RCP8.5 scenario giving a 4°C rise by 2100, while the ambitious RCP2.6, with CO_2 actually declining after mid-century, manages to hold the twenty-first temperature rise to about 1°C. That's a total of about 2°C from preindustrial times—a target that most climate experts agree should not be exceeded if we're to avoid dangerous climate change.

REGIONAL CLIMATE PROJECTIONS

Figure 15.11 shows projected increases for global average temperature over the twenty-first century. But, as we've already discovered when looking at past climates, we can't expect that change to be distributed uniformly across the globe. Although there's more uncertainty in regional details than in global average projections, some regional effects emerge consistently enough that we can have considerable confidence in them. Figure 15.12 maps regional projections of the average temperature change expected under RCP4.5 by the late twenty-first century. Other scenarios show qualitatively similar patterns. For example, all scenarios agree with Figure 15.12 in showing the greatest projected temperature increase in the Arctic. That's entirely consistent with the rapid change observed recently in Arctic climate, as discussed in Chapter 14, and it's a change that's explained by processes such as ice–albedo feedback. Another

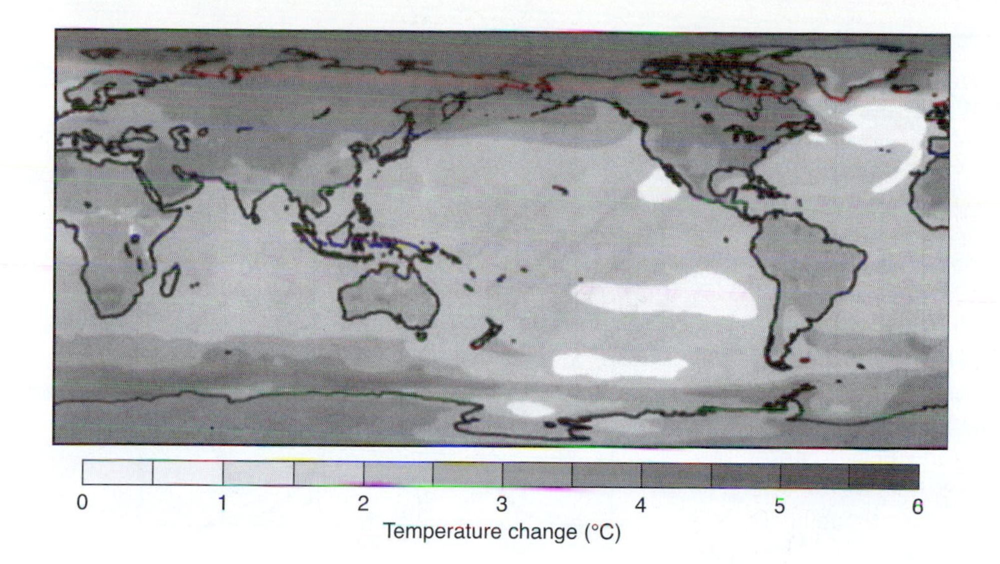

FIGURE 15.12
Projected temperature increases for the period 2081–2100, compared with 1986–2005, for the RCP4.5 scenario. Note that increases are largest in the Arctic and generally greater on land than over the oceans.

obvious pattern visible in Figure 15.12 is that the continents warm more than the oceans. Again, there's a good explanation: The oceans' vast heat capacity means it takes a lot of energy to change ocean temperature. The result is that a global average warming of, say, 3°C translates into considerably greater warming over nearly all land areas.

15.3 Consequences of Global Climate Change

Global warming won't just make Earth warmer. It will also bring changes in sea level, altered precipitation patterns, changes in soil moisture content, increases in some extreme weather events, more flooding and more drought, alteration of natural climate cycles such as El Niño, changes in species ranges and the composition of ecosystems, advances of tropical diseases into formerly temperate regions, changes in ocean circulation, melting of Arctic permafrost, and a host of other effects that have an impact on Earth and its inhabitants. Here we'll take a quick look at the more important of these impacts.

EXTREME TEMPERATURE EVENTS

Obviously, we should expect more hot spells in a warming world. Most of Earth's land area has already seen such increases, and model projections suggest a continued increase in the number of abnormally hot days and in higher maximum temperatures nearly everywhere. At the other end of the temperature spectrum, most land areas can expect fewer exceptionally cold days and longer frost-free growing seasons. Surprisingly, it takes only a modest rise in global average temperatures to increase substantially the frequency of high temperature extremes. Box 15.2 examines the reason.

BOX 15.2 | Means and Extremes

Many varying quantities, from temperatures to heights of individuals to exam grades, are naturally distributed on a bell-shaped curve. Values of the given quantity cluster around the mean, which, for a symmetric curve, is also the most probable value. The probability that the quantity in question will take a given value decreases with that value's deviation from the mean.

A slight change in the mean value—for example, a degree or so of global warming—can nevertheless cause large changes in the probability of rare extreme values. Figure 15.13 shows why. The figure depicts two hypothetical temperature distributions with slightly different means. The probability of a given temperature occurring is proportional to the height of the curve at that temperature, and the probability of temperatures in a given *range* depends on the relative area under the curve in that range. A slight increase in the mean has little effect on the distribution of the more probable values near the mean, but it substantially raises the curve at the extreme tail end of the distribution. The greater area under the raised curve means that extremely high temperatures, although still rare, increase disproportionately. At the left-hand end of the bell-shaped curve, similar reasoning shows that the likelihood of extremely cold days decreases substantially with a slight upward shift in the mean.

It's difficult to argue that a particular extreme temperature occurrence is the result of a shift in the mean temperature—just as it's difficult to argue that a given case of emphysema, for example, results from air pollution, or that a case of cancer is caused by nuclear radiation. Nevertheless, some events are so far along the tail of the bell curve that they are almost impossibly unlikely absent an increase in the mean. Such is the case with the summer 2003 heat wave in Europe, which killed tens of thousands of people. A statistical analysis suggests that human influence—especially greenhouse gas emissions—at least doubled the probability of this extreme temperature event.

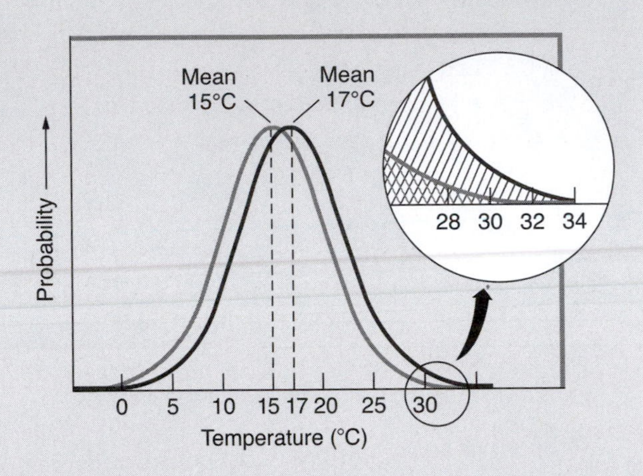

FIGURE 15.13

Bell-shaped curves representing temperature distributions with means of 15°C and 17°C. The two curves don't differ much at the highly probable temperature values near the mean, but the enlargement shows that high-temperature extremes become much more likely with the higher mean.

THE HYDROLOGIC CYCLE AND WEATHER

A warming Earth means a moister atmosphere, as increased thermal energy leads to more evaporation and therefore a more intense hydrologic cycle. So we can generally expect more precipitation with global warming. Again, though, this effect won't be evenly distributed in either space or time. The spatial distribution of projected changes in precipitation is especially wide, with drier areas seeing substantial declines despite the overall trend of an average increase. In general, precipitation and river flows are expected to increase by some 10% to 40% at high latitudes and in some parts of the tropics, while decreasing by up to 30% over the mid-latitudes and dry tropical regions. In addition, more precipitation is likely to occur in short, intense events. One result of these changes in precipitation patterns will be increased flooding in some areas. Elsewhere—especially in many continental interiors—warmer soils, greater evaporation, and less precipitation will lead to increased drought.

The intense tropical storms known as hurricanes, cyclones, and typhoons get their energy from warm surface water, so it's not surprising that tropical storms are expected to grow more intense, at least in some regions. Indeed, as Figure 14.13 indicates, we've seen an increase in hurricane intensity over the past few decades, an increase that correlates well with rising sea-surface temperatures. However, the formation and strength of tropical storms depends on a number of complex factors, and the observed correlation between strength and temperature is actually stronger than theory and models suggest. So we're less certain in attributing hurricane intensity to anthropogenic climate change than we are, say, in making a similar attribution for heat waves. Furthermore, it's not at all clear whether the number of hurricanes should be expected to change significantly.

SEA LEVEL RISE

Twenty thousand years ago so much water was locked up in continental ice sheets that sea level was some 120 m (nearly 400 feet) below where it is now. As Earth emerged from the ice age, sea level rose at the rapid rate of about 1 cm per year. By 6,000 years ago this rate had slowed to about 0.5 mm per year, and by 3,000 years ago it was down to 0.1 to 0.2 mm per year. But by the start of the twentieth century the rate had increased to 1 mm per year; it then tripled to today's 3.5 mm per year.

Measuring sea level isn't easy. The sea surface is rarely smooth, and it's subject to tidal variations. Even weather affects sea level, with the sea rising beneath regions of low atmospheric pressure. The land itself doesn't stay still either, with some continental edges sinking and others rising—the latter sometimes an ongoing rebound from the time when the weight of massive ice sheets depressed the land surface. Earthquakes and slower tectonic motions also change land elevation. Although we think of water as seeking its own level, across the vastness of the world's oceans there's no single unambiguous value for sea level. Sea level rise, like other consequences of climate change, isn't distributed evenly around the globe. Nevertheless we have enough information, from a worldwide network of tide-gauging stations and more recently from satellite altimetry, to reconstruct the history of sea level rise since the late 1800s.

You might expect that sea level rise is associated largely with the melting ice described in Chapter 14, but in fact the single most significant contribution in the past few decades is thermal expansion of water as its temperature increases. In effect, the ocean acts like a giant thermometer, its level rising with temperature. However, it takes decades for warming to penetrate into the ocean depths, so sea level doesn't respond immediately. As a result, it's not easy to pin down thermal expansion accurately, either in attributing observed sea level rise or in projecting future sea levels.

Melting ice makes the second most important contribution, but here, too, things are complicated. First of all, only land-based ice melt is significant. Ice that's already floating in the water makes essentially no change in sea level when it melts, because the greater density of water means that the melted ice occupies the same volume as the portion of ice that was submerged. Second, melting isn't the only warming-related change in land-based ice cover. Because global warming intensifies the hydrologic cycle, the result can be more snow. In very cold regions, such as the interior of Antarctica, this may lead to an increase in ice accumulation. Over the twentieth century, this effect may have given Antarctica a *negative* contribution to sea level rise, although here the uncertainty is so great that we can't even be sure of that negative sign. By 2010, though, it had became clear that melting near Antarctica's margins makes the frozen continent a positive contributor to sea level rise, with West Antarctica alone responsible for some 10% of the global rise. Table 15.1 details ice-related contributions to sea level rise.

Other factors that contribute to a rise in sea level include melting permafrost, the deposition of sediments by river flows, and, significantly, human alteration of the hydrologic cycle. For example, our depletion of groundwater aquifers moves water from natural underground storage to the oceans. The opposite effect occurs with dams and agricultural irrigation that block the natural flow of water back to the oceans and instead send much of it to the atmosphere via evaporation and transpiration. Although these effects aren't a direct result of climate change, they're another example of human influence on the global environment that happens to coincide with anthropogenic climate change.

Uncertainty about these processes leads to a range of projections for twenty-first-century sea level rise, typically about half a meter to over 1.5 m. These estimates have increased substantially over those in the 2007 IPCC report, whose projections range from 0.35 m for the more optimistic A1B scenario to 0.43 m for

TABLE 15.1 | SEA LEVEL RISE FROM LAND ICE

Ice source	Potential sea level rise equivalent*	Estimated contribution to twentieth-century sea level rise
Glaciers	0.24 m	} 3 cm
Ice caps	0.27 m	
Greenland	7.2 m	0.5 mm
Antarctica	61.1 m	–1 cm

* Rise if all the ice melted.

the business-as-usual A1FI scenario. Figure 15.14 shows more recent projections for the RCP4.5 concentration pathway, along with year 2100 projections for the comparable IPCC A1B scenario. Part of the reason for the increase in estimates of sea level rise lies with better modeling, but there's also empirical evidence for more rapid melting of mountain glaciers and the Greenland ice cap—both of which contribute directly to sea level rise.

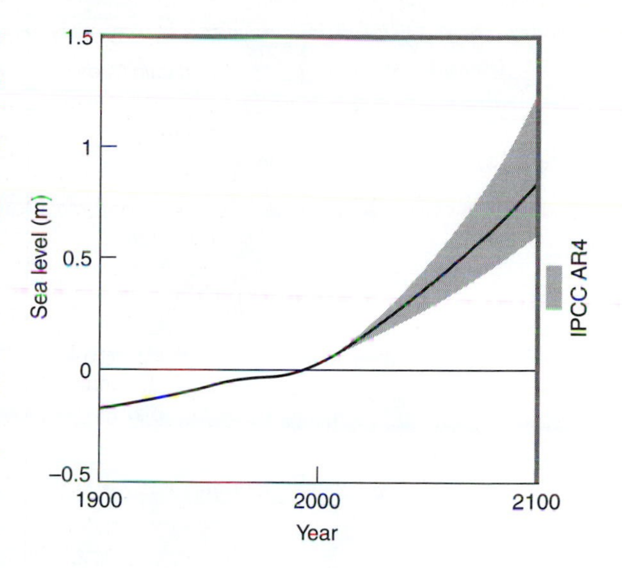

FIGURE 15.14
Graph showing twentieth-century sea level rise of some 20 cm, along with a twenty-first-century projection using the RCP4.5 scenario. Black curve is the mean from a range of model runs, represented by the gray area. Also shown is a projected range at 2100 from the IPCC 2007 Fourth Assessment Report (gray band at right), now considered unrealistically low. The IPCC projection is for the A1B scenario, whose 2100 forcing of 4.2 W/m² is comparable to the 4.5 W/m² of RCP4.5.

A 1-meter rise in sea level—about 3 feet—may not sound like much; after all, it's much less than typical tidal variation or wave action. So why the fuss? Because half the human population lives in coastal areas—some in river deltas like much of Bangladesh, some already below sea level as in the Netherlands or New Orleans, and some on low-lying islands. In Bangladesh, a 1-meter rise in sea level would submerge nearly 20% of the country's land and displace some 20 million people. In addition, an overall sea level rise adds to the highest tides and storm surges, increasing the risk of flooding over much greater areas. Globally, such flooding could affect hundreds of millions of people.

A more subtle manifestation of sea level rise is the intrusion of salt water into the underground water table. Already a problem in such low-lying areas as Bangladesh, Florida, and the Nile delta, salt intrusion leads to reduced agricultural productivity and exacerbates the growing scarcity of freshwater resources. Finally, rising sea level may inundate coastal wetlands, where, as we saw in Chapter 8's discussion of tidal energy, the interplay of saltwater and freshwater helps to nourish rich ecosystems that act as the nurseries for much of the world's marine life.

Even with the significant sea level rise expected this century, you can probably forget apocalyptic visions of New York, London, Tokyo, Cape Town, Rio de Janeiro, and other great coastal cities going under water. But because it takes a long time for a changing climate to melt land ice and for rising surface temperatures to penetrate the ocean depths, the two major causes of rising sea level—melting ice and thermal expansion—will continue for a very long time. Table 15.1 shows that land ice, in particular, has a huge potential for increasing sea level. We don't expect the many-meter rise implicit in Table 15.1 any time soon; however, it's worrisome that paleoclimate studies show our planet was essentially ice-free about 35 million years ago, when atmospheric CO_2 was at 450 ppm—a value we'll almost certainly reach in a few decades. So our greenhouse emissions may be setting the stage for a really huge sea level rise several thousand years hence. And in the near term, other nasty surprises are possible. Land ice needn't melt to raise sea level; the same rise occurs if the ice falls bodily into the sea, as its submerged portion displaces seawater and pushes up the ocean surface. Scientists worry that water from even modest melting can act as a lubricant between the bottommost ice layer and the land, thus allowing large ice sheets to slide into the sea. Of particular concern is the West Antarctic ice

FIGURE 15.15
Collapse of the West Antarctic ice sheet would raise sea level by some 3 m; as a result, the Florida peninsula would lose a substantial part of its land area (shaded in dark gray).

sheet, which could raise sea level by about 3 meters, or 10 feet. This amount would be enough to flood many coastal cities, and it would significantly reduce the land area of low-lying regions (Fig. 15.15).

SPECIES RANGES

Every species has a natural range of climatic conditions it can tolerate. As Earth warms, species ranges are expected to shift toward the poles and to higher elevations. Recent studies confirm this: On average, species ranges in the Northern Hemisphere have been shifting northward at an average rate of some 6 km per decade. For alpine species, the shift was 6 m upward in elevation each decade. You might think this effect would simply redistribute species about the globe, but many climate-induced changes are actually detrimental to species' populations, and even to their long-term survival. For one thing, the warming of the twenty-first century will be more rapid than any natural climate change occurring at least since the end of the last ice age. Some species, especially trees, may not be able to move poleward fast enough to match the changing climate. Thus they may die off at the warmer end of their range, without having had time to advance and populate newly suitable poleward regions. The result could be a decline in some forest ecosystems, including animal species that require particular forest types.

Another detrimental effect of climate change follows from the so-called species–area relationship. This long-established principle of ecology says that smaller areas can sustain fewer different species. As global warming moves species ranges upward in both latitude and altitude, the effect is often to shrink the size of regions with particular climatic conditions. The result, according to the species–area relationship, should be fewer species. This climate-induced effect may soon rival the loss of biodiversity resulting from direct human-induced habitat destruction. In addition, outright extinction looms for many species. A study covering some 20% of Earth's land area suggests that somewhere between 15% and 37% of all species will be on the road to extinction by 2050, depending on just how rapidly global temperatures increase.

Shifts in geographical range aren't the only evidence of climate change to be found in ecological systems. Events such as budding of leaves and flowers, mating, nest building, and migration are getting earlier at an average rate of 2.3 days per decade. (These are statistical results obtained by averaging data from hundreds of species.) Climate change isn't the only factor influencing range and timing in natural systems, and not all species share in the general trends of northward advance and earlier timing. However, the vast majority of species do, leading to a high level of confidence that we're seeing the effects of climate change on natural systems.

It isn't just birds, frogs, butterflies, and trees that are advancing toward higher latitudes; so are many pathogens and their insect hosts. This movement could make tropical diseases more common in temperate regions. A variety of factors contribute to the spread of disease in our global society, but there's already some indication that illnesses such as malaria, Lyme disease, West Nile virus, and dengue fever are becoming more widespread with increasing temperatures.

OCEAN CIRCULATION

Ocean currents play a major role in determining regional climate. Perhaps the best-known example is the Gulf Stream, which carries warm surface water from the Gulf of Mexico toward northern Europe. Its warming influence gives Europe mild weather at the same latitudes that are near-Arctic in North America. One Scottish town situated at the same latitude as Juneau, Alaska, has palm trees lining its main street!

However, water that flows northward must eventually return to keep the oceans in balance. In the case of the Gulf Stream, the return flow begins north of Iceland, between Greenland and Scandinavia. The Gulf Stream water isn't only warm; it also has a higher salt content as a result of evaporation in the warm tropics. The water's warmth tends to decrease its density, while the high salt concentration tends to increase it. As the Gulf Stream flows northward and cools, its density therefore increases. Finally, north of Iceland, the Gulf water becomes denser than the surrounding water and therefore sinks. The cool water then flows southward at a depth of several kilometers. Together, the Gulf Stream and its cold, deep return flow constitute part of a global pattern of **thermohaline circulation**, whose name derives from its two drivers, heat (*thermo*) and salt (*haline*).

As Europe's present climate attests, thermohaline circulation in the North Atlantic represents a significant energy flow—comparable, at high latitudes in the northeastern Atlantic, to the energy delivered by sunlight. A reduction in the North Atlantic thermohaline circulation could therefore cool the climate of northwestern Europe. Ironically, global warming could bring about just such an effect, because increased precipitation and melting ice would dilute the salty North Atlantic surface with freshwater, reducing its density and inhibiting the sinking that drives thermohaline circulation. With the amount of global warming projected for the twenty-first century, this effect could reduce the North Atlantic thermohaline flow by as much as 60%. Nevertheless, even this reduction in Gulf Stream warmth won't be enough to keep Europe from warming in response to greenhouse forcing. And thermohaline circulation isn't the only driver of the Gulf Stream; wind-induced currents also contribute. So visions of a deep freeze in Europe or elsewhere seem far-fetched. However, our observational and theoretical knowledge of the thermohaline circulation is incomplete, and this is one area where nonlinearity in the climate system could bring a nasty surprise.

OCEAN ACIDIFICATION

Chapter 13 described how land and oceans together absorb slightly more than half of all anthropogenic CO_2 emissions, with the oceans taking a greater share. This is a good thing, since it slows the buildup of the most important greenhouse gas in the atmosphere and thus limits the rate of climate change. But is it a good thing for the oceans? The answer depends on the complex chemistry of oceanic carbon, and recent research suggests that the oceans' CO_2 uptake is adversely affecting marine organisms.

When CO_2 dissolves in water, it forms carbonic acid (H_2CO_3), a mild acid that gives carbonated sodas their acidity. Carbonic acid dissociates into hydrogen ions (H^+) and bicarbonate ions (HCO_3^-). The hydrogen ions combine with carbonate (CO_3^-) to form additional bicarbonate. But carbonate is what marine organisms use to build shells and other structures, like the rigid skeleton of coral. So adding CO_2 to the oceans has the net effect of decreasing the amount of carbonate available to marine organisms.

Already the effect of anthropogenic carbon has been to lower the oceans' average pH by 0.1. (Recall from Chapter 6's discussion of acid rain that pH measures the concentration of hydrogen ions in solution, with a decrease of 1 in pH being a tenfold increase in H^+.) With a doubling of atmospheric CO_2, the oceans' pH could fall by another 0.3 to 0.4 units. Under such acidic conditions, some species of high-latitude marine plankton may have trouble surviving. Because such organisms are at the base of marine food chains, the populations of larger marine species could suffer as well. Although these problems will be limited at first to polar and high-latitude oceans, they'll spread toward the tropics as the acidification increases.

15.4 Climate Change and Society

Climate change will have broad impacts on human society. Changes projected to occur in this century won't have the apocalyptic, civilization-destroying character of, say, an all-out nuclear war or a direct hit from a large asteroid. But climate change will be disruptive in many ways to a civilization whose planetary home is already under stress from the demands of billions of human beings. Although the focus of this book is on the scientific and technical issues of energy, environment, and climate, I do want to end this chapter with a brief survey of the major societal impacts expected from twenty-first-century climate change.

We've already seen that the physical impacts of climate change won't be evenly distributed around the globe. Low-lying coastal regions, for example, will suffer more from sea level rise; continental interiors will see more drought; and high latitudes will warm disproportionately. For that reason alone, impacts will vary from country to country. But social, political, and economic considerations also alter the impact of a changing climate. That's because wealthier countries are better able to adapt to and counter deleterious effects. Wealthy countries can build protective seawalls to ward off the rising oceans; they can engineer massive water projects to ensure continuing supplies of freshwater; they can outbid poorer countries if declines in domestic agriculture require increased food imports; their advanced health care systems can better cope with spreading tropical diseases; and their robust economies can speed recovery from extreme weather events. We're forced to the unfortunate conclusion that climate change will exacerbate the gap between rich and poor nations, as those with the fewest resources will be the most vulnerable. This conclusion should be especially troubling because anthropogenic climate change is almost entirely the result of fossil fuel consumption, deforestation, and other activities carried out by or for the benefit of the developed world.

Not all climate change will be harmful. For example, melting of Arctic sea ice is opening access to undersea resources and new routes for shipping—already, circumpolar nations are squabbling about ownership. And if warming is limited to one or two additional degrees, some regions will experience enhanced agricultural productivity resulting from increased precipitation, more rapid growth associated with higher levels of atmospheric CO_2, and a longer growing season. But beyond 2°C to 3°C of global average warming, essentially all impacts become deleterious, and the chance of surprising nonlinear changes occurring increases significantly. A sobering example is the European summer heat wave of 2003, which reduced plant growth so much that the plants actually became net sources of CO_2, emitting more CO_2 to the atmosphere than they took in through photosynthesis. In this extreme event, any positive effect of CO_2-enhanced plant growth was overwhelmed by the negative effect of heat and drought.

So what are the expected impacts on society? Some, such as increased disease and mortality rates among the elderly and the urban poor, will result directly from increased temperatures—especially from extreme heat waves, whose probability increases disproportionately as described in Box 15.2. Demand for air conditioning will stress already frail electric power infrastructures, reducing reliability and increasing brownouts as well as fuel consumption. And tropical diseases such as malaria may spread to higher latitudes. But the most significant impacts on humans will probably come from changes in the hydrologic cycle. More intense precipitation will lead to additional soil erosion and landslides. The former accelerates the loss of agricultural soils while the latter threaten burgeoning hillside populations, both rich and poor. At the opposite extreme, drier summers will decrease crop yields, make scarce freshwater resources even scarcer, and increase the risk of forest fires. These fires also will be more likely to destroy homes, as residential development at the urban fringes spreads into forested areas. In the Arctic, melting permafrost has already caused the collapse of roads, buildings, and pipelines, while the loss of sea ice alters the local ecology, threatening the livelihood of indigenous peoples. Both rising sea level and increased storm intensity will increase damage to coastal regions and may make some entirely uninhabitable. Tourist destinations will shift, precipitating both economic downturns and new opportunities.

No single impact of climate change—barring such surprises as the disintegration of the West Antarctic ice sheet—will in itself be insurmountable. But coming as the world approaches what will probably be its peak population, in a time at once of increasing economic interdependence and growing contrast and hostility between the developed and developing world, climate change will exacerbate economic, political, and humanitarian stresses that we already handle, at best, imperfectly.

For most people, climate change remains low on the list of potential threats to future well-being. But some, such as those in the insurance industry, are seriously worried. Losses from natural disasters have skyrocketed in recent decades. Most of this increased loss results from more people choosing to live in more vulnerable places, but some of it is likely attributable to climate change. Reinsurance companies—those that back up conventional insurance in case of catastrophic losses—are particularly concerned. Indeed, the Swiss company Swiss Re is contributing to the development of carbon-free energy technologies, and along with other European reinsurers is helping to sponsor public education programs on climate change. In the United States, the ski industry has joined the Natural Resources Defense Council in a campaign entitled "Keep Winter Cool." Even the U.S. military establishment, concerned about the security implications of sudden climate change, commissioned a study of low-probability but high-impact climate events.

Climate change presents humankind with myriad challenges, but it also presents opportunities. These include opportunities for technological and lifestyle changes that wean us from fossil fuels and thus reduce the greenhouse emissions that drive climate change; opportunities for creative adaptation to a changing planet; and opportunities for wiser, more efficient use of energy, water, and other natural resources. Chapter 16 takes a look at how humankind might move forward with energy technologies, practices, and policies that don't have to damage the local environment or upset the global climate.

CHAPTER REVIEW

BIG IDEAS

15.1 Climate models project future climates based on assumptions about greenhouse gas concentrations and other factors. They range from the simplest energy-balance models to large, three-dimensional, time-dependent versions that tax the fastest computers. Models are validated by projecting present-day climate from past conditions, by their successful reproduction of spatial and temporal distributions of climatic conditions, and by comparison with natural experiments such as volcanic eruptions.

15.2 Projections of future climates depend on assumptions about human behavior. Scenarios ranging from business as usual to concentrated efforts at reducing greenhouse emissions suggest a wide range in projected greenhouse gas concentrations by the year 2100. All scenarios show a global temperature rise through the twenty-first century. The greatest rise is expected at higher latitudes, especially in the Arctic.

15.3 Likely impacts of anthropogenic climate change include more heat waves and other high-temperature extremes; increased precipitation, both in amount and intensity; sea level rise due to thermal expansion and melting land ice; species range shifts and extinctions; changes in growing seasons; and the spread of tropical diseases. Less likely are nonlinear "surprise" events such as abrupt changes in ocean circulation or large land-based ice caps sliding into the sea. Ocean acidification is another effect of increased atmospheric CO_2.

15.4 Climate change will affect human society in a variety of ways. Sea level rise will threaten low-lying areas; changes in temperature and precipitation will alter agricultural production, positively in some places and negatively in others. Heat waves and the spread of tropical diseases will have an impact on human health. In general, wealthy countries will be better able to adapt to a changing climate than developing countries.

TERMS TO KNOW

Earth system model (ESM) (p. 388)
equilibrium model (p. 388)
general circulation model (GCM) (p. 387)
global climate model (GCM) (p. 387)
grid cell (p. 388)
one-dimensional model (p. 386)
representative concentration pathways (RCPs) (p. 394)
resolution (p. 388)
subgrid parameterization (p. 389)
thermohaline circulation (p. 403)
three-dimensional model (p. 386)
time-dependent model (p. 388)
two-box model (p. 386)
two-dimensional model (p. 386)

GETTING QUANTITATIVE

Global climate model, typical grid cell: 50 to 200 km

Global warming by 2100, projected: 2°C to 5°C over preindustrial

Sea level rise by 2100, projected: 0.5 to 1.5 m

QUESTIONS

1. What was the approximate increase in global average temperature during the twentieth century? What is the range of increases projected for the twenty-first century?
2. Why does halving the size of a climate model's grid cells more than double the computer time required to calculate the equations?
3. Explain what is meant by a coupled climate model.
4. Given the same resolution, does a three-dimensional climate model require just three times more computer time than a one-dimensional model? Explain.
5. What are subgrid phenomena, and how are they handled in climate models?
6. What are some ways modelers validate their climate models?
7. Figure 15.10 shows a wide range in CO_2 emissions under various RCP scenarios. Figure 15.9 shows much less variation in projected CO_2 concentrations. Why the difference?
8. What factor currently makes the greatest contribution to sea level rise?
9. Antarctica might have made a negative contribution to sea level rise during the twentieth century. How is that consistent with global warming?
10. Future climate depends in part on human choices. What are some choices we might make that could explain the difference between the RCP8.5 and RCP2.6 scenarios for future climate forcings?

EXERCISES

1 In Chapter 13 I gave an alternate definition of climate sensitivity as the temperature increase per unit (W/m²) increase in forcing. Given that a forcing of 3.75 W/m² is associated with a doubling of atmospheric CO_2, use Figure 15.4 to estimate both the transient and long-term climate sensitivity of the models used to produce that figure.

2 At the global average temperature of about 15°C, water increases in volume by about 0.02% for each degree Celsius of temperature increase. Use this result to make a rough estimate of the contribution to sea level rise from an ocean temperature increase of 2°C, assuming that only the top 500 m initially experiences the warming.

3 The volume of the Greenland ice sheet is some 2.85 million km³. If this ice melts, it will become water occupying about 92% of this volume. Given that the oceans cover 71% of Earth's surface, estimate the sea level rise that would result if all of Greenland's ice melted. Compare with Table 15.1.

4 (a) Calculate the number of grid cells in the atmospheric component of the HadGEM2-ES climate model described in Box 15.1. (b) At the equator, HadGEM2-ES's ocean component has grid cells measuring 1° in longitude and 0.33° in latitude. What's the actual size of such a cell?

5 Verify that the 1% per year increase in atmospheric CO_2 used in the model runs shown in Figure 15.4 results in CO_2 doubling in about 70 years.

RESEARCH PROBLEMS

1 In this problem you'll build a two-box climate model based on the energy flows shown in Figures 12.5 and 15.1. The table here gives values of the various energy flows represented by the arrows in Figure 15.1. For example, arrow 1, representing incident solar energy, is taken from Figure 12.5 as 341 W/m². Arrows that involve radiation are described by the familiar expression $e\sigma T^4$ for the energy emitted per square meter from an object at absolute temperature T. For example, arrow 8, the infrared emitted upward from the surface, has the value $e_s\sigma T_s^4$, where the subscript *s* stands for "surface" and *e* is the emissivity, the property that describes an object's effectiveness as an emitter and absorber of radiation. Similarly, arrow 4, infrared emitted from the atmosphere to space, is $e_a\sigma T_a^4$. The warm atmosphere emits radiation in both directions, so the downward arrow 9 has the same value as arrow 4 in this simplified model. Arrow 3 is the infrared from the surface (arrow 8) that manages to escape to space. Since *e* describes both emission and absorption, the atmosphere absorbs a fraction e_a of the arrow 8 energy flow; this is the effect of the greenhouse gases. So the fraction of the surface radiation that gets through the atmosphere is $1 - e_a$; hence the entry for arrow 3 (this point was made earlier in connection with Figure 12.3). For this simplified model, ignore convection and evapotranspiration (arrow 7), as indicated in the table.

Arrow	Physical meaning	Value (W/m²)
1	Incident solar energy	341
2	Sunlight reflected from atmosphere	
3	Infrared transmitted through atmosphere to space	$(1 - e_a)\, e_s\sigma T_s^4$
4	Infrared emitted by atmosphere to space	$e_a\sigma T_a^4$
5	Sunlight absorbed by surface	
6	Sunlight reflected from surface	
7	Convection and evapotranspiration	Ignore
8	Infrared emitted upward from surface	$e_s\sigma T_s^4$
9	Infrared emitted downward from atmosphere	$e_a\sigma T_a^4$

Here's your job: First, consult Figure 12.5 and fill in the remaining quantities in the table. You should be able to read the values right off the figure. Next, set up equations representing energy balance for both the surface and atmosphere. On the left-hand side of each equation should be all the flows *to* the surface or atmosphere; on the right-hand side should be all the flows *away* from the surface or atmosphere. The equal sign between them expresses energy balance. The terms in your equations are the numbers or expressions in the table.

You're almost ready to solve your equations, but first you need values for the surface and atmospheric emissivities: $e_s = 0.83$ and $e_a = 0.92$. These values aren't necessarily the most accurate, but they've been chosen so this oversimplified model gives reasonable results. The relatively high value of e_a represents the effect of greenhouse gases, which make the atmosphere nearly opaque to infrared. You also need the constant σ, introduced in Chapter 4; it's $5.67 \times 10^{-8}\,\mathrm{W/m^2 \cdot K^4}$.

You now have two equations in the two unknowns T_s and T_a. Since these appear to the fourth power, it's far easier to solve for T_s^4, and T_a^4, treating each of these quantities as a single unknown. Then take the fourth root (take the square root twice, or raise to the power 0.25). The result is the surface and atmospheric temperatures as predicted by this simple two-box model. Are your answers reasonable? (Don't expect them to be exactly right.) Is the atmospheric temperature lower, as we discussed in Chapter 12?

2 Now include the effect of convection and evapotranspiration, which is approximately proportional to the temperature difference between the surface and atmosphere. So use the term $c(T_s - T_a)$ for arrow 7, with $c = 4\,\mathrm{W/m^2 \cdot K}$ (again, not a particularly accurate value, but chosen for this simplified model). (a) Modify your equations to include this new term. (b) Solve for the two temperatures. Now you have both T and T^4 terms, so this is harder. You'll need to use computer software or a calculator that can solve nonlinear equations. Compare your answer with the result of Research Problem 1 and comment. (c) Calculate the flow represented by arrow 7. Is it roughly comparable to the convection and evapotranspiration arrows in Figure 12.5?

3 Get involved in climate modeling! Join the climate*prediction*.net project by signing up at the web site www.climateprediction.net. In its spare time, your computer will join tens of thousands of others in sharing the effort of running large-scale climate models. You'll be kept up to date on the progress of climate*prediction*'s model experiments, and you'll make a real contribution to climate research.

ARGUE YOUR CASE

1 "GIGO," for "Garbage In, Garbage Out" is a derogatory term for computer programs that produce meaningless output because their input isn't realistic or correct. You'll find the term widely used by climate skeptics in dismissing the projections of global climate models. Formulate an argument to suggest that what goes into climate models isn't garbage.

2 A friend argues that we can't have confidence in climate model projections because we have no way to verify that the models correctly describe future climate. Criticize this argument.

Chapter 16

ENERGY AND CLIMATE: BREAKING THE LINK

This final chapter asks what we can do to break the link between our energy demand and the deleterious impact our energy consumption has on Earth's environment. In many ways, the same remedies apply both to conventional impacts, such as air pollution and water pollution, and to climate change. But in other ways the truly global problem of climate change presents unique challenges that require new ways of thinking and acting about energy and the environment.

As I explained at the beginning of Chapter 6, there's an important distinction between fossil emissions of traditional pollutants and the greenhouse gas CO_2. Traditional pollutants are undesirable and largely unnecessary by-products of fossil fuel combustion. With enough money and technology we can reduce these pollutant emissions to arbitrarily low levels. For that reason, traditional pollution presents no barriers *in principle* to our continued use of fossil fuels or even to increased fossil fuel consumption. I say *in principle* because we have to be willing to pay what it takes to achieve our desired air- and water-quality standards, and to restore environmental quality damaged in fossil fuel extraction and transportation. But we know how to do that if we, as a society, so choose.

Carbon dioxide is different. Along with water, it's a necessary product of fossil fuel consumption. To burn fossil fuels is to make CO_2, so on a certain level we *want* to turn fossil fuels into CO_2. Whereas traditional pollutants occur in relatively small quantities, we produce huge amounts of CO_2 because efficient fossil fuel combustion converts essentially all the fuel carbon into CO_2. For example, the data in Table 7.3 show that a 1-GWe coal-fired power plant produces 1,000 tons of CO_2 per hour, compared with some 75 tons of air pollutants and ash. So climate-changing CO_2 emissions are more than an order of magnitude greater than conventional pollutants. And the impact of CO_2 emissions is truly global, while most pollution has primarily local and regional impacts.

16.1 Carbon Emissions: Where We're Going and Where We Need to Be

The U.S. National Oceanic and Atmospheric Administration recently made headlines with an announcement that last year's atmospheric CO_2 concentration set a new record. Actually, this isn't the least bit surprising, and it doesn't matter what year "last year" was. As Figure 13.7 shows, atmospheric CO_2 has been rising steadily throughout the industrial era of the past few centuries. This means every year sees a new record concentration of atmospheric CO_2. In the early twenty-first century, the annual increase has ranged from about 1.5 ppm to 3 ppm, as you can verify in Research Problem 1. If this trend continues, we're looking at a rise in atmospheric CO_2 of something like 200 ppm by 2100, reaching nearly 600 ppm at that time. However, the rate of increase has itself been increasing—again, try Research Problem 1 to verify this—so we're likely to reach that concentration somewhat sooner.

"DANGEROUS ANTHROPOGENIC INTERFERENCE"

The ultimate objective of the **United Nations Framework Convention on Climate Change (UNFCCC)**—an international agreement signed at the 1992 Earth Summit in Rio de Janeiro—is "stabilization of greenhouse gas concentrations in the atmosphere at a level that would prevent dangerous anthropogenic interference with the climate system." As Chapter 15 suggests, there's no clear threshold at which point our interference with the climate system (read greenhouse emissions) can be considered "dangerous." There used to be general agreement that avoiding dangerous interference with the climate system meant limiting the twenty-first-century atmospheric CO_2 concentration to no more than a doubling of its preindustrial value—that is, to 560 ppm or less. But a growing understanding of the perils we face from climate change, sea level rise, and ocean acidification has led to a more stringent and physically meaningful criterion: limiting the global temperature rise for the industrial era to under 2°C. An accord reached at the 2009 United Nations Climate Change Conference in Copenhagen formalized this 2°C goal, and called for a possible lowering to 1.5°C following discussions to take place by 2015. The 2010 UN Climate Conference in Cancun reaffirmed the 2°C goal and set forth clear objectives and mechanisms for meeting that goal. However, the Cancun conference did not result in any binding requirements, and emissions-reduction pledges made by the assembled nations fell far short of what's needed to keep the rise under 2°C. Given political and economic realities, it's not at all clear that even those inadequate pledges are realistic.

Limiting the industrial-era temperature rise to 2°C would not be easy. We're already close to 1°C above the preindustrial climate, so the 2°C goal allows only a little more than 1°C in the remainder of the twenty-first century. Of the four RCP scenarios discussed in Chapter 15, only RCP2.6 keeps the industrial-era temperature rise under 2°C—and it requires a net removal of CO_2 from the atmosphere near the end of the century. Figure 16.1 makes this clear, by plotting century-end CO_2 emissions, concentrations, and industrial-era temperature rises for the four RCPs. With nations squabbling about modest reductions in emissions, or in some cases only reductions in the growth of emissions, we're far from an international consensus or the technology needed to turn those emissions negative and suck CO_2 from the atmosphere.

In terms of CO_2 concentration, Figure 16.1 shows RCP4.5 at 580 ppm—somewhat over a CO_2 doubling from its preindustrial concentration and no longer considered acceptable. RCPs 6.0 and 8.5 are even higher. RCP2.6, in contrast, ends the century with CO_2 at about 425 ppm—only modestly above today's level. Still, even that may be too high. Increasingly, climate scientists and activists are embracing 350 ppm as the maximum safe value for atmospheric CO_2 in the long run—although acknowledging that it may peak at well above today's 390 ppm before dropping to 350 ppm. Box 16.1 explores several scientific reasons in support of "350."

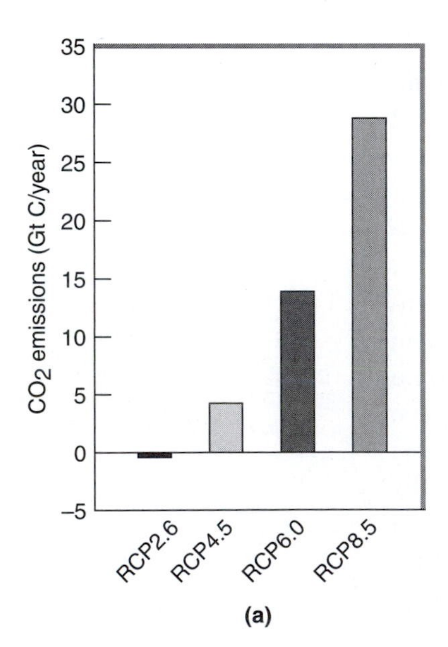

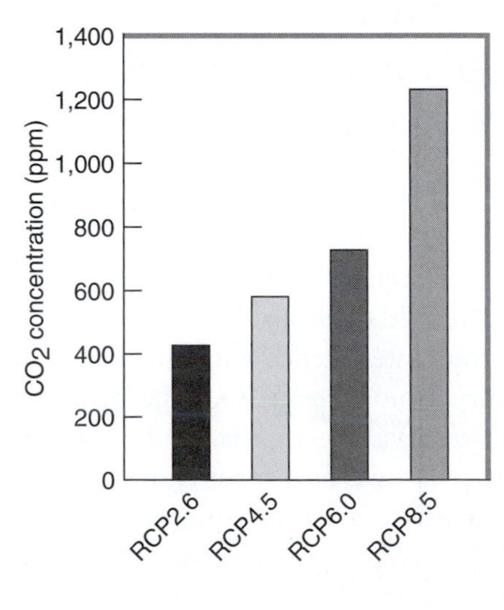

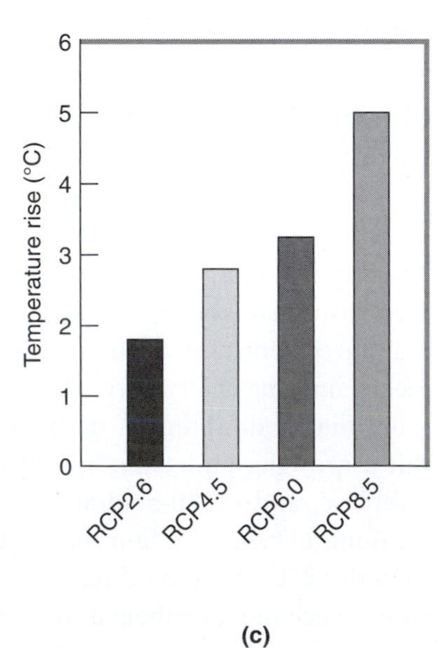

FIGURE 16.1

Graphs of (a) CO_2 emissions, (b) CO_2 concentrations, and (c) projected temperature increases from preindustrial times, at year 2100 for the four representative concentration pathways (RCPs) introduced in Chapter 15. Numbers on RCPs give the approximate radiative forcing due to all industrial-era forcing agents.

BOX 16.1 | 350: The Science behind the Number

"350" has become a rallying cry for climate activists hoping to preserve a stable climate for future generations. What's the scientific basis for this number? Several lines of reasoning point to 350 ppm as a prudent safe level:

- Earth's energy imbalance today is most likely a little over 0.5 W/m²—a result of industrial-era forcings that have not yet come to equilibrium. Doubled CO_2 from preindustrial times (a 280-ppm increase) results in a forcing of about 4 W/m², so we'd need a reduction of somewhat more than 1/8 of that level, some 35–40 ppm, to bring our planet into energy balance. With 390 ppm today, that puts us in the vicinity of 350 ppm.
- Terrestrial ecosystems are under threat at today's 390 ppm, and ocean acidification is already damaging marine ecosystems. We need lower atmospheric CO_2 to reverse these threats.
- Most significantly, today's 390 ppm and the higher concentrations projected for 2100 risk upsetting long-term ice-sheet dynamics—which could result in catastrophic sea level rise of tens of meters.

Evidence for ice-sheet effects comes from paleoclimate studies summarized in Figure 16.2. These studies show that Earth was ice-free until about 35 million years ago, when atmospheric CO_2 dropped below 450 ppm and the Antarctic ice sheet formed. This implies that, over thousands of years, a CO_2 concentration above 450 ppm will result in melting of all the planet's ice. The result, as Table 15.1 suggests, would be a sea level rise of nearly 70 m or about 200 feet.

A second line of evidence involves climate sensitivity, defined in Chapter 13 as the temperature increase associated with a doubling of atmospheric CO_2; there the best estimate was about 3°C. But this value neglects long-term feedbacks associated with ice-sheet dynamics. A simple way to estimate the overall sensitivity with these feedbacks is to compare the paleoclimate temperature and greenhouse-gas forcing. In a seminal 2008 paper entitled "Target Atmospheric CO_2: Where Should Humanity Aim?", climatologist James Hansen and colleagues plotted Vostok ice-core data for temperature and greenhouse forcing together on the same graph, adjusting the scaling so both curves lie right on top of each

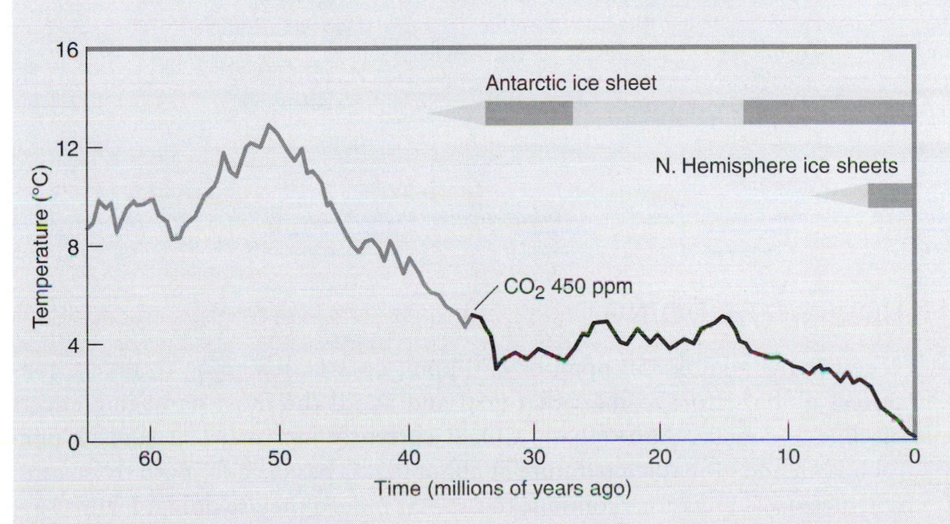

FIGURE 16.2 Global deep ocean temperature going back 65 million years, determined from oxygen isotope ratios (see Box 14.2). Bars represent presence of polar ice sheets, with darker bands indicating heavier glaciation. Earth was essentially ice-free before 35 million years ago, when the atmospheric CO_2 concentration was above 450 ppm.

(continued)

BOX 16.1 | 350: The Science behind the Number (Continued)

other (Figure 16.3). This shows immediately that a temperature rise of 3°C corresponds to a forcing of 2 W/m². Since doubled CO_2 gives 4 W/m² of forcing, the long-term climate sensitivity is 6°C, or twice the climate sensitivity without ice-sheet feedbacks. Thus in the long term, the climate may be twice as sensitive to anthropogenic forcing as we had thought—again arguing for a lower ultimate atmospheric CO_2 concentration.

The work of Hansen and colleagues provides additional insights into climate change, especially the extra warming "in the pipeline" that we can expect even if we stopped our greenhouse emissions today. Figure 16.3 also explores this phenomenon.

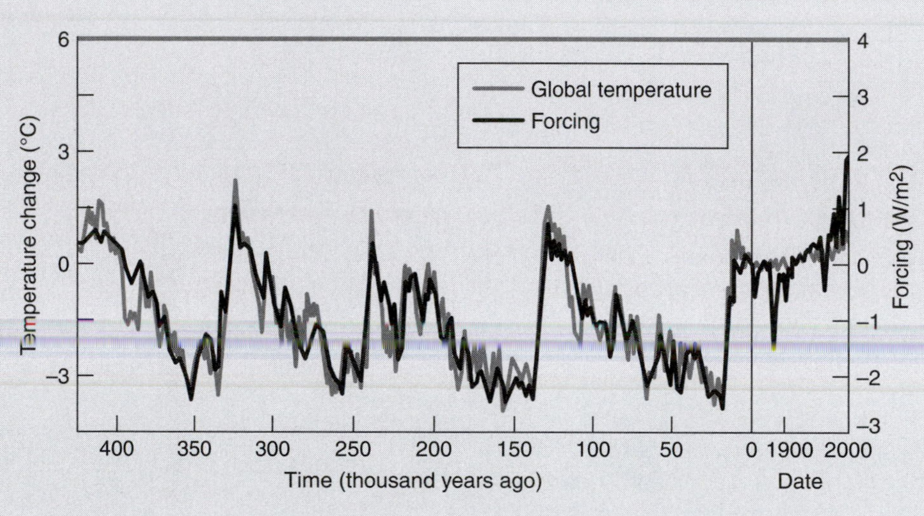

FIGURE 16.3
Global temperature and forcing, going back 420,000 years or four ice-age cycles. Curves have been scaled so they lie on top of each other; dashed line then shows that a 3°C warming corresponds to a forcing increase of 2 W/m². Since 4 W/m² is the forcing for doubled CO_2, this implies a climate sensitivity of 6°C for 2×CO_2. Expanded horizontal (time) scale at right shows industrial-era temperature and forcing. Here the temperature curve lies below the forcing curve, implying about 1.4°C of warming still "in the pipeline"—warming that would occur even if we stopped climate-changing emissions today.

HOW WE'RE DOING

Whether the goal is 350 ppm, or 450 ppm, or 2°C, it's going to take a concerted, global effort to meet that goal and avoid the most damaging effects of climate change. Atmospheric CO_2 is currently increasing at some 2 ppm each year, and global temperature at about 0.2°C per decade. Both those rates will increase if emissions continue to rise—as indeed they're doing. Figure 16.4 shows actual carbon emissions for 2000–2010, along with two RCP scenarios and a projection by the U.S. Department of Energy of future emissions. If the

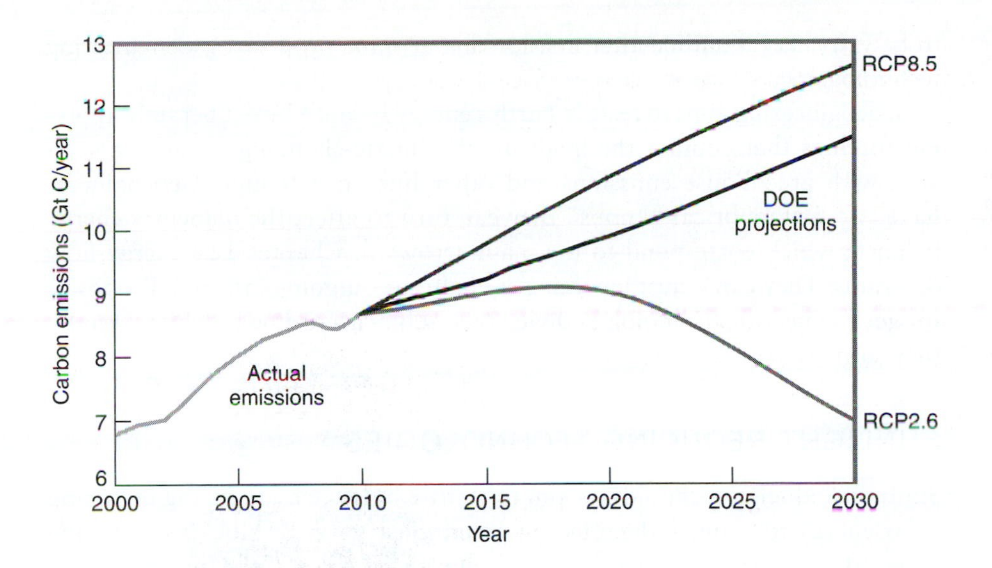

FIGURE 16.4
Actual global carbon emissions for 2000–2010, followed by emissions in two RCP scenarios and a projection by the U.S. Department of Energy. RCP2.6 is the only RCP scenario that meets the 2°C goal in 2100. RCP and projection data have been rescaled so all agree in 2010.

DOE projections prove accurate, then we're between business-as-usual RCP8.5 and the emissions-reducing RCP2.6—whose really serious emissions cuts don't kick in until late in the current decade. By that time the DOE projections parallel RCP8.5, although at somewhat lower levels. As Figure 16.1 showed, that puts us on target for a century-end CO_2 concentration and temperature rise that are well above any current estimates of maximum safe levels. So the bottom line is that we—the whole world, that is—are not doing particularly well, although we still have a few years to get our emissions under control.

I can think of four approaches we might take, individually or in combination, to minimize anthropogenic climate change. At one extreme is **geoengineering**: global-scale engineering projects that alter Earth's energy balance to counter anthropogenic warming. Although many scientists are decidedly unenthusiastic about geoengineering, it's increasingly seen as a possible last-ditch solution in the face of threateningly rapid global warming. Three other approaches to the climate problem involve changing our energy behavior: We could continue producing CO_2 through a fossil-based energy system, but get serious about treating CO_2 as a pollutant and try to keep it out of the atmosphere. We could switch to energy sources that entail less CO_2 production. Finally, we could simply use less energy—either by changing our lifestyles to require less or by using energy more efficiently. The remainder of this chapter explores these four options.

16.2 Geoengineering

Geoengineering is a cop-out: A successful geoengineering technology would let humanity continue its profligate CO_2 emissions, while technologically countering the associated climate change. For that reason alone, many scientists and policymakers are wary of geoengineering. But there are plenty of other reasons to be wary, as I'll outline after first looking at some proposed geoengineering technologies.

Geoengineering aims to restore Earth's energy balance by deliberately applying forcings that counter the inadvertent, climate-changing forcings associated with greenhouse emissions and other human activities. Geoengineers have two metaphorical "knobs" they can turn to affect the planetary energy balance, which correspond to the main arrows in Chapter 12's energy-flow diagrams: They can adjust incoming sunlight, or outgoing infrared. Proposals for geoengineering technologies divide themselves according to which "knob" they turn.

SUNLIGHT-REDUCING TECHNOLOGIES

Sunlight-reducing technologies aim to restore balance by reducing incoming solar energy to counter the effect of greenhouse gases in blocking outgoing infrared. The simplest approach is to reflect a fraction of the incoming sunlight back to space. The most widely studied, and probably the least expensive, of such approaches is to inject sulfur dioxide into the stratosphere. Like SO_2 emitted in coal burning, the injected SO_2 would react to create highly reflective sulfate aerosols. Like the sulfate aerosols from fossil pollution, these particles would exhibit a negative forcing and thus produce a cooling effect.

We have some quantitative understanding of the climate impacts of stratospheric aerosols. The 1991 Pinatubo eruption injected some 20 million tons of sulfur dioxide into the upper atmosphere, causing a clearly measurable global cooling that was shown in Figure 15.6. So we can, in principle, calculate how much SO_2 would be required to counter a given amount of greenhouse warming. Such calculations suggest quantities in the millions of tons per year to counter the degree of warming we expect in the coming century. The actual injection could be done with balloons, by a fleet of specially equipped aircraft, or by launching SO_2-filled projectiles from the surface.

An alternative proposal calls for placing trillions of dinner-plate-sized reflectors at a point in space a million miles sunward of Earth, where they would orbit the Sun with a one-year period and thus remain always between Sun and Earth. Equipping each with a navigation system would allow for control of this vast space-based "sunshade" as needed to adjust Earth's temperature. The cost? Prohibitive.

A more down-to-Earth suggestion has specially designed ships spraying seawater into the lower atmosphere, where it evaporates and leaves salt crystals on which water can condense to form clouds. With more condensation nuclei, the condensed droplets would be smaller and more reflective. Again, more reflection means negative forcing and a cooling effect.

One disadvantage of all sunlight-reducing geoengineering proposals is that they don't touch the problem of ocean acidification. Rather, by countering only the warming effect of CO_2, they permit ever rising atmospheric CO_2 concentration. As a result, CO_2 continues to dissolve in the oceans, increasing acidity.

INFRARED-ALTERING APPROACHES

Greenhouse warming occurs as atmospheric greenhouse gases absorb outgoing infrared radiation. Geoengineering schemes would counter this absorption by removing greenhouse gases from the atmosphere—something we'll need to do if we're to implement the RPC2.6 scenario's negative emissions in the final decades of this century. Removing greenhouse gases is a more direct solution to the climate-change problem than is tampering with incoming sunlight, because it directly reverses the immediate problem we're causing, namely emission of greenhouse gases.

One widely discussed approach is to seed the oceans with iron, a nutrient that currently restricts the growth of plankton. More plankton performing photosynthesis would increase the downward flux of carbon from atmosphere to oceans, helping counter the upward flux from fossil fuel combustion. Alternatives include using chemical reactions to capture atmospheric carbon dioxide, perhaps with "artificial trees" that expose their reactive "leaves" to the air, or altering ocean chemistry so the oceans absorb more CO_2. Simpler CO_2-absorption technologies range from planting more trees to grinding up rocks and spreading them over vast land areas to accelerate the natural weathering that, on the longest timescales, takes CO_2 from the atmosphere.

GEOENGINEERING: A CRITIQUE

The first edition of this book dismissed geoengineering in a single paragraph. But increasingly, geoengineering is getting a serious airing in the scientific community. Scientific conferences have geoengineering sessions, scientific societies are commissioning reports, and climate scientists are using their models to explore the impacts of geoengineering technologies. So should we take geoengineering seriously?

Most geoengineering proposals entail a host of other consequences besides mitigating climate change. Some of these are known, some not. Space-based reflectors would make the skies darker, with implications for, among other things, solar energy generation. Sulfate-aerosol schemes would put a permanent global haze into the atmosphere. Ocean fertilization and chemistry-altering schemes might affect marine ecosystems in unpredictable ways. Many of the more advanced schemes depend on as-yet-undeveloped technologies, and may have significant materials and energy requirements. All have costs, which in some cases dwarf any economic impact resulting from emissions reductions. And the "geo" in geoengineering means we're talking about global projects designed to affect the entire planet. Unless we know precisely what we're doing, we risk unexpected consequences that could include plunging Earth into a period of global cooling. Not knowing what we were doing—in the case of inadvertently geoengineering Earth's atmosphere with greenhouse emissions—is what got us into our present climate predicament. I personally doubt that we can get out of it by further tinkering with our planet's energy balance. Better that we turn our attention to reducing greenhouse emissions.

In this view I'm echoing an authoritative report on geoengineering done for Britain's Royal Society. At the start of its thorough evaluation of proposed geoengineering technologies, the report cautions: "The safest and most predictable method of moderating climate change is to take early and effective action to reduce emissions of greenhouse gases. No geoengineering method can provide an easy or readily acceptable alternative solution to the problem of climate change."

Nevertheless, there may be a place for geoengineering. What if we fail to reign in greenhouse emissions, and rapid climate change begins to pose a serious threat to human society? What if we find ourselves on the verge of a dangerous nonlinear "tipping point"? Wouldn't it be good to have researched and developed some last-ditch geoengineering schemes that we could deploy in a real climate emergency? Maybe. But would having such schemes weaken the political will needed for immediate emissions reductions? On the other hand, might geoengineering—especially CO_2-removal technologies—help us buy time to for the transition to a low-carbon energy economy? Geoengineering presents thorny dilemmas, and in my view it does not, by itself, qualify as a viable solution for mitigating climate change.

16.3 Carbon Capture and Sequestration

Carbon capture and sequestration (CCS) involves collecting CO_2 from fossil energy production and sequestering it underground so it won't enter the atmosphere. This is a daunting task, given the vast amounts of CO_2 produced from fossil fuels. For the fossil-fueled engines that power today's cars, trucks, and airplanes, CCS is for all practical purposes impossible. But it might be feasible for large, stationary sources like power plants and industrial boilers. Worldwide, such sources account about half of today's 9 Gt of carbon emissions from fossil fuels. That's still a vast quantity of carbon dioxide; you can show in Exercise 1 that the 1,000 tons per hour released from a single 1-GW power plant amounts to half a million cubic meters.

CAPTURING CARBON DIOXIDE

Combustion in a conventional fossil-fueled power plant or boiler uses air, which is roughly 20% oxygen and 80% nitrogen. Oxygen combines with the fuel to make CO_2 and H_2O, but very little of the nitrogen does (although it's the source of NO_x pollution). So the resulting mix—called flue gas—is mostly nitrogen, with only about 15% CO_2. Separating the CO_2 from the diffuse flue gas isn't easy.

One approach is to react the flue gas with chemicals that absorb CO_2, and then heat the chemicals to release concentrated CO_2 for sequestration while making the chemicals available for reuse. This is similar to geoengineering schemes for chemical removal of CO_2 from the atmosphere. In either case the vast quantities of CO_2 rule out one-time use of chemicals.

Another approach is to remove carbon before combustion. Chapter 5 introduced efficient combined-cycle power plants, some of which use coal. Normally, a combined-cycle coal plant gasifies coal in the presence of limited oxygen, producing a mixture of hydrogen (H_2) and carbon monoxide (CO) that burns in gas turbines. Using more oxygen, however, produces a mix of H_2 and CO_2. Capturing the CO_2 at this precombustion state is relatively easy, leaving hydrogen as the combustible fuel.

A final approach is to use pure oxygen in place of air to support combustion. The flue gas is then mostly CO_2, readily available for sequestration. But extracting oxygen from air is expensive and energy intensive, and combustion with pure oxygen occurs at temperatures too high for most materials.

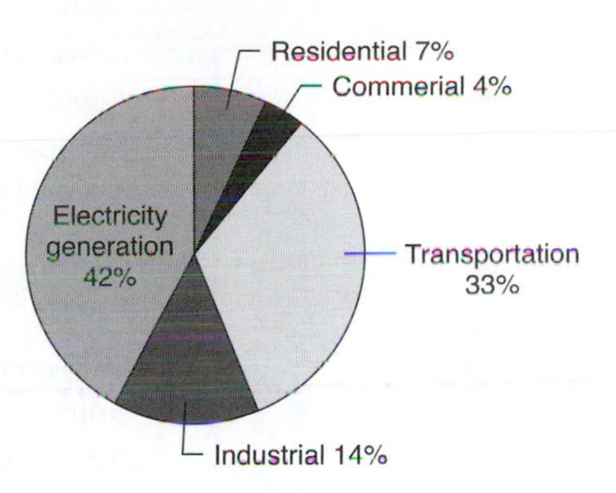

FIGURE 16.5
Sources of CO_2 emissions for the United States. Capture of CO_2 is practical only for electricity generation and some industrial facilities; together, these account for only about half of all CO_2 emissions.

I emphasize that CO_2 capture is practical only for large stationary sources. CO_2 removal is virtually impossible with mobile sources, and impractical with small stationary sources like home heating systems. Yet large sources account for only about half of all global CO_2 emissions. Figure 16.5 quantifies this point for U.S. emissions.

SEQUESTRATION

Once we've captured CO_2, we have to sequester it where it can't escape to the atmosphere. But where? The most promising sequestration sites are porous rock deposits permeated with brine—very salty water—at high pressure. Injecting CO_2 at these pressures produces a dense fluid that combines properties of gas and liquid. If impermeable layers cap the porous rock, then the injected CO_2 is likely to remain in place.

Another place to sequester CO_2 is in the very wells that produce fossil petroleum. Drillers routinely pump CO_2 into old wells to enhance petroleum yields, and such yield enhancement could be combined with sequestration. A pilot project at a Canadian oil field sequestered 5 million tons of CO_2 from a North Dakota coal-gasification plant while increasing oil production by 10,000 barrels per day (see Exercises 3 and 4). But oil-field sequestration isn't ideal, because oil fields are riddled with thousands of oil-well penetrations, each offering a potential escape route for CO_2.

Yet another place to put CO_2 is in the deep ocean. Figure 13.10 shows that the oceans already store nearly 40,000 Gt of carbon. Exercise 5 will convince you that it would take more than a century for the carbon from all the world's power plants to increase oceanic carbon by a mere 1%. But most oceanic carbon isn't in the form of CO_2, so we don't have experience with deep-ocean CO_2. What effects might it have on marine life? And what's the cost—in dollars and energy—of compressing CO_2 to force it into the deep ocean? And is it legal, under the Law of the Sea, to dump a waste product into international waters? Questions like these make deep-ocean sequestration a tentative proposition at best.

A carbon-sequestration scheme is technically viable if we can ensure that it will retain at least 90% to 95% of the stored CO_2 for the first century, and 65%

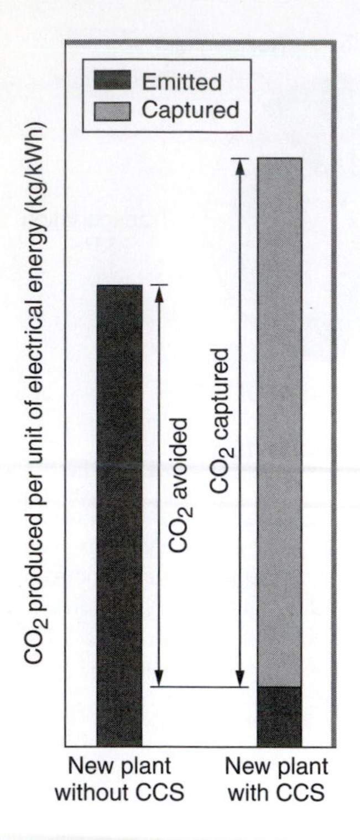

FIGURE 16.6

A new power plant with CCS consumes more fuel than a comparable plant without CCS. Nevertheless, carbon capture results in a significant reduction in CO_2 emissions, by the amount labeled "CO_2 avoided." Figure is for new power plants, built with CCS from the ground up; retrofitting existing plants is less effective in reducing emissions.

or more for 500 years. Those numbers appear feasible; the IPCC report *Carbon Capture and Storage* suggests that well-engineered geologic sequestration facilities are likely to retain 99% of their CO_2 for at least a century and maybe a millennium or more. But this is far from a proven technology.

PROSPECTS FOR CCS

CCS could buy us several decades to ease the transition to carbon-free energy sources, especially with coal-generated electricity. But, again, CCS only works with large stationary sources, which account for only half of our carbon emissions. So it's not a complete solution. And construction of CCS facilities is a massive undertaking, not likely to be accomplished quickly. In fact, the IPCC estimates that only 20% to 40% of fossil carbon emissions might be amenable to CCS by 2050. Everything you know from Chapters 12 to 15 suggests that we can't wait that long for major reductions in CO_2 emissions.

Furthermore, CCS carries significant energy and economic costs. CO_2 capture is energy intensive, and it takes even more energy to pressurize CO_2 for injection into sequestration facilities. Adding CCS to existing coal-fired plants would result in a 24% to 40% increase in fuel consumption for the same electrical output. This increases emissions of pollution and solid waste, and raises substantially the cost of electric power. Although CCS might capture some 95% of CO_2, increased fuel consumption somewhat offsets this reduction (see Figure 16.6).

Like other technological fixes, CCS hasn't yet proven effective or economical on a large scale. A $1.3 billion demonstration CCS power plant, called FutureGen, is scheduled for construction in the United States; meanwhile, China has built a coal-fired power plant that reportedly sequesters CO_2 more economically than earlier estimates. But even if these projects succeed, we're a long way from large-scale CCS deployment. CCS is no panacea, and it alone won't stop anthropogenic climate change. So the bottom line remains: We must move away from fossil fuels.

16.4 Alternative Energy Sources

Chapters 7 to 10 introduced alternative energy sources that don't involve fossil fuels. Some, such as solar energy and its derivatives, wind and hydropower, are renewable—they come in a steady stream that doesn't run out. Another solar derivative, biomass, is renewable if it's harvested sustainably. Other sources, such as nuclear energy, aren't strictly renewable but promise nearly unlimited fuel resources in their more advanced manifestations, particularly nuclear fusion. Geothermal and tidal energy can make only modest contributions to our global energy supply but could be significant in localized regions. All these alternatives have far lower emissions of CO_2 and other greenhouse gases than do fossil fuels.

Far lower—but not zero. From our survey of energy sources, it's clear that none is completely benign. But some are better than others, and how much better depends on which measures you apply and how you evaluate such factors as risks of catastrophic nuclear accidents or dam failures. So how do the climate impacts of different energy sources compare? It's easy to measure greenhouse emissions from fossil energy, whose emissions result almost entirely from combustion. Fossil fuel is a commercial commodity, so we know pretty accurately how much we're burning. But greenhouse emissions from other energy sources are harder to quantify, because they don't involve ongoing energy production as much as processes like manufacturing energy facilities, producing cement, mining nuclear fuels, growing biomass, transporting materials, and decommissioning. Chapter 7 mentioned a study of greenhouse emissions associated with nuclear power; Chapter 10 gave the surprising result that building hydroelectric dams in tropical regions may cause methane emissions with greater climate impact than comparable fossil-fueled facilities. A comprehensive study for the International Atomic Energy Agency examined nearly all the alternative energy sources and gave estimated high and low ranges for greenhouse emissions; Figure 16.7 summarizes the results. In most cases, the high estimate includes existing technologies, whereas the low estimate refers to technologies expected to be available by the year 2020. For hydropower, the high category is for tropical dams and the low category is for run-of-the river power plants that require far less concrete and don't entail large, stagnant reservoirs. For other sources, the high and low estimates are simply ranges of quantities that are difficult to determine accurately. Much of the CO_2 emission implied in Figure 16.7 comes from fossil fuels consumed in the manufacture and installation of energy facilities. Solar energy ranks fairly high in CO_2 emissions, because present-day semiconductor fabrication is very energy intensive (although this is changing as techniques for manufacturing thin-film photovoltaic cells become widespread). CO_2 emissions for some alternative energy sources also result from cement production; that's the case for nuclear power plants, hydro dams, and the massive concrete foundations of large-scale wind turbines. With nuclear power,

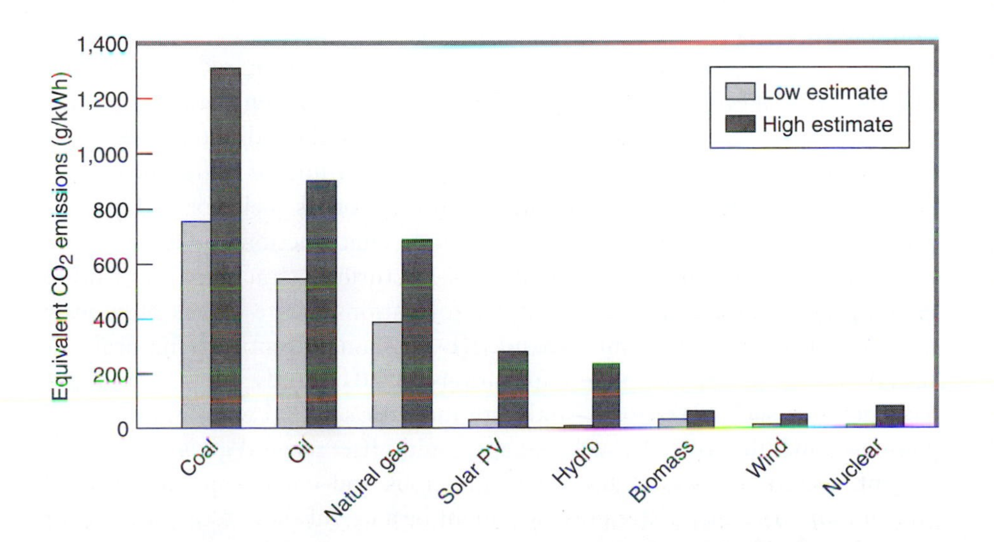

FIGURE 16.7
Estimates of CO_2 emissions from various energy sources. Emissions come from combustion (fossil fuels only), as well as construction, fuel production, transportation, and other relevant processes. High estimates for fossil fuels are based on typical power plants; low on the best technologies currently available. High solar PV estimate assumes crystalline silicon manufactured as for the semiconductor industry, and is increasingly unrepresentative. Ranges on other energy sources reflect assumptions about manufacturing, fuel production (nuclear), location (hydro), and other factors.

carbon emissions also result from the considerable fossil energy used in producing nuclear fuel; the level of emissions depends significantly on the quality of uranium ore. Once nonfossil energy sources are deployed on a large scale, however, the energy they produce can be used in manufacturing additional nonfossil energy facilities, thus lowering the associated carbon emissions. Although alternative-energy advocates may be disappointed to find that their favorite alternative isn't necessarily "pure" from a climate-change standpoint, it's nevertheless clear from Figure 16.7 that all the alternatives have far less climate impact than fossil fuels.

So why not switch to one or more energy alternatives and be done with the climate-change problem? That may well be what's needed, but it's not an easy switch. Of the energy alternatives considered in Chapters 7 to 10, only hydro and nuclear power currently make significant contributions to the global energy supply. Both are best suited to producing electrical energy, so they're not good for powering vehicles. Hydropower potential in the developed world is largely tapped already, and nuclear power continues to face an uncertain future. Geothermal and tidal energy, though useful in some local situations, just aren't sufficient to make a major contribution. Direct solar, wind, and biomass have the potential to supply far more energy than we use. Although their contribution is growing rapidly, it will be some time before they can achieve their full potential. Direct solar and wind, too, aren't generally adaptable for transportation. And although wind and some biomass fuels can now compete with conventional energy sources, solar photovoltaics remain economically uncompetitive.

Substituting nonfossil sources in transportation is particularly challenging. Biofuels can help, but it will be decades, if ever, before we have enough biofuel capacity to supply the world's burgeoning vehicle fleet. Hydrogen has promise for transportation, but remember that there's no hydrogen energy resource here on Earth. So unless we make hydrogen with some nonfossil means—for example, using nuclear, hydro, wind, or photovoltaic-generated electricity—then hydrogen doesn't help much with climate change. Nuclear fusion could offer virtually unlimited energy, including producing hydrogen for transportation, but fusion is at best decades in the future.

Figure 16.7 suggests one more modest approach to carbon reduction within the fossil-fueled economy—switching from coal and oil to natural gas (which is mostly methane). You've seen how the chemistry of these different fuels gives roughly a factor-of-2 spread in carbon emissions per unit of energy produced. Natural gas has become more prominent in the world's fuel mix, today comprising some 27% of fossil fuel consumption. That fraction could rise, especially with the increased use of hydraulic fracturing ("fracking") and other techniques to extract gas from geologic formations where it was previously uneconomical to do so. But expanded use of unconventional natural gas sources carries its own climate implications: A 2011 study suggests high gas leakage from such operations—and with methane's global warming potential some 30 times that of CO_2, such leakage could offset natural gas's greenhouse advantage relative to coal and oil. A judicious and environmentally careful switch to natural gas, where possible, might be a useful short-term strategy for reducing the climate impact of fossil fuels, but it's not an answer for the long term. At best, natural gas can act as a bridge to a future economy that's not based on fossil fuels.

So which energy source is the long-term solution? Over the next half-century or so, there simply isn't one answer. Maybe someday we'll live in a society powered entirely by direct solar energy or nuclear fusion—the two alternatives whose availability far exceeds humanity's energy demand. We ought to be working toward that day, but in the meantime we need to adopt an energy strategy that can reduce carbon emissions over the next few decades. One study suggests that we could complete a transition to nonfossil energy sources by 2030—and with the added benefit of lowering our total energy consumption by replacing fossil-fueled transportation with more efficient electric vehicles. Whether or not that optimistic date is possible, we're going to need to act soon to begin our energy transition. In the short term, that transition can include strategies like CCS and substitution of natural gas for coal—but in the long term it must mean breaking our dependence on fossil fuels.

16.5 Using Less Energy

Phrases such as *using less energy* and *energy conservation* conjure up images of people clad in bulky sweaters shivering in dimly lit rooms or crammed into impossibly tiny vehicles. But that's not what I mean by using less energy. I mean something a lot smarter—namely, using energy more efficiently so that we get the same benefits and living standards we're used to, but with a lot lower energy consumption.

My comments in this section are aimed primarily at the developed world, and particularly at North Americans, whose energy consumption is, by any measure, excessive. There are plenty of countries in the developing world whose energy use is *below* what's needed for a decent standard of living, and they have every right to strive for greater energy use and the better life it can bring. But even developing countries will benefit from increased energy efficiency, which will make their path to development easier, more economical, and "greener."

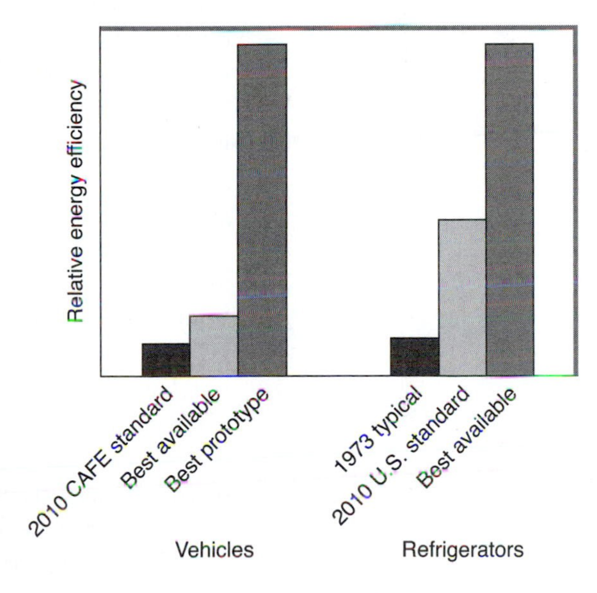

FIGURE 16.8
Potential efficiency improvements for light-duty vehicles and household refrigerators. The best prototype vehicle is a Volkswagen that goes 282 miles on a gallon of diesel fuel; it's compared with the Toyota Prius (best available) and the 2010 U.S. Corporate Average Fuel Economy (CAFE) standard. Best refrigerator is a Sunfrost R-19, designed for solar-powered homes and other applications demanding maximum efficiency.

Using less energy is the simplest, most straightforward, and often the most economical way to reduce climate-changing CO_2 emissions, not to mention many other deleterious environmental impacts. And the "resource" here is huge, thanks to the inefficiencies inherent in most of today's energy technologies. Even the most advanced technologies now in commercial use have energy efficiencies far below their theoretical maxima, and well below the efficiencies of the best prototype models. Figure 16.8 suggests the energy efficiency improvements available with two common technologies, namely vehicles and refrigerators.

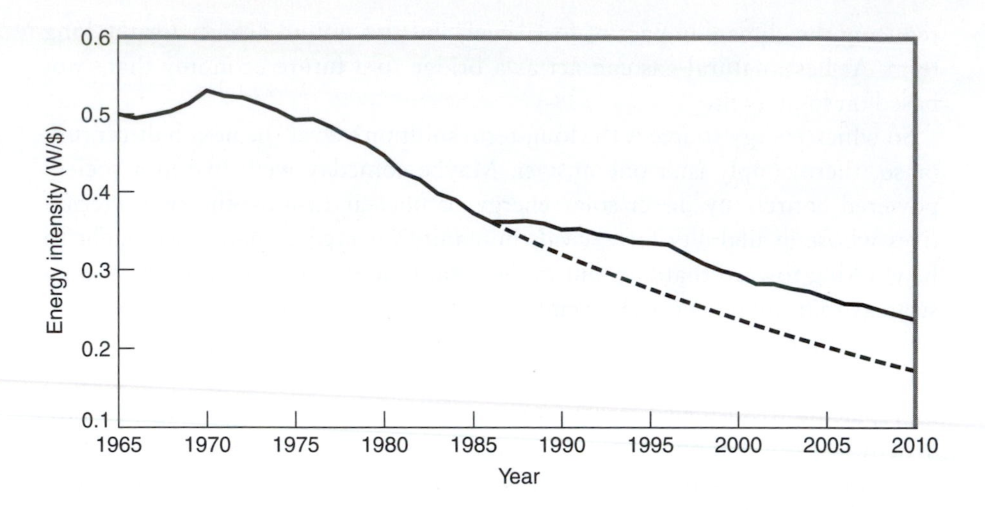

FIGURE 16.9
U.S. energy intensity (energy consumption per inflation-adjusted dollar of gross domestic product) since 1965. The actual intensity dropped by half between 1970 and 2004, but continuation of the steep drop during the late 1970s and early 1980s (dashed curve) would have resulted in a greater decline, which could have displaced more than two-thirds of our current oil imports.

Unlike some of the "greenest" alternative energies, which have yet to demonstrate that they're economically or technologically feasible on a large scale, energy efficiency is a proven way to reduce carbon emissions. We've done it before, and we continue—sometimes modestly, sometimes dramatically—to improve efficiency in some energy-use sectors. In Chapter 2 I introduced the notion of energy intensity, the ratio of a country's energy consumption to its gross domestic product (GDP). Figure 2.8 indicates almost steady reductions in U.S. energy intensity throughout the twentieth century, with a drop of nearly 50% since 1975. Many industrialized countries show similar trends. A reduction in energy intensity is like discovering an equivalent amount of oil, or natural gas, or other fuel—usually at a price far below what it would have cost to purchase the actual fuel.

It's particularly instructive to consider the times when we consciously tried to reduce energy consumption under pressure from market forces, shortages, or government incentives. Figure 16.9 focuses on the period since 1965 in the United States, revealing some 15 years of rapid decline in consumption as the energy shortages of the early 1970s drove Americans to smaller cars, better home insulation, more efficient appliances and heating systems, and even solar energy. I've fit an exponential curve to this data, suggesting that U.S. energy intensity was then decreasing annually at about 3%, and I've extrapolated that curve to show where we would be today had the trend continued. Although it may not look huge, the difference between the actual and extrapolated declines would be the energy equivalent of some 3 billion barrels of oil per year, more than two-thirds of our net annual imports—as you can verify in Exercise 6.

You might argue that we surely couldn't have continued those downward trends indefinitely, and you'd probably be right. But it wasn't technological barriers that slowed Figure 16.9's trend toward greater energy efficiency—it was a drop in the price of oil. And it wasn't technological failures that derailed the nascent solar-energy industry of the late 1970s and early 1980s—it was a conscious political decision to eliminate tax credits and research funding designed to grow the solar alternative. Had that not happened, we might now have a robust solar industry whose contribution to the U.S. energy mix could greatly exceed its pathetic one-tenth of one percent (recall Fig. 2.5). On a more positive note, California's progressive energy policies—including the renewable portfolio standards mentioned in Chapter 9—have resulted in that state's substantial deviation from the national rate of growth in per capita electricity consumption, as shown in Figure 16.10.

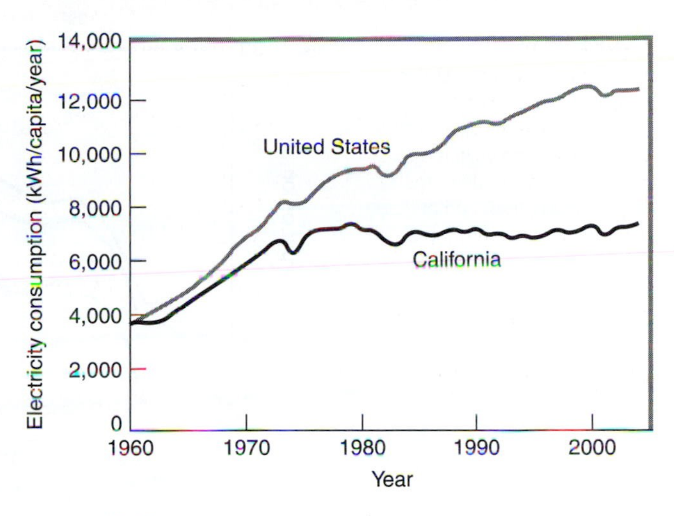

FIGURE 16.10

Per capita electrical energy consumption in California compared with the rest of the United States. The two were closely correlated in the 1960s, but subsequent policies encouraging energy efficiency leveled California's per capita electricity consumption while the rest of the country's consumption continued to grow. Curves show per capita consumption and don't account for actual consumption growth associated with population increases.

A look at specific energy uses shows how we can achieve substantial reductions in energy intensity or, equivalently, increases in efficiency. Refrigerators, for example, are among the greatest energy consumers in a typical home, after heating and cooling systems and hot-water heaters. Their energy intensity has fallen by nearly 80% since the 1970s. Figure 16.11 shows this trend and makes clear that the implementation of state and national standards has been a significant factor; you can explore the associated energy savings in Exercise 7. Airplanes are another example. Cost pressures have driven airlines to scrap older models for newer, more fuel-efficient ones; as a result, the U.S. airline fleet today uses only about one-third of the energy it did in 1970 per passenger per mile flown. Similarly, the most energy-efficient cars available have been getting better over time, although here the picture is more mixed as we—especially we in the United States—have chosen to drive larger cars and the so-called light trucks that include most SUVs (Fig. 16.12).

One caution here: Energy intensity is not a direct indicator of energy consumption. We can make refrigerators more efficient, but if we become so affluent that we put two or three refrigerators in each home, then we defeat those gains. We can make cars twice as efficient, but if we drive them twice as far then we've gained nothing. In fact, economists have found that that's what often happens: gains in efficiency are offset or even reversed by increased usage that follows the efficiency gains.

But even if we succeed in reducing per capita energy consumption, our total consumption may still increase with increasing population. For the entire period shown in Figure 16.9, in fact, U.S. per capita energy consumption did rise, by a modest 11%. The per capita GDP, in contrast,

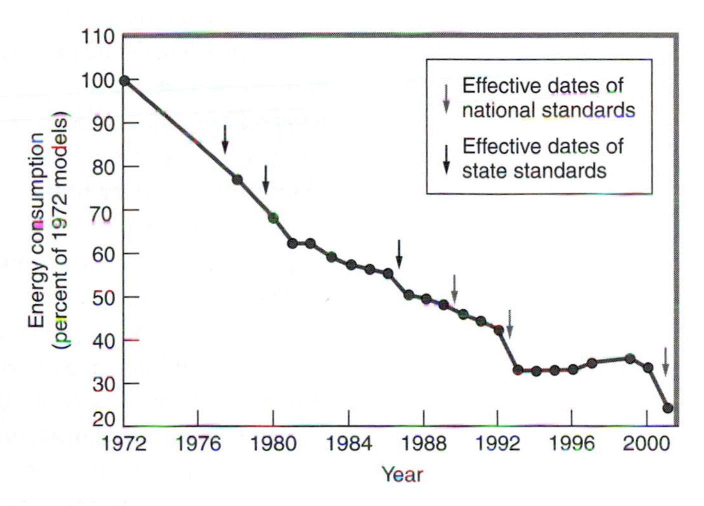

FIGURE 16.11

Energy consumption of household refrigerators in the United States has dropped dramatically in recent decades. Arrows mark the introduction of state and federal efficiency standards, which have played a significant role in decreasing energy consumption.

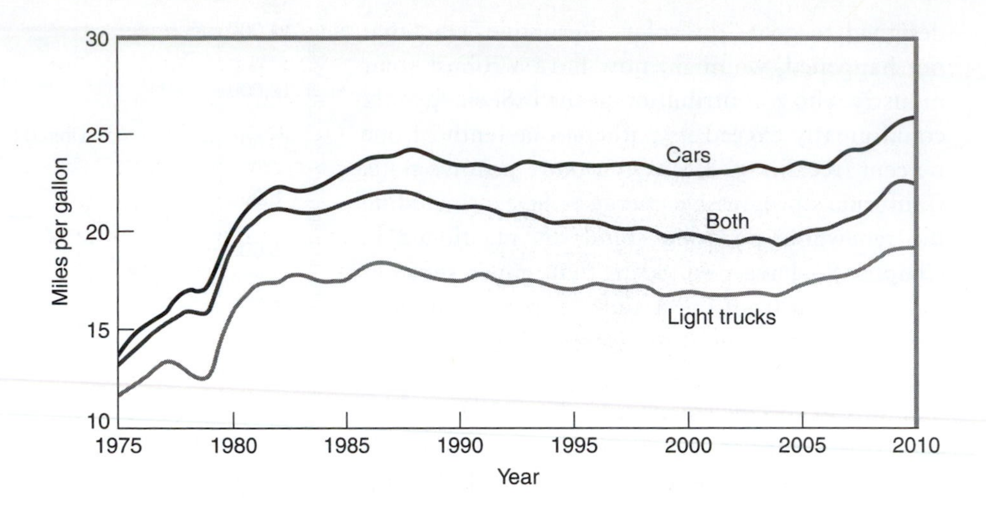

FIGURE 16.12
Fuel efficiency of the U.S. car and light-truck fleet. Light trucks include many SUV models. The trend toward larger cars and SUVs caused the decrease in combined mileage from about 1990 through 2004.

rose 128 percent, which is why our energy intensity declined significantly. Our overall energy consumption nevertheless increased substantially because the population increased, but it would have gone up a lot more if our energy efficiency hadn't improved.

COMPOUNDING EFFICIENCY GAINS

Energy efficiency has a way of compounding. Take home heating, for example: A homeowner achieves modest energy savings by installing insulating window shades and by caulking to eliminate air infiltration. This reduces the amount of natural gas or oil burned to heat the house, but it's still burned in the same furnace. Building a new home presents greater opportunities for savings—a lot more insulation goes into the structure, windows are double-glazed argon-filled units with substantial *R* values, and air infiltration is minimized. So the new house can use a smaller, less-expensive furnace—one that not only burns less fuel but that also took less material and energy to manufacture. The energy-efficient house maintains a more even temperature, so it needs less piping or ductwork to deliver heat throughout the house, which means more savings in costs and manufacturing energy. Insulate the house to an even greater extent, and you may be able to eliminate the heating system altogether, relying entirely on solar gain, the metabolism of the occupants (remember, that's about 100 W each), and the heat generated by appliances, lighting, computers, and other energy-consuming devices. Down, down, down go the home's energy consumption, the amount of energy and materials that go into making it, and the costs. You can make similar arguments for virtually every other energy-consuming product, from cars to airplanes to appliances to entire factories.

ECONOMICS OF ENERGY EFFICIENCY

Imagine that you're the owner of an electric utility faced with a 10% increase in power demand because of population growth, the trend toward larger homes, the proliferation of electrical gadgets, and the increasing "leakage" of power from supposedly "smart" electronic devices that consume energy even when they're turned off (see Exercise 8). So what do you do? You could crank up your generating facilities to maximum output, if they aren't already there. You could buy power from other utilities, but this solution is subject to the vagaries of the market, because your neighboring utilities, too, face increasing demand. You could ration your power through brownouts of individual towns or neighborhoods on a rotating basis—a draconian approach that's common in developing countries and a rare last resort in the developed world. You could build another power plant, at huge capital cost and with a long lead time before your new generators come online. Or you could pay your customers to use less electricity by providing them with compact fluorescent lamps that use only 20% as much energy as incandescents; giving them extra insulation for their water heaters, reducing so-called standby losses from the hot water stored in the tank; providing rebates for purchasing newer, more efficient appliances; or subsidizing individual wind and solar electric installations. If these energy-saving measures obviate the need to build that expensive new power plant, then you've saved money. You've also prevented more carbon emissions, or radioactive waste, or the disruption of a river ecosystem with dams.

The idea that reducing demand for a product can actually save money for its producer is a radical one; after all, it's unusual to find a manufacturer who doesn't want to increase sales. But it's an idea whose time has come, and one that benefits all parties involved: the power company, the consumer, and the environment.

The same arguments apply to other sectors of the energy industry. Take oil, for example: We've cut our energy intensity for oil in half since 1975, meaning that we now get twice as much GDP out of each barrel of oil. We can probably cut it in half again using currently available energy technologies. The cost of applying those technologies would be about $12 per barrel of oil saved, which is a huge bargain. As I write these words, the cost of oil stands at $106 per barrel. If you run a company that's looking to invest in a new piece of equipment, you'd happily pay the extra $12 for the more efficient version: This expenditure would save you $106 in energy costs, so you'd pocket the $94 difference.

What we're talking about here are *negawatts* and *negabarrels*—watts of power we don't need to generate and barrels of oil we don't need to burn. We can ascribe a cost to them just as we do for the "real thing." This cost is associated with buying more-efficient energy technologies rather than more energy. If the cost of greater efficiency is less than the cost to produce real energy, then it's more economical to buy the *negenergy*. And, because of the compounding effect described in the preceding section, moving from modest efficiency gains to super-efficient designs may actually lower the cost of negenergy.

ENERGY AND LIVING STANDARDS

Let me emphasize again that increased energy efficiency is a very different proposition than simply reducing our energy consumption. The former is a smart approach that buys us the same energy benefits at a lower cost; the latter requires that we give up some of the comforts provided by our energy consumption. But

BOX 16.2 | Negawatts: The Rocky Mountain Institute

The term *negawatt* is the brainchild of Amory Lovins, cofounder of the Rocky Mountain Institute (RMI) in Snowmass, Colorado. Lovins champions the view that dramatic gains in energy efficiency are both technologically possible and economically preferable to business as usual. Lovins and and his colleagues at RMI make their case in numerous well-documented publications, such as *Winning the Oil Endgame: Innovation for Profits, Jobs, and Security*. Such efficiency gains would not only buy us reduced greenhouse emissions and less conventional pollution, but also security from the economic, political, and even military upheavals that could accompany the final decades of a fossil-fueled economy running on empty.

RMI's physical facilities carry the philosophy of energy efficiency to its extreme. The 4,000-square-foot headquarters building is nestled into a south-facing slope high in Colorado's Rocky Mountains (Fig. 16.13), and it doesn't have a conventional heat source. Solar gain through advanced windows and the glazing of a built-in greenhouse provides all the energy needed. The glazing boasts R values as high as 8, while selective coatings let visible light through but block outgoing infrared to reduce losses even further. The building retains its solar-supplied energy with walls insulated to $R = 40$ and its roof to $R = 80$, and some 500 tons of concrete and soil provide a thermal mass that maintains an even temperature despite fluctuations in sunlight. The RMI building cost only about 1% more than a conventional structure, and its energy savings paid for that excess in a mere three years. What was state of the art when RMI was constructed in the 1980s is now widely available commercial technology. Lovins cites contemporary examples of RMI's philosophy applied to large office buildings that use 80% to 90% less energy than conventional structures while taking less time and money to build. For the transportation sector, RMI champions lightweight carbon-composite materials whose use could result in greatly increased vehicle fuel efficiency.

Are RMI's visions pie-in-the-sky dreams? Or are they grounded in a solid reality of physics, technology, and economics? RMI's emphasis on working with the market economy suggests a mix of reality and radical but perhaps achievable vision. RMI has already helped major corporations redesign their energy facilities, saving money in the process. Time will tell whether RMI's super-efficient designs become the standard. If they do, we'll have gone a long way toward breaking the link between energy and climate.

FIGURE 16.13
The headquarters building of the Rocky Mountain Institute has no internal heat source, despite its location in the cold climate of the Colorado Rockies. Technology that was cutting-edge when the building was constructed is now widely available.

might we do some of each? Could we have a comfortable, technologically advanced, economically prosperous, and culturally rich society that offers its citizens satisfying lives with, say, half the energy consumption we're now used to? Figure 16.14 suggests that, for North Americans, the answer is yes. We would hardly consider the countries of Europe to be technologically, economically, or culturally backward, yet Figure 16.14 shows that Europeans use only about half as much energy per capita as North Americans do and that their greenhouse emissions are comparably lower. There are, of course, some legitimate reasons why North Americans consume more energy. Some of these I discussed in Chapter 2, including geography and North America's role as a world food supplier. But others are simply a matter of societal choice. European countries have chosen high gasoline taxes to help curb consumption (recall Fig. 5.25). Even many smaller European cities have vibrant public transportation, and reliable, convenient rail systems link European communities. Zoning regulations encourage city-, town-, and village-based development over suburban and exurban sprawl. As a result of such tax, public-transportation, and community-development policies, Europeans have fewer cars per capita and their cars are smaller and more fuel efficient (see Research Problem 5).

In Chapter 2 you did deep knee bends to experience physically your body's typical 100-W power output, and I tied subsequent energy-consumption figures to that 100-W human equivalent. Figure 2.3, for example, shows that it would take 100 energy servants working around the clock to satisfy the average American's energy demand. You've now seen how greater energy efficiency, coupled with sensible public policies that discourage profligate energy consumption, might make that figure considerably lower. How much lower? And how does lower per capita energy use translate into future global energy consumption and greenhouse emissions?

Projections of future energy consumption range widely, as suggested in Figure 16.15. Energy expert Vaclav Smil, in his book *Energy at the Crossroads*, explains how simple assumptions bracket a fivefold range in energy projections for the year 2100. A conservative projection, in Smil's view, would be doubled

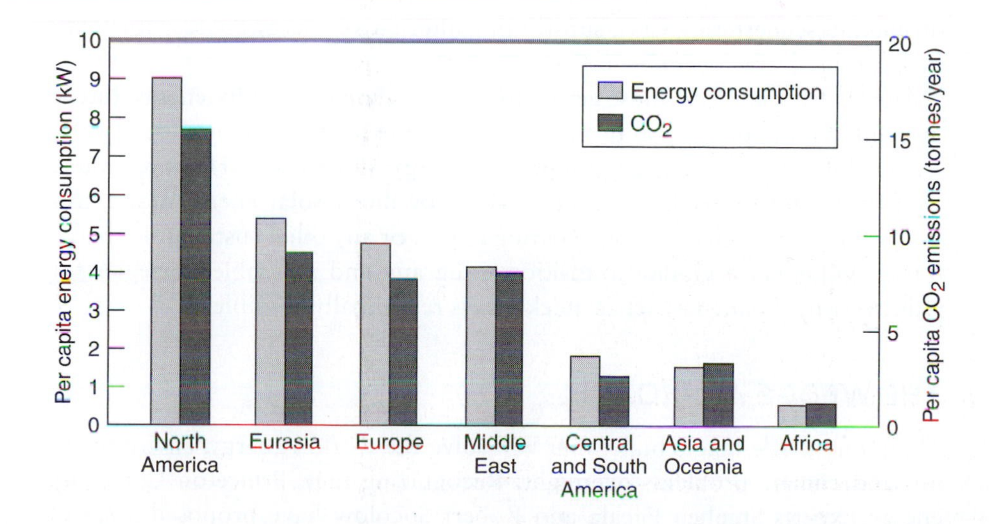

FIGURE 16.14
Per capita energy consumption (light gray) and CO_2 emissions (dark gray) for different world regions.

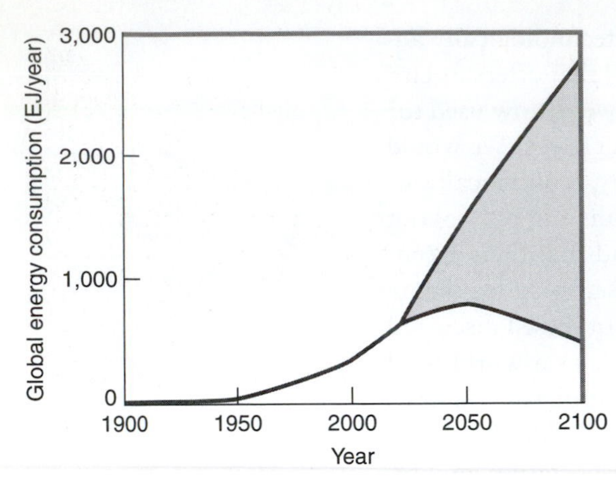

FIGURE 16.15
Projections of future world energy use show a fivefold uncertainty (shaded area), depending on assumptions made about population growth, energy efficiency gains, and per capita energy consumption.

energy efficiency (conservative because we did much better during the twentieth century) and a world population of 7 billion averaging today's per capita energy consumption. This scenario, he argues, leads to a global energy-consumption rate in 2100 that's just a little over what it is today, and it provides an energy-based living standard for everyone that's equivalent to what Italy enjoys today. That requires significant reductions in per capita consumption among the more profligate energy users, particularly the United States. But the potential for energy savings described earlier in this chapter suggests that it's certainly feasible to halve the number of energy servants working for each of us, and that with a concerted effort we could go a lot lower. But what's technically feasible may not be politically feasible, and the past few decades' history of both international and domestic attempts to limit energy consumption and carbon emissions shows that achieving what's possible may in fact be very difficult.

On the other hand, if per capita energy consumption continues the twentieth century's nearly fourfold increase, and if population is still climbing in 2100, then the world could be using some five times as much energy at the end of the century as it does today. Everything you know about available energy resources and their environmental and climatic impacts says this scenario is one to be avoided, if not actually impossible.

16.6 Strategy for a Sustainable Future

I started this chapter by asking what we need to do to stabilize atmospheric CO_2 at levels that will prevent "dangerous anthropogenic interference" with Earth's climate. The quick answer is that we need to reduce our CO_2 emissions, most of which come from burning fossil fuels. But how? We've now seen four distinct approaches: geoengineering, continuing to burn fossil fuels while capturing and sequestering CO_2, substituting alternative energy sources for fossil fuels, and reducing overall energy consumption.

Any realistic strategy must get us to a more climate-friendly energy future without massive disruption of the world economy or reliance on theoretically wonderful but technologically unproven energy alternatives. However much we might want to live in a world powered by direct solar energy alone, this isn't going to happen overnight. Getting to that or any other sustainable energy future will entail a gradual transition using any and all viable strategies for reducing our climate impact as quickly as is realistically possible.

THE WEDGE APPROACH

There's no single magic bullet that will solve the world's energy, environmental, and climate problems overnight. Recognizing that, Princeton University energy experts Stephen Pacala and Robert Socolow have proposed a set of interchangeable "wedges," each of which eliminates some of the projected increase in future greenhouse emissions. Their idea is to break a seemingly overwhelming problem into manageable pieces, and to provide enough such pieces to allow some choice in how we solve the problem.

Pacala and Socolow begin with a 50-year extrapolation of global anthropogenic carbon emissions, which projects an approximate doubling over that period. Their goal is to prevent that doubling by holding emissions to their current level for the next 50 years. As Figure 16.16 shows, the discrepancy between projected growth and the flat-path goal is a triangular region in the emissions-versus-time graph. The beauty of this idea is that it doesn't require a major upheaval at first, because the emissions we're trying to eliminate don't yet exist. That is, the difference between the growing emissions projection and the flat emissions goal starts out small and manageable. But over 50 years the flat path results in far less carbon in the atmosphere. Pacala and Socolow call the missing emissions region the *stabilization triangle*.

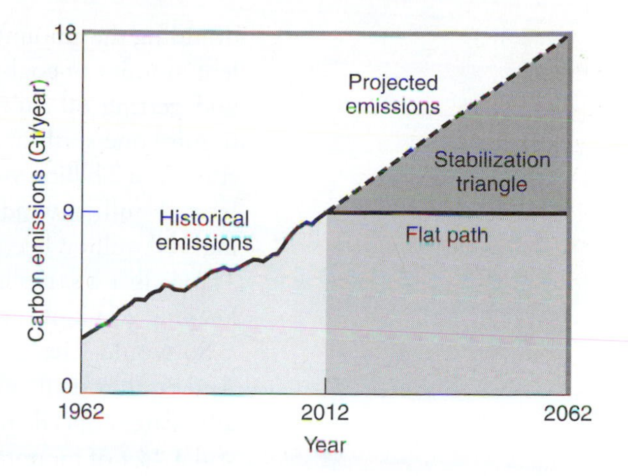

FIGURE 16.16
The stabilization triangle represents emissions to be avoided over 50 years if we're to stabilize atmospheric CO_2 at less than a doubling from its preindustrial concentration. The shaded areas represent total CO_2 emissions (in gigatonnes of carbon).

Global anthropogenic CO_2 emissions from fossil fuel combustion are currently about 9 Gt of carbon per year, so the projected 50-year doubling would add another 9 Gt per year. Socolow and Pacala would prevent these additional emissions with "wedges," each of which would eliminate 1 Gt per year of carbon emissions by 50 years from now (Fig. 16.17). But again, the wedges' initial job is more modest; in the first year, each wedge has to displace a relatively small amount of carbon emissions. But over 50 years, each wedge will have kept some 25 Gt of carbon out of the atmosphere. Pacala and Socolow made their proposal in 2004, when annual emissions were 7 Gt; therefore they needed 7 wedges. Today's emissions are close to 9 Gt annually, so I've updated to 9 wedges. You can explore the Pacala–Socolow idea quantitatively in Exercise 9.

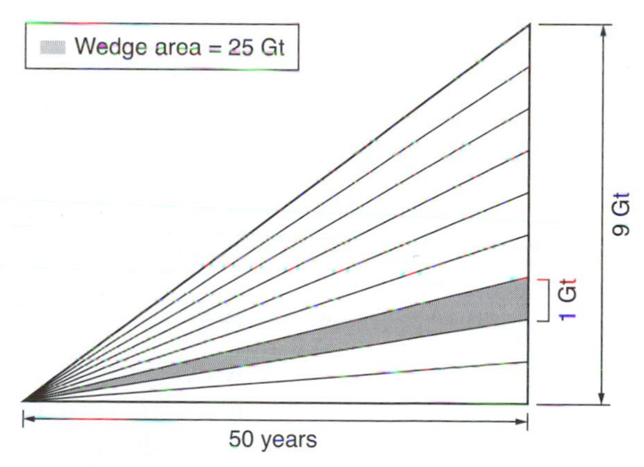

FIGURE 16.17
Stabilization can be achieved with nine wedges, each representing a different energy strategy, and each of which grows to the point where it eliminates 1 Gt of carbon emissions per year 50 years from now. Over 50 years, each wedge keeps 25 Gt of carbon out of the atmosphere.

So what are the wedges? Pacala and Socolow identify more than the 9 wedges required, giving a range of choices. A sampling of their proposed wedges includes energy efficiency; renewable energy sources; nuclear energy; substituting natural gas for coal; carbon capture and sequestration; biofuels and carbon-free hydrogen for transportation; afforestation (planting trees to remove atmospheric CO_2); and better management of methane from agriculture, landfills, and natural gas operations. They quantify each wedge, showing the magnitude of the effort involved after 50 years. If we choose to include a CCS wedge, for example, we would need to have CCS installed at 800 1-GW coal-fired power plants by the end of the 50-year period. If we choose a wind-power wedge, we'd have to build 2 million 1-MW wind turbines over 50 years to replace coal power. A solar photovoltaic wedge would mean 2,000 GW of peak photovoltaic power in 50 years, requiring globally about 2 million hectares of land—roughly the area of New Jersey. A nuclear wedge would have us doubling the amount of nuclear-generated electricity and displacing an equivalent amount of coal-generated electricity. Two billion cars running on biofuels and getting 60 miles per gallon would provide another wedge, and would require one-sixth of the world's cropland for biofuel production. Or we could run those 2 billion vehicles on hydrogen produced from fossil fuels with CCS or from 4 million wind turbines. Stopping tropical deforestation and rehabilitating 300 million hectares of tropical forest would give us a forest-based wedge. This is just a sampling of the list of well-documented choices, each capable of keeping 25 Gt of carbon out of the atmosphere over the next 50 years.

So would Pacala and Socolow's wedges solve our energy-climate problem? And do they really make the problem manageable? On the second point, wedge advocates concede that their approach breaks a "heroic challenge" into what's still a "set of monumental tasks." But those tasks are achievable: For example, although it may seem daunting to manufacture 4 million wind turbines or a New Jersey's worth of solar panels over the next 50 years, consider that the world today produces nearly 100 million motor vehicles *each year*. We as a species already have the tools needed for each wedge, and there are enough possible wedges that we can pick and choose among them. Each wedge represents the same interchangeable unit, namely 25 Gt of eliminated carbon, which makes it straightforward to compare their relative costs, risks, and environmental impacts.

The wedge approach stabilizes carbon *emissions*, but does it achieve the UN goal of "stabilization of greenhouse gas *concentrations* in the atmosphere at a level that would prevent dangerous anthropogenic interference with the climate system" (my italics)? Not in itself: Fifty years of CO_2 emissions at today's levels will still bring us to nearly 500 ppm of atmospheric CO_2—above what's now regarded as "safe." At that point we'd have to begin not only reducing emissions but deploying technologies to remove CO_2 from the atmosphere. But if we don't implement something like the wedge approach, and do it soon, then we'll face an even more difficult emissions-reduction challenge and CO_2 levels far above the danger point.

PARTING WORDS

Here, at the end of what has been a mostly scientific textbook, I'll wrap up with some personal thoughts. I've stressed from the beginning the huge magnitude of humankind's energy consumption, and I've tried throughout to get you thinking quantitatively about energy and its environmental impacts. You can now talk fluently of gigawatts, exajoules, and quads with a feel for those energy units' colossal size. You can also quantify conventional pollutants, and you know how gigatonnes of anthropogenic carbon emissions alter the natural carbon cycle. You can scrutinize an alternative energy proposal and decide whether it's quantitatively and technologically realistic. In the end, though, all of these energy numbers come down to one quantitative fact: The magnitude of humanity's energy enterprise has become large enough to have a global impact on our planet. Reducing that impact and mitigating its effects will require people who, like you, can undertake a serious quantitative analysis of the problem and its potential solutions. But it also will take visionaries with new technological, economic, and social ideas that can reconcile energy-consuming humanity with a healthy planet. And it will need leaders, activists, and policymakers with the optimism, courage, and confidence to guide us through what, by any measure, will be some challenging decades. I hope you'll play a part.

CHAPTER REVIEW

BIG IDEAS

16.1 The goal of the **United Nations Framework Convention on Climate Change** is "stabilization of greenhouse gas concentrations in the atmosphere at a level that would prevent dangerous anthropogenic interference with the climate system." This goal probably means limiting the industrial-era temperature rise to about 2°C, and atmospheric CO_2 concentration to 450 ppm and possibly much less. We've considered four possible paths to that goal.

16.2 **Geoengineering** would alter Earth's energy balance, either by reducing incoming sunlight or by removing atmospheric carbon dioxide. Geoengineering schemes are unproven and involve unknown risks, and most should be regarded as last resorts.

16.3 **Carbon capture and sequestration** captures the carbon from fossil fuels, either before or after combustion, and sequesters it underground or in the deep ocean. Its large-scale technological or economic viability has not yet been demonstrated, but if proven it would allow continued use of fossil fuels in stationary sources with reduced climate impact.

16.4 The alternative energy sources described in Chapters 7 to 11 offer the prospect of greatly reduced—although not zero—carbon emissions.

16.5 Using energy more efficiently results in less energy consumption and lower greenhouse emissions without reducing the benefits that energy provides.

16.6 Achieving a sustainable future with a stable climate will require using many of the different approaches available for reducing greenhouse emissions.

TERMS TO KNOW

carbon capture and sequestration (CCS) (p. 418)
geoengineering (p. 415)
United Nations Framework Convention on Climate Change (UNFCCC) (p. 441)

GETTING QUANTITATIVE

Estimated maximum industrial-era temperature rise to avoid "dangerous anthropogenic interference with the climate system": 2°C; corresponding atmospheric CO_2 concentration: 350–450 ppm

CCS requirement for typical coal-fired power plant: 1,000 tons of CO_2 every hour

Estimated range of carbon emissions from different energy sources: high estimate = 755 g/kWh (coal); low estimate = 4 g/kWh (hydro)

QUESTIONS

1 Why isn't carbon capture a straightforward problem in pollution control—as is, for example, the removal of sulfur emissions from flue gases?

2 The oceans already hold nearly 40,000 Gt of carbon. So why do we worry about the effects of adding a little more in the form of CO_2, as we might do if we decided to capture CO_2 and sequester it in the oceans?

3 Why aren't energy alternatives such as solar photovoltaics and nuclear power totally free of greenhouse emissions?

4 Why might the "wedge" approach be easier to implement than some other strategies for stabilizing atmospheric CO_2?

5 Explain how Figure 16.3 shows that Earth would experience additional warming even if we stopped greenhouse gas emissions today.

6 What would *you* consider to be "dangerous anthropogenic interference with the climate system"?

EXERCISES

1 (a) Estimate the volume occupied by the 1,000 tons of CO_2 produced each hour in a 1-GWe coal-fired power plant, assuming it's at approximately normal atmospheric temperature and pressure. Under these conditions, 1 mole of any gas occupies 22.4 liters. (b) Use your answer to determine the volume occupied by the power plant's yearly CO_2 production, expressed in cubic kilometers.

2 If you drive a car getting 22 miles per gallon 11,000 miles per year, how much CO_2 do you produce in a year? See Exercise 1 in Chapter 6.

3 Section 16.3 mentions a pilot project that's sequestered 5 million tons of CO_2 in a Canadian oil field. How much time would it take a 1-GWe coal-fired power plant to produce that much CO_2?

4 The pilot sequestration project mentioned in Section 16.3 increased the oil field's output by 10,000 barrels per day. How much oil-fired electric power could that excess generate, assuming 40% efficiency?

5 How long could we inject CO_2 from one thousand 1-GWe power plants into the deep ocean before we increased the stored carbon (now at 39,000 Gt) by 1%? (That 39,000 Gt is *carbon*, and the emissions from a 1-GWe power plant is given in this chapter in tons of *carbon dioxide*.)

6 Assuming a 100-EJ annual U.S. energy consumption, and a 2010 GDP of $13 trillion, show that the difference between actual and extrapolated energy intensities at the end of the curve in Figure 16.9 is the energy equivalent

of between 4 and 5 billion barrels of oil annually, and compare this quantity with our annual oil imports.

7 Figures 16.8 and 16.11 reflect dramatic gains in the efficiency of U.S. household refrigerators since the 1970s. In 1973 the average refrigerator used 1,800 kWh of electricity per year; in 2010 the average was 430 kWh. (a) Estimate the total yearly energy savings in the United States today that is associated with this drop in refrigerator energy consumption, and estimate how many large electric power plants this improvement in efficiency displaced.

8 Today's "smart" electronic devices and appliances typically consume about 5 W continuously, even when they're off. Suppose the typical household has ten such devices. Estimate the total of this standby power consumption in the United States. How many large power plants are at work just to supply this standby power?

9 Estimate the total carbon emissions associated with each of the three shaded regions in Figure 16.16—that is, the historical emissions from 1962 to 2012, the emissions from 2012 to 2062 under the flat-path scenario, and the emissions associated with the stabilization triangle.

10 A 20-W compact fluorescent lamp replaces a 100-W incandescent bulb. How much energy does it save over its 10,000-hour lifetime? At 10¢ per kilowatt-hour, how much money does it save?

11 (a) Using data found elsewhere in this book, determine what forcing from stratospheric SO_2 injection would be needed to counter all industrial-era forcings. (b) Would the eventual result be a stable climate, continued warming, or global cooling?

RESEARCH PROBLEMS

1 Consult the Mauna Loa CO_2 data used for Figure 13.7, and make a table or graph of the annual *change* in atmospheric CO_2 concentration since 1960.

2 Has California continued to hold its per capita electrical energy consumption steady beyond the time shown in Figure 16.10? Find the data to continue the graph as far as you can.

3 What has happened to the U.S. vehicle fleet's average fuel efficiency since the last data point shown in Figure 16.12? Continue this graph as far as you can.

4 Find data to plot either U.S. or world oil consumption from approximately 1970 to the present. Identify any periods of declining consumption that last a few years or more. Use a spreadsheet's trendline or curve-fitting capabilities to fit an exponential decay, and extrapolate to find what consumption would be today if consumption had continued to decline exponentially.

5 Find data and prepare a graph showing the number of motor vehicles per 1,000 people in about 10 countries. Include the United States, at least one country in Western Europe, and at least one developing country.

ARGUE YOUR CASE

1 Your friend is involved with 350.org, a worldwide organization of climate activists advocating 350 ppm as a CO_2-concentration goal. But your friend can't articulate forcefully why that's a desirable goal. Formulate some scientific arguments that will help your friend.

2 Formulate your own scenario, and sketch (don't calculate!) graphs analogous to Figures 15.9 and 15.10, for likely CO_2 concentrations and emissions through the year 2100. Argue for some policies that could bring about your scenario.

APPENDIX

//

PROPERTIES OF MATERIALS

TABLE A.1 | AIR AND WATER

Material	Density (kg/m³)	Specific heat (kJ/kg·K)	Melting point (K)	Boiling point (K)	Heat of fusion (kJ/kg)	Heat of vaporization (kJ/kg)
Air	1.2	1.01	O_2: 54.8 N_2: 63.3	O_2: 90.2 N_2: 77.4		
Water	Liquid: 1,000 Ice: 917	Liquid: 4.184 Ice: 2.05	273 (0°C)	373 (100°C)	334	2,257

TABLE A.2 | THERMAL CONDUCTIVITIES OF SELECTED MATERIALS

Material	Thermal conductivity, (W/m·K)	Thermal conductivity, (Btu·inch/hour·ft²·°F)
Air	0.026	0.18
Aluminum	237	1,644
Concrete (typical)	1	7
Fiberglass	0.042	0.29
Glass (typical)	0.8	5.5
Rock (granite)	3.37	23.4
Steel	46	319
Styrofoam (extruded polystyrene foam)	0.029	0.2
Water	0.61	4.2
Wood (pine)	0.11	0.78
Urethane foam	0.22	0.15

TABLE A.3 | *R* VALUES ($ft^2 \cdot °F \cdot h/Btu$) OF SOME COMMON BUILDING MATERIALS

Material	*R* value
Air layer:	
Adjacent to inside wall	0.68
Adjacent to outside wall, no wind	0.17
Concrete, 8-inch	1.1
Fiberglass:	
3.5-inch (fits "4-inch" lumber)	12
5.5-inch (fits "6-inch" lumber)	19
Glass, 1/8-inch single pane	0.023
Gypsum board (Sheetrock), 1/2-inch	0.45
Polystyrene foam, 1-inch	5
Urethane foam, 1-inch	6.6
Window (*R* values include adjacent air layer):	
Single-glazed wood	0.9
Standard double-glazed wood	2.0
Argon-filled double-glazed with low-E coating	2.9
Argon-filled triple-glazed with low-E coating	5.5
Best commercially available windows	11.1
Wood:	
1/2-inch cedar	0.68
3/4-inch oak	0.74
1/2-inch plywood	0.63
3/4-inch white pine	0.96

TABLE A.4 | SPECIFIC HEATS (J/kg·K) OF SOME COMMON MATERIALS

Material	Specific heat
Aluminum	900
Concrete	880
Glass	753
Steel	502
Stone (granite)	840
Water:	
Liquid	4,184
Ice	2,050
Wood	1,400

TABLE A.5 | GLOBAL WARMING POTENTIALS

Gas	Atmospheric lifetime (years)	Global warming potential relative to CO_2 — Time frame: 20 years	100 years	500 years
Carbon dioxide (CO_2)	~1,000*	1	1	1
Methane (CH_4)	11	67	23	6.9
Nitrous oxide (N_2O)	114	291	298	153
CFC-11 (CCl_3F)	45	6,700	4,750	1,620
HCFC-22 ($CHClF_2$)	12	5,200	1,800	550
HFC-23 (CHF_3)	270	12,000	14,800	12,200

*The lifetime of CO_2 is ambiguous; see Section 13.5.

TABLE A.6 | SOME IMPORTANT RADIOACTIVE ISOTOPES

Isotope	Half-life (approximate)
Carbon-14	5,730 years
Iodine-131	8 days
Potassium-40	1.25 billion years
Plutonium-239	24,000 years
Radon-222	3.8 days
Cesium-137	30 years
Strontium-90	29 years
Tritium (hydrogen-3)	12 years
Uranium-235	704 million years
Uranium-238	4.5 billion years

Periodic Table of the Elements

Key:
- 1 — Atomic number
- H — Symbol
- Hydrogen — Name
- 1.00794 — Average atomic mass (for elements with no stable or long-lived isotopes, number in brackets is mass number of most stable isotope).

Metals | Metalloids | Nonmetals

1 H Hydrogen 1.00794																	2 He Helium 4.002602
3 Li Lithium 6.941	4 Be Beryllium 9.012182											5 B Boron 10.811	6 C Carbon 12.0107	7 N Nitrogen 14.0067	8 O Oxygen 15.994	9 F Fluorine 18.9984032	10 Ne Neon 20.1797
11 Na Sodium 22.98976928	12 Mg Magnesium 24.3050											13 Al Aluminum 26.9815386	14 Si Silicon 28.0855	15 P Phosphorus 30.973762	16 S Sulfur 32.065	17 Cl Chlorine 35.453	18 Ar Argon 39.948
19 K Potassium 39.0983	20 Ca Calcium 40.078	21 Sc Scandium 44.955912	22 Ti Titanium 47.867	23 V Vanadium 50.9415	24 Cr Chromium 51.9961	25 Mn Manganese 54.938045	26 Fe Iron 55.845	27 Co Cobalt 58.933195	28 Ni Nickel 58.6934	29 Cu Copper 63.546	30 Zn Zinc 65.38	31 Ga Gallium 69.723	32 Ge Germanium 72.64	33 As Arsenic 74.92160	34 Se Selenium 78.96	35 Br Bromine 79.904	36 Kr Krypton 83.798
37 Rb Rubidium 85.4678	38 Sr Strontium 87.62	39 Y Yttrium 88.90585	40 Zr Zirconium 91.224	41 Nb Niobium 92.90638	42 Mo Molybdenum 95.96	43 Tc Technetium [98]	44 Ru Ruthenium 101.07	45 Rh Rhodium 102.90550	46 Pd Palladium 106.42	47 Ag Silver 107.8682	48 Cd Cadmium 112.411	49 In Indium 114.818	50 Sn Tin 118.710	51 Sb Antimony 121.760	52 Te Tellurium 127.60	53 I Iodine 126.90447	54 Xe Xenon 131.293
55 Cs Cesium 132.9054519	56 Ba Barium 137.327	57 La Lanthanum 138.90547	72 Hf Hafnium 178.49	73 Ta Tantalum 180.94788	74 W Tungsten 183.84	75 Re Rhenium 186.207	76 Os Osmium 190.23	77 Ir Iridium 192.217	78 Pt Platinum 195.084	79 Au Gold 196.966569	80 Hg Mercury 200.59	81 Tl Thallium 204.3833	82 Pb Lead 207.2	83 Bi Bismuth 208.98040	84 Po Polonium [209]	85 At Astatine [210]	86 Rn Radon [222]
87 Fr Francium [223]	88 Ra Radium [226]	89 Ac Actinium [227]	104 Rf Rutherfordium [261]	105 Db Dubnium [262]	106 Sg Seaborgium [266]	107 Bh Bohrium [264]	108 Hs Hassium [277]	109 Mt Meitnerium [268]	110 Ds Darmstadtium [271]	111 Rg Roentgenium [272]	112 Cn Copernicium [285]	113 [284]	114 [289]	115 [288]	116 [292]	117 [294]	118 [294]

Lanthanides	58 Ce Cerium 140.116	56 Pr Praseodymium 140.90765	60 Nd Neodymium 144.242	61 Pm Promethium [145]	62 Sm Samarium 150.36	63 Eu Europium 151.964	64 Gd Gadolinium 157.25	65 Tb Terbium 158.92535	66 Dy Dysprosium 162.500	67 Ho Holmium 164.93032	68 Er Erbium 167.259	69 Tm Thulium 168.93421	70 Yb Ytterbium 173.05	71 Lu Lutetium 174.967
Actinides	90 Th Thorium 232.03806	91 Pa Protactinium 231.03588	92 U Uranium 238..02891	93 Np Neptunium [237]	94 Pu Plutonium [244]	95 Am Americium [243]	96 Cm Curium [247]	97 Bk Berkelium [247]	98 Cf Californium [251]	99 Es Einsteinium [252]	100 Fm Fermium [257]	101 Md Mendelevium [258]	102 No Nobelium [259]	103 Lr Lawrencium [262]

Elements 113–118 were synthesized so recently that they don't yet have names. They're so short-lived that their chemical properties haven't been firmly established.

GLOSSARY

//

absolute zero the lowest possible temperature, corresponding to a system's state of minimum energy

absorber plate the black-coated element in a solar collector that absorbs solar radiation

acid mine drainage seepage of acidic water from coal mines into surface streams

acid rain precipitation of higher than normal acidity, resulting from the reaction of sulfur dioxide pollution with water

active solar heating a solar heating system that uses active devices such as pumps or fans to move heated air or liquid

aerobic respiration respiration that uses oxygen and produces carbon dioxide

albedo a number between 0 and 1 representing the fraction of incident sunlight that a planet reflects back to space

alpha decay radioactive decay by emission of an alpha particle; occurs in nuclei that have excess protons

alpha particle a high-energy helium-4 nucleus (${}^{4}_{2}H$), emitted when a radioactive nucleus undergoes alpha decay

amorphous silicon silicon that lacks a crystal structure; it is inexpensive to produce, but solar photovoltaic cells made from amorphous silicon are less efficient than those made from crystalline silicon

ampere the SI unit of electric current, equal to 1 coulomb per second

anaerobic respiration respiration in the absence of oxygen, often producing methane

anthropogenic of human origin

atmospheric lifetime the time that a molecule or particle remains in the atmosphere before being removed by physical or chemical processes

atomic nucleus the inner core of the atom, consisting of protons and neutrons and containing nearly all the atomic mass

atomic number the number of protons in a nucleus; determines the chemical element

background radiation the level of radiation from both natural and anthropogenic sources to which a population is typically exposed

baghouse a chamber filled with fabric bags used for filtering particulates from an exhaust stream

barrel of oil equivalent an energy unit equal to 6.12 GJ; the energy contained in one 42-gallon barrel of oil

baseload the average minimum electrical energy use in a given region; baseload power plants such as large nuclear and coal-fired facilities generally run continually to supply this load

battery a device that uses chemical reactions to separate electric charge and thus supply electrical energy

becquerel the SI unit of radioactivity, equal to 1 decay per second

beta decay radioactive decay by emission of a beta particle; occurs in nuclei that have excess neutrons

beta particle a high-energy electron, emitted when a radioactive nucleus undergoes beta decay

binary system a geothermal power system that uses geothermal fluids to heat a working fluid, which in turn boils to vapor and drives a turbine

binding energy the energy needed to disassemble a nucleus; equivalently, the energy released when the nucleus forms from individual nucleons; usually expressed as binding energy per nucleon

biodiesel a biofuel made from plant oils that can be used in diesel engines

biofuel a fuel made from biomass; examples include ethanol, biodiesel, and methane

biological pump the process whereby dead organisms and organic waste sink into the deep ocean thus removing carbon from the surface waters

biomass matter derived from living things and useful as fuel; dry biomass has an energy content of typically 15 to 20 MJ/kg

boiling-water reactor (BWR) a light-water reactor in which water boils within the reactor vessel and the resulting steam drives a turbine

borehole a hole drilled into the ground to produce a profile of temperature with depth, which is useful for assessing climate history

breeder reactor a reactor designed to breed plutonium-239 from uranium-238, thus producing more fissile fuel than it consumes

British thermal unit an energy unit equal to 1,054 joules; the energy needed to raise 1 pound of water by 1°F

brownout a reduction in voltage on the power grid, resulting in less power to end users

calorie an energy unit equal to 4.184 joules; the energy needed to raise the temperature of 1 gram of water by 1°C

carbon capture and sequestration (CCS) the process of capturing carbon dioxide and sequestering it underground or in the deep ocean to prevent fossil carbon emissions from adding to atmospheric greenhouse gases

carbon cycle the system of flows and reservoirs whereby carbon moves through the Earth-ocean-atmosphere-biosphere system

carbon neutral a system that produces zero net carbon emissions

carbon sink any system that removes carbon from the atmosphere

Carnot engine a heat engine that provides the theoretical maximum efficiency

catalytic converter a device that uses a catalyst to accelerate chemical reactions that render pollutants less harmful; for example, catalytic converters in cars convert nitrogen oxides into nitrogen and oxygen; carbon monoxide into carbon dioxide; and unburned hydrocarbons into carbon dioxide and water

chain reaction a nuclear fission reaction wherein neutrons released in one fission go on to cause additional fissions

chemical energy energy stored in the chemical bonds that join atoms into molecules

chlorofluorocarbon (CFC) a class of organic compounds that include fluorine and chlorine

Clean Air Act a law, first enacted in 1963 and subsequently amended substantially, that has been instrumental in improving air quality in the United States

climate the average conditions prevailing in Earth's atmosphere

climate forcing an upset of Earth's energy balance, generally considered to occur at the tropopause; generally leads to a temperature change

climate sensitivity a measure of the response of the climate system to a change in forcing; expressed either as a temperature change for a doubling of atmospheric carbon dioxide or as a temperature change per watt per square meter of forcing

closed-loop system a geothermal energy system in which geothermal fluids heat a secondary working fluid that is confined to a closed loop

coal a black, solid fossil fuel high in carbon

coefficient of performance (COP) in a heat pump, the ratio of heat delivered to high-quality energy required to run the pump

cogeneration the process whereby fuel energy is used to produce both electricity and heat for industrial use or space heating, thus putting to use the waste heat required by the second law of thermodynamics

combined-cycle power plant a power plant using gas turbines whose waste heat then drives a conventional steam cycle, resulting in high overall efficiency

combined heat and power (CHP) a system that produces both electric power and heat for industrial processes or space heating

compound parabolic concentrator a solar concentrator that boosts light intensity without focusing the light to a point or line

compression-ignition engine an intermittent-combustion engine in which fuel ignites as the fuel–air mixture in the cylinder is heated by compression to ignition temperature, as in diesel engines

concentration ratio the ratio by which a solar concentrator boosts the light intensity

condenser in a steam power plant, the device in which steam is recondensed to water

conduction heat flow by physical contact, when energy is transferred by collisions among molecules

continuous combustion engine a heat engine in which combustion takes place continuously, as in a power plant, in contrast to the intermittent combustion in gasoline and diesel engines

control rods neutron-absorbing rods that move in and out of a reactor to control the chain reaction

convection heat transfer by the bulk motion of a fluid

coolant a material used to carry heat away from the core of a nuclear reactor or other heat engine

cooling tower a structure used in power plants to cool the water that flows through the plant's condenser; extracts waste heat and discharges it to the atmosphere

cornucopian one who believes that human ingenuity will alleviate the problem of finite resources through new discoveries and technologies

crankshaft a rotating shaft in an engine that converts the back-and-forth motion of pistons into rotary motion

criteria pollutant one of six substances—carbon monoxide, lead, nitrogen dioxide, ozone, sulfur oxides, and two classes of particulates—for which the Clean Air Act sets maximum concentrations in ambient air

critical mass the amount of nuclear material needed for a self-sustaining chain reaction

crude oil oil as it comes from the ground, before refining or other processing

curie an older unit of radioactivity, equal to 37 billion decays per second, and approximately the radioactivity of 1 gram of radium-226

curve of binding energy a graph of binding energy per nucleon versus mass number; the curve peaks near mass number 56 (iron), indicating that nuclear energy release is possible by fusion of nuclei lighter than this mass and by fission of heavier nuclei

cyclone a device that swirls exhaust gases in a high-speed vortex to extract particulate pollution

demand charge a cost that electric utilities charge institutional users based on their peak power demand

deuterium hydrogen-2 ($^{2}_{1}H$), the stable heavy hydrogen isotope whose nucleus contains 1 proton and 1 neutron

diffuse insolation solar energy that has been scattered in the atmosphere before reaching Earth's surface

direct insolation solar energy coming directly from the Sun

dry cooling tower structure designed to transfer waste heat from a power plant's cooling water to the atmosphere; dry towers keep the water in a closed system where it can't evaporate

D–T fusion nuclear fusion involving the hydrogen isotopes deuterium and tritium

Earth system model (ESM) a complex computer model that includes not only atmospheric and ocean models, but also other aspects of the Earth system, including chemical and biological processes

elastic potential energy energy stored in the altered configuration of elastic materials such as rubber bands or springs

electric charge a fundamental property of matter that comes in two kinds (positive and negative) and that gives rise to electric fields

electric current a flow of electric charge

electric field the influence of electric charge on the space around it, giving rise to electric forces on other charges; also arises from a changing magnetic field

electric power grid the interconnected system of electric power plants, transmission lines and related infrastructure, and end users

electric vehicle a car or other vehicle powered entirely by electrical energy

electrolysis the use of electric current to dissociate water into hydrogen and oxygen

electromagnetic force a force associated with electric charge (electric force) or moving electric charge (magnetic force)

electromagnetic induction a fundamental phenomenon whereby a changing magnetic field produces an electric field; the basis of electric generators and essential for the existence of electromagnetic waves

electromagnetic radiation a flow of electromagnetic energy in the form of electromagnetic waves

electromagnetic wave a wave of self-regenerating electric and magnetic fields that propagates through empty space, carrying electromagnetic energy; light is one example of an electromagnetic wave

electrostatic precipitator a device that uses a strong electric field to extract particulates from an exhaust stream

emissivity a number between 0 and 1 that gives the efficiency with which a material radiates electromagnetic waves; equal to the efficiency of absorption at a given wavelength

end use the final use to which energy is being put

end-use energy energy delivered to end users, as distinguished from primary energy

end user an individual, institution, or other agent who uses energy supplied from a utility or other source

energy one of the two basic "substances" that make up the universe (the other is matter), energy is associated with motion, in stored form with forces, and with what we call, loosely, "heat"

energy balance a state in which a system maintains a constant temperature because its energy input balances its energy loss

energy carrier a substance, such as hydrogen (H_2), which contains energy but isn't available naturally as an energy resource

energy density the energy per unit volume at any point in space

energy flow a movement of energy from one place to another, as in the flows of solar and geothermal energy that bring energy to Earth's surface

energy intensity a measure of energy consumption per unit of gross domestic product

energy payback time the time it takes a system to produce as much energy as was required to construct it

enrichment the process of enhancing the fraction of fissile uranium-235 in nuclear fuel to levels above its natural concentration of 0.7%

entropy a measure of disorder; systems with higher entropy have lower energy quality

equilibrium model a climate model that determines the equilibrium climate after all changes have stopped occurring

ethanol ethyl alcohol, used as a biofuel and generally produced by fermentation of plant matter

evacuated tube collector a solar collector for hot-water systems that utilizes a heat transfer fluid insulated from its surroundings by vacuum

evapotranspiration the two processes of evaporation and plant transpiration, both of which emit water vapor into the atmosphere

external combustion engine a heat engine in which combustion takes place outside the cylinders or other chambers holding the engine's working fluid; an example is a steam power plant

feedback effect an additional effect that follows as the result of some change in a system; a feedback is positive if it tends to enhance the original change and negative if it tends to diminish it

feed-in tariff (FIT) an arrangement whereby utilities agree to pay a fixed rate, generally above prevailing market rates, for energy from renewable sources

filter a simple device for removing particulates via filtration

first law of thermodynamics a statement of energy conservation; the change in a system's internal energy is the sum of the work done on the system and the heat that flows into it

fissile capable of fission when struck by low-energy neutrons; uranium-235 and plutonium-239 are the most important fissile isotopes

fission an energy-releasing nuclear reaction in which a heavy nucleus splits into two smaller nuclei; the source of energy in nuclear power reactors

fission product a substance produced as a result of nuclear fission; typical fission products span a range of middleweight nuclei and are highly radioactive

fissionable capable of fission when struck by neutrons; the required neutron energy is high except in the subcategory of fissile isotopes

flat-plate collector a solar collector with a flat absorber plate

flue gas the mix of gaseous and particulate effluent that emerges from the combustion chamber of a fossil- or biomass-fueled power plant or industrial boiler

flue gas desulfurization a chemical process for removing sulfur compounds from flue gas before it is emitted to the atmosphere

fly ash the ash produced in coal-burning power plants

focus the point or line on which a solar concentrator focuses sunlight

force a push or a pull

forced convection convection driven by a mechanical device such as a fan or pump, or by wind

fossil fuel a hydrocarbon fuel formed over tens to hundreds of millions of years from once-living matter, containing stored solar energy originally captured by photosynthetic plants; coal, oil, and natural gas are the common fossil fuels

fractional distillation the process used in oil refineries to separate crude oil into different substances based on their molecular weights

frictional force a force that acts between two surfaces to oppose their relative motion

fuel a substance that stores energy

fuel cell a device that uses a steady stream of fuel to produce electricity directly, without combustion; hydrogen fuel cells, which combine hydrogen and oxygen are particularly promising and produce only water as a by-product

fusion an energy-releasing nuclear reaction in which two light nuclei join to make a heavier nucleus; the source of energy that powers the Sun and other stars, and may someday be harnessed for terrestrial energy generation

gamma decay radioactive decay by emission of a gamma ray; occurs in nuclei with excess energy
gamma ray a high-energy photon, emitted as a nucleus, undergoes gamma decay
gas–electric hybrid a vehicle propulsion system that uses both a gasoline engine and an electric motor; the batteries powering the electric motor are charged directly by the gasoline engine or by the vehicle's motion
gas turbine a continuous-combustion rotary engine operating at high temperature; used in advanced power plants and jet aircraft engines
general circulation model (**GCM**) a large, complex model that includes details of atmospheric circulation
generation-I reactor early nuclear reactors deployed in the 1950s
generation-II reactor nuclear reactors deployed mostly during the 1960s through 1980s, when the use of nuclear power was growing rapidly; most operating reactors today are still gen-II designs
generation-III reactor nuclear reactors representing incremental improvements on gen-II designs, featuring greater inherent safety and economic advantages; deployed in the 1990s and early 2000s
generation-III+ reactor nuclear reactors being deployed today; represent modest improvements on gen-III designs
generation-IV reactor refers to any of a number of advanced designs that feature greater safety, higher efficiency, proliferation resistance, and less waste production. Some gen-IV reactors will be able to utilize much more of the nonfissile uranium-238, and some may "burn" nuclear waste
geoengineering processes that would modify Earth's environment on a large scale, generally with the goal of reducing or countering global warming
geoexchange heat pump another name for a geothermal heat pump
geopressured system a geothermal system in which energy is stored in pressurized water
geothermal energy energy from Earth's interior
geothermal gradient the rate at which temperature increases with depth below Earth's surface
geothermal heat pump a heat pump that moves thermal energy from the ground into a building
global climate model (**GCM**) a large, complex climate model capable of projecting future climate with detailed spatial resolution; often used interchangeably with general circulation model
global warming potential (**GWP**) a measure of a substance's effectiveness in causing global warming, measured relative to that of carbon dioxide; depends on the time frame involved
gravitational force the force that every piece of matter exerts on every other piece
gravitational potential energy potential energy associated with objects that have been lifted against gravity
gray a unit of radiation dose, equal to 1 joule of radiation energy per kilogram of absorbing material
greenhouse effect the warming of a planet's surface due to the absorption of outgoing infrared radiation by atmospheric greenhouse gases
greenhouse gas an atmospheric gas that absorbs infrared radiation; water vapor and carbon dioxide are the most important greenhouse gases
greenhouse term a term in the energy-balance equation at the top of the atmosphere that represents the greenhouse effect; shows that the greenhouse effect depends on the upper atmosphere being cooler than the surface; the greenhouse term is $e_a\sigma(T_s^4 - T_a^4)$
grid shorthand for the electric power grid
grid cell the fundamental unit of atmosphere, ocean, or surface in which a climate model computes a single value for each quantity
gross domestic product (**GDP**) a measure of the total economic value of a country's goods and services
gross primary productivity see *primary productivity*
ground-source heat pump a heat pump that takes its energy from underground, where the temperature is roughly constant year round

half-life the time it takes half the nuclei of a given radioactive substance to decay
head the vertical distance through which water drops to drive a hydropower turbine
heat energy that is flowing as a result of a temperature difference
heat capacity a measure of the energy required to raise a particular object's temperature; in SI, its units are J/K or J/°C
heat engine a device that extracts thermal energy from a hot source and delivers mechanical energy;

the second law of thermodynamics states that this process cannot be 100% efficient

heat exchanger a device that transfers energy from one fluid to another

heat of fusion the energy per unit mass required to change a material from the solid to the liquid state

heat of transformation the energy per unit mass required to effect a phase transition

heat of vaporization the energy per unit mass required to change a material from a liquid to a gas

heat pump a refrigerator operated so as to transfer heat from exterior air or the ground into a building for purposes of heating; can also operate in reverse to provide summer cooling

heavy water water made with deuterium (D) instead of ordinary hydrogen (H) and used as a moderator and coolant in some nuclear reactor designs, especially Canadian reactors

heliostat a Sun-tracking mirror

higher heating value the energy content of a fuel that includes the heat of vaporization of the water vapor produced in combustion

hole the absence of an electron in an interatomic bond in a semiconductor; acts like a positive charge that can move through the material

horsepower a unit of power equal to 746 watts

Hubbert's peak the point at which production of a resource peaks, after which there may be a discrepancy between falling production and rising demand

hydride a chemical compound of hydrogen and another light element; possibly useful for storing hydrogen fuel in hydrogen-powered vehicles

hydrologic cycle the natural cycle whereby water evaporates from oceans and lakes, forms clouds, precipitates, and runs back into surface waters or underground aquifers

hydropower power produced by running or falling water

hydrothermal system a geothermal resource in which water and steam circulate freely

ice age a period of cool climate marked by the advance of polar ice sheets

ice–albedo feedback a positive feedback associated with global warming in which reduced ice cover decreases Earth's reflectivity and thus results in more solar energy being absorbed, enhancing the original warming

ignition temperature the temperature at which a plasma undergoes nuclear fusion; the lowest ignition temperature, for deuterium–tritium fusion, is still 50 million kelvins

indirect gain a passive solar design in which solar radiation warms a thermal mass and the energy is then circulated through the building by natural convection

inertial confinement a fusion scheme for confining plasma at high density for such short times that the inertia of the plasma keeps its particles from leaving the fusion site; usually induced with high-power laser or particle beams

infiltration seepage of outside air into a building, resulting in additional heating or cooling loads

insolation the intensity of sunlight at a given location and time, measured in watts per square meter

interglacial the warm, relatively short period between ice ages

Intergovernmental Panel on Climate Change (IPCC) an international body established by the World Meteorological Organization and the United Nations Environment Program; issues major reports on climate change approximately every 6 years

intermittent combustion engine refers to a heat engine in which combustion takes place intermittently, as in the spark-initiated ignition of a gasoline engine

internal combustion engine a heat engine in which combustion takes place within a closed chamber, such as the cylinders of a gasoline engine

internal energy energy associated with random thermal motion of molecules; also called *thermal energy*

inverter a device that converts direct current, such as that produced by photovoltaic cells, into standard alternating current

irreversible process a thermodynamic process that converts some mechanical energy into random thermal energy, resulting in a loss of energy quality

isotope a particular nuclear species determined by both its atomic number (Z) and mass number (N); different isotopes of the same element have the same atomic number but different mass numbers

joule the SI unit of energy, equal to 1 watt-second or 1 newton-meter

Kelvin scale the SI temperature scale; a temperature difference of 1 K is the same as a difference of 1°C, but the zero in the Kelvin scale is absolute zero (–273 K)

kerogen a waxy substance formed from organic material that is an intermediate stage in the formation of petroleum

kinetic energy energy of motion

lapse rate the rate at which temperature changes with altitude, normally a decrease of about 6.5°C per kilometer

latent heat the energy stored in a material as a result of its having changed from a lower- to a higher-energy phase

Lawson criterion a criterion for successful fusion; the product of plasma density and confinement time; must exceed 10^{20} s/m^3 for deuterium–tritium fusion

light-water reactor (**LWR**) a nuclear reactor using ordinary water as coolant and moderator; the most common reactor design, especially in the United States

linear response a response that's directly proportional to its cause

liquefied natural gas (**LNG**) natural gas that has been compressed and cooled to a liquid state that occupies only 1/600 the volume of the gaseous product

load management schemes or mechanisms to manage electrical energy demand so as to help match supply and demand

lower heating value the energy content of a fuel that does not include the heat of vaporization of the water vapor produced in combustion

magnetic confinement a fusion scheme that uses magnetic fields to confine plasma at relatively low density and for relatively long times

magnetic field the influence of moving electric charge on the space around it, giving rise to magnetic forces on other moving charges; also arises from a changing electric field

mass number the total number of nucleons (neutrons plus protons) in a nucleus; a rough measure of the nuclear mass

matter one of the two basic "substances" that make up the universe (the other is energy), matter is made up of elementary particles, atoms, and molecules

mesosphere the atmospheric layer above the stratosphere, characterized by a temperature that decreases with altitude

methane clathrate a compound formed underwater at great depth, containing methane (CH_4; natural gas) trapped in an ice-like structure

microwave sounding unit (**MSU**) a satellite-based instrument that measures microwave radiation from atmospheric oxygen; the data can be used to calculate temperature at different levels in the atmosphere

mixed-oxide fuel (**MOX**) nuclear reactor fuel made from a mix of uranium and plutonium oxides

moderator a material used in a nuclear reactor to slow down neutrons, making them more likely to cause fission

mountaintop removal a coal-mining technique in which entire mountaintops are removed to get at coal; the residue is often dumped into adjacent valleys

multiplication factor the factor by which the number of fissioning nuclei is multiplied in each successive generation of fission events; should be exactly 1 for a controlled reaction as in a nuclear reactor; a multiplication factor much greater than 1 leads to an explosive chain reaction, as in a fission bomb

National Ambient Air-Quality Standards (**NAAQS**) the maximum allowed concentrations of the criteria pollutants under the Clean Air Act

natural convection convection that occurs naturally when a fluid is heated and the warm fluid then rises, carrying energy upward

natural gas a gaseous fossil fuel, consisting mostly of methane (CH_4)

natural greenhouse effect the greenhouse effect caused by naturally occurring greenhouse gases, especially water vapor and carbon dioxide; for Earth, this effect increases the surface temperature by some 33°C

negative feedback a feedback effect that tends to diminish the original change

net metering a scheme for electrical energy billing in which distributed producers—like individual homeowners' photovoltaic systems or wind turbines—can generate power that goes to the grid, in the process running the electric meter backward and earning credits for the distributed producer

net primary productivity see *primary productivity*

neutron a neutral elementary particle that, along with the proton, makes up the atomic nucleus

neutron-induced fission fission that results when a neutron strikes a nucleus

newton the SI unit of force, equal to 1 kg·m/s^2

nonlinear response a response that's not in direct proportion to its cause; in extreme cases, a nonlinear response may be an abrupt jump in some quantity in response to a very small change in another quantity

***N*-type semiconductor** a semiconductor that's been doped with impurities so that the dominant carriers of electric current are negative electrons

nuclear difference the factor of about 10 million difference between the energy released in nuclear reactions compared with chemical reactions

nuclear force the strong, short-range force that binds neutrons and protons in the atomic nucleus

nuclear fuel cycle the entire sequence of steps from mining of nuclear fuel through disposal of nuclear waste

nuclear radiation high-energy particles produced in the decay of radioactive nuclei

nucleon a constituent of the atomic nucleus—a neutron or proton

ocean thermal energy conversion (OTEC) a technological scheme using the temperature difference between warm, surface ocean water and cool, deep water to drive a heat engine

ohm the unit of electrical resistance (symbol Ω), equal to 1 volt per ampere

Ohm's law the statement that voltage and current are proportional: $I = V/R$. An empirical law that applies, approximately, to some electrical conductors, especially metals

oil shale rock containing kerogen that can be extracted and processed into oil

one-dimensional model a climate model in which conditions vary in only one direction, often height

open-loop system a geothermal energy system in which geothermal fluids themselves constitute the working fluid

ozone the oxygen compound O_3; a greenhouse gas; in the stratosphere, ozone is an absorber of solar ultraviolet radiation, but it is a highly reactive, toxic pollutant near ground level

ozone layer the region of the stratosphere containing ozone that absorbs solar ultraviolet radiation

parabolic dish a dish-shaped solar concentrator with a parabolic shape

parabolic reflector a solar concentrator with a parabolic shape that provides the optimum focusing of light

parabolic trough concentrator a trough-shaped parabolic concentrator that focuses light to a line

parallel hybrid a hybrid vehicle in which a gasoline engine provides most of the motive power, assisted occasionally by the electric motor

particulate pollution fine particles of solid material suspended in the air

passive solar heating solar heating technology that does not make use of active devices such as pumps and fans to move energy around

peaking power smaller power facilities that can be turned on and off as needed to provide extra power during times of peak demand; hydropower is an excellent form of peaking power

peat a crumbly, brown material formed from decaying vegetation; a first step in the formation of coal, peat itself is burned as a fuel in some parts of the world

petroleum strictly speaking, a liquid or gaseous fossil fuel, but often used as a synonym for oil

pH scale the scale used to designate acidity, with 7 being neutral and lower values being acidic

photochemical reaction a reaction driven by the energy of sunlight; important in forming smog from other air pollutants

photochemical smog a mix of harmful chemicals formed by the action of sunlight on air pollutants

photolysis a process using sunlight to break water into hydrogen and oxygen

photon a particle-like bundle of electromagnetic wave energy

photosynthesis the process whereby green plants take in carbon dioxide and solar energy to make carbohydrates that serve as energy sources for life

photovoltaic cell a semiconductor device that converts light energy directly into electricity

photovoltaic module an assemblage of photovoltaic cells

Planck's constant a fundamental constant of nature that quantifies the essential "graininess" of nature in quantum physics

plankton tiny marine organisms at the base of the marine food chain; plankton are important precursors of petroleum

plasma a gas that's so hot it's ionized, so its particles are positively charged ions and negative electrons

plug-in hybrid a gasoline–electric hybrid vehicle whose battery can be recharged either with energy from its gasoline engine or by plugging into the electric power grid

***PN* junction** a junction between *P*-type and *N*-type semiconductors; such a junction contains a strong electric field that accelerates charges in a photovoltaic cell; a *PN* junction conducts electricity in only one direction

pollution traditionally, an unwanted by-product of combustion or other human processes that is toxic to people or damaging to the environment

polycrystalline silicon silicon that consists of many distinct crystals; less expensive to make than single-crystal silicon

positive feedback a feedback effect that tends to enhance the original change

potential energy stored energy, associated with work done against a force

power the rate of energy production or consumption, measured in watts

power coefficient the fraction of the wind's power that a turbine actually extracts; the theoretical maximum is 59%

power curve a plot of power output versus wind speed for a wind turbine

power tower a solar thermal system in which a field of Sun-tracking mirrors concentrates light on an absorber high in a central tower

pressure vessel the heavy steel chamber that contains the nuclear fuel in a light-water reactor

pressurized-water reactor (PWR) a light-water reactor in which water in the reactor vessel is under such high pressure that it doesn't boil, but instead transfers its heat to a secondary loop in which water boils and drives a turbine

primary energy the energy input to a system such as a power plant, including any that ends up as waste

primary productivity the rate at which plants capture solar energy through photosynthesis; gross primary productivity is the total amount, whereas net primary productivity—typically about half of gross—measures what's left over after plants use energy themselves

primary standards air-quality standards aimed at protecting human health

proton a positively charged elementary particle that, along with the neutron, makes up the atomic nucleus

proton-exchange membrane fuel cell (PEMFC) a fuel cell using a special membrane that permits only protons to pass through

proxy a quantity that "stands in" for another, as when tree rings or isotope ratios serve as proxies for climatic conditions

***P*-type semiconductor** a semiconductor that's been doped with impurities so that the dominant carriers of electric current are positive holes

pumped storage a scheme for storing energy by pumping water from a lower source to a higher reservoir; running the water downhill to turn turbines then recovers the stored energy as electricity

quad 1 quadrillion (10^{15}) Btu

rad an older unit of radiation dose, equal to 0.01 Gy

radiation energy flow by electromagnetic waves

radiative forcing forcing due to an upset in the balance of radiation energy flows

radioactive decay the process whereby a radioactive nucleus emits one or more high-energy particles, thus turning into a different nuclear species (or, with gamma decay, a lower-energy state of the same nucleus)

radiosonde a balloon-borne suite of instruments for measuring conditions at different altitudes in the atmosphere and transmitting the results via radio

reactor-year a unit of nuclear power operation equivalent to 1 nuclear reactor operating for 1 year

rechargeable battery a battery capable of storing electrical energy via chemical reactions when current is run "backward" through the battery

refine the process of chemically altering crude oil to form a variety of products, including heavy oils, aviation fuels, and gasoline

refrigerator a device that transfers heat from cooler objects to warmer surroundings; the second law of thermodynamics says this can't happen without an input of high-quality energy

regenerative braking braking in a hybrid vehicle that uses an electric generator to slow the wheels; the generator charges the battery, thus recovering some of the energy of the vehicle's motion

rem an older unit of absorbed radiation dose, equal to 0.01 Sv

renewable portfolio standard (RPS) a standard requiring an entity—typically a country, state, or utility—to produce a certain portion of its electrical energy using renewable sources

repair mechanism a cellular mechanism that repairs radiation damage to DNA

representative concentration pathways (RCPs) scenarios that depict hypothetical future evolution of the concentrations of greenhouse gases and other climate-forcing agents; used with computer modeling to explore future climate and possibilities for mitigation of climate change

reprocessing processing of spent nuclear fuel to extract different substances, usually plutonium-239

reserve the amount of a fuel that is known or reasonably certain to exist

reserve/production (R/P) ratio the ratio of a fuel or other natural resource's reserves to the rate at which it's being extracted from the ground; has the units of time and is a rough estimate of the time the given resource will last

reservoir a repository where materials reside temporarily or permanently, as in the atmospheric carbon reservoir or the carbon reservoir associated with biomass

resistance the property of an electrical conductor that describes the relation between voltage and current in the conductor; for a given voltage, higher resistance means lower current

resolution a measure of the fineness in which a climate model divides the physical domain it is modeling; smaller grid cells give higher resolution

resource the total amount of a fuel in the ground, whether or not it's been discovered or even estimated to exist

rolling blackout a strategy used during times when electric power production is insufficient to meet demand, and neighborhoods, regions, or institutions are denied power for periods of time before the next region faces its blackout

***R* value** a measure of the insulating value of a material; in English units, an *R* value of 1 corresponds to an energy loss rate of 1 Btu/hour through every square foot for every degree Fahrenheit temperature difference across the material

saturation effect the complete absorption of all infrared radiation by greenhouse gases in certain wavelength bands, meaning that further increases in greenhouse gas concentrations don't result in more absorption in those bands; makes greenhouse gas forcing a nonlinear function of concentration

scrubbing removal of chemical pollutants, especially sulfur compounds, from flue gas by reacting it with another substance

second law of thermodynamics the statement that systems naturally evolve toward states of lower organization and therefore lower energy quality; applied to heat engines, the second law says it's impossible to extract thermal energy from a hot source and convert it to mechanical energy with 100% efficiency

secondary standards air-quality standards aimed at protecting the general welfare, including property and environmental quality

selective surface a solar absorber surface whose properties are wavelength dependent, typically absorbing in the visible but not in the infrared, and therefore minimizing infrared radiation loss

semiconductor a material that conducts electricity by virtue of either negative electrons or positive holes, and whose electrical properties can be manipulated precisely; at the heart of all modern electronics, including photovoltaic cells

sensible heat thermal energy associated with an object's being at a given temperature; as opposed to latent heat

sequester the process of storing substances—especially carbon dioxide from fossil fuel combustion—away from Earth's surface

series hybrid a hybrid vehicle in which the gasoline engine is used solely to generate electricity that runs the electric motor; not in widespread use

sievert a unit of radiation dose, similar to the gray but adjusted for the biological impact of different types of radiation

single-axis concentrator a concentrator that tracks the Sun in only one direction and focuses light to a line

smart grid an electric power grid that senses power usage by individual end users and, in some cases, automatically manages loads inside homes or institutions

solar chimney a system that uses the flow of solar-heated air rising up a huge chimney-like structure to power turbines

solar collector a device consisting of one or more glass covers and an absorber plate, used in active solar heating systems

solar constant the rate at which solar energy crosses a square meter of area oriented perpendicular to the Sun's rays, at Earth's distance from the Sun; equal to approximately 1,364 W/m^2

solar lighting use of sunlight for illumination inside buildings

solar oven a cooking device that works by concentrating sunlight

solar panel a photovoltaic module

solar pond a pond that functions as a solar collector; typically uses layers of salty water to inhibit convective energy loss

solar still a device that uses sunlight to distill water, making freshwater from saltwater

solar thermal power system a system in which sunlight is concentrated to boil a fluid and drive a turbine or other engine

spark-ignition engine an intermittent-combustion engine in which a spark initiates combustion, as in gasoline engines

specific heat a measure of the energy per unit mass required to raise a material's temperature; in SI, its units are J/kg·K or J/kg·°C

stagnation temperature the maximum temperature attainable in a solar collector under given conditions; equal to the temperature that would be reached under stagnation conditions—that is, if the heat-transfer fluid stops flowing

steam generator the device in a pressurized-water reactor in which heat from the primary coolant water transfers energy to boil water in the secondary loop

steam reforming a reaction of high-temperature steam with natural gas to produce hydrogen and carbon monoxide; the resulting mixture can be burned as a fuel

Stefan–Boltzmann constant The constant in the Stefan–Boltzmann law, given by $\sigma = 5.67 \times 10^{-8}$ W/m^2·K^4

Stefan–Boltzmann law the statement that the power radiated from an object with surface area A and emissivity e at temperature T is given by $P = e\sigma AT^4$

step-down transformer a transformer whose secondary coil produces a lower voltage than the voltage supplied to the transformer's primary coil

step-up transformer a transformer whose secondary coil produces a higher voltage than the voltage supplied to the transformer's primary coil

stratification in a solar hot-water tank, the desirable situation in which hotter water is at the top of the tank

stratosphere the second level of the atmosphere, extending from the tropopause to some 50 km

strip mining extracting coal by stripping off overlying surface layers

subgrid parameterization a procedure for dealing with phenomena that occur on size scales smaller than the grid size of a climate model

sulfate aerosol sulfur-based particulates, particularly from coal burning; these are highly reflective and thus have a cooling effect on climate

superconductor a material that exhibits zero electrical resistance when cooled below a critical temperature

tar sands sand containing a heavy tar from which oil can be extracted

temperature a measure of the average thermal energy of the molecules in a substance

temperature anomalies deviations in temperature from established values or averages

temperature inversion a situation in which atmospheric temperature increases with altitude from the surface upward

thermal conductivity a measure of a material's ability to conduct heat

thermal energy energy associated with random thermal motion of molecules; also called *internal energy*

thermal mass a large, massive structure whose heat capacity helps store solar energy and even out temperature fluctuations in a building

thermal pollution waste heat, typically from a power plant, discharged to the environment and possibly disrupting the local ecology

thermal power plant a power plant that extracts energy from a hot source, usually a fossil-fueled boiler or nuclear fission reactor

thermal splitting the use of high temperatures to dissociate water into hydrogen and oxygen

thermodynamic efficiency limit the theoretical maximum efficiency for a heat engine, given by

$$e = 1 - \frac{T_c}{T_h}$$

where T_h and T_c are the maximum and minimum temperatures, respectively

thermohaline circulation ocean circulation driven by variations in temperature and salinity

thermosiphon a solar hot-water system in which heat-transfer fluid circulates by natural convection

thermosphere the atmospheric layer above the meso sphere, characterized by hot, diffuse gas whose temperature increases with altitude

three-dimensional model a climate model that includes variations in all three spatial dimensions

tidal energy energy originating in the motions of Earth and Moon, resulting in ocean tides

time-dependent model a model that computes climatic conditions as a function of time

tokamak a toroidal plasma device that appears to be the most promising candidate for a nuclear fusion reactor

ton, tonne, metric ton units of mass; 1 ton is 1 English ton, or 2,000 pounds; 1 tonne (or metric ton) is

1,000 kg or 2,200 pounds, roughly the same as 1 ton

tonne oil equivalent an energy unit equal to 41.9 GJ; the energy contained in 1 metric ton of oil

total primary energy supply (**TPES**) the total primary energy supplied to a country or the world

transformer a device incorporating two coils, working by electromagnetic induction, that changes voltage levels in alternating-current systems

transpiration the process whereby plants take in water from their roots and release it to the atmosphere

transuranic an element heavier than uranium; not found in nature but formed by neutron capture in fission reactors

Trombe wall an indirect-gain solar heating scheme in which a massive wall serves as both absorber and thermal mass

tropopause the boundary between the troposphere and stratosphere

troposphere the lowest level of the atmosphere, extending 8 to 18 km above the surface

turbine a fanlike arrangement of blades driven by a fluid flow; includes steam turbines in power plants and wind turbines

two-axis concentrator a concentrator, such as a parabolic dish, that tracks the Sun in two directions and focuses light to a point

two-box model a simple climate model that treats Earth's surface and atmosphere as separate systems, with each characterized by a single value for relevant physical quantities

two-dimensional model a climate model in which conditions vary in two directions, usually height and latitude

United Nations Framework Convention on Climate Change (**UNFCCC**) the framework, originally forged at the Rio Earth Summit in 1992, that establishes international mechanisms for dealing with climate change; the Kyoto Protocol is a subsidiary agreement to the UNFCCC

urban heat island effect the apparent warming observed over time at a fixed location due not to global climate change but to increasing urbanization at that location

vapor-dominated system a geothermal system saturated with steam at high pressure; the most valuable type of geothermal resource

volt the SI unit of voltage, equal to 1 joule per coulomb

voltage a measure of the energy per unit charge in electrical situations

water vapor feedback a positive feedback effect associated with global warming; warming results in more evaporation, which increases atmospheric water vapor, and since water vapor is a greenhouse gas, this enhances the original warming

watt the SI unit of power, equal to 1 J/s

weight the force that gravity exerts on an object; near Earth's surface, weight is the product of mass and the strength of gravity (*mg*)

well-mixed greenhouse gas a greenhouse gas that remains in the atmosphere long enough to reach an approximately uniform global concentration; these include carbon dioxide, methane, and most other common greenhouse gases

wet cooling tower structure designed to transfer waste heat from a power plant's cooling water to the atmosphere; wet towers do so by evaporation from water exposed directly to the air

work a measure of the mechanical energy supplied to an object; equal to the product of applied force times the distance moved in the direction of the force

work–energy theorem the statement that the net work done on a system goes into changing the system's kinetic energy

zero-dimensional energy-balance model an energy-balance climate model that treats Earth as a single point and yields only a single global average temperature

SUGGESTED READINGS

BOOKS/PUBLICATIONS

Annual Review of Environment and Resources. Palo Alto, CA: Annual Reviews. Each volume of this ongoing series features papers on energy and environmental issues by leading scholars in these fields; emphasis varies with volume. Note: Volumes 1 to 15 (1976–1990) were published as the *Annual Review of Energy,* and volumes 16 to 27 (1991–2002) were published as the *Annual Review of Energy and the Environment.* Relevant to the entire book.

Bent, Robert, Lloyd Orr, and Randall Baker, eds. *Energy: Science, Policy, and the Pursuit of Sustainability*. Washington, DC: Island Press, 2002. In seven chapters, individual experts look at technological and policy aspects of energy and its relationship to the environment. Relevant to Chapters 5 to 11 and 16.

Bodansky, David. *Nuclear Energy: Principles, Practices, and Prospects.* 2nd ed. New York: American Institute of Physics Press, 2004. A detailed scholarly account of all aspects of contemporary nuclear energy applications. Relevant to Chapter 7.

Boeker, Egbert, and Rienk van Grondelle. *Environmental Physics: Sustainable Energy and Climate Change.* 3rd ed. Chichester, UK: John Wiley & Sons, 2011. Scientific analysis of energy sources and technologies, climate change, pollution, and environmental analysis. Not for the mathematically challenged! Relevant to the entire book, for those who want quantitative depth and solid grounding in the underlying physics.

Brown, Robert C. *Biorenewable Resources: Engineering New Products from Agriculture.* Ames, IA: Iowa State Press, 2003. An overview of biofuels and related agricultural products. Readable, but assumes some math and chemistry. Relevant to Chapter 10.

Deffeyes, Kenneth S. *Beyond Oil: The View from Hubbert's Peak.* New York: Hill and Wang, 2006. Deffeyes, an oil geologist who worked with M. King Hubbert, is a firm believer that global peak oil production is upon us. Relevant to Chapter 5.

Ferguson, Charles. *Nuclear Energy: What You Need to Know.* New York: Oxford University Press, 2011. Using a question-and-answer format, this brief book looks at the major issues of nuclear power. Especially strong on nuclear security. Relevant to Chapter 7.

Garwin, Richard L., and Georges Charpak. *Megawatts and Megatons: The Future of Nuclear Power and Nuclear Weapons.* Chicago: University of Chicago Press, 2002. Long-time nuclear science and policy expert Garwin teams up with Nobel laureate Charpak in this up-to-date account of nuclear power and nuclear weapons. The authors make a reasoned argument for their stand in support of nuclear power while urging nuclear disarmament. Relevant to Chapter 7.

Gipe, Paul. *Wind Power: Renewable Energy for Home, Farm, and Business.* White River Junction, VT: Chelsea Green Publishing, 2004. A good survey of wind technology and a practical guide to those who are thinking seriously about installing small- and medium-scale wind systems, written by a tireless wind energy enthusiast. Relevant to Chapter 10.

Glantz, Michael. *Climate Affairs: A Primer*. Washington, DC: Island Press, 2003. A discussion of climate change and its impacts written for a general audience by a social scientist who is the former head of the Environmental and Societal Impacts

Group at the National Center for Atmospheric Research. Complements Chapters 12 to 15 from a social science perspective.

Hansen, James. *Storms of My Grandchildren.* Bloomsbury USA, 2009. Subtitled *The Truth about the Coming Climate Catastrophe and Our Last Chance to Save Humanity,* this is a very personalized account of climate science and climate politics by a leading and outspoken climatologist. Tends to the alarmist side, but given Hansen's solid scientific reputation, his concerns should be taken seriously. Relevant to Chapters 12 to 16.

Hoffmann, Peter. *Tomorrow's Energy: Hydrogen, Fuel Cells, and the Prospects for a Cleaner Planet.* Cambridge, MA: MIT Press, 2001. A clear and readable account of hydrogen energy that is especially strong on the history of hydrogen technology. Relevant to Chapter 11.

Houghton, John. *Global Warming: The Complete Briefing.* 4th ed. Cambridge, UK: Cambridge University Press, 2009. A thorough, authoritative, yet eminently readable account of global climate change. Sir John Houghton was co-chair of the science working group for the Intergovernmental Panel for Climate Change's Third Assessment Report. Relevant to Chapters 12 to 15.

Intergovernmental Panel on Climate Change. *Climate Change 2007: The Physical Science Basis.* IPCC 4. Cambridge, UK: Cambridge University Press, 2007. This is the science portion of the IPCC's Fourth Assessment Report—a comprehensive survey that synthesizes a vast array of theory, data, and modeling results to describe the state of climate science. It also is available at www.ipcc.ch. Relevant to Chapters 12 to 15.

Intergovernmental Panel on Climate Change, *Special Report on Renewable Energy Sources and Climate Change Mitigation.* IPCC, 2011. A detailed assessment of the prospects for renewable energy development in the context of mitigating climate change. Available at http://srren.ipcc-wg3.de/report. Relevant to the entire book.

International Energy Agency. *Biofuels for Transport: An International Perspective.* Paris: Chirat, 2004. A well-organized wealth of technical information and statistics on biofuels. Relevant to Chapter 10.

Lovins, Amory B., E. K. Datta, O.-E. Bustnes, J. G. Koomey, and N. J. Glasgow. *Winning the Oil Endgame: Innovation for Profits, Jobs, and Security.* Snowmass, CO: Rocky Mountain Institute, 2005. This elaborately documented report shows how humankind might end its dependence on climate-changing fossil fuels. It also is available in full at www.oilendgame.com. Relevant to Chapters 6 to 11 and 16.

Mann, Michael, and Lee Krump. *Dire Predictions: Understanding Global Warming.* New York: Pearson Education, 2008. Despite its alarmist title, this book contains solid science, clothed in a series of illustrated vignettes that make the book read like a magazine article. The lead author is the originator of the famous "hockey stick" reconstruction of millennial climate, and the book is largely a distillation and explanation of findings from the IPCC Fourth Assessment Report. Relevant to Chapters 12 to 16.

Manwell, J. F., J. G. McGowan, and A. L. Rogers. *Wind Energy Explained: Theory, Design, and Application.* Chichester, UK: Wiley, 2002. A thorough and mathematically detailed account of wind turbine engineering, this book also covers economics and environmental impacts. Relevant to Chapter 10.

National Research Council, and National Academy of Engineering. *The Hydrogen Economy: Opportunities, Costs, Barriers, and R&D Needs.* Washington, DC: National Academies Press, 2004. A thoughtful, serious study of the prospects for hydrogen energy produced by a diverse group ranging from industry executives to environmentalists, this publication contains a good mix of technology, economics, and policy. Unfortunately, there's no index. Relevant to Chapter 11.

Pahl, Greg. *Biodiesel: Growing a New Energy Economy.* White River Junction, VT: Chelsea Green Publishing, 2005. A look at the technology, policies, and worldwide spread of biodiesel fuel. Pahl is a biodiesel enthusiast, but he presents a realistic analysis of the challenges that biodiesel faces. Relevant to Chapter 10.

Pepper, Ian, Charles Gerba, and Mark Brusseau, eds. *Pollution Science.* San Diego: Academic Press, 1996. This textbook gives a thorough survey of pollution science at about the same level of *Energy, Environment, and Climate.* Especially relevant to Chapter 6.

Romm, Joseph J. *The Hype about Hydrogen: Fact and Fiction in the Race to Save the Planet.* Washington, DC: Island Press, 2004. A realistic assessment of hydrogen's prospects (it's far off) by a former assistant secretary in the U.S. Department of Energy's Office of Energy Efficiency and Renewable Energy. Romm sees fossil fuel shortages and climate change as looming events, and he warns that a hydrogen economy won't develop soon enough to avoid them. Relevant to Chapters 11 and 16.

Ruddiman, William. *Earth's Climate: Past and Future*, 2nd ed. New York: W. H. Freeman, 2008. A beautifully illustrated textbook that is especially strong on paleoclimate. Relevant to Chapters 12 to 15.

Schewe, Phillip. *The Grid: A Journey through the Heart of Our Electrified World*. Washington, DC: Joseph Henry Press, 2007. A historical description of the evolution of the electric power grid, including descriptions of some of the major power failures. Light on technical details, but a nice introduction to something we're all connected to but don't often think about. Relevant to Chapter 11.

Schmidt, Gavin, and Joshua Wolfe. *Climate Change: Picturing the Science*. New York: W.W. Norton, 2009. A lavishly illustrated, scientifically accurate but non-quantitative introduction to contemporary climate change issues. The author team comprises noted climatologist Schmidt and photographer Wolfe. Relevant to Chapters 12 to 16.

Schneider, Stephen. *Science as a Contact Sport: Inside the Battle to Save Earth's Climate*. Washington, DC: National Geographic Society, 2009. A personal account of the history of climate science and policy by a noted climatologist whose death in 2010 deprived climate research and public outreach of a vigorous spokesperson. Relevant to Chapters 12 to 16.

Smil, Vaclav. *Energy Transitions: History, Requirements, Prospects*. Santa Barbara, CA: Praeger, 2010. In this brief volume, Smil traces the history of energy transitions—from wood to coal, coal to oil, etc.—for different societies, and uses the results to speculate on the prospects for a transition to climate-neutral energy sources. Relevant to the entire book.

Smil, Vaclav. *Energy: A Beginner's Guide*. Oxford, UK: Oneworld Publications, 2006. A quick guide to energy that's particularly strong on energy history. Relevant to Chapters 1, 2, 6 to 11, and 16.

Smil, Vaclav. *Energy at the Crossroads: Global Perspectives and Uncertainty*. Cambridge, MA: MIT Press, 2003. A thorough, scholarly, but readable account of energy history, alternatives, and energy futures by an independent thinker. Relevant to Chapters 1, 2, 6 to 11, and 16.

Sørensen, Bent. *Renewable Energy: Physics, Engineering, Environmental Impacts, Economics and Planning*. 4th ed. Burlington, MA: Academic Press, 2010. A thorough, quantitative, and highly detailed account of the basic physics and engineering of renewable energy sources and technologies. Relevant to Chapters 7 to 11.

U.S. Department of Energy. *Solar Energy: Complete Guide to Solar Power and Photovoltaics, Practical Information on Heating, Lighting, and Concentrating, Energy Department Research*. Progressive Management, 2007. This set of two CD-ROMs packs some 51,000 pages containing practical tips on solar energy as well as a survey of U.S. Department of Energy research on solar technologies. Relevant to Chapter 9.

Weart, Spencer R. *The Discovery of Global Warming*. Cambridge, MA: Harvard University Press, 2004. An authoritative history of the science that has led to our present-day understanding of climate change. Relevant to Chapters 12 to 15.

Wolfson, Richard. *Nuclear Choices: A Citizen's Guide to Nuclear Technology*. Rev. ed. Cambridge, MA: MIT Press, 1993. Your author describes nuclear technology and policy issues associated with both nuclear power and nuclear weapons. Although somewhat dated, especially on weapons, it's useful for a simple explanation of nuclear technologies. Relevant to Chapter 7.

World Meteorological Organization. *Climate into the 21st Century*. Cambridge, UK: Cambridge University Press, 2003. A quick introduction to climate issues in a profusely illustrated coffee-table–style book. Relevant to Chapters 12 to 15.

FILM/VIDEO

Kohn, Walter, and Alan Heeger. *Power of the Sun*. DVD. Santa Barbara, CA: University of California at Santa Barbara, 2005. Nobel laureates Kohn and Heeger explore the promise of photovoltaic technology. The DVD includes a 22-minute film on the physics of photovoltaic devices and a 57-minute feature on photovoltaics and their applications.

National Geographic. *A Way Forward: Facing Climate Change*. A series of short videos documenting impacts of and potential solutions to global warming. Available at http://video.nationalgeographic.com/video/player/environment/global-warming-environment/way-forward-climate.html

Wolfson, Richard. *Earth's Changing Climate*. DVD. Chantilly, VA: The Teaching Company, 2007. In this twelve-lecture video course (also on audio), your author explores the science of climate and the role of humankind in the ongoing changes in Earth's climate. It's less mathematical than this book, with more emphasis on climate and less on energy.

AUTHORITATIVE WEB SITES

http://cdiac.ornl.gov The Carbon Dioxide Information Analysis Center at Oak Ridge National Laboratory makes available data on concentrations of CO_2 and other greenhouse gases, as well as historical temperature records from a host of sources such as ice cores.

www.cru.uea.ac.uk The Climatic Research Unit of the University of East Anglia in England maintains one of the most comprehensive and accessible datasets on global temperature.

www.eia.doe.gov/emeu/aer/ The U.S. Department of Energy's Energy Information Administration posts the *Annual Energy Review* at this site; a new version appears each summer. The site also contains a wealth of tables on all aspects of energy production and use, mostly in the United States but also with a section on international energy.

www.eia.gov The portal to the U.S. Department of Energy's Energy Information Administration gives access to a wealth of energy statistics including, but not limited to, those in the *Annual Energy Review*.

www.eia.gov/countries/ This page gives access to international energy statistics. The DATA link gets you to pages that let you build and download your own tables of detailed energy data over time.

www.eia.gov/cneaf/electricity/page/capacity/capacity.html This page gives access to yearly spreadsheets entitled "Existing Electric Generating Units in the United States." These offer a fascinating and detailed look at where our electricity comes from, and are used in a number of research problems throughout the book.

www.epa.gov The home page of the U.S. Environmental Protection Agency. The "Clean Air Act" and "Climate Change" links are especially relevant; follow the latter link to find the latest inventory of greenhouse gas emissions.

www.iea.org The home page of the International Energy Agency offers links to a wide range of energy information. The "Statistics" link lets you extract desired data for any country or region, and the "Publications/Surveys Free for Download" link therein gets you to a downloadable version of the IEA's *Key World Energy Statistics*, published yearly.

www.ipcc.ch The home page of the Intergovernmental Panel on Climate Change has links to the IPCC's reports, both the regular climate assessment reports and a variety of special reports.

www.ncdc.noaa.gov/oa/ncdc.html The National Climatic Data Center, operated by the National Oceanographic and Atmospheric Administration, has a host of climate data ranging from individual weather stations to national and global averages.

www.noaa.gov The U.S. National Oceanic and Atmospheric Administration posts frequent updates and studies on climate and weather at this site.

www.nrel.gov The U.S. Department of Energy's National Renewable Energy Laboratory describes its many research programs on this site.

http://nsidc.org The National Snow and Ice Data Center, located at the University of Colorado's Cooperative Institute for Research in Environmental Sciences, is an affiliate of the National Oceanographic and Atmospheric Administration. It maintains data archives on snow and ice coverage as measured from satellites and ground surveys.

www.pewclimate.org Here the Pew Center on Climate Change posts its many reports on anthropogenic climate change and strategies for dealing with it. The Pew Center is especially strong on getting the business community involved with climate issues.

www.realclimate.org This site contains running commentaries by reputable climate scientists on climate change, including refutations of claims by "climate skeptics."

www.undp.org The home page of the United Nations Development Programme, which publishes the annual *Human Development Report* and makes it available on this site. The section on Human Development Data includes the Human Development Index, an alternative to gross national product.

ANSWERS TO ODD-NUMBERED QUESTIONS AND EXERCISES

CHAPTER 1

Questions

1. Life has changed Earth's atmosphere
3. Oxygen is highly reactive
5. Volcanoes emitted CO_2
7. There were more people reproducing in 1988

Exercises

1. 170 PW
3. 1985 (83 million people)
5. About 10 TW

CHAPTER 2

Questions

1. Megawatts are not energy units—they measure power, or the *rate* of energy use
3. Laptop charger AC input (120 V × 1.5)/2 A = 90 W, DC output 20 V × 2.3 A = 46 W; electric teapot = 1,500 W; blender = 450 W (max); refrigerator (115 VAC × 7.1 A)/2 = 409 W; freezer (115 VAC × 5 A)/2 = 288 W
5. No
7. Spending billions of dollars to try to clean up a blowout of a deep water oil drilling platform in the Gulf of Mexico, for example, could increase the GDP, though the accident could decrease the quality of life for countless citizens (and end it for some)

Exercises

1. $80
3. 45%
5. 470 W
7. $\frac{3.7\ \text{k\$/yr}}{\text{kW}}$ is the per capita GDP/per capita energy consumption, or the inverse energy intensity

CHAPTER 3

Questions

1. When walking uphill, one does work against gravity
3. Natural gas (CH_4) is more dense than hydrogen (H_2)
5. (a) 10^{12}, (b) 1, (c) 1 million

Exercises

1. 97 W
3. 15.7 PW
5. (a) 32 kW; (b) 4.8 kW
7. 4 million barrels per day
9. 26 minutes
11. 82 kW = 110 hp
13. 302 W
15. (a) 30 PJ; (b) about 3 hours

CHAPTER 4

Questions

1. Heat is a form of energy flow due to temperature gradients. Thermal energy is due to small scale kinetic motions; it depends on an object's temperature and composition
3. Fiberglass traps air and prevents convection
5. Stefan–Boltzmann law
7. Systems tend toward higher entropy, or disorder. Useful energy dissipates into thermal energy

9. When we turn useful energy into thermal energy, it is ultimately wasted
11. Energy systems have a higher heating value when they capture the latent energy of water vapor, e.g., via a condensation system. Systems that simply exhaust water vapor have lower heating values

Exercises

3. 100 W
5. 35 minutes
7. (b) 750 kJ; (c) 700 kJ. The primary energy required in either case is more than 7 times that required to heat the water alone
9. (a) 173 PJ; (b) 18 days
11. The original house (a) loses heat at a rate of 40,340 Btu/hr and (b) burns 263 gallons of oil per month, (c) at a cost of $723/month. (d) Upgrades save $564/month on the heating bill
13. 2.2 mm^2
15. Summer: e = 48%, P_{max} = 1.2 GWe; winter: e = 53%, P_{max} = 1.3 GWe

CHAPTER 5

Questions

1. Natural gas has relatively more H and less C, so for a given amount of energy (H_2) it produces less CO_2
3. Combined-cycle power plants use waste heat from the gas cycle to power the steam cycle, significantly increasing the overall efficiency
5. Resources are greater than reserves. Resources include all fuels potentially discoverable, whether they are recoverable or not. Reserves include only those fuels that we know about and can easily recover
7. Cornucopians trust in human ingenuity to stay one step ahead of resource limitations, by means of technological innovations and clever adaptations, such as efficiencies

Exercises

1. 12.5 molecules of O_2 (well, 13)
3. 79%
11. (b) 24 years; (c) 30 years; (d) 39 years
13. 10 years

CHAPTER 6

Questions

1. Combustion adds O_2 to C. That adds extra mass. See Exercise 1 for details
3. CO binds to hemoglobin and reduces blood's ability to carry oxygen. This can cause tissue damage, heart irregularities, and impaired brain function, leading to accidents or direct system failure
5. Primary air quality standards are designed to protect human health. Secondary standards are designed to protect general welfare, including environmental quality and property
7. The catalytic converter combines incompletely combusted *toxic* CO with O_2 to produce CO_2 (which is safer to breathe in exhaust)
9. UV can break down exhaust constituents and give them the energy to recombine into chemicals such as ozone (O_3)
11. (i) Natural gas has a higher H/C ratio than coal, so it produces more energy/CO_2. (ii) Natural gas is lighter, so it doesn't require as much energy to extract and transport

Exercises

1. 8.5 kg = 19 pounds CO_2 per gallon of gas
3. (a) About 10 billion tonnes/yr; (b) 75 million tonnes
5. 1.57 kg CO_2
7. (a) 5.6 billion gallons; (b) 27 *Deepwater Horizon* blowouts
9. About 50 kilotonnes SO_2
11. Tier 1: 0.96 kg; Tier 2: 0.12 kg
13. Approximately 10 tonnes
15. 6 GJ per tonne, or about 15% of the energy content of the oil transported

CHAPTER 7

Questions

1. The moderator's role is to slow down the fast neutrons emitted by uranium so they can split another U-235 nucleus. One nice thing about a light-water reactor is that the water, which cools the reactor, also acts as the moderator. So if the reactor loses its coolant, it also loses the moderator, and that stops the reaction. Fail safe?
3. (i) One product of uranium fission in power plants is plutonium, which is useful in nuclear weapons. (ii) Spent nuclear fuel or long-lived waste from a nuclear reactor included in a "dirty bomb" can have severe environmental and health consequences
5. The nuclear energy released by the splitting or formation of one atom in a fission or fusion

event is more than a million times greater than the chemical energy released by the combustion of one CO_2 molecule. Therefore we burn (and waste) millions of times more carbon-based mass to get the same energy as from a little nuclear fuel

Exercises

1. Cu-65
3. 0.08%
5. Using 29 MJ/kg for the energy content of coal, the 3 trainloads carry 8.7×10^{14} J
7. ${}^{238}_{92}\mathrm{U} \rightarrow {}^{234}_{90}\mathrm{Th} + {}^{4}_{2}\mathrm{He}$, ${}^{234}_{90}\mathrm{Th} \rightarrow {}^{234}_{91}\mathrm{Pa} + {}^{0}_{-1}e + \bar{\nu}$, ${}^{234}_{91}\mathrm{Pa} \rightarrow {}^{234}_{92}\mathrm{U} + {}^{0}_{-1}e + \bar{\nu}$, ${}^{234}_{92}\mathrm{U} \rightarrow {}^{230}_{90}\mathrm{Th} + {}^{4}_{2}\mathrm{He}$, ${}^{230}_{90}\mathrm{Th} \rightarrow {}^{226}_{88}\mathrm{Ra} + {}^{4}_{2}\mathrm{He}$, with energy emissions all along the way
9. (a) 108,000 cases from the background radiation; (b) natural: about 88,000; artificial: about 20,000 (mostly from medical procedures)
11. Roughly 30 billion years
13. 10^{20} particles/m^3

CHAPTER 8

Questions

1. Tidal effects on the Earth's core and radioactivity left over from the formation of the solar system
3. Ground source heat is mostly solar, not geothermal
5. (i) Loss of water due to steam extraction. (ii) Cooling of the source, or heating of the environment
7. Gravitational interaction of the Earth–Moon system (with a little help from the Sun)

Exercises

1. 44 TW
3. 84 mW/m^2
5. (a) 29%; (b) 76%
7. (a) 6.7; (b) 2.2 kW
9. 62% greater
11. (a) Gas: 1.18 MBtu; heat pump: 0.92 MBtu; (b) gas: \$4.53/MBtu; heat pump: \$29.71/MBtu. The gas furnace uses more energy, but costs less

CHAPTER 9

Questions

1. The whole spherical area of the Earth radiates energy away ($4\pi R_E^2$), and the effective area of the Earth receiving solar energy is that of a flat disk (πR_E^2). The ratio is our familiar geometric factor of 4
3. Cooler water tends to sink. Hot water is taken from the top, and replaced by cold water. With the heat exchanger at the bottom of the tank, the evacuated-tube collectors can stay cooler and operate more efficiently
5. (i) The trough system can move on just one axis instead of two. (ii) The heated fluid can turn turbines at a remote location instead of requiring a Stirling engine at each focus, e.g., of a parabolic dish system
7. Once a stone house heats up, it can stay warm for a long time. Once it gets cold, it can stay cold for a long time. This can work in some climates if you fine-tune the thermal mass with the home design. If you live in a region with extreme seasons, or if you expect the climate to change over the lifetime of the house, insulation is an especially good idea
9. Light with a short enough wavelength (i.e., a high enough energy) can eject an electron from the crystal structure of the PV cell. That light has the "band gap" energy, but longer wavelength light has lower energy and can't do the trick.

   However, if the light's wavelength is too short, that is, its energy is too high, it can heat the cell, causing the ejected electron to recombine with the "hole" it left behind, and short circuiting the creation of PV electricity

Exercises

1. 1,361–1,372 W/m^2
3. 48 minutes
5. 8.8 W/m^2·°C
7. 16 m^2
9. 76 minutes
11. (a) 24%; (b) 62%
13. About 12 years
15. (a) 450 MW; (b) 2,360 ha, about 34% the area of the Cheviot mine

CHAPTER 10

Questions

1. Hydropower and wind: yes, in theory. Biomass: not really
3. If a wind turbine extracted all of the wind's kinetic energy, it would stop the air passing through the turbine, deflect the air around it, and produce no power

5. Ethanol is more volatile than gasoline. The highest VOC losses occur during handling, transport, and dispensing of ethanol

Exercises

1. Energy to lift water ~ 1% energy to vaporize water
3. About 3 billion W
5. 29 m
7. 39%
9. 42%
13. (a) 260 billion kWh, or 7% of the total U.S. annual electric energy production; (b) 29 GW, or 29 coal-fired plants
15. 6.7%

CHAPTER 11

Questions

1. Lower power losses at higher voltages
3. Losses due to resistance of power lines and limited efficiencies

Exercises

1. 230 kV
3. 27,381
5. (a) 11.3 kV; (b) 3.8 kV
7. (a) Atmosphere's volume ~ $3 \times 10^{19} m^3$; (b) chemical energy ~ 200 EJ; (c) < ½ year
9. (a) 2.2 billion J; (b) 0.73 million J, which is 0.033% as much
11. The fuel-cell car is responsible for twice as much CO_2 emission

CHAPTER 12

Questions

1. The insolation is inversely proportional to distance squared, and proportional to the temperature to the fourth power. Therefore, the temperature decreases proportionally to just the square root of the distance. This is a weak dependence
3. If the atmosphere had the same temperature as the surface of Earth, then the atmosphere would radiate energy to space at the same rate at which it absorbs it, thus nullifying the greenhouse effect. A cooler atmosphere, however, is a less effective radiator (radiating at a slower rate than it absorbs energy), so it can trap the infrared radiation, permitting a greenhouse effect
5. Without greenhouse gases (GHG), Mars, Earth, and Venus would have rather similar temperatures, because their distances from the Sun do not vary greatly (see Question 1). The extreme differences in the planets' atmospheres account for most of their temperature variations. The atmosphere of Venus is mostly CO_2, and its greenhouse effect makes the planet very hot. Earth's atmosphere is partly made of GHG, and it is warm. Mars has almost no atmosphere, and it is cold

Exercises

1. 278 K
3. (a) 328 K; (b) 232 K
5. 189 K
7. It would rise by 1.8 K
9. 700 W/m^2

CHAPTER 13

Questions

1. *Positive feedbacks are destabilizing.*
   •A warming positive feedback: Water vapor feedback: climate warms → water evaporates → atmospheric H_2O is a greenhouse gas → climate warms more . . .
   •A cooling positive feedback: Earth cools rapidly (e.g., by Milankovitch mechanism or volcanoes) → snow and ice advance → higher albedo → less solar radiation absorbed → more cooling → more snow and ice . . . snowball Earth
   Negative feedbacks are stabilizing.
   •Earth warms slowly → radiates more energy → Earth cools back toward equilibrium temperature
   •Earth warms → more evaporation → more clouds → higher albedo → less sunlight absorbed → Earth cools
3. The cooling effect of aerosols is small compared to the large global warming potential of GHG emitted by coal plants, and aerosols do not mix well or survive long in the atmosphere
5. CO_2 that's added to the atmosphere cycles rapidly through the atmosphere-biosphere-surface ocean system, but is removed to the deep ocean only on much longer timescales. Thus additional anthropogenic CO_2 is with us—some of it in the atmosphere—for centuries to thousands of years

7. Both have lifetimes greater than 100 years but less than the 500 years at which their GWPs are dropping. N_2O has an atmospheric lifetime of about 120 years, and HFC-23 has an atmospheric lifetime of 260 years

Exercises

1. $T_1 = 254.8$ K, $T_2 = 256.1$ K, $\Delta T = 1.3$ K
3. $\Delta T = 1.33$ K
5. (a) $\Delta F = 1.77\ \text{W/m}^2$, higher than the IPCC's earlier reported forcing of 1.5W/m^2; (b) $\Delta F = 3.71\ \text{W/m}^2$ for CO_2 doubling
9. 27.5 tons of CO_2 per person per year

CHAPTER 14

Questions

1. By about 0.8°C. It could have been 0.5°C higher, if not for the effects of volcanoes
3. Milankovitch cycles: changes in Earth's position and/or tilt with respect to the Sun, in combination with other dynamics that push these small solar forcings over the edge
5. It shows the stratosphere (the upper atmosphere) cooling and the troposphere (near Earth's surface) warming, just as predicted for greenhouse warming models in Chapter 12

Exercises

There are no exercises for Chapter 14.

CHAPTER 15

Questions

1. 0.8°C; 1–4°C more
3. Ocean and atmosphere dynamics are modeled together, tracking interactions between flows of heat, energy, water, carbon, chemistry, and more
5. Features and dynamics on smaller scales than a model's grid are parametrized
7. All scenarios have *nonzero emissions* throughout most of the century. Therefore, CO_2 concentrations continue to grow, even in the scenario with decreasing emissions
9. Global warming → more evaporation of seawater → more precipitation → more snowfall in Antarctica → more ice accumulation

Exercises

1. Transient climate sensitivity: $\approx 0.64 \dfrac{°\text{C}}{\text{W/m}^2}$; long-term: $\approx 0.93 \dfrac{°\text{C}}{\text{W/m}^2}$
3. 7.2 m

CHAPTER 16

Questions

1. Much carbon is emitted from moving sources, such as vehicles. It is easier to capture emissions from stationary sources, such as coal plants. Even so, separating CO_2 from mixed gaseous exhaust is tricky
3. It takes energy to manufacture and install solar and wind power facilities, and most of that energy is currently supplied by fossil fuels
5. Over long timescales, temperature and CO_2 curves on this graph are in lockstep—showing that, in equilibrium, the two curves should be right on top of each other. But on the short timescales shown for the modern era, the curves deviate because the climate hasn't had time to equilibrate to the CO_2. So, there would be additional warming as the climate equilibrates

Exercises

1. (a) 0.5 million m^3 CO_2/ hour; (b) 4.5 km^3/year
3. 7 months
5. 163 years
7. About 140 billion kWh per year of electrical energy; almost 16 large (1 GW) power plants
9. Historical (1962–2012) emissions: ≈275 Gt C; flat path rectangle (2012–2062): 450 Gt C; stabilization triangle: 225 Gt C (total ~ 950 Gt C)
11. (a) −1.6 W/m²; (b) depends on CO_2 and other emissions; could be continued warming, global cooling, or oscillations between warming and cooling

CREDITS AND DATA SOURCES

//

ABBREVIATIONS: AGENCIES

CDIAC: Carbon Dioxide Information Analysis Center, Oak Ridge National Laboratory, Oak Ridge, TN

DOE: U.S. Department of Energy, Washington, DC

EIA: U.S. Department of Energy, Energy Information Administration

EPA: U.S. Environmental Protection Agency, Washington, DC

IAEA: International Atomic Energy Agency, Vienna, Austria

IEA: International Energy Agency, Paris, France

ABBREVIATIONS: PUBLICATIONS

AER: U.S. Department of Energy, Energy Information Administration, *Annual Energy Review*.

IPCC 2: Intergovernmental Panel on Climate Change, *Climate Change 1995: The Science of Climate Change*. Cambridge: Cambridge University Press, 1996.

IPCC 3: Intergovernmental Panel on Climate Change, *Climate Change 2001: The Scientific Basis*. Cambridge: Cambridge University Press, 2001.

IPCC 4: Intergovernmental Panel on Climate Change, *Climate Change 2007: The Physical Science Basis*. Cambridge: Cambridge University Press, 2007.

MER: U.S. Department of Energy, Energy Information Administration, *Monthly Energy Review*.

SOURCES

Preface

p. xix: Tad Merrick.

Chapter 1

Fig. 1.3 © John Sibbick.

Fig. 1.5 Adapted from L. D. Danny Harvey, *Global Warming: The Hard Science* (Harlow, England: Prentice Hall/Pearson Education, 2000), Fig. 1.1a; based on L. A. Frakes, *Climates throughout Geologic Time* (New York: Elsevier, 1979).

Fig. 1.7 Adapted from Paul F. Hoffman and Daniel P. Schrag, "Snowball Earth," *Scientific American* 282 (Jan. 2000): 72.

Fig. 1.8 Adapted from Robert Romer, *Energy Facts and Figures* (Amherst, MA: Spring Street Press, 1985); Brent Sørensen, *Renewable Energy: Its Physics, Engineering, Use, Environmental Impacts, Economy and Planning Aspects*, 2nd ed. (London: Academic Press, 2000).

Fig. 1.9 U.S. Census Bureau, "International Programs," www.census.gov/ipc/www/idb/worldpopinfo.php (accessed 4/18/2011), links to World Population, World Population Growth Rates, and Historical Estimates of World Population.

Chapter 2

Fig. 2.1 © tomroster.com.

Fig. 2.3 World energy from Vaclav Smil, *Energy Transitions: History, Requirements, Prospects* (Santa Barbara, CA: Praeger, 2010), Appendix. World population from PBL Netherlands Environmental Assessment Agency, "History Database of the Global Environment," www.pbl.nl/en/themasites/hyde/basicdrivingfactors/population/index.html (accessed 4/18/2011), link to summary of historical population data. U.S. energy through 1945 from *Annual Energy Review* 2005, Appendix E, Table E1. U.S. energy for 1950 on from AER 2009, Table 1.3. U.S. population through 1970 from U.S. Census Bureau, U.S. Department of Commerce, "1800–1970 from Historical Statistics, Colonial Times to 1970" (U.S. Government Printing Office: Washington, D.C., 1975); Series A 6-8. U.S.

population 1970–2008 from the 2010 *Statistical Abstract of the United States*. Energy data from *Annual Review of Energy and the Environment*, vol. 25 (2000), DOE *International Energy Annual* 2003, Table E1; Vaclav Smil, *Energy at the Crossroads: Global Perspectives and Uncertainty* (Cambridge, MA: MIT Press, 2003), Table 1, p. 24. Population data from Population Division, Department of Economic and Social Affairs, United Nations Secretariat, "The World at Six Billion," www.un.org/esa/population/publications/sixbillion/sixbilpart1.pdf (accessed 4/18/2011); inset data from World Bank via Google Public Data.

Fig. 2.4 IEA, Energy Balance Data for selected countries, www.iea.org/stats/index.asp (accessed 4/18/2011).

Fig. 2.5 (a) Ibid.; (b) AER 2008, Fig. 2.1a.

Fig. 2.6 (a) AER 2009, Table 1.3; (b) EIA, "International Energy Annual 2006," www.eia.doe.gov/iea/wecbtu.html (accessed 4/2/2011), Table 18.

Fig. 2.7 IEA, *Key World Energy Statistics 2009*, Section 8: Selected Energy Indicators for 2007, using GDP PPP.

Fig. 2.8 Data for U.S. energy 1900–1945 from AER 2005, Appendix E, Table E1. 1950–2009 from AER 2009, Table 1.1. U.S. GDP from Measuring Worth, "What Was the U.S. GDP Then?" www.measuringworth.com/usgdp/?q=hmit/gdp/ (accessed 8/16/2011), requesting real GDP (G$ in $2005). World energy and GWP (in $1990) from Vaclav Smil, "Energy in the Twentieth Century: Resources, Conversions, Costs, Uses, and Consequences," *Annual Review of Energy and the Environment* 25 (2000): 24, Table 1 (adjusted to $2005 using conversion factor 1.23 obtained from Economic History Services calculator at Measuring Worth, "Seven Ways to Compute the Relative Value of a U.S. Dollar Amount - 1774 to Present," www.measuringworth.com/calculators/uscompare/ [accessed 8/16/2011], using GDP deflator); 2010 scaled from 2000 using Enerdata energy-intensity figures.

Fig. 2.9 Energy data from IEA, *Key World Energy Statistics 2009*, Section 8: Selected Energy Indicators for 2007, using GDP PPP. Life expectancy from United Nations Development Programme, "Human Development Report 2009," hdr.undp.org/en/reports/global/hdr2009/ (accessed 4/18/2011), Table H.

Chapter 3

Fig. 3.4 (b) Peter Bowater/Photo Researchers, Inc.

Fig. 3.5 MER April 2011, Table 7.2a, 2010 full year data.

Fig. 3.8 G.R. "Dick" Roberts © Natural Sciences Image Library.

Chapter 4

Fig. 4.5 Courtesy Dow Building Solutions.

Fig. 4.13 © Comet, Zurich.

Fig. 4.14 Courtesy Michael Moser, Assistant Director of Facilities Services, Middlebury College.

Fig. 4.15 Lawrence Livermore National Laboratory and U.S. Department of Energy, "Estimated U.S. Energy Use in 2009: ~94.6 Quads," publicaffairs.llnl.gov/news/news_releases/2010/images/energy-flow-annotated.pdf (accessed 4/18/2011).

Chapter 5

Fig. 5.1 Ted Clutter/Photo Researchers, Inc.

Fig. 5.2 Adapted from web-based animation associated with Stephen Marshak, *Essentials of Geology* (New York: W. W. Norton, 2006).

Fig. 5.3 AER 2009, Fig. 5.11.

Fig. 5.5 Bloomberg via Getty Images.

Fig. 5.6 Data for 1950–1965: Worldwatch Institute, *Vital Signs 2005* (New York: W. W. Norton, 2005), p. 31. Data for 1970 forward from AER 2009, Table 11.1.

Fig. 5.7 Natural Gas Supply Association, "Natural Gas and the Environment," www.naturalgas.org/environment/naturalgas.asp (accessed 4/18/2011).

Fig. 5.8 Colin Garratt/Milepost 92 1/2 /Corbis.

Fig. 5.9 (stack, pollutants out, cooling tower): G.R. "Dick" Roberts © Natural Sciences Image Library; (turbine): G. Bowater/Corbis; (coal train): © Earl Richardson; (plant surrounded by water): Lowell Georgia/Corbis.

Fig. 5.10 Courtesy Chevron Corporation.

Fig. 5.11 Adapted from Marc Ross, "Fuel Efficiency and the Physics of Automobiles" (revised paper, Department of Physics, University of Michigan), available at sitemaker.umich.edu/mhross/files/fueleff_physicsautossanders.pdf (accessed 4/18/2011), pp. 16–17.

Fig. 5.12 Data from DOE and EPA, www.fueleconomy.gov (accessed 4/18/2011), 2010 models.

Fig. 5.13 (a) Toyota 14 Engine and Electric-Drive Motor, © Hatsukari715, 2008 http://en.wikipedia.org/wiki/Public_domain; (b) adapted from Toyota Motor Sales, USA.

Fig. 5.14 Adapted from Rolls-Royce plc.

Fig. 5.16 Data from AER 2009, Table 11.13.

Fig. 5.17 Data from AER 2009, Table 11.4.

Fig. 5.18 Data from AER 2009, Table 5.1.

Fig. 5.20 G.R. "Dick" Roberts © Natural Sciences Image Library.

Fig. 5.21 Sarah Leen/National Geographic Image Collection.

Fig. 5.24 Adapted from Ibrahim Sami Nashawi, Adel Malallah, and Mohammed Al-Bisharah, "Forecasting World Crude Oil Production Using Multicyclic Hubbert Model," *Energy Fuels* 24 (2010): 1788–1800, Fig. 27; doi:10.1021/ef901240p.

Fig. 5.25 Data from Deutsche Gesellschaft für Technische Zusammenarbeit (GTZ), "International Fuel Prices 2009," www.gtz.de/de/dokumente/gtz2009-en-ifp-part-2.pdf (accessed 8/16/2011), Section 2 tables by continent. Taxes from "International Fuel Prices 2005," www.gtz.de/de/dokumente/en_International_Fuel_Prices_2005.pdf (accessed 8/16/2011), Table 12.5.

Chapter 6

Fig. 6.1 Adapted from Volkswagen of America, Inc.

Fig. 6.2 NASA.

Fig. 6.3 Archive Photos/Getty Images.

Fig. 6.4 (b) Courtesy Rees-Memphis, Inc.

Fig. 6.5 (b) Photo courtesy FLSmidth & Co.

Fig. 6.6 National Atmospheric Deposition Program (NRSP-3), 2006. NADP Program Office, Illinois State Water Survey, 2204 Griffith Drive, Champaign, IL 61820 (2004 data), nadp.sws.uiuc.edu/isopleths/maps2004/phfield.pdf (web site no longer accessible).

Fig. 6.7 Data from EPA, "Carbon Monoxide," www.epa.gov/air/emissions/co.htm (accessed 4/18/2011).

Fig. 6.8 Data from EPA, "Nitrogen Oxides," www.epa.gov/air/emissions/nox.htm (accessed 4/18/2011).

Fig. 6.10 Peggy & Yoram Kahana/Peter Arnold/Photolibrary.

Fig. 6.11 California Air Resources Board, data for Thursday, January 5, 2006, North Main Street location in Los Angeles County, with downloads for specific pollutants, www.arb.ca.gov/aqmis2/paqdselect.php (web site no longer accessible).

Fig. 6.12 The Image Bank/Getty Images.

Fig. 6.13 EPA, "Mercury Maps: A Quantitative Spatial Link between Air Deposition and Fish Tissue," EPA-823-R-01-009, water.epa.gov/scitech/datait/models/maps/upload/2006_12_27_models_maps_report.pdf (accessed 4/18/2011), Fig. 1.

Fig. 6.14 Photo C Vivian Stockman/www.ohvec.org. Flyover courtesy South Wings.

Fig. 6.15 US Coast Guard/Handout/Corbis.

Fig. 6.16 Corbis.

Fig. 6.17 Sarah Leen/National Geographic Image Collection.

Fig. 6.18 (a) Anchorage Daily News/MCT/Landov; (b) AP Photo.

Fig. 6.19 International Tanker Owners Pollution Federation Limited, "Statistics," www.itopf.com/information-services/data-and-statistics/statistics/#major (accessed 5/23/2011), Table 2.

Table 6.1 EPA, "National Ambient Air Quality Standards (NAAQS)," www.epa.gov/air/criteria.html (accessed 4/18/2011).

Table 6.2 EPA.

Fig. 6.20 Data from EPA Office of Mobile Sources, "Emission Facts," EPA420-F-99-017, www.epa.gov/oms/consumer/f99017.pdf (accessed 4/18/2011).

Fig. 6.21 Data from EPA, "National Emissions Inventory (NEI) Air Pollutant Emissions Trends Data," www.epa.gov/ttn/chief/trends/ (accessed 4/18/2011).

Fig. 6.22 EPA, www.epa.gov/oaqps001/greenbk/map/mapnpoll.pdf (accessed 5/24/2011).

Chapter 7

Fig. 7.1 National data from IAEA, "Energy, Electricity, and Nuclear Power," www-pub.iaea.org/MTCD/publications/PDF/IAEA-RDS-1-30_web.pdf (accessed 4/18/2011), 2010 edition, Fig. 1. State data downloaded from Nuclear Energy Institute, "Resources and Stats," www.nei.org/resourcesandstats/documentlibrary/reliableandaffordableenergy/graphicsandcharts/stateelectricitygenerationfuelshares (accessed 4/18/2011).

Fig. 7.2 Data from IAEA, "Energy, Electricity, and Nuclear Power Estimates for the Period up to 2030," July 2003 edition, Table 1.

Fig. 7.7 (a) Courtesy Entergy Vermont Yankee; (b) © Eric Richardson.

Fig. 7.10 Courtesy of General Electric Nuclear Energy.

Fig. 7.12 Courtesy of Pennsylvania Power & Light Corporation.

Fig. 7.15 Adapted from Bernard L. Cohen, "The Disposal of Radioactive Wastes from Fission Reactors,"

Scientific American 236 (June 1977): 21–31. Reproduced with permission.

Fig. 7.16 Adapted from Richard Garwin and Georges Charpak, *Megawatts and Megatons: The Future of Nuclear Power and Nuclear Weapons* (Chicago: University of Chicago Press, 2001), Fig. 5.9, p. 150.

Fig. 7.17 National Council on Radiation Protection and Measurements, Report No. 93, *Ionizing Radiation Exposure of the Population of the United States*.

Fig. 7.18 Data for 1980–2000 from DOE, *International Energy Annual* 2006, Tables 6.3 (world electricity) and 2.7 (nuclear electricity). Data for 2010 (using 2009 data) and projections from IAEA, *Energy, Electricity and Nuclear Power Estimates for the Period up to 2050*, Table 4.

Fig. 7.21 Photo courtesy of the United States Department of Energy (DOE).

Fig. 7.22 Photo courtesy of Princeton Plasma Physics Laboratory (PPPL).

Chapter 8

Fig. 8.1 Adapted from Wendell A. Duffield and John H. Sass, "Geothermal Energy: Clean Power from the Earth's Heat," *U.S. Geological Survey Circular* 1249 (2003).

Fig. 8.2 DOE, " U.S. Geothermal Resource Map," www1.eere.energy.gov/geothermal/geomap.html (web site no longer accessible).

Fig. 8.3 Bo Zaunders/Corbis.

Fig. 8.4 Courtesy Geo-Heat Center, Oregon Institute of Technology.

Fig. 8.5 Data from John W. Lund, "Direct Heat Utilization of Geothermal Resources Worldwide 2005," Geo-Heat Center, Oregon Institute of Technology, geoheat.oit.edu/pdf/directht.pdf (accessed 4/18/2011), Table 2.

Fig. 8.7 Courtesy of United States Department of Energy.

Fig. 8.9 Geothermal Resources Council.

Table 8.1 Geothermal capacities from Alison Holm, Leslie Blodgett, Dan Jennejohn and Karl Gawell, "Geothermal Energy: International Market Update," Geothermal Energy Association, www.geo-energy.org/pdf/reports/GEA_International_Market_Report_Final_May_2010.pdf (accessed 8/16/2011), percentages from above were listed in discussions of individual countries; otherwise from above source's geothermal generating capacity, divided by total electrical generating capacity. California data from Dan Jennejohn, "2010 US Geothermal Power Production and Development Update," Geothermal Energy Association, geo-energy.org/pdf/reports/April_2010_US_Geothermal_Industry_Update_Final.pdf (accessed 4/18/2011).

Fig. 8.13 (left): Photolibrary; (right): John Elk III/Alamy.

Fig. 8.14 Bent Sørensen, *Renewable Energy*, 2nd ed. (San Diego: Academic Press, 2000), p. 280.

Fig. 8.15 JP5\ZOB/WENN/Newscom.

Fig. 8.16 Courtesy Ocean Power Delivery, Ltd.

Chapter 9

Fig. 9.2 Courtesy of Bill Marshall.

Fig. 9.4 Theoretical curves adapted from John A. Duffie and William A. Beckman, *Solar Energy Thermal Processes* (New York: John Wiley & Sons, 1974), Fig. 3.4.3, p. 44. City values from Whitlock, C. E., et al., "Release 3 NASA Surface Meteorology and Solar Energy Data Set for Renewable Energy Industry Use" (Rise & Shine 2000, the 26th Annual Conference of the Solar Energy Society of Canada Inc. and Solar, Oct. 21–24, 2000, Halifax, Nova Scotia, Canada), www.apricus-solar.com/html/solar_collector_insolation.htm (accessed 4/18/2011).

Fig. 9.5 Data from Whitlock, C. E., et al., "Release 3 NASA Surface Meteorology and Solar Energy Data Set for Renewable Energy Industry Use" (Rise & Shine 2000, the 26th Annual Conference of the Solar Energy Society of Canada Inc. and Solar, Oct. 21–24, 2000, Halifax, Nova Scotia, Canada), www.apricus-solar.com/html/solar_collector_insolation.htm (accessed 4/18/2011).

Table 9.1 National Renewable Energy Laboratory, *Solar Radiation Data Manual for Flat-Plate and Concentrating Collectors*, rredc.nrel.gov/solar/old_data/nsrdb/redbook/sum2/ (accessed 4/18/2011).

Fig. 9.6 Adapted from National Renewable Energy Laboratory.

Fig. 9.7 Peter Foukal, *Solar Astrophysics* (New York: Wiley, 1990), Table 3.1, p. 68.

Fig. 9.11 Roberto Mettifogo/Getty Images.

Fig. 9.13 Tad Merrick.

Fig. 9.16 Brian Green/Alamy.

Fig. 9.17 Image courtesy of Torreso.

Fig. 9.18 MCT/HO/Landov.

Fig. 9.21 Tommaso Guicciardini/Photo Researchers, Inc.

Fig. 9.22 Courtesy Solar World.

Fig. 9.23 www.solarhouse.com.

Fig. 9.24 First Solar.

Fig. 9.25 IEA, Photovoltaic Power Systems Programme, *Report IEA-PVPS T1-19:2010*, www.iea-pvps.org/index.php?id-92 (accessed 4/18/2011), 2010 datum extrapolated.

Fig. 9.27 Adapted from www.solarbuzz.com/Module-prices.htm (web site no longer accessible).

Fig. 9.28 DOE, "FY 2011 Congressional Budget Request: Budget Highlights" www.cfo.doe.gov/budget/11budget/Content/FY2011Highlights.pdf (accessed 4/18/2011).

Chapter 10

Fig. 10.1 AER 2009, Tables 1.2 and 10.1.

Fig. 10.2 EIA International Energy Statistics, www.eia.gov/emeu/international/contents.html (accessed 4/18/2011).

Fig. 10.3 Data from UNESCO, www.unesco.org/science/waterday2000/Cycle.htm (web site no longer accessible).

Fig. 10.4 Data from World Energy Council, www.worldenergy.org/wec-geis/publications/reports/ser/hydro/hydro.asp (web site no longer accessible).

Fig. 10.6 Photo: Jonas N. Jordan/United States Army Corps of Engineers.

Fig. 10.7 (a) Bill Johnson/United States Army Corps of Engineers; (b) Corbis.

Fig. 10.10 National Renewable Energy Laboratory, "Wind Energy Resource Atlas of the United States," rredc.nrel.gov/wind/pubs/atlas/maps/chap2/2-01m.html (accessed 4/18/2011).

Fig. 10.11 © tomroster.com.

Fig. 10.14 William Manning Photography/Alamy.

Fig. 10.15 World Wind Energy Association, *World Wind Energy Report 2009*, www.wwindea.org/home/index.php?option=com_content&task=view&id=266&Itemid=43 (accessed 4/11/2011).

Fig. 10.16 Data from Michael Pidwirny, *Fundamentals of Physical Geography*, 2nd ed. (online textbook, 2006), Table 9l-1, www.physicalgeography.net/fundamentals/9l.html (accessed 8/16/2011). Corn data from J. Johnston and M. Bowman, "Performance of Conventional, Dwarf, Grazing, and Silage Corn at New Liskeard," New Liskeard Agricultural Research Station, Ontario, Canada, assuming 18.2 MJ/kg for dry corn.

Fig. 10.17 Courtesy of Burlington Electric Company.

Fig. 10.18 EIA International Energy Statistics, www.eia.gov/emeu/international/contents.html (accessed 4/18/2011).

Fig. 10.19 Ibid.

Fig. 10.20 Photo by Steve Jurvetson.

Fig. 10.21 Van D. Bucher/Photo Researchers, Inc.

Chapter 11

Fig. 11.1 Data from EIA International Energy Statistics, *International Energy Outlook 2010*, www.eia.doe.gov/oiaf/ieo/pdf/0484(2010).pdf (accessed 4/18.2011), Table 11.

Fig. 11.2 World and country data from EIA, "International Energy Statistics," tonto.eia.doe.gov/cfapps/ipdbproject/IEDIndex3.cfm (accessed 4/18/2011), following links to Electricity, Generation (2008 data). State data from EIA, *Electric Power Annual*, www.eia.doe.gov/fuelelectric.html (accessed 4/18/2011), link to Electricity Databases at bottom, then State-level Spreadsheets 1990–2008, Net Generation by State by Type of Producer by Energy Source (2008 data).

Fig. 11.4 Photos: Shutterstock.

Fig. 11.7 Data from NationalGrid, www.nationalgrid.com/uk/Electricity/Data/Demand+Data/ (accessed 4/18/2011).

Fig. 11.8 Data from multiple sources; data points represent averages where ranges are given in source.

Fig. 11.10 Courtesy First Light Power Resources.

Fig. 11.13 Transtock Inc./Alamy.

Fig. 11.14 Adapted from *The Hydrogen Economy: Opportunities, Costs, Barriers, and R&D Needs* (Washington, DC: National Academies Press, 2004), Fig. 6.1, p. 67.

Chapter 12

Fig. 12.4 NASA/NOAA/SSEC.

Fig. 12.5 Adapted from J. Kiehl and K. Trenberth, *Bulletin of the American Meteorological Society* 78 (1997): 197, as reproduced in IPCC 3, Fig. 1.2; updated with numbers from K. Trenberth, J. Fasullo, and J. Kiehl, *Bulletin of the American Meteorological Society* (March 2009): 314, with downward IR flow adjusted for balance.

Fig. 12.6 Courtesy of Lunar and Planetary Institute, NASA.

Chapter 13

Fig. 13.4 IPCC 4, Fig. SPM-2 and following text.

Fig. 13.5 Carbon dioxide data from CDIAC, "Global

Fossil-Fuel CO_2 Emissions," cdiac.ornl.gov/trends/emis/tre_glob.html (accessed 8/16/2011). Land use data from CDIAC, "Annual Net Flux of Carbon to the Atmosphere from Land-Use Change: 1850–2005," cdiac.esd.ornl.gov/trends/landuse/houghton/1850-2005.txt (accessed 2/3/2011).

Fig. 13.6 Methane data: 1750–1992 from CDIAC, "Historical CH4 Mixing Ratios from Law Dome (Antarctica) and Summit (Greenland) Ice Cores," cdiac.esd.ornl.gov/ftp/trends/atm_meth/EthCH498B.txt (accessed 2/2/2011). 1992 forward from Mauna Loa data at CDIAC, "Atmospheric Methane Record from Mauna Loa, Hawaii, USA," cdiac.esd.ornl.gov/trends/atm_meth/csiro/csiro-mloch4.html (accessed 2/2/2011), 1992 is average of both datasets. HCFC-22 data from Montzka, S. A., et al., "Recent Increases in Global HFC-23 Emissions," *Geophysical Research Letters* 37 (2010): L02808, data included in Supplement, Fig. 1; doi: 10.1029/2009GL041195.

Fig. 13.7 Main graph data through 1953 from CDIAC, "Historical CO2 Record from the Siple Station Ice Core," cdiac.ornl.gov/ftp/trends/co2/siple2.013 (accessed 2/25/2011). 1959 forward from Mauna Loa yearly average data and inset from Mauna Loa monthly data, at National Oceanic & Atmospheric Administration, "Trends in Atmospheric Carbon Dioxide," www.esrl.noaa.gov/gmd/ccgg/trends (accessed 2/25/2011).

Fig. 13.8 (a) Data for 1010–1975 from CDIAC, "Historical CO2 Record from the Law Dome DE08, DE08-2, and DSS Ice Cores," cdiac.ornl.gov/ftp/trends/co2/lawdome.combined.dat (accessed 2/25/2011). 1976 forward from Mauna Loa, National Oceanic & Atmospheric Administration, "Trends in Atmospheric Carbon Dioxide," www.esrl.noaa.gov/gmd/ccgg/trends (accessed 2/25/2011). (b) Vostok ice core data from CDIAC, "Historical CO2 Record from the Vostok Ice Core," cdiac.ornl.gov/ftp/trends/co2/vostok.icecore.co2 (accessed 2/25/2011).

Fig. 13.9 NASA.

Fig. 13.10 Data (rounded to integers) assembled from IPCC 3 and IPCC 4; John Houghton, *Global Warming*, 3rd ed. (Cambridge, UK: Cambridge University Press, 2004); L. D. Danny Harvey, *Global Warming: The Hard Science* (Harlow, UK: Prentice-Hall/Pearson Education, 2000); Global Carbon Project, www.globalcarbonproject.org/carbonbudget/09/hl-full.htm (accessed 2/26/2011); CDIAC, "Global Fossil-Fuel CO_2 Emissions," cdiac.ornl.gov/trends/emis/tre_glob.html (accessed 8/16/2011); CDIAC, "Annual Net Flux of Carbon to the Atmosphere from Land-Use Change: 1850–2005," cdiac.esd.ornl.gov/trends/landuse/houghton/1850-2005.txt (accessed 2/3/2011); private communication from Dr. Peter Vitousek, Stanford University; Woods Hole Oceanographic Institute, "Global Ecology," www.whrc.org/global/carbon/index.html (accessed 8/16/2011). Data have been adjusted for consistency.

Fig. 13.11 Data for 1850–2007 from CDIAC, "Global Fossil-Fuel CO_2 Emissions," cdiac.ornl.gov/trends/emis/tre_glob.html (accessed 2/26/2011). Data for 2008–2009 from CDIAC, "Fossil-Fuel CO_2 Emissions," cdiac.ornl.gov/trends/emis/meth_reg.html (accessed 2/26/2011). 2010 data from estimate calculated in P. Friedlingstein, R. A. Houghton, G. Marland, J. Hackler, T. A. Boden, T. J. Conway, J. G. Canadell, M. R. Raupach, P. Ciais, and C. Le Quéré, "Update on CO_2 emissions," *Nature Geoscience* 3 (2010): 8110812; doi: 10.1038/ngeo1022.

Fig. 13.12 Plotted using CO_2 pulse decay equation from footnote (a) in IPCC 4, Table TS-2.

Chapter 14

Fig. 14.1 Data from University of East Anglia, Climatic Research Unit, "1: Global Temperature Record," www.cru.uea.ac.uk/cru/info/warming/ (accessed 4/18/2011). Error bars estimated from IPCC 3, Fig. 2.1.

Fig. 14.2 Data for CRU from University of East Anglia, Climatic Research Unit, "1: Global Temperature Record," www.cru.uea.ac.uk/cru/info/warming/ (accessed 4/18/2011); National Climatic Data Center, www.ncdc.noaa.gov/cmb-faq/anomalies.html#mean (accessed 8/16/2011), link to Annual Global (land and ocean combined). GISS: NASA Goddard Institute for Space Studies, data.giss.nasa.gov/gistemp/tabledata/GLB.Ts.txt (accessed 4/18/2011), J-D annual averages, adjusted for 1960–1991 average used in other two datasets.

Fig. 14.3 NASA Goddard Institute for Space Studies, "GISS Surface Temperature Analysis," data.giss.nasa.gov/gistemp/maps/ (accessed 4/18/2011), using data sources GISS analysis for land, Had/Reyn_v2 for oceans, map type: anomalies, mean period: annual (Jan.–Dec.), time interval: 2010–2010, base period: 1961–1990, smoothing radius: 1200 km, projection type: regular.

Fig. 14.4 Data from National Oceanographic Data Center, ftp://ftp.nodc.noaa.gov/pub/data.nodc/woa/DATA_ANALYSIS/3M_HEAT_CONTENT/DATA/basin/yearly/h22-w0-700m.dat (accessed 4/18/2011), from www.nodc.noaa.gov/OC5/3M_HEAT_CONTENT/basin_data.html.

Fig. 14.5 Upper two graphs from Remote Sensing Systems, "Description of MSU and AMSU Data Products," www.ssmi.com/msu/msu_data_description.html#msu_amsu_time_series (accessed 3/14/2011), bottom curve is a portion of Fig. 14.1.

Fig. 14.6 Graph produced by Global Warming Art Project, from reconstructions in P. D. Jones, K. R. Briffa, T. P. Barnett, and S.F.B. Tett, "High-resolution Palaeoclimatic Records for the last Millennium: Interpretation, Integration and Comparison with General Circulation Model Control-run Temperatures," *The Holocene* 8 (1998): 455–471; M. E. Mann, R. S. Bradley, and M. K. Hughes, "Northern Hemisphere Temperatures during the Past Millennium: Inferences, Uncertainties, and Limitations," *Geophysical Research Letters* 26, no. 6 (1999): 759–762; T. Crowley and T. Lowery, "Northern Hemisphere Temperature Reconstruction," *Ambio* 29 (2000): 51–54. Modified as published in T. Crowley, "Causes of Climate Change over the Past 1000 Years," *Science* 289 (2000): 270–277; K. R. Briffa, T. J. Osborn, F. H. Schweingruber, I. C. Harris, P. D. Jones, S. G. Shiyatov, S. G. and E. A. Vaganov, "Low-frequency Temperature Variations from a Northern Tree-ring Density Network," *Journal of Geophysical Research* 106 (2001): 2929–2941; J. Esper, E. R. Cook, and F. H. Schweingruber, "Low-Frequency Signals in Long Tree-Ring Chronologies for Reconstructing Past Temperature Variability," *Science* 295 (2002): 2250–2253; M. E. Mann and P. D. Jones, "Global Surface Temperatures over the Past Two Millennia," *Geophysical Research Letters* 30, no. 15 (2003): 1820; P. D. Jones and M. E. Mann, "Climate over Past Millennia," *Reviews of Geophysics* 42 (2004): RG2002; S. Huang, "Merging Information from Different Resources for New Insights into Climate Change in the Past and Future," *Geophysical Research Letters* 31 (2004): L13205; A. Moberg, D. M. Sonechkin, K. Holmgren, N. M. Datsenko and W. Karlén, "Highly Variable Northern Hemisphere Temperatures Reconstructed from Low- and High-resolution Proxy Data," *Nature* 443 (2005): 613–617; J. H. Oerlemans, "Extracting a Climate Signal from 169 Glacier Records," *Science* 308 (2005): 675–677.

Fig. 14.7 Data from J. R. Petit, D. Raynaud, C. Lorius, J. Jouzel, G. Delaygue, N. I. Barkov, and V. M. Kotlyakov, "Historical Isotopic Temperature Record from the Vostok Ice Core," *Trends: A Compendium of Data on Global Change* (CDIAC, 2000), cdiac.ornl.gov/trends/temp/vostok/jouz_tem.htm (accessed 3/15/2011).

Fig. 14.8 Adapted from United States Geological Survey, pubs.usgs.gov/gip/continents/map.jpg (accessed 3/15/2011).

Fig. 14.9 CDIAC, "Historical CO2 Record from the Vostok Ice Core," cdiac.ornl.gov/ftp/trends/co2/vostok.icecore.co2 (accessed 2/25/2011); J. R. Petit, D. Raynaud, C. Lorius, J. Jouzel, G. Delaygue, N. I. Barkov, and V. M. Kotlyakov, "Historical Isotopic Temperature Record from the Vostok Ice Core," *Trends: A Compendium of Data on Global Change* (CDIAC, 2000), cdiac.ornl.gov/trends/temp/vostok/jouz_tem.htm (accessed 3/15/2011).

Fig. 14.10 Data from R. B. Alley, "GISP2 Ice Core Temperature and Accumulation Data," IGBP PAGES/World Data Center for Paleoclimatology, Data Contribution Series #2004-013, NOAA/NGDC Paleoclimatology Program, Boulder, CO, ftp.ncdc.noaa.gov/pub/data/paleo/icecore/greenland/summit/gisp2/isotopes/gisp2_temp_accum_alley2000.txt (accessed 3/15/2011).

Fig. 14.11 Courtesy of United States Geological Survey.

Fig. 14.12 (a) NASA/Goddard Space Flight Center Scientific Visualization Studio; (b) data provided by R. Gersten.

Fig. 14.13 Data from Kerry Emanuel, "Papers, Data, and Graphics Pertaining to Tropical Cyclone Trends and Variability (Updated through 2009)," wind.mit.edu/~emanuel/Papers_data_graphics.htm (accessed 8/16/2011), from download link to Excel spreadsheet; both curves smoothed with 3-point running average.

Fig. 14.14 Adapted from Gerald A. Meehl, Warren M. Washington, Caspar M. Ammann, Julie M. Arblaster, T.M.L. Wigley, and Claudia Tebaldi, "Combinations of Natural and Anthropogenic Forcings in Twentieth-Century Climate," *Journal of Climate* 17, no. 19. Reprinted with permission.

Chapter 15

Fig. 15.2 Adapted from William F. Ruddiman, *Earth's Climate: Past and Future* (New York: W. H. Freeman, 2001), p. 74.

Fig. 15.3 Ibid.

Fig. 15.4 Adapted from IPCC 3, Fig. 9.1.

Fig. 15.5 From IPCC 3, Fig. 8.8a.

Fig. 15.6 From IPCC 2, p. 258.

Fig. 15.7 From IPCC 3, Fig. 9.3.

Fig. 15.8 From IPCC 3, Fig. 9.16.

Fig. 15.9 Data from International Institute for Applied Systems Analysis, Laxenburg, Austria, RCP (Representative Concentration Pathways) database version 2.05, www.iiasa.ac.at/web-apps/tnt/RcpDb/dsd?Action=htmlpage&page=download (accessed 3/23/2011).

Fig. 15.10 Ibid.

Fig. 15.11 Adapted from Gerald A. Meehl, Warren M. Washington, Julie M. Arblaster, Aixue Hu, Haiyan Teng, Claudia Tebaldi, Warren G. Strand, and James B. White III, "Climate System Response to External Forcings and Climate Change Projections in CCSM4," submitted to *Journal of Climate CCSM4 Special Collection*, Fig. 9.

Fig. 15.12 Adapted from ibid., Fig. 10d.

Fig. 15.14 Adapted from J. C. Moore, S. Jevrejeva, and A. Grinsted, "Efficacy of Geoengineering to Limit 21st Century Sea-level Rise," *PNAS (Proceedings of the National Academy of Sciences)* 107 (Sept. 7, 2010): 15699–15703, Fig. 3.

Fig. 15.15 Adapted from University of Arizona, Department of Geosciences, Environmental Studies Laboratory, "Mapping Areas Potentially Impacted by Sea Level Rise," www.geo.arizona.edu/dgesl/research/other/climate_change_and_sea_level/sea_level_rise/florida/slr_usafl_i.htm (accessed 8/16/2011), using +3 meters map.

Chapter 16

Fig. 16.1 From International Institute for Applied Systems Analysis, Laxenburg, Austria, RCP (Representative Concentration Pathways) database version 2.05, www.iiasa.ac.at/web-apps/tnt/RcpDb/dsd?Action=htmlpage&page=download (accessed 3/23/2011); Gerald A. Meehl, Warren M. Washington, Julie M. Arblaster, Aixue Hu, Haiyan Teng, Claudia Tebaldi, Warren G. Strand, and James B. White III, "Climate System Response to External Forcings and Climate Change Projections in CCSM4," submitted to *Journal of Climate CCSM4 Special Collection*, Fig. 9.

Fig. 16.2 Adapted from James Hansen, Makiko Sato, Pushker Kharecha, David Beerling, Valerie Masson-Delmotte, Mark Pagani, Maureen Raymo, Dana L. Royer, and James C. Zachos, "Target Atmospheric CO_2: Where Should Humanity Aim?" *Open Atmospheric Science Journal* 2 (2008): 217–231, Fig. 3.

Fig. 16.3 Adapted from ibid., Fig. 2.

Fig. 16.4 Data sources: actual emissions (2000–2010) from CDIAC, "Global Fossil-Fuel CO_2 Emissions," cdiac.ornl.gov/trends/emis/tre_glob.html (accessed 2/26/2011); CDIAC, "Fossil-Fuel CO_2 Emissions," cdiac.ornl.gov/trends/emis/meth_reg.html (accessed 2/26/11); P. Friedlingstein, R. A. Houghton, G. Marland, J. Hackler, T. A. Boden, T. J. Conway, J. G. Canadell, M. R. Raupach, P. Ciais, and C. Le Quéré, "Update on CO_2 emissions," *Nature Geoscience* 3 (2010): 8110812; doi: 10.1038/ngeo1022. RCPs from International Institute for Applied Systems Analysis, Laxenburg, Austria, RCP (Representative Concentration Pathways) database version 2.05, www.iiasa.ac.at/web-apps/tnt/RcpDb/dsd?Action=htmlpage&page=download (accessed 3/23/2011). DOE projections from DOE EIA, "International Energy Outlook 2010, Energy-Related Carbon Dioxide Emissions," www.eia.doe.gov/oiaf/ieo/emissions.html (accessed 4/8/2011), figure source and data link for Fig. 104.

Fig. 16.5 Data from EPA, "Inventory of U.S. Greenhouse Gas Emissions and Sinks: 1990–2009," www.epa.gov/climatechange/emissions/downloads11/US-GHG-Inventory-2011-Executive-Summary.pdf (accessed 4/8/2011), Table ES-3.

Fig. 16.6 Adapted from Intergovernmental Panel on Climate Change, *Carbon Dioxide Capture and Storage* (Geneva: IPCC, 2005), Fig. SPM.2.

Fig. 16.7 All data except nuclear high estimate from J. V. Spadaro, L. Langlois, and B. Hamilton, "Greenhouse Gas Emissions of Electricity Generating Chains: Assessing the Difference," *IAEA Bulletin* 42, no. 2 (2000): 21. Nuclear high estimate is based on an assessment of several studies, including U. R. Fritsche, "Comparing Greenhouse-Gas Emissions and Abatement Costs of Nuclear and Alternative Energy Options from a Life-Cycle Perspective," www.oeko.de/service/gemis/files/info/nuke_co2_en.pdf (accessed 4/11/2011), p. 5.

Fig. 16.8 Data for cars: EPA, Volkswagen; refrigerators: EPA.

Fig. 16.9 Data for U.S. energy from AER 2009, Table 1.1. U.S. GDP from Measuring Worth, "What Was the U.S. GDP Then?" www.measuringworth.com/usgdp/?q=hmit/gdp/ (accessed 8/16/2011), requesting real GDP (G$ in $2005).

Fig. 16.10 Data from Pat McAuliffe, California Energy Commission, e-mail message to author, 4/26/2006.

Fig. 16.11 Adapted from S. Nadel, American Council for an Energy-Efficient Economy, from European Council for an Energy-Efficient Economy 2003 Summer Study, in Arthur Rosenfeld, Pat McAuliffe, and John Wilson, "Energy Efficiency and Climate Change," *Encyclopedia of Energy*, Vol. 2 (Boston: Elsevier Press, 2004), Fig. 3, p. 373, www.fypower.org/pdf/CEC_EE_and_Climate_Change.pdf (accessed 4/18/2011).

Fig. 16.12 Data from EPA, "Light-Duty Automotive Technology, Carbon Dioxide Emissions, and Fuel Economy Trends: 1975 through 2010," www.epa.gov/oms/fetrends.htm#summary (accessed 4/13/2011), following download link to Full Report Tables (420r10023-main-rpt-tables.xls; data from Table 1, Adj Comp MPG columns).

Fig. 16.13 Courtesy of the Rocky Mountain Institute.

Fig. 16.14 Data from EIA, "International Energy Statistics," tonto.eia.doe.gov/cfapps/ipdbproject/IEDIndex3.cfm (accessed 8/16/2011), following links to (1) Total Energy, Total Primary Energy Consumption, data for Total Primary Energy Consumption Per Capita, year 2008 and (2) Indicators, CO2 Emissions, data for Per Capita Carbon Dioxide Emissions from the Consumption of Energy, year 2008.

Fig. 16.15 Adapted from Vaclav Smil, *Energy at the Crossroads: Global Perspectives and Uncertainties* (Cambridge, MA: MIT Press, 2003), Fig. 3.27, p. 171.

Fig. 16.16 Adapted and updated for approximately 2012 start date using data through 2007 from CDIAC, "Global Fossil-Fuel CO_2 Emissions," cdiac.ornl.gov/trends/emis/tre_glob.html (accessed 2/26/2011). Data for 2008–2009 from CDIAC, "Fossil-Fuel CO_2 Emissions," cdiac.ornl.gov/trends/emis/meth_reg.html (accessed 2/26/2011). 2010 data from estimate calculated in P. Friedlingstein, R. A. Houghton, G. Marland, J. Hackler, T. A. Boden, T. J. Conway, J. G. Canadell, M. R. Raupach, P. Ciais, and C. Le Quéré, "Update on CO_2 emissions," *Nature Geoscience* 3 (2010): 8110812; doi: 10.1038/ngeo1022.

Modified from Robert Socolow, Princeton University, "Stabilization Wedges: Mitigation Tools for the Next Half-Century" (lecture at Avoiding Dangerous Climate Change: A Scientific Symposium on Stabilisation of Greenhouse Gases, Met Office, Exeter, UK, 2/3/2005), and based on S. Pacala and R. Socolow, *Science* 305, no. 5686 (2004): 968–972.

Fig. 16.17 Ibid.

INDEX

Page numbers in italics refer to figures; *t* indicates a table.

PHYSICAL CONSTANTS

Earth's radius: $R_E = 6.37 \times 10^6$ m
Planck's constant: $h = 6.63 \times 10^{-34}$ J·s
Speed of light: $c = 3.00 \times 10^8$ m/s
Gravitational acceleration: $g = 9.8$ m/s^2
Stefan–Boltzmann constant: $\sigma = 5.67 \times 10^{-8}$ W/m^2·K^4
Solar constant: $S = \sim 1{,}364$ W/m^2

IMPORTANT EQUATIONS

Energy of a photon: $E = hf = \frac{hc}{\lambda}$ (Equation 3.1; p. 44)
Work: $W = Fd$ (Equation 3.2; p. 51)
Force of gravity: $F_g = mg$ (Equation 3.3; p. 52)
Work done lifting object of mass m a distance h: $W = mgh$ (Equation 3.4; p. 52)
Kinetic energy: $K = \frac{1}{2}mv^2$ (Equation 3.5; p. 54)
Conductive heat flow: $H = kA\frac{T_h - T_c}{d}$ (Equation 4.1; p. 66)
Stefan–Boltzmann radiation law: $P = e\sigma AT^4$ (Equation 4.2; p. 73; also Equation 12.1, p. 320)
Heat required for temperature change: $Q = mc\Delta T$ (Equation 4.3; p. 76)
Efficiency of an engine: $e = \frac{\text{mechanical energy delivered}}{\text{energy extracted from fuel}}$ (Equation 4.4; p. 82)
Thermodynamic efficiency limit: $e = 1 - \frac{T_c}{T_h}$ (Equation 4.5; p. 82)
Exponential growth: $N = N_0 e^{rt}$ (Equation 5.1; p. 118)
Cumulative consumption: $\frac{N_0}{r}(e^{rt} - 1)$ (Equation 5.2; p. 118)
Mass-energy equivalence: $E = mc^2$ (Box 7.1; p. 170)
Coefficient of performance: $\text{COP} = \frac{\text{heat delivered}}{\text{electrical energy required}}$ (Equation 8.1; p. 215)
Maximum possible COP: $\text{COP}_{\text{max}} = \frac{T_h}{T_h - T_c}$ (Equation 8.2; p. 216)
Wind power per unit area: $\frac{1}{2}\rho v^3$ (Equation 10.1; p. 269)
Electric power: $P = IV$ (Equation 11.1; p. 295)
Ohm's law: $I = \frac{V}{R}$ (Equation 11.2; p. 296)
Earth's energy balance, no atmosphere: $\frac{S}{4} = e\sigma T^4$ (Equation 12.2; p. 321)
Power radiated to space: $\sigma T_s^4 - e_a\sigma(T_s^4 - T_a^4)$ (Equation 12.3; p. 328)

CLIMATE QUANTITIES

Natural greenhouse effect: 33°C
Earth's albedo: 0.30
Anthropogenic CO_2 from fossil fuels: ~9 Gt carbon/year
Pre-industrial CO_2 concentration: 280 ppm
2010 CO_2 concentration: ~390 ppm; increasing ~2 ppm/year
Global warming over the past one hundred years: ~0.8°C
Temperature difference between present day and ice age: ~6°C
Climate sensitivity: 2°C to 4.5°C for CO_2 doubling
Projected temperature rise by 2100: 1.5°C to 6°C